COMPUTATIONAL AEROACOUSTICS

Computational Aeroacoustics (CAA) is a relatively new research area. CAA algorithms have developed rapidly, and the methods have been applied in many areas of aeroacoustics. The objective of CAA is not simply to develop computational methods, but also to use these methods to solve practical aeroacoustics problems and to perform numerical simulations of aeroacoustic phenomena. By analyzing the simulation data, an investigator can determine noise generation mechanisms and sound propagation processes. This is both a textbook for graduate students and a reference for researchers in CAA and, as such, is self-contained. No prior knowledge of numerical methods for solving partial differential equations is needed; however, a general understanding of partial differential equations and basic numerical analysis is assumed. Exercises are included and are designed to be an integral part of the chapter content. In addition, sample computer programs are included to illustrate the implementation of the numerical algorithms.

Dr. Christopher K. W. Tam is the Robert O. Lawton Distinguished Professor in the Department of Mathematics at Florida State University. His research in computational mathematics and numerical simulation involves the development of low dispersion and dissipation computation schemes, numerical boundary conditions, and the mathematical analysis of the scheme's computational properties for use in large-scale numerical simulation of a number of real-world problems. In addition, he is developing jet as well as other aircraft noise theories, and prediction codes in NASA and the US Aircraft Industry's noise reduction effort as well.

Cambridge Aerospace Series

Editors:
Wei Shyy
and
Vigor Yang

1. J. M. Rolfe and K. J. Staples (eds.): *Flight Simulation*
2. P. Berlin: *The Geostationary Applications Satellite*
3. M. J. T. Smith: *Aircraft Noise*
4. N. X. Vinh: *Flight Mechanics of High-Performance Aircraft*
5. W. A. Mair and D. L. Birdsall: *Aircraft Performance*
6. M. J. Abzug and E. E. Larrabee: *Airplane Stability and Control*
7. M. J. Sidi: *Spacecraft Dynamics and Control*
8. J. D. Anderson: *A History of Aerodynamics*
9. A. M. Cruise, J. A. Bowles, C. V. Goodall, and T. J. Patrick: *Principles of Space Instrument Design*
10. G. A. Khoury (ed.): *Airship Technology*, 2nd Edition
11. J. P. Fielding: *Introduction to Aircraft Design*
12. J. G. Leishman: *Principles of Helicopter Aerodynamics*, 2nd Edition
13. J. Katz and A. Plotkin: *Low-Speed Aerodynamics*, 2nd Edition
14. M. J. Abzug and E. E. Larrabee: *Airplane Stability and Control: A History of the Technologies that Made Aviation Possible*, 2nd Edition
15. D. H. Hodges and G. A. Pierce: *Introduction to Structural Dynamics and Aeroelasticity*, 2nd Edition
16. W. Fehse: *Automatic Rendezvous and Docking of Spacecraft*
17. R. D. Flack: *Fundamentals of Jet Propulsion with Applications*
18. E. A. Baskharone: *Principles of Turbomachinery in Air-Breathing Engines*
19. D. D. Knight: *Numerical Methods for High-Speed Flows*
20. C. A. Wagner, T. Hüttl, and P. Sagaut (eds.): *Large-Eddy Simulation for Acoustics*
21. D. D. Joseph, T. Funada, and J. Wang: *Potential Flows of Viscous and Viscoelastic Fluids*
22. W. Shyy, Y. Lian, H. Liu, J. Tang, and D. Viieru: *Aerodynamics of Low Reynolds Number Flyers*
23. J. H. Saleh: *Analyses for Durability and System Design Lifetime*
24. B. K. Donaldson: *Analysis of Aircraft Structures*, 2nd Edition
25. C. Segal: *The Scramjet Engine: Processes and Characteristics*
26. J. F. Doyle: *Guided Explorations of the Mechanics of Solids and Structures*
27. A. K. Kundu: *Aircraft Design*
28. M. I. Friswell, J. E. T. Penny, S. D. Garvey, and A. W. Lees: *Dynamics of Rotating Machines*
29. B. A. Conway (ed): *Spacecraft Trajectory Optimization*
30. R. J. Adrian and J. Westerweel: *Particle Image Velocimetry*
31. G. A. Flandro, H. M. McMahon, and R. L. Roach: *Basic Aerodynamics*
32. H. Babinsky and J. K. Harvey: *Shock Wave–Boundary-Layer Interactions*
33. C. K. W. Tam: *Computational Aeroacoustics: A Wave Number Approach*

Computational Aeroacoustics

A WAVE NUMBER APPROACH

Christopher K. W. Tam
Florida State University

CAMBRIDGE
UNIVERSITY PRESS

CAMBRIDGE
UNIVERSITY PRESS

32 Avenue of the Americas, New York NY 10013-2473, USA

Cambridge University Press is part of the University of Cambridge.

It furthers the University's mission by disseminating knowledge in the pursuit of education, learning and research at the highest international levels of excellence.

www.cambridge.org
Information on this title: www.cambridge.org/9781107656338

© Cambridge University Press 2012

First published 2012
First paperback edition 2014

A catalogue record for this publication is available from the British Library

Library of Congress Cataloguing in Publication data

Tam, Christopher K. W.
Computational aeroacoustics : a wave number approach / Christopher K. W. Tam.
    p.   cm. – (Cambridge aerospace series)
ISBN 978-0-521-80678-7 (hardback)
1. Aerodynamic noise – Mathematical models.   2. Sound-waves – Mathematical models.   I. Title.
TL574.N6T36   2012
629.132´3–dc23   2011044020

ISBN 978-0-521-80678-7 Hardback
ISBN 978-1-107-65633-8 Paperback

Contents

Preface   *page* xi

1. **Finite Difference Equations** 1

   1.1 Order of Finite Difference Equations: Concept of Solution   1
   1.2 Linear Difference Equations with Constant Coefficients   2
   1.3 Finite Difference Solution as an Approximate Solution of a
        Boundary Value Problem   4
   1.4 Finite Difference Model for a Surface of Discontinuity   8
   EXERCISES   17

2. **Spatial Discretization in Wave Number Space** 21

   2.1 Truncated Taylor Series Method   21
   2.2 Optimized Finite Difference Approximation in Wave Number
        Space   22
   2.3 Group Velocity Consideration   25
   2.4 Schemes with Large Stencils   30
   2.5 Backward Difference Stencils   32
   2.6 Coefficients of Several Large Optimized Stencils   34
   EXERCISES   36

3. **Time Discretization** 38

   3.1 Single Time Step Method: Runge-Kutta Scheme   38
   3.2 Optimized Multilevel Time Discretization   40
   3.3 Stability Diagram   41
   EXERCISES   43

4. **Finite Difference Scheme as Dispersive Waves** 45

   4.1 Dispersive Waves of Physical Systems   45
   4.2 Group Velocity and Dispersion   47
   4.3 Origin of Numerical Dispersion   48
   4.4 Numerical Dispersion Arising from Temporal Discretization   53
   4.5 Origin of Numerical Dissipation   57

4.6 Multidimensional Waves   59
EXERCISES   60

5. **Finite Difference Solution of the Linearized Euler Equations** 61

5.1 Dispersion Relations and Asymptotic Solutions of the Linearized
     Euler Equations   61
5.2 Dispersion-Relation-Preserving (DRP) Scheme   67
5.3 Numerical Stability   69
5.4 Group Velocity for Finite Difference Schemes   70
5.5 Time Step Δt: Accuracy Consideration   72
5.6 DRP Scheme in Curvilinear Coordinates   73
EXERCISES   75

6. **Radiation, Outflow, and Wall Boundary Conditions** 80

6.1 Radiation Boundary Conditions   81
6.2 Outflow Boundary Conditions   81
6.3 Implementation of Radiation and Outflow Boundary Conditions   83
6.4 Numerical Simulation: An Example   84
6.5 Generalized Radiation and Outflow Boundary Conditions   88
6.6 The Ghost Point Method for Wall Boundary Conditions   89
6.7 Enforcing Wall Boundary Conditions on Curved Surfaces   103
EXERCISES   106

7. **The Short Wave Component of Finite Difference Schemes** 112

7.1 The Short Waves   112
7.2 Artificial Selective Damping   113
7.3 Excessive Damping   118
7.4 Artificial Damping at Surfaces of Discontinuity   122
7.5 Aliasing   124
7.6 Coefficients of Several Large Damping Stencils   125
EXERCISES   127

8. **Computation of Nonlinear Acoustic Waves** 130

8.1 Nonlinear Simple Waves   130
8.2 Spurious Oscillations: Origin and Characteristics   132
8.3 Variable Artificial Selective Damping   137
EXERCISES   142

9. **Advanced Numerical Boundary Treatments** 144

9.1 Boundaries with Incoming Disturbances   144
9.2 Entrainment Flow   146
9.3 Outflow Boundary Conditions: Further Consideration   149
9.4 Axis Boundary Treatment   152
9.5 Perfectly Matched Layer as an Absorbing Boundary Condition   158
9.6 Boundaries with Discontinuities   167
9.7 Internal Flow Driven by a Pressure Gradient   170
EXERCISES   171

10. **Time-Domain Impedance Boundary Condition** 180

    10.1 A Three-Parameter Broadband Model   182
    10.2 Stability of the Three-Parameter Time-Domain Impedance
        Boundary Condition   182
    10.3 Impedance Boundary Condition in the Presence of a Subsonic
        Mean Flow   185
    10.4 Numerical Implementation   187
    10.5 A Numerical Example   189
    10.6 Acoustic Wave Propagation and Scattering in a Duct with
        Acoustic Liner Splices   192
    EXERCISES   201

11. **Extrapolation and Interpolation** 203

    11.1 Extrapolation and Numerical Instability   203
    11.2 Wave Number Analysis of Extrapolation   205
    11.3 Optimized Interpolation Method   213
    11.4 A Numerical Example   219
    EXERCISES   226

12. **Multiscales Problems** 229

    12.1 Spatial Stencils for Use in the Mesh-Size-Change Buffer Region   231
    12.2 Time Marching Stencil   235
    12.3 Damping Stencils   238
    12.4 Numerical Examples   241
    12.5 Coefficients of Several Large Buffer Stencils   252
    12.6 Large Buffer Selective Damping Stencils   256
    EXERCISES   261

13. **Complex Geometry** 263

    13.1 Basic Concept of Overset Grids   263
    13.2 Optimized Multidimensional Interpolation   266
    13.3 Numerical Examples: Scattering Problems   280
    13.4 Sliding Interface Problems   284
    EXERCISES   294

14. **Continuation of a Near-Field Acoustic Solution to the Far Field** 298

    14.1 The Continuation Problem   299
    14.2 Surface Green's Function: Pressure as the Matching Variable   301
    14.3 Surface Green's Function: Normal Velocity as the Matching
        Variable   305
    14.4 The Adjoint Green's Function   308
    14.5 Adjoint Green's Function for a Conical Surface   313
    14.6 Generation of a Random Broadband Acoustic Field   321
    14.7 Continuation of Broadband Near Acoustic Field on a Conical
        Surface to the Far Field   323
    EXERCISE   328

15. **Design of Computational Aeroacoustic Codes** 329

   15.1 Basic Elements of a CAA Code   329
   15.2 Spatial Resolution Requirements   333
   15.3 Mesh Design: Body-Fitted Grid   344
   15.4 Example I: Direct Numerical Simulation of the Generation of
        Airfoil Tones at Moderate Reynolds Number   350
   15.5 Computation of Turbulent Flows   379
   15.6 Example II: Numerical Simulation of Axisymmetric Jet Screech   385

APPENDIX A: Fourier and Laplace Transforms 395
APPENDIX B: The Method of Stationary Phase 397
APPENDIX C: The Method of Characteristics 398
APPENDIX D: Diffusion Equation 400
APPENDIX E: Accelerated Convergence to Steady State 403
APPENDIX F: Generation of Broadband Sound Waves with a Prescribed
             Spectrum by an Energy-Conserving Discretization Method . . . 407
APPENDIX G: Sample Computer Programs 410

References   471
Index   477

Preface

Computational Aeroacoustics (CAA) is a relatively young research area. It began in earnest fewer than twenty years ago. During this time, CAA algorithms have developed rapidly. These methods soon found applications in many areas of aeroacoustics.

The objective of CAA is not simply to develop computational methods, but also to use these methods to solve real practical aeroacoustics problems. It is also a goal of CAA to perform numerical simulation of aeroacoustic phenomena. By analyzing the simulation data, an investigator can determine noise generation mechanisms and sound propagation processes. Hence, CAA offers a way to obtain a better understanding of the physics of a problem.

Computational Aeroacoustics is not the same as Computational Fluid Dynamics (CFD). In fact, CAA faces a different set of computational challenges, because aeroacoustics problems are intrinsically different from standard aerodynamics and fluid mechanics problems. By definition, aeroacoustics problems are time dependent, whereas aerodynamics and fluid mechanics problems are, in general, time independent or involve only low-frequency unsteadiness. The following list outlines some of the major computational challenges facing CAA:

1. Aeroacoustics problems typically involve a broad range of frequencies. Numerical resolution of the high-frequency waves with extremely short wavelengths becomes a formidable obstacle to accurate numerical simulation.
2. The amplitudes of acoustic waves are usually small, in particular, when compared with the mean flow. Oftentimes, the sound intensity of the acoustic waves is five to six orders smaller than that of the mean flow. To compute sound waves accurately, a numerical scheme must have high resolution and extremely low numerical noise.
3. In most aeroacoustics problems, interest is in the sound waves radiated to the far field. This requires a solution that is uniformly valid from the source region all the way to the measurement point many acoustic wavelengths away. Because of the long propagation distance, computational aeroacoustics schemes must have minimal numerical dispersion and dissipation. They should also propagate the waves at the correct wave speeds independent of the orientation of the computation mesh.

4. A computation domain is inevitably finite in size. For aerodynamics or fluid mechanics problems, flow disturbances tend to decay very quickly away from a body or their source of generation; therefore, the disturbances are usually small at the boundary of the computation domain. Acoustic waves, on the other hand, decay very slowly and actually reach the boundaries of a finite computation domain. To avoid the reflection of outgoing sound waves back into the computation domain and thus contaminating the solution, specially developed radiation and outflow or absorbing boundary conditions must be imposed at the artificial exterior boundaries to assist the waves in exiting smoothly. For standard CFD problems, such boundary conditions are usually not required.

5. Aeroacoustics problems are archetypical examples of multiscales problems. The length scale of the acoustic source is usually very different from the acoustic wavelength. That is, the length scale of the source region and that of the acoustic region can be vastly different. CAA methods must be designed to handle problems involving tremendously different length scales in different parts of the computational domain.

Many of these major computational issues and challenges to CAA have now been resolved or at least partly resolved. This book offers an overview of the methods, analysis, and new ideas introduced to overcome such challenges.

It should be clear, as elaborated above, that the nature of aeroacoustics problems is substantially different from that of traditional fluid dynamics and aerodynamics problems. As a result, standard CFD schemes, designed for applications to fluid mechanics problems, are generally not adequate for computing or simulating aeroacoustics problems accurately and efficiently. For this reason, there is a need for the independent development of CAA.

At the outset of CAA development, it was clear to investigators that a fundamental need in CAA was to develop algorithms that could resolve high-frequency short waves with the minimum number of mesh points per wavelength. Standard CFD second-order schemes often require 18 to 25 mesh points per wavelength to ensure adequate accuracy. This large number of mesh points is clearly not acceptable if the method is to be adopted for practical computation. It was soon recognized that a solution to the problem was to use large-stencil high-resolution CAA schemes. At present, high-resolution algorithms implemented on stencils of a reasonable size can resolve waves using about six to seven mesh points per wavelength.

The small amplitude of acoustic waves in comparison with that of the mean flow has raised a good deal of initial apprehension as to whether the inherent noise level of a numerical scheme used to compute the mean flow would overwhelm the actual radiated sound. Such apprehension is well founded, for experience has indicated that some CFD schemes do have a high intrinsic noise level. The development of new high-resolution CAA methods has proven that it is, indeed, possible to capture sound waves of minute amplitude with good accuracy.

One of the most significant differences between traditional CFD and CAA methodology is the method of error analysis. In CFD, the standard way to assess the quality of a scheme is by the order of Taylor series truncation. In general, it is assumed that a fourth-order scheme is better than a second-order scheme, which, in turn, is better than a first-order scheme, but all such assessments are qualitative not quantitative. There is no way to find out by order-of-magnitude analysis how many mesh points per wavelength are needed to achieve, say, a half-percent accuracy in a computation. Furthermore, traditional numerical analysis does not provide a way to quantify wave propagation errors. Dispersion and dissipation errors are often erroneously linked to the phase velocity and amplification factor.

The development of wave number analysis, through the use of Fourier-Laplace transforms, has provided a firm mathematical foundation for error analysis in CAA. Wave number analysis shows that the order of a scheme is not the most important factor in achieving high-quality results. Instead, the resolved bandwidth of a scheme in wave number space is the more important issue. As far as numerical wave propagation error is concerned, wave number analysis shows that the phase velocity is totally irrelevant. Rather, it is the dependence of the group velocity of the scheme on wave number that is important. In wave propagation, space and time play an important partnership role. The relationship is all encoded in the dispersion function. Thus, a dispersion-relation-preserving scheme (a numerical scheme having formally the same dispersion relation as the original partial differential equations) would not only automatically guarantee a numerically accurate solution, but it would also replicate the number of wave modes (acoustic, vorticity, and entropy) and the characteristics supported by the original partial differential equations. Wave number analysis provides an understanding of the existence and characteristics of spurious short waves. Such knowledge allows the design of very effective artificial selective damping stencils and filters. Wave number analysis, together with the dispersion relation of the discretized equations, offers a simple quantitative method for analyzing the numerical stability of CAA algorithms. Such an analysis is crucial when selecting the size of time marching step.

The development of numerical boundary conditions is also an integral part of CAA, and the importance of high-quality numerical boundary conditions for CAA can never be overemphasized. There is no exaggeration in saying that the development of high-quality numerical boundary conditions is as important as the development of a high-quality time marching scheme. The construction and analysis of numerical boundary conditions form a core part of the materials covered in this book.

A large difference between the size of a noise source and the acoustic wavelength would invariably result in a multiscale problem. However, very often, large length scale disparity arises because of a change in the dominant physics in different parts of the computational domain. For example, for the problem of sound waves dissipation by a resonant acoustic liner, the dominant physics adjacent to the solid surface and hole openings of the liner is viscous effect. An oscillatory Stokes layer, driven by the sound field, would develop. The thickness of the Stokes layer (the scale to be resolved) is very small even when excited by moderately low-frequency sound. Away from the wall, compressibility effect dominates. The length scale is the acoustic wavelength, which can be hundreds of times the thickness of the viscous Stokes layer. Another instance that would lead to severe length scale disparity is when sound waves propagate against a flow. The wavelength of an upstream propagating sound wave is drastically reduced over the region where the mean flow is transonic. Numerical treatment of multiscale problems is an important part of this book.

Once the appropriate physical model and computational methods have been developed to study a real phenomenon computationally, it is necessary to design a simulation code. Designing a computer code is very different from developing a computational algorithm. A computer code is formed by synthesizing many elements of basic computational methods. A good code should be stable, accurate, and efficient. Because many students of CAA may not have experience in designing computer code for real problems, an effort is made in this book to provide some guidance on code design. As a part of this effort, examples at a modest level of complexity are provided.

This book is written both as a text for graduate students and as a reference for researchers using CAA. Every effort has been made to make this book self-contained. No prior knowledge of numerical methods for solving partial differential equations is needed; however, a general understanding of partial differential equations and basic numerical analysis is assumed. Exercises are included at the end of many chapters; they are designed to be an integral part of the chapter content. In addition, sample computer programs are included to illustrate the implementation of the numerical algorithms.

This book has benefited from the work and publications of many investigators. In particular, the contributions of Drs. Jay C. Webb, Konstantin K. Kurbatskii, Nikolai N. Pastouchenko, Hongbin Ju, Hao Shen, Fang Q. Hu, Tom Dong, Laurent Auriault, Andrew T. Thies, and Philip P. LePoudre should be mentioned. Their work and that of others have made this book possible. The assistance of Dr. Sarah A. Parrish is greatly appreciated.

Tallahassee, Florida, 2011

1 Finite Difference Equations

In this chapter, the exact analytical solution of linear finite difference equations is discussed. The main purpose is to identify the similarities and differences between solutions of differential equations and finite difference equations. Attention is drawn to the intrinsic problems of using a high-order finite difference equation to approximate a partial differential equation. Since exact analytical solutions are used, the conclusions of this chapter are not subjected to numerical errors.

1.1. Order of Finite Difference Equations: Concept of Solution

Domain: In this chapter the domain considered consists of the set of integers $k = 0$, $\pm 1, \pm 2, \pm 3, \ldots$. The general member of the sequence $\ldots, y_{-2}, y_{-1}, y_0, y_1, y_2, \ldots$ will be denoted by y_k.

An ordinary difference equation is an algorithm relating the values of different members of the sequence y_k. In general, a finite difference equation can be written in the form

$$y_{k+n} = F(y_{k+n-1}, y_{k+n-2}, \ldots, y_k, k), \tag{1.1}$$

where F is a general function.

The order of a difference equation is the difference between the highest and lowest indices appearing in the equation. For linear difference equations, the number of linearly independent solutions is equal to the order of the equation.

A difference equation is linear if it can be put in the following form:

$$y_{k+n} + a_1(k) y_{k+n-1} + a_2(k) y_{k+n-2} + \cdots + a_{n-1}(k) y_{k+1} + a_n(k) y_k = R_k, \tag{1.2}$$

where $a_i(k), i = 1, 2, 3, \ldots, n$ and R_k are given functions of k.

EXAMPLES

(a) $y_{k+1} - 3y_k + y_{k-1} = 6e^{-k}$ (second-order, linear)
(b) $y_{k+1} = y^2{}_k$ (first-order, nonlinear)
(c) $y_{k+2} = \sin(y_k)$ (second-order, nonlinear)

The solution of a difference equation is a function $y_k = \phi(k)$ that reduces the equation to an identity.

1.2 Linear Difference Equations with Constant Coefficients

Linear difference equations with constant coefficients can be solved in much the same way as linear differential equations with constant coefficients. The characteristics of the two types of solutions are similar but not identical.

Consider the nth-order homogeneous finite difference equation with constant coefficients:

$$y_{k+n} + a_1 y_{k+n-1} + a_2 y_{k+n-2} + \cdots + a_n y_k = 0, \tag{1.3}$$

where $a_1, a_2, \ldots, a_n$ are constants. The general solution of such an equation has the form:

$$y_k = cr^k, \tag{1.4}$$

where c and r are constants. Substitution of Eq. (1.4) into Eq. (1.3) yields, after factoring out the common factor cr^k,

$$f(r) \equiv r^n + a_1 r^{n-1} + a_2 r^{n-2} + \cdots + a_{n-1} r + a_n = 0. \tag{1.5}$$

Here, $f(r)$ is an nth-order polynomial and thus has n roots r_i, $i = 1, 2, \ldots, n$. For each r_i we have a solution:

$$y_k = c_i r_i^k, \tag{1.6}$$

where c_i is an arbitrary constant. The most general solution may be found by superposition.

1.2.1 Distinct Roots

If the characteristic roots of Eq. (1.5) are distinct, then a fundamental set of solutions is

$$y_k^{(i)} = r_i^k, \qquad i = 1, 2, \ldots, n$$

and the general solution of the homogeneous equation is

$$y_k = c_1 r_1^k + c_2 r_2^k + \cdots + c_n r_n^k, \tag{1.7}$$

where $c_1, c_2, \ldots, c_n$ are n arbitrary constants.

EXAMPLE. Find the general solution of

$$y_{k+3} - 7y_{k+2} + 14y_{k+1} - 8y_k = 0.$$

Let $y_k = cr^k$. Substitution into the difference equation yields the characteristic equation

$$r^3 - 7r^2 + 14r - 8 = 0$$

or

$$(r - 1)(r - 2)(r - 4) = 0.$$

The characteristic roots are $r = 1, 2,$ and 4. Therefore, the general solution is

$$y_k = A + B2^k + C4^k,$$

where A, B, and C are arbitrary constants.

1.2.2 Repeated Roots

Now consider the case where one or more of the roots of the characteristic equation are repeated. Suppose the root r_1 has multiplicity m_1, the root r_2 has multiplicity m_2, and the root r_ℓ has multiplicity m_ℓ such that

$$m_1 + m_2 + \cdots + m_\ell = n. \tag{1.8}$$

The characteristic equation can be written as

$$(r - r_1)^{m_1} (r - r_2)^{m_2} \cdots (r - r_\ell)^{m_\ell} = 0. \tag{1.9}$$

Corresponding to a repeated root of the characteristic polynomial (1.9) of multiplicity m, the solution is

$$y_k = \left(A_1 + A_2 k + A_3 k^2 + \cdots A_m k^{m-1}\right) r^k, \tag{1.10}$$

where $A_1, A_2, \ldots, A_m$ are arbitrary constants.

EXAMPLE. Consider the general solution of the equation

$$y_{k+2} - 6y_{k+1} + 9y_k = 0.$$

The characteristic equation is

$$r^2 - 6r + 9 = 0 \quad \text{or} \quad (r - 3)^2 = 0.$$

Thus, there is a repeated root $r = 3, 3$. The general solution is

$$y_k = (A + Bk)\, 3^k.$$

1.2.3 Complex Roots

Since the coefficients of the characteristic polynomial are real, complex roots must appear as complex conjugate pairs. Suppose r and $r*$ ($* = $ complex conjugate) are roots of the characteristic equation; then, corresponding to these roots the solutions may be written as

$$y_k^{(1)} = r^k, \qquad y_k^{(2)} = (r^*)^k.$$

If a real solution is desired, these solutions can be recasted into a real form. Let $r = Re^{i\theta}$, then an alternative set of fundamental solutions is

$$y_k^{(1)} = R^k \cos(k\theta), \qquad y_k^{(2)} = R^k \sin(k\theta).$$

If r and $r*$ are repeated roots of multiplicity m, then the set of fundamental solutions corresponding to these roots is

$$
\begin{aligned}
y_k^{(1)} &= R^k \cos(k\theta) & y_k^{(m+1)} &= R^k \sin(k\theta) \\
y_k^{(2)} &= kR^k \cos(k\theta) & y_k^{(m+2)} &= kR^k \sin(k\theta) \\
&\;\;\vdots & &\;\;\vdots \\
y_k^{(m)} &= k^{m-1} R^k \cos(k\theta) & y_k^{(2m)} &= k^{m-1} R^k \sin(k\theta).
\end{aligned}
\tag{1.11}
$$

EXAMPLE. Find the general solution of

$$y_{k+2} - 4y_{k+1} + 8y_k = 0.$$

The characteristic equation is

$$r^2 - 4r + 8 = 0.$$

The roots are $r = 2 \pm 2i = 2\sqrt{2} \ e^{\pm i(\pi/4)}$. Therefore, the general solution (can be verified by direct substitution) is

$$y_k = A(2\sqrt{2})^k \cos\left(\frac{\pi}{4}k\right) + B(2\sqrt{2})^k \sin\left(\frac{\pi}{4}k\right),$$

where A and B are arbitrary constants.

1.3 Finite Difference Solution as an Approximate Solution of a Boundary Value Problem

A concrete example will now illustrate the inherent difficulties of using the finite difference solution to approximate the solution of a boundary value problem governed by partial differential equations.

Suppose the frequencies of the normal acoustic wave modes of a one-dimensional tube of length L as shown in Figure 1.1 is to be determined. The tube has two closed ends and is filled with air. The governing equations of motion of the air in the tube are the linearized momentum and energy equations, as follows:

$$\rho_0 \frac{\partial u}{\partial t} + \frac{\partial p}{\partial x} = 0 \tag{1.12}$$

$$\frac{\partial p}{\partial t} + \gamma p_0 \frac{\partial u}{\partial x} = 0, \tag{1.13}$$

where ρ_0, p_0, and γ are, respectively, the static density, the pressure, and the ratio of specific heats of the air inside the tube; and u is the velocity. The boundary conditions are

$$\text{At} \quad x = 0, L; \quad u = 0. \tag{1.14}$$

Upon eliminating p from (1.12) and (1.13), the equation for u is

$$\frac{\partial^2 u}{\partial t^2} - a^2 \frac{\partial^2 u}{\partial x^2} = 0, \tag{1.15}$$

where $a = (\gamma p_0/\rho_0)^{1/2}$ is the speed of sound.

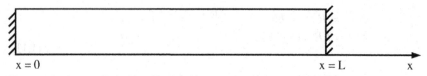

$x = 0$ $x = L$ x

Figure 1.1. A one-dimensional tube with closed ends.

1.3.1 Analytical Solution

To find the normal acoustic modes of the tube, consideration will be given to solutions of the form:

$$u(x, t) = \mathrm{Re}[\hat{u}(x)e^{-i\omega t}], \tag{1.16}$$

where $\mathrm{Re}[\ ]$ is the real part of $[\ ]$. Substitution of Eq. (1.16) into Eqs. (1.15) and (1.14) yields the following eigenvalue problem:

$$\frac{d^2\hat{u}}{dx^2} + \frac{\omega^2}{a^2}\hat{u} = 0 \tag{1.17}$$

$$\hat{u}(0) = \hat{u}(L) = 0. \tag{1.18}$$

The two linearly independent solutions of Eq. (1.17) are

$$\hat{u}(x) = A\sin\left(\frac{\omega x}{a}\right) + B\cos\left(\frac{\omega x}{a}\right). \tag{1.19}$$

On imposing boundary conditions (1.18), it is found that

$$B = 0, \quad \text{and} \quad A\sin\left(\frac{\omega L}{a}\right) = 0.$$

For a nontrivial solution A cannot be zero, this leads to,

$$\sin\left(\frac{\omega x}{a}\right) = 0 \quad \text{or} \quad \frac{\omega L}{a} = n\pi \quad (n = \text{integer}).$$

Therefore,

$$\omega_n = \frac{n\pi a}{L}, \quad (n = 1, 2, 3, \dots) \tag{1.20}$$

is the eigenvalue or eigenfrequency. The eigenfunction or mode shape is obtained from Eq. (1.19); i.e.,

$$\hat{u}_n(x) = \sin\left(\frac{n\pi x}{L}\right), \quad n = 1, 2, 3, \dots. \tag{1.21}$$

1.3.2 Finite Difference Solution

Now consider solving the normal mode problem by finite difference approximation. For this purpose, the tube is divided into M equal intervals with a spacing of $\Delta x = L/M$ as shown in Figure 1.2. ℓ is the spatial index ($\ell = 0$ to M). Both second- and fourth-order standard central difference approximation will be used.

$$\left(\frac{\partial^2 u}{\partial x^2}\right)_\ell = \frac{u_{\ell+1} - 2u_\ell + u_{\ell-1}}{(\Delta x)^2} + O(\Delta x^2) \tag{1.22}$$

$$\left(\frac{\partial^2 u}{\partial x^2}\right)_\ell = \frac{-u_{\ell+2} + 16u_{\ell+1} - 30u_\ell + 16u_{\ell-1} - u_{\ell-2}}{12(\Delta x)^2} + O(\Delta x^4). \tag{1.23}$$

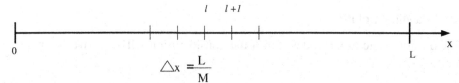

Figure 1.2. The computation grid for finite difference solution.

1.3.2.1 Second-Order Approximation

On replacing the spatial derivative of Eq. (1.15) by Eq. (1.22), the finite difference equation to be solved is

$$\frac{d^2 u_\ell}{dt^2} - \frac{a^2}{(\Delta x)^2}(u_{\ell+1} - 2u_\ell + u_{\ell-1}) = 0. \tag{1.24}$$

Eq. (1.24) is a second-order finite difference equation, the same order as the original partial differential equation. For a unique solution, two boundary conditions are required. This is given by the boundary conditions of the physical problem, Eq. (1.14); i.e.,

$$u_0 = 0, \qquad u_M = 0. \tag{1.25}$$

On following Eq. (1.16), a separable solution of a similar form is sought,

$$u_\ell(t) = \mathrm{Re}\left[\tilde{u}_\ell e^{-i\omega t}\right]. \tag{1.26}$$

Substitution of Eq. (1.26) into Eqs. (1.24) and (1.25) leads to the following eigenvalue problem:

$$\tilde{u}_{\ell+1} + \left[\frac{\omega^2 (\Delta x)^2}{a^2} - 2\right]\tilde{u}_\ell + \tilde{u}_{\ell-1} = 0 \tag{1.27}$$

$$\tilde{u}_0 = 0, \qquad \tilde{u}_M = 0. \tag{1.28}$$

Two linearly independent solutions of finite difference equation (1.27) in the form of Eq. (1.4) can easily be found. The characteristic equation is

$$r^2 + \left[\frac{\omega^2 (\Delta x)^2}{a^2} - 2\right]r + 1 = 0. \tag{1.29}$$

The two roots of Eq. (1.29) are complex conjugates of each other. The absolute value is equal to unity. Thus, the general solution of Eq. (1.27) may be written in the following form:

$$\tilde{u}_\ell = A \sin(\Theta\ell) + B \cos(\Theta\ell), \tag{1.30}$$

where

$$\Theta = \cos^{-1}\left[1 - \frac{\omega^2 (\Delta x)^2}{2a^2}\right]. \tag{1.31}$$

Upon imposition of boundary conditions (1.28), it is easy to find

$$B = 0, \quad A \sin(\Theta M) = 0.$$

For a nontrivial solution, it is required that $\sin(\theta M) = 0$. Hence,

$$\Theta M = n\pi, \qquad n = 1, 2, 3, \ldots$$

or

$$\cos^{-1}\left[1 - \frac{\omega^2 (\Delta x)^2}{2a^2}\right] = \frac{n\pi}{M}.$$

This yields

$$\omega_n = \frac{2^{\frac{1}{2}}a}{\Delta x}\left(1 - \cos\left(\frac{n\pi}{M}\right)\right)^{\frac{1}{2}}, \qquad n = 1, 2, 3, \ldots \tag{1.32}$$

and from solution (1.30), the eigenfunction or mode shape is

$$\tilde{u}_\ell = \sin(\Theta\ell) = \sin\left(\frac{n\pi\ell}{M}\right). \tag{1.33}$$

Now, it is instructive to compare finite difference solutions (1.32) and (1.33) with the exact solution of the original partial differential equations (1.20) and (1.21). One obvious difference is that the exact solution has infinitely many eigenfrequencies and eigenfunctions, whereas the finite difference solution supports only a finite number ($2M$) of such modes. Furthermore, ω_n of Eq. (1.32) is a good approximation of the exact solution only for $n\pi/M \ll 1$. In other words, a second-order finite difference approximation provides good results only for the low-order long-wave modes. The error increases quickly as n increases.

1.3.2.2 Fourth-Order Approximation
If the fourth-order approximation of Eq. (1.23) is used instead of Eq. (1.22), it is easy to show that the governing finite difference equation for $\tilde{u}_\ell$ is

$$\tilde{u}_{\ell+2} - 16\tilde{u}_{\ell+1} + \left(30 - \frac{12\omega^2 (\Delta x)^2}{a^2}\right)\tilde{u}_\ell - 16\tilde{u}_{\ell-1} + \tilde{u}_{\ell-2} = 0. \tag{1.34}$$

The two physical boundary conditions of Eq. (1.15) are

$$\tilde{u}_0 = 0, \qquad \tilde{u}_M = 0. \tag{1.35}$$

Now, Eq. (1.34) is a fourth-order finite difference equation. There are four linearly independent solutions. In order to have a unique solution, four boundary conditions are necessary. However, only two physical boundary conditions are available. To ensure a unique solution of the fourth-order finite difference equation, two extra (nonphysical) boundary conditions need to be created. Also, two of the four solutions of Eq. (1.34) are spurious solutions unrelated to the physical problem. Therefore, the use of high-order approximation will result in

(A) Possible generation of spurious numerical solutions.
(B) A need for extra boundary conditions or special boundary treatment.

These are definite disadvantages in the use of a high-order scheme to approximate partial differential equations. Are there any advantages? To show that there could be an advantage, note that the eigenfunction of the finite difference equation (1.33) is identical to the exact eigenfunction (1.21). As it turns out, the eigenfunction (1.33)

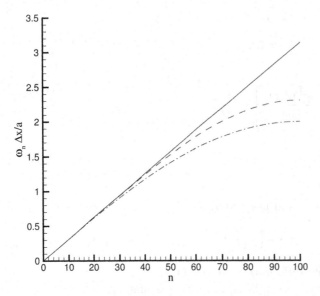

Figure 1.3. Comparison of normal mode frequencies. _____, exact, Eq. (1.20); ――――, fourth-order, Eq. (1.37); ·······, second-order, Eq. (1.32).

of the second-order approximation is also the eigenfunction of the fourth-order approximation, namely, the solution of Eqs. (1.34) and (1.35) is

$$\tilde{u}_\ell = \sin\left(\frac{n\pi\ell}{M}\right), \qquad n = 1, 2, 3, \ldots. \tag{1.36}$$

These eigenfunctions satisfy boundary conditions (1.35). On the substitution of solution (1.36) into Eq. (1.34), it is easy to find that the corresponding eigenfrequency is given by

$$\omega_n = \frac{a}{(\Delta x)\,6^{\frac{1}{2}}}\left[15 - 16\cos\left(\frac{n\pi}{M}\right) + \cos\left(\frac{2n\pi}{M}\right)\right]^{\frac{1}{2}}. \tag{1.37}$$

It is straightforward to find that frequency formula (1.37) is a much improved approximation to the exact eigenfrequency of formula (1.20) than formula (1.32) of the second-order method. Figure 1.3 shows a comparison for the case $M = 100$. This result illustrates the fact that, when the problems of spurious waves and extra boundary conditions are adequately taken care of, a high-order method does give more accurate numerical results.

1.4 Finite Difference Model for a Surface of Discontinuity

How best to transform a boundary value problem governed by partial differential equations into a computation problem governed by difference equations is not always obvious. The task is even more difficult if the original problem involves a surface of discontinuity. There is a lack of discussion in the literature about how to model a discontinuity in the context of finite difference. The purpose of this section is to show how one such model may be set up. At the same time, this model will demonstrate that the finite difference formulation of boundary value problems may support spurious boundary modes. These modes might have no counterpart in the original problem.

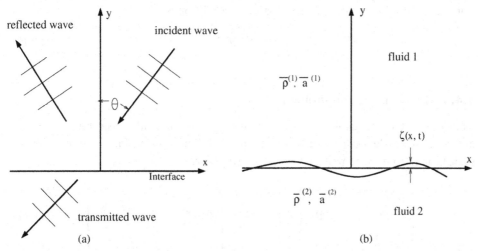

Figure 1.4. Schematic diagram showing (a) the incident, reflected, and transmitted sound waves at a fluid interface, (b) the slightly deformed fluid interface.

They are not generally known or expected. If one of these spurious boundary modes grows in time, then this could lead to numerical instability or a divergent solution.

Consider the transmission of sound through the interface of two fluids of densities $\bar{\rho}^{(1)}$ and $\bar{\rho}^{(2)}$ and sound speeds $\bar{a}^{(1)}$ and $\bar{a}^{(2)}$, respectively, as shown in Figure 1.4a. It is known that refraction takes place at such an interface.

Superscripts (1) and (2) will be used to denote the flow variables above and below the interface. For small-amplitude incident sound waves, it is sufficient to use the linearized Euler equations and interface boundary conditions. Let $y = \varsigma(x, t)$ be the location of the interface. The governing equations are

$$y \geq 0, \qquad \frac{\partial u^{(1)}}{\partial t} = -\frac{1}{\bar{\rho}^{(1)}} \frac{\partial p^{(1)}}{\partial x} \tag{1.38}$$

$$\frac{\partial v^{(1)}}{\partial t} = -\frac{1}{\bar{\rho}^{(1)}} \frac{\partial p^{(1)}}{\partial y} \tag{1.39}$$

$$\frac{\partial p^{(1)}}{\partial t} + \gamma \bar{p}^{(1)} \left(\frac{\partial u^{(1)}}{\partial x} + \frac{\partial v^{(1)}}{\partial y} \right) = 0 \tag{1.40}$$

$$y \leq 0 \qquad \frac{\partial u^{(2)}}{\partial t} = -\frac{1}{\bar{\rho}^{(2)}} \frac{\partial p^{(2)}}{\partial x} \tag{1.41}$$

$$\frac{\partial v^{(2)}}{\partial t} = -\frac{1}{\bar{\rho}^{(2)}} \frac{\partial p^{(2)}}{\partial y} \tag{1.42}$$

$$\frac{\partial p^{(2)}}{\partial t} + \gamma \bar{p}^{(2)} \left(\frac{\partial u^{(2)}}{\partial x} + \frac{\partial v^{(2)}}{\partial y} \right) = 0. \tag{1.43}$$

The dynamic and kinematic boundary conditions at the interface are

$$y = 0, \qquad p^{(1)} = p^{(2)} \tag{1.44}$$

$$\frac{\partial \varsigma}{\partial t} = v^{(1)} = v^{(2)}. \tag{1.45}$$

For static equilibrium, the pressure balance condition is

$$\bar{p}^{(1)} = \bar{p}^{(2)} \quad \text{or} \quad \bar{\rho}^{(1)}(\bar{a}^{(1)})^2 = \bar{\rho}^{(2)}(\bar{a}^{(2)})^2. \tag{1.46}$$

1.4.1 The Transmission Problem

Consider a plane acoustic wave of angular frequency ω incident on the interface at an angle of incidence θ as shown in Figure 1.4. The appropriate solution of Eqs. (1.38) to (1.40) may be written in the following form:

$$\begin{bmatrix} u^{(1)} \\ v^{(1)} \\ p^{(1)} \end{bmatrix}_{\text{incident}} = \text{Re}\left\{ A \begin{bmatrix} -\frac{\sin\theta}{\bar{\rho}^{(1)}\bar{a}^{(1)}} \\ -\frac{\cos\theta}{\bar{\rho}^{(1)}\bar{a}^{(1)}} \\ 1 \end{bmatrix} e^{-i\omega(\sin\theta x + \cos\theta y + \bar{a}^{(1)}t)/\bar{a}^{(1)}} \right\}, \tag{1.47}$$

where A is the amplitude and Re{} is the real part of. The reflected wave in region (1) has a form similar to Eq. (1.47), which may be written as

$$\begin{bmatrix} u^{(1)} \\ v^{(1)} \\ p^{(1)} \end{bmatrix}_{\text{reflected}} = \text{Re}\left\{ R \begin{bmatrix} -\frac{\sin\theta}{\bar{\rho}^{(1)}\bar{a}^{(1)}} \\ \frac{\cos\theta}{\bar{\rho}^{(1)}\bar{a}^{(1)}} \\ 1 \end{bmatrix} e^{-i\omega(\sin\theta x - \cos\theta y + \bar{a}^{(1)}t)/\bar{a}^{(1)}} \right\}, \tag{1.48}$$

where R is the amplitude of the reflected wave. The transmitted wave in region (2) must have the same dependence on x and t as the incidence wave. Let

$$\begin{bmatrix} u^{(2)} \\ v^{(2)} \\ p^{(2)} \end{bmatrix}_{\text{transmitted}} = \text{Re}\left\{ \begin{bmatrix} \hat{u}(y) \\ \hat{v}(y) \\ \hat{p}(y) \end{bmatrix} e^{-i\omega(\sin\theta x + \bar{a}^{(1)}t)/\bar{a}^{(1)}} \right\}. \tag{1.49}$$

By substituting Eq. (1.49) into Eqs. (1.41) to (1.43), it is easy to find after some simple elimination,

$$\frac{d^2\hat{p}}{dy^2} + \frac{\omega^2}{\left(\bar{a}^2\right)^2}\left[1 - \left(\frac{\bar{a}^{(2)}}{\bar{a}^{(1)}}\right)^2 \sin^2\theta\right]\hat{p} = 0. \tag{1.50}$$

On solving Eq. (1.50), the transmitted wave with an amplitude T may be written as

$$\begin{bmatrix} u^{(2)} \\ v^{(2)} \\ p^{(2)} \end{bmatrix}_{\text{transmitted}}$$

$$= \text{Re}\left\{ T \begin{bmatrix} -\frac{\sin\theta}{\bar{\rho}^{(2)}\bar{a}^{(1)}} \\ -\frac{[1-(\bar{a}^{(2)}/\bar{a}^{(1)})^2 \sin^2\theta]^{1/2}}{\bar{\rho}^{(2)}\bar{a}^{(2)}} \\ 1 \end{bmatrix} e^{-i\omega\{\sin\theta x + (\bar{a}^{(1)}/\bar{a}^{(2)})[1-(\bar{a}^{(2)}/\bar{a}^{(1)})^2 \sin^2\theta]^{1/2}y + \bar{a}^{(1)}t\}/\bar{a}^{(1)}} \right\}. $$

$$\tag{1.51}$$

Now the amplitudes of the reflected and the transmitted wave may be found by enforcing boundary conditions (1.44) and (1.45) at $y = 0$. This yields

$$A + R = T \tag{1.52}$$

$$-A\frac{\cos \theta}{\overline{\rho}^{(1)}\overline{a}^{(1)}} + R\frac{\cos \theta}{\overline{\rho}^{(1)}\overline{a}^{(1)}} = -T\frac{(1 - (\overline{a}^{(2)}/\overline{a}^{(1)})^2 \sin^2 \theta))^{1/2}}{\overline{\rho}^{(2)}\overline{a}^{(2)}}. \tag{1.53}$$

Therefore, the amplitudes of the transmitted and reflected waves are

$$T = \frac{2A}{1 + \frac{\overline{a}^{(2)}}{\overline{a}^{(1)}} \frac{[1 - (\overline{\rho}^{(1)}/\overline{\rho}^{(2)}) \sin^2 \theta]^{1/2}}{\cos \theta}} \tag{1.54}$$

$$R = T - A. \tag{1.55}$$

Eq. (1.46) has been used to simplify these expressions.

1.4.2 Finite Difference Model

It is easy to eliminate $u^{(1)}$ and $v^{(1)}$ from Eqs. (1.38) to (1.40) to obtain a single equation for $p^{(1)}$,

$$\frac{\partial^2 p^{(1)}}{\partial t^2} - (\overline{a}^{(1)})^2 \left(\frac{\partial^2 p^{(1)}}{\partial x^2} + \frac{\partial^2 p^{(1)}}{\partial y^2}\right) = 0. \tag{1.56}$$

Once $p^{(1)}$ is found, $v^{(1)}$ may be calculated by Eq. (1.39) or

$$\frac{\partial v^{(1)}}{\partial t} = -\frac{1}{\overline{\rho}^{(1)}} \frac{\partial p^{(1)}}{\partial y}. \tag{1.57}$$

These equations are valid for $y \geq 0$. Similarly for $y \leq 0$, the equations are

$$\frac{\partial^2 p^{(2)}}{\partial t^2} - (\overline{a}^{(2)})^2 \left(\frac{\partial^2 p^{(2)}}{\partial x^2} + \frac{\partial^2 p^{(2)}}{\partial y^2}\right) = 0 \tag{1.58}$$

$$\frac{\partial v^{(2)}}{\partial t} = -\frac{1}{\overline{\rho}^{(2)}} \frac{\partial p^{(2)}}{\partial y}. \tag{1.59}$$

A rectangular mesh with mesh size Δx and Δy, as shown in Figure 1.5, will be used for the finite difference model. Let ℓ and m be the mesh indices in the x and y directions. The fluid interface is at $m = 0$.

The time dependence of all variables is in the form $e^{-i\omega t}$. This may be factored out from the problem. For simplicity, let

$$u_{\ell,m}^{(1)}(t) = \text{Re}\left\{\hat{u}_{\ell,m}^{(1)} e^{-i\omega t}\right\} \tag{1.60}$$

and similarly for all the other variables. Suppose the standard second-order central difference is used to approximate the spatial derivatives in Eqs. (1.56) to (1.59), it is straightforward to obtain

$$\omega^2 \hat{p}_{\ell,m}^{(1)} + (\overline{a}^{(1)})^2 \left[\frac{1}{(\Delta x)^2}\left(\hat{p}_{\ell+1,m}^{(1)} - 2\hat{p}_{\ell,m}^{(1)} + \hat{p}_{\ell-1,m}^{(1)}\right)\right.$$
$$\left. + \frac{1}{(\Delta y)^2}\left(\hat{p}_{\ell,m+1}^{(1)} - 2\hat{p}_{\ell,m}^{(1)} + \hat{p}_{\ell,m-1}^{(1)}\right)\right] = 0 \tag{1.61}$$

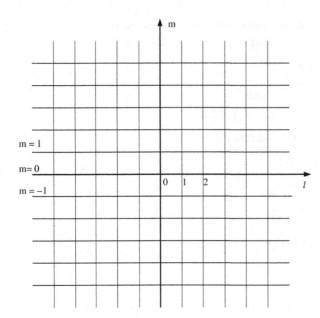

Figure 1.5. Computation mesh.

$$i\omega\hat{v}^{(1)}_{\ell,m} = \frac{1}{\overline{\rho}^{(1)}}\frac{\hat{p}^{(1)}_{\ell,m+1} - \hat{p}^{(1)}_{\ell,m-1}}{2\Delta y} \tag{1.62}$$

$$\omega^2 \hat{p}^{(2)}_{\ell,m} + (\overline{a}^{(2)})^2 \left[\frac{1}{(\Delta x)^2}\left(\hat{p}^{(2)}_{\ell+1,m} - 2\hat{p}^{(2)}_{\ell,m} + \hat{p}^{(2)}_{\ell-1,m} \right) \right.$$
$$\left. + \frac{1}{(\Delta y)^2}\left(\hat{p}^{(2)}_{\ell,m+1} - 2\hat{p}^{(2)}_{\ell,m} + \hat{p}^{(2)}_{\ell,m-1} \right) \right] = 0 \tag{1.63}$$

$$i\omega\hat{v}^{(2)}_{\ell,m} = \frac{1}{\overline{\rho}^{(2)}}\frac{\hat{p}^{(2)}_{\ell,m+1} - \hat{p}^{(2)}_{\ell,m-1}}{2\Delta y}. \tag{1.64}$$

Now, because a 3-point central difference stencil is used, Eqs. (1.61) and (1.62) should be applicable to values of $m \geq 1$. However, if it is insisted that they are to hold true at $m = 0$, just above the interface discontinuity, then a problem arises. These equations will include the nonphysical variable $\hat{p}^{(1)}_{\ell,-1}$. Similarly, Eqs. (1.63) and (1.64) should be applicable to $m \leq -1$. But if they are to be satisfied at $m = 0$, just below the interface discontinuity, then there is the problem of nonphysical variable $\hat{p}^{(2)}_{\ell,1}$ appearing in the equations. $\hat{p}^{(1)}_{\ell,-1}$ and $\hat{p}^{(2)}_{\ell,1}$ will be referred to as ghost values and the corresponding row of points at $m = -1$ and $m = +1$ are the ghost points. In a time-domain computation, the values $\hat{p}^{(1)}_{\ell,0}$, $\hat{v}^{(1)}_{\ell,0}$ are given by the time marching scheme of Eqs. (1.61) and (1.62). Similarly, $\hat{p}^{(2)}_{\ell,0}$ and $\hat{v}^{(2)}_{\ell,0}$ are given by those of Eqs. (1.63) and (1.64). Boundary conditions (1.44) and (1.45), however, require the continuity of pressure and vertical velocity component at the interface. This would not be possible in general. Here, it is important to recognize that, although the pressure is continuous at the interface, the pressure gradient is not. In fact, the pressure gradient should be discontinuous. The jump at the discontinuity is determined by boundary conditions (1.44) and (1.45). In the finite difference model, one may accept $\hat{p}^{(1)}_{\ell,-1}$ and $\hat{p}^{(2)}_{\ell,1}$, the ghost values, as the variables controlling the pressure gradients. These ghost values of pressure are to be chosen such that the discretized version of boundary conditions (1.44) and (1.45), namely,

$$\hat{p}^{(1)}_{\ell,0} = \hat{p}^{(2)}_{\ell,0} \tag{1.65}$$

$$\hat{v}^{(1)}_{\ell,0} = \hat{v}^{(2)}_{\ell,0} \tag{1.66}$$

are satisfied. Notice that there are two boundary conditions. They can be enforced by choosing the two ghost values properly.

The finite difference model of sound transmission as formulated, consisting of Eqs. (1.61) to (1.66), can be solved exactly analytically. The exact solution may be found as follows.

The incoming wave at an angle of incidence θ and having a total wave number k may be written in the form

$$\hat{p}^{(\text{incoming})}_{\ell,m} = A e^{ik(\sin\theta \, \Delta x \ell + \cos\theta \, \Delta y m)}. \tag{1.67}$$

By substitution of Eq. (1.67) into Eq. (1.61), i.e., replacing $\hat{p}^{(1)}_{\ell,m}$ by $\hat{p}^{(\text{incoming})}_{\ell,m}$, the equation is satisfied if k is given by the root of

$$\frac{\cos(k\sin\theta\,\Delta x)}{(\Delta x)^2} + \frac{\cos(k\cos\theta\,\Delta y)}{(\Delta y)^2} - \left(\frac{1}{(\Delta x)^2} + \frac{1}{(\Delta y)^2}\right) + \frac{\omega^2}{2(\bar{a}^{(1)})^2} = 0. \tag{1.68}$$

The cosine functions are symmetric with respect to k, so that if k is a positive real root, then $-k$ is also a root. The positive real root is the total wave number of the incident wave. If ω is complex, then k is also complex. It is to be determined by analytic continuation in the complex plane. The reflected wave has the same dependence on ℓ but propagates in the positive m direction. It is straightforward to find that the total wave field in region (1), being the sum of the incident and the reflected waves, may be written as

$$\hat{p}^{(1)}_{\ell,m} = A e^{-ik(\sin\theta\,\Delta x\ell + \cos\theta\,\Delta y m)} + \overline{R} e^{-ik(\sin\theta\,\Delta x\ell - \cos\theta\,\Delta y m)}$$
$$(m = 0, 1, 2, \ldots). \tag{1.69}$$

$\overline{R}$ is the reflected wave amplitude. By substituting Eq. (1.69) into Eq. (1.62), the vertical velocity component is found, as follows:

$$\hat{v}^{(1)}_{\ell,m}$$
$$= \begin{cases} \dfrac{\sin(k\cos\theta\,\Delta y)}{\omega\overline{\rho}^{(1)}\Delta y}\left[-A e^{-ik(\sin\theta\,\Delta x\ell + \cos\theta\,\Delta y m)} + \overline{R}\,e^{ik(-\sin\theta\,\Delta x\ell + \cos\theta\,\Delta y m)}\right], & m = 1, 2, 3, \ldots \\[2ex] \dfrac{-i}{2\omega\overline{\rho}^{(1)}\Delta y}\left[A e^{-ik(\sin\theta\,\Delta x\ell + \cos\theta\,\Delta y)} + \overline{R}\,e^{ik(-\sin\theta\,\Delta x\ell + \cos\theta\,\Delta y)} - \hat{p}^{(1)}_{\ell,1}\right], & m = 0 \end{cases}$$
$$\tag{1.70}$$

The transmitted wave in region (2) must have the same dependence on ℓ as the incident and reflected waves. Thus, let

$$\hat{p}^{(2)}_{\ell,m} = \overline{T} e^{-ik\sin\theta\,\Delta x\ell - i\beta\,\Delta y m}$$
$$(m = 0, -1, -2, -3, \ldots). \tag{1.71}$$

The transmitted wave must satisfy (1.63). On substituting (1.71) into (1.63), it is found that the equation is satisfied if β is the root of

$$\cos(\beta \Delta y) = (\Delta y)^2 \left[\frac{1}{(\Delta x)^2} + \frac{1}{(\Delta y)^2} - \frac{\cos(k \sin \theta \Delta x)}{(\Delta x)^2} - \frac{\omega^2}{2(\overline{a}^2)^2} \right]. \quad (1.72)$$

For the outgoing wave, β is the positive real root. If ω is complex, then β is obtained by analytic continuation.

The vertical velocity component of the transmitted wave is given by the substitution of (1.71) into (1.64). This yields

$$\hat{v}_{\ell,m}^{(2)} = \begin{cases} -\frac{\overline{T}}{\omega \overline{\rho}^{(2)} \Delta y} \sin(\beta \Delta y) e^{-ik \sin \theta \Delta x \ell - i\beta \Delta y m}, & m = -1, -2, -3, \ldots \\[2mm] \frac{-i}{2\omega \overline{\rho}^{(2)} \Delta y} \left[-\overline{T} e^{-ik \sin \theta \Delta x \ell + i\beta \Delta y} + \hat{p}_{\ell,1}^{(2)} \right], & m = 0. \end{cases} \quad (1.73)$$

It should be obvious that the ghost value of pressure must have the same dependence on ℓ as the incident, reflected, and transmitted waves. Hence, we may write

$$\hat{p}_{\ell,-1}^{(1)} = \hat{p}_{-1} e^{-ik \sin \theta \Delta x \ell} \quad (1.74)$$

$$\hat{p}_{\ell,1}^{(2)} = \hat{p}_1 e^{-ik \sin \theta \Delta x \ell}. \quad (1.75)$$

There are four boundary conditions that need to be satisfied at $m = 0$. The first two are from Eqs. (1.61) and (1.63). By setting $m = 0$, these equations give

$$\frac{\omega^2}{(\overline{a}^{(1)})^2} \hat{p}_{\ell,0}^{(1)} + \frac{1}{(\Delta x)^2} \left(\hat{p}_{\ell+1,0}^{(1)} - 2\hat{p}_{\ell,0}^{(1)} + \hat{p}_{\ell-1,0}^{(1)} \right) + \frac{1}{(\Delta y)^2} \left(\hat{p}_{\ell,1}^{(1)} - 2\hat{p}_{\ell,0}^{(1)} + \hat{p}_{\ell,-1}^{(1)} \right) = 0 \quad (1.76)$$

$$\frac{\omega^2}{(\overline{a}^{(2)})^2} \hat{p}_{\ell,0}^{(2)} + \frac{1}{(\Delta x)^2} \left(\hat{p}_{\ell+1,0}^{(2)} - 2\hat{p}_{\ell,0}^{(2)} + \hat{p}_{\ell-1,0}^{(2)} \right) + \frac{1}{(\Delta y)^2} \left(\hat{p}_{\ell,1}^{(2)} - 2\hat{p}_{\ell,0}^{(2)} + \hat{p}_{\ell,-1}^{(2)} \right) = 0. \quad (1.77)$$

The other two boundary conditions are the dynamic and kinematic boundary conditions (1.65) and (1.66). The incident, reflected, and transmitted wave solutions as found consist of Eqs. (1.69), (1.70), (1.71), and (1.73). There are four constants in the solution; they are $\overline{R}$, $\overline{T}$, $\hat{p}_{-1}$, and $\hat{p}_1$. By substituting the solution into the four boundary conditions, it is straightforward to find that the transmitted and reflected wave amplitudes are given by

$$\frac{\overline{T}}{A} = \frac{2}{1 + \left(\frac{\overline{\rho}^{(1)}}{\overline{\rho}^{(2)}} \right) \frac{\sin(\beta \Delta y)}{\sin(k \cos \theta \Delta y)}} \quad (1.78)$$

$$\overline{R} = \overline{T} - A. \quad (1.79)$$

It is easy to show that, in the limit $\Delta x, \Delta y \to 0$, $k \to \omega/\overline{a}^{(1)}$, $\beta \to \omega/\overline{a}^{(2)} [1 - (\overline{\rho}^{(1)}/\overline{\rho}^{(2)}) \sin^2 \theta]^{1/2}$, so that Eq. (1.78) is identical to the exact solution (1.54) of the continuous model.

Figure 1.6 shows the transmitted wave amplitude computed by Eq. (1.78) for the case $\overline{\rho}^{(1)}/\overline{\rho}^{(2)} = 1.2$ with $\omega \Delta x/\overline{a}^{(1)} = \pi/6$ and $\pi/4$, $\Delta x = \Delta y$ as a function of the angle of incidence. There is good agreement with the exact solution (1.54). For the given density ratio, total internal refraction occurs at $\theta = 65.91°$. For an

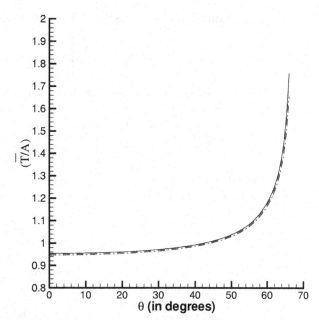

Figure 1.6. Transmitted wave amplitude as a function of the angle of incidence. _____, exact solution; – – –, $\omega\Delta x/\bar{a}^{(1)} = \pi/6$; ······, $\omega\Delta x/\bar{a}^{(1)} = \pi/4$.

incident angle greater than this value, the incident wave is totally reflected at the interface and there is no transmitted wave. T becomes complex, and the transmitted wave becomes damped exponentially in the negative y direction. The cutoff angle as determined by Eq. (1.78) is found to be very close to the exact value.

1.4.3 Boundary Modes

It turns out that the discretized interface problem consisting of finite difference equations (1.61) to (1.64) and boundary conditions (1.65) and (1.66) supported nontrivial homogeneous solutions. These are eigensolutions. They will be referred to as boundary modes. These solutions are bounded spatially, but they may grow or decay in time. A numerical solution is stable only if all boundary modes decay in time. If there is one growing boundary mode, then, in time, this mode will dominate the entire solution leading to numerical instability.

To find the boundary modes, let them be of the form,

$$\hat{p}_{\ell,m}^{(1)} = Ae^{i\alpha\ell\Delta x + i\beta_1 m\Delta y}, \quad m = -1, 0, 1, 2, \ldots \tag{1.80}$$

$$\hat{p}_{\ell,m}^{(2)} = Be^{i\alpha\ell\Delta x - i\beta_2 m\Delta y}, \quad m = 1, 0, -1, -2, \ldots \tag{1.81}$$

where α, the wave number, is real and arbitrary. β_1 and β_2 may be complex. For spatially bounded boundary modes, it is required that

$$\text{Im}(\beta_1) \geq 0; \quad \text{Im}(\beta_2) \geq 0. \tag{1.82}$$

If $\text{Im}(\beta_1) = 0$, then $\text{Re}(\beta_1) > 0$ and, similarly, if $\text{Im}(\beta_2) = 0$, then $\text{Re}(\beta_2) > 0$ to satisfy the outgoing wave condition.

Substitution of Eq. (1.80) into Eq. (1.61) yields

$$\cos(\beta_1 \Delta y) = \left(\frac{\Delta y}{\Delta x}\right)^2 \left[1 - \cos(\alpha \Delta x) - \frac{1}{2}\left(\frac{\omega \Delta x}{\overline{a}^{(1)}}\right)^2\right] + 1. \qquad (1.83)$$

Similarly, substitution of Eq. (1.81) into Eq. (1.63) leads to

$$\cos(\beta_2 \Delta y) = \left(\frac{\Delta y}{\Delta x}\right)^2 \left[1 - \cos(\alpha \Delta x) - \frac{1}{2}\left(\frac{\omega \Delta x}{\overline{a}^{(1)}}\right)^2 \left(\frac{\overline{a}^{(1)}}{\overline{a}^{(2)}}\right)^2\right] + 1. \qquad (1.84)$$

The corresponding $\hat{v}_{\ell,m}^{(1)}$ and $\hat{v}_{\ell,m}^{(2)}$ may easily be found by solving Eqs. (1.62) and (1.64). They are found to be

$$\hat{v}_{\ell,m}^{(1)} = A \frac{\sin(\beta_1 \Delta y)}{\omega \overline{\rho}^{(1)} \Delta y} e^{i\alpha\ell\Delta x + i\beta_1 m\Delta y} \qquad (1.85)$$

$$\hat{v}_{\ell,m}^{(2)} = -B \frac{\sin(\beta_2 \Delta y)}{\omega \overline{\rho}^{(2)} \Delta y} e^{i\alpha\ell\Delta x - i\beta_2 m\Delta y}. \qquad (1.86)$$

Upon imposing the dynamic and kinematic boundary condition (1.65) and (1.66) on solutions (1.80), (1.81), (1.85), and (1.86), it is straightforward to obtain

$$A - B = 0$$

$$\frac{\sin(\beta_1 \Delta y)}{\overline{\rho}^{(1)}} A + \frac{\sin(\beta_2 \Delta y)}{\overline{\rho}^{(2)}} B = 0.$$

For nontrivial solutions, the determinant, D, of the coefficient matrix of the above 2×2 linear system for A and B must be equal to zero; i.e.,

$$D \equiv \frac{\sin(\beta_1 \Delta y)}{\overline{\rho}^{(1)}} + \frac{\sin(\beta_2 \Delta y)}{\overline{\rho}^{(2)}} = 0. \qquad (1.87)$$

The eigenvalues, $\omega \Delta x / \overline{a}^{(1)}$, of the boundary modes are the roots of Eq. (1.87) for a given $\alpha \Delta x$.

To determine the eigenvalues numerically, the following grid search method may be used. To start the search, the complex $\omega \Delta x / \overline{a}^{(1)}$ plane is divided into a grid. The intent is to calculate the complex value of D at each grid point. Once the values of D on the grid are known, a contour plot is used to locate the curves $\text{Re}(D) = 0$ and $\text{Im}(D) = 0$ (where $\text{Re}(\ )$ and $\text{Im}(\ )$ denote the real and imaginary part). The intersections of these curves give $D = 0$ or the locations of the boundary mode frequencies. To improve accuracy, one may use a refined grid or use a Newton or similar iteration method.

To calculate the value of D at a grid point in the complex $\omega \Delta x / \overline{a}^{(1)}$ plane for a prescribed density and mesh size ratio and $\alpha \Delta x$, the values of $\beta_1 \Delta y$ and $\beta_2 \Delta y$ are first computed by solving Eqs. (1.83) and (1.84). All these values are then substituted into Eq. (1.87) to compute D. Figure 1.7 shows the contours $\text{Re}(D) = 0$ and $\text{Im}(D) = 0$ for the case $\overline{\rho}^{(1)} / \overline{\rho}^{(2)} = 2.0$, $\Delta x = \Delta y$, and $\alpha \Delta x = \pi/10$. These two curves intercept each other on the real $\omega \Delta x / \overline{a}^{(1)}$ axis at $\omega \Delta x / \overline{a}^{(1)} = (0.536, 0.0)$. Since $\omega \Delta x / \overline{a}^{(1)}$ is real, this root corresponds to a neutral mode. With the addition of artificial selective damping to the finite difference scheme, which is a standard procedure and will be

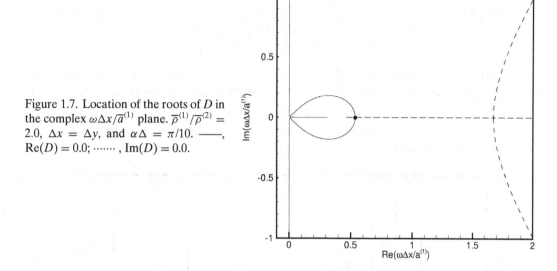

Figure 1.7. Location of the roots of D in the complex $\omega \Delta x / \overline{a}^{(1)}$ plane. $\overline{p}^{(1)} / \overline{p}^{(2)} = 2.0$, $\Delta x = \Delta y$, and $\alpha \Delta = \pi/10$. ———, $\mathrm{Re}(D) = 0.0$; ⋯⋯⋯ , $\mathrm{Im}(D) = 0.0$.

discussed in a later chapter, the mode will be damped. Thus, the finite difference scheme is not subjected to numerical instability arising from boundary modes.

EXERCISES

1.1. The governing equation and boundary conditions for a rectangular vibrating membrane of width W and breadth B are

$$\frac{\partial^2 u}{\partial t^2} = c^2 \left(\frac{\partial^2 u}{\partial x^2} + \frac{\partial^2 u}{\partial y^2} \right),$$

$$\text{At } x = 0, \quad x = W, \quad y = 0, \quad y = B, \quad u = 0,$$

where u is the displacement of the membrane. It is easy to show that the normal mode frequencies of the vibrating membrane are

$$\omega_{jk} = c \left[\left(\frac{j\pi}{W} \right)^2 + \left(\frac{k\pi}{B} \right)^2 \right]^{1/2}, \quad \text{where } j \text{ and } k \text{ are integers.}$$

Now divide the width into N subdivisions and the breadth into M subdivisions so that $\Delta x = W/N$ and $\Delta y = B/M$. Suppose a second-order central difference is used to approximate the x and y derivatives. This converts the partial differential equation into a partial difference equation, as follows:

$$\frac{\partial^2 u_{\ell,m}}{\partial t^2} = c^2 \left[\frac{u_{\ell+1,m} - 2u_{\ell,m} + u_{\ell-1,m}}{(\Delta x)^2} + \frac{u_{\ell,m+1} - 2u_{\ell,m} + u_{\ell,m+1}}{(\Delta y)^2} \right],$$

where ℓ, m are the spatial indices in the x and y directions ($\ell = 0, 1, 2, \ldots, N$; $m = 0, 1, 2, \ldots, M$). The boundary conditions for the finite difference equations are

$$\text{at } \ell = 0, \quad \ell = N, \quad m = 0, \quad m = M, \quad u_{\ell,m} = 0.$$

Look for the normal mode solution of the finite difference equation of the following form:

$$u_{\ell,m}(t) = A \sin\left(\frac{j\pi\ell}{N}\right) \sin\left(\frac{k\pi m}{M}\right) e^{-i\omega t},$$

where j and k are integers. Find the normal mode frequencies and mode shapes of the discretized system and compare them with the exact eigensolution of the original problem.

1.2. Consider a one-dimensional tube with an end wall at $x = 0$ as shown. Acoustic disturbances inside the tube are governed by the simple wave equation,

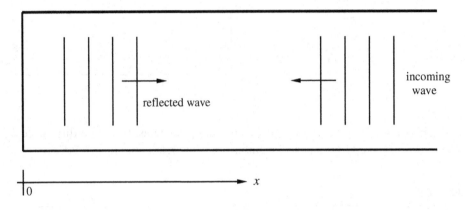

$$\frac{\partial^2 u}{\partial t^2} = c^2 \frac{\partial^2 u}{\partial x^2},$$

where u is the fluid velocity and c is the sound speed. The boundary condition at the wall is

$$\text{at } x = 0, \quad u = 0.$$

An incoming wave (a solution of the simple wave equation) of angular frequency ω and amplitude A of the form

$$u_{\text{incident}} = A e^{-i\omega(x/c+t)}$$

impinges on the wall at $x = 0$. This creates a reflected wave. The frequency and amplitude of the reflected wave (also a solution of the simple wave equation) must be such that the wall boundary condition is satisfied; i.e., at $x = 0$

$$u_{\text{incident}} + u_{\text{reflected}} = 0.$$

It is easy to find that the reflected wave is given by

$$u_{\text{reflected}} = -A e^{i\omega(x/c-t)}.$$

Now, suppose the simple wave equation is discretized by using a second-order central difference approximation. This gives

$$\frac{d^2 u_\ell}{dt^2} = \frac{c^2}{(\Delta x)^2}(u_{\ell+1} - 2u_\ell + u_{\ell-1}),$$

where ℓ is the spatial index.

Consider an incident wave of the form

$$u_\ell(t) = Ae^{-i(\kappa\ell\Delta x+\omega t)},$$

where $\kappa\Delta x = \cos^{-1}(1 - \frac{\omega^2\Delta x^2}{2c^2})$. Note: the value of κ, the wave number, is determined by the requirement that the incident wave satisfies the governing finite difference equation.

Find the reflected wave of the discretized system and show that the reflected wave has the same amplitude as the incident wave. Notice that the wave number κ and, hence, the phase velocity, ω/κ, and group velocity, $d\omega/d\kappa$, of the waves supported by the finite difference approximation are not the same as those of the original simple wave equation. Is the approximation good at high or low frequency?

1.3. Acoustic waves in a medium at rest are governed by the linearized Euler and energy equations. Let (u,v) be the velocity components in the x and y directions and ρ_0 and p_0 be the static density and pressure. The governing equations are

$$\frac{\partial u}{\partial t} + \frac{1}{\rho_0}\frac{\partial p}{\partial x} = 0$$

$$\frac{\partial v}{\partial t} + \frac{1}{\rho_0}\frac{\partial p}{\partial y} = 0$$

$$\frac{\partial p}{\partial t} + \gamma p_0\left(\frac{\partial u}{\partial x} + \frac{\partial v}{\partial y}\right) = 0,$$

where γ is the ratio of specific heats. Upon eliminating u and v, it is easy to find that the governing equation for p is

$$\frac{\partial^2 p}{\partial t^2} = c^2\left(\frac{\partial^2 p}{\partial x^2} + \frac{\partial^2 p}{\partial y^2}\right), \quad c^2 = \frac{\gamma p_0}{\rho_0}. \tag{A}$$

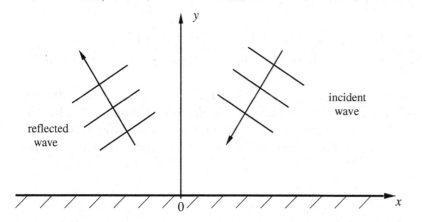

Consider plane acoustic waves propagating at an angle θ with respect to the y-axis incident on a wall lying on the x-axis. The incident wave is given by

$$p_{\text{incident}} = Ae^{-\frac{i\omega}{c}(x\sin\theta+y\cos\theta+ct)}. \tag{B}$$

The boundary condition at the wall is

$$y = 0, \qquad v = 0.$$

By the y-momentum equation, the wall boundary condition for pressure p is

$$\text{At } y = 0, \qquad \frac{\partial p}{\partial y} = 0.$$

In order to satisfy the wall boundary condition, a reflected wave is generated at the wall. This reflected wave cancels the pressure gradient of the incident wave in the y direction at the wall, namely,

$$y = 0, \qquad \frac{\partial p_{\text{incident}}}{\partial y} + \frac{\partial p_{\text{reflected}}}{\partial y} = 0 \qquad \text{(C)}$$

It is easy to find that the solution of Eq. (A) that satisfies Eq. (C) is,

$$p_{\text{reflected}} = A e^{i\frac{\omega}{c}(-x\sin\theta + y\cos\theta - ct)} \qquad \text{(D)}$$

By formulas (B) and (D), it is straightforward to show that the following well-known properties of acoustic wave reflection are true.

  (i) The angle of reflection is equal to the angle of incidence.
  (ii) The reflected wave has the same intensity as the incident wave.
  (iii) There is a pressure doubling at the wall, that is,

$$\overline{p_{\text{wall}}^2} = 2\overline{p_{\text{incident}}^2} = 2\overline{p_{\text{reflected}}^2},$$

where the overbar denotes a time average.

Suppose the problem is solved by finite difference approximation with a mesh size of Δx and Δy. On using a second-order central difference approximation, Eqs. (A) and (C) become

$$\frac{\partial^2 p_{\ell,m}}{\partial t^2} = c^2 \left[\frac{p_{\ell+1,m} - 2p_{\ell,m} + p_{\ell-1,m}}{(\Delta x)^2} + \frac{p_{\ell,m+1} - 2p_{\ell,m} + p_{\ell,m-1}}{(\Delta y)^2} \right] \qquad \text{(E)}$$

$$m = 0, \qquad \left(p_{\ell,m+1}^{\text{incident}} - p_{\ell,m-1}^{\text{incident}} \right) + \left(p_{\ell,m+1}^{\text{reflected}} - p_{\ell,m-1}^{\text{reflected}} \right) = 0, \qquad \text{(F)}$$

where ℓ, m are the spatial indices in the x and y directions and the wall is located at $m = 0$.

Find the incident wave with frequency ω and angle of incidence θ. Find the reflected wave so that boundary condition (F) is satisfied. Determine whether the acoustic wave reflection properties (i), (ii), and (iii) are satisfied. Would you be able to find the incident and reflected waves if a fourth-order central difference scheme is used?

2 Spatial Discretization in Wave Number Space

In this chapter, how to form finite difference approximations to partial derivatives of the spatial coordinates is considered. The standard approach assumes that the mesh size goes to zero, i.e., $\Delta x \to 0$, in formulating finite difference approximations. However, in real applications, Δx is never equal to zero. It is, therefore, useful to realize that it is possible to develop an accurate approximation to $\partial/\partial x$ by a finite difference quotient with finite Δx. Many finite difference approximations are based on truncated Taylor series expansions. Here, a very different approach is introduced. It will be shown that finite difference approximation may be formulated in wavenumber space. There are advantages in using a wave number approach. They will be elaborated throughout this book.

2.1 Truncated Taylor Series Method

A standard way to form finite difference approximations to $\partial/\partial x$ is to use Taylor series truncation. Let the x-axis be divided into a regular grid with spacing Δx. Index ℓ (integer) will be used to denote the ℓth grid point. On applying Taylor series expansion as $\Delta x \to 0$, it is easy to find $(x_\ell = \ell \Delta x)$ as follows:

$$u_{\ell+1} = u(x_\ell + \Delta x) = u_\ell + \left(\frac{\partial u}{\partial x}\right)_\ell \Delta x + \frac{1}{2}\left(\frac{\partial^2 u}{\partial x^2}\right)_\ell \Delta x^2 + \frac{1}{3!}\left(\frac{\partial^3 u}{\partial x^3}\right)_\ell \Delta x^3 + \cdots$$

$$u_{\ell-1} = u(x_\ell - \Delta x) = u_\ell - \left(\frac{\partial u}{\partial x}\right)_\ell \Delta x + \frac{1}{2}\left(\frac{\partial^2 u}{\partial x^2}\right)_\ell \Delta x^2 - \frac{1}{3!}\left(\frac{\partial^3 u}{\partial x^3}\right)_\ell \Delta x^3 + \cdots$$

$$(2.1)$$

Thus,

$$u_{\ell+1} - u_{\ell-1} = 2\Delta x \left(\frac{\partial u}{\partial x}\right)_\ell + \frac{1}{3}\left(\frac{\partial^3 u}{\partial x^3}\right)_\ell \Delta x^3 + \cdots.$$

On neglecting higher-order terms, a second-order central difference approximation is obtained,

$$\left(\frac{\partial u}{\partial x}\right)_\ell = \frac{u_{\ell+1} - u_{\ell-1}}{2\Delta x} + O(\Delta x^2) \qquad (2.2)$$

This is a 3-point second-order stencil. By keeping a larger number of terms in (2.1) before truncation, it is easy to derive the following 5-point fourth-order stencil and the 7-point sixth-order stencil as follows:

$$\left(\frac{\partial u}{\partial x}\right)_\ell = \frac{-u_{\ell+2} + 8u_{\ell+1} - 8u_{\ell-1} + u_{\ell-2}}{12\Delta x} + O(\Delta x^4) \tag{2.3}$$

$$\left(\frac{\partial u}{\partial x}\right)_\ell = \frac{u_{\ell+3} - 9u_{\ell+2} + 45u_{\ell+1} - 45u_{\ell-1} + 9u_{\ell-2} - u_{\ell-3}}{60\Delta x} + O(\Delta x^6). \tag{2.4}$$

From the order of the truncation error, one expects that the 7-point stencil is probably a more accurate approximation than the 5-point stencil, which is, in turn, more accurate than the 3-point stencil. However, how much better, and in what sense it is better, are not clear. These are the glaring weaknesses of the Taylor series truncation method.

2.2 Optimized Finite Difference Approximation in Wave Number Space

Suppose a central $(2N+1)$-point stencil is used to approximate $\partial/\partial x$; i.e.,

$$\left(\frac{\partial f}{\partial x}\right)_\ell \simeq \frac{1}{\Delta x} \sum_{j=-N}^{N} a_j f_{\ell+j}. \tag{2.5}$$

At one's disposal are the stencil coefficients a_j; $j = -N$ to N. These coefficients will now be chosen such that the Fourier transform of the right side of Eq. (2.5) is a good approximation of that of the left side, for arbitrary function f. Fourier transform is defined only for functions of a continuous variable. For this purpose, Eq. (2.5), which relates the values of a set of points spaced Δx apart, is assumed to hold for any similar set of points Δx apart along the x-axis. The generalized form of Eq. (2.5), applicable to the continuous variable x, is

$$\frac{\partial f}{\partial x} \simeq \frac{1}{\Delta x} \sum_{-N}^{N} a_j f(x + j\Delta x). \tag{2.6}$$

Eq. (2.5) is a special case of Eq. (2.6). By setting $x = \ell\Delta x$ in Eq. (2.6), Eq. (2.5) is recovered.

The Fourier transform of a function $F(x)$ and its inverse are related by

$$\tilde{F}(\alpha) = \frac{1}{2\pi} \int_{-\infty}^{\infty} F(x)e^{-i\alpha x} dx$$

$$F(x) = \int_{-\infty}^{\infty} \tilde{F}(\alpha)e^{i\alpha x} d\alpha. \tag{2.7}$$

The following useful theorems concerning Fourier transform are proven in Appendix A.

2.2.1 Derivative Theorem

$$\text{Transform of } \frac{\partial f(x)}{\partial x} = i\alpha \tilde{f}(\alpha).$$

2.2.2 Shifting Theorem

$$\text{Transform of } f(x + \lambda) = e^{i\alpha\lambda}\tilde{f}(\alpha).$$

Now, by applying the Fourier transform to Eq. (2.6) and making use of the Derivative and Shifting theorems, it is found that

$$i\alpha\tilde{f}(\alpha) \simeq \frac{1}{\Delta x}\left[\sum_{j=-N}^{N} a_j e^{ij\alpha\Delta x}\right]\tilde{f}(\alpha) \equiv i\bar{\alpha}\tilde{f}(\alpha), \tag{2.8}$$

where

$$\bar{\alpha} = \frac{-i}{\Delta x}\sum_{j=-N}^{N} a_j e^{ij\alpha\Delta x}. \tag{2.9}$$

α on the left of (2.8) is the wave number of the partial derivative. By comparing the left and right sides of this equation, it becomes clear that $\bar{\alpha}$ on the right side is effectively the wave number of the finite difference scheme. $\bar{\alpha}\,\Delta x$ is a periodic function of $\alpha\Delta x$ with a period of 2π. To ensure that the Fourier transform of the finite difference scheme is a good approximation of that of the partial derivative over the wave number range of, say, waves with wavelengths longer than four Δx or $\alpha\Delta x \leq (\pi/2)$, it is required that a_j be chosen to minimize the integrated error, E, over this wave number range, where

$$E = \int_{-\frac{\pi}{2}}^{\frac{\pi}{2}} |\alpha\Delta x - \bar{\alpha}\Delta x|^2 d\,(\alpha\Delta x). \tag{2.10}$$

In order that $\bar{\alpha}$ is real, the coefficients a_j must be antisymmetric; i.e.,

$$a_0 = 0, \qquad a_{-j} = -a_j. \tag{2.11}$$

On substituting $\bar{\alpha}$ from Eq. (2.9) into Eq. (2.10) and taking Eq. (2.11) into account, E may be rewritten as,

$$E = \int_{-\frac{\pi}{2}}^{\frac{\pi}{2}} \left[k - 2\sum_{j=1}^{N} a_j \sin(kj)\right]^2 dk. \tag{2.12}$$

The conditions for E to be a minimum are

$$\frac{\partial E}{\partial a_j} = 0, \qquad j = 1, 2, \ldots, N. \tag{2.13}$$

Eq. (2.13) provides N equations for the N coefficients $a_j, j = 1, 2, \ldots, N$.

Now consider the case of $N = 3$ or a 7-point stencil.

$$\frac{\partial f}{\partial x} \simeq \frac{1}{\Delta x}\sum_{j=-3}^{3} a_j f(x + j\Delta x); \qquad a_{-j} = -a_j. \tag{2.14}$$

There are three coefficients $a_1, a_2,$ and a_3 to be determined. It is possible to combine the truncated Taylor series method and the Fourier transform optimization method.

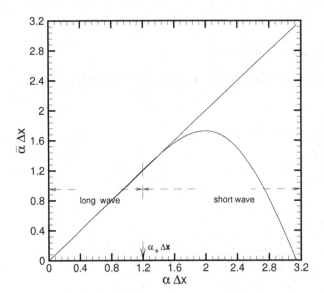

Figure 2.1. $\bar{\alpha}\Delta x$ versus $\alpha\Delta x$ for the 7-point fourth-order optimized central difference scheme.

Let Eq. (2.14) be required to be accurate to $O(\Delta x^4)$. This can be done by expanding the right side of Eq. (2.14) as a Taylor series and then equating terms of order Δx and $(\Delta x)^3$. This gives

$$a_2 = \frac{9}{20} - \frac{4a_1}{5}$$

$$a_3 = -\frac{2}{15} + \frac{a_1}{5}.$$

On substitution into Eq. (2.12), the integrated error E is a function of a_1 alone. The value a_1 may be found by solving

$$\frac{\partial E}{\partial a_1} = 0.$$

This gives the following values

$$a_0 = 0 \qquad\qquad a_1 = -a_{-1} = 0.79926643$$
$$a_2 = -a_{-2} = -0.18941314 \quad a_3 = -a_{-3} = 0.02651995\,.$$

The relationship between $\bar{\alpha}\,\Delta x$ and $\alpha\Delta x$ over the interval 0 to π for the optimized 7-point stencil is shown graphically in Figure 2.1. For $\alpha\Delta x$ up to about 1.2, the curve is nearly the same as the straight line $\bar{\alpha}\,\Delta x = \alpha\Delta x$. Thus, the finite difference scheme can provide reasonably good approximation for $\alpha\Delta x < 1.2$ or $\lambda > 5.2\Delta x$, where λ is the wavelength.

For $\alpha\Delta x$ greater than 1.6, the $\bar{\alpha}(\alpha)\Delta x$ curve deviates increasingly from the straight-line relationship. Thus, the wave propagation characteristics of the short waves of the finite difference scheme are very different from those of the partial differential equation. There is an in-depth discussion of short waves later.

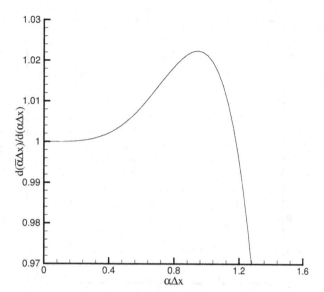

Figure 2.2. $\frac{d(\bar{\alpha}\Delta x)}{d(\alpha\Delta x)}$ versus $\alpha\Delta x$ for $N = 3$, fourth-order scheme optimized over the range of $-\pi/2$ to $\pi/2$.

2.3 Group Velocity Consideration

There is no question that a good finite difference scheme should have $\bar{\alpha}\Delta x$ equal to $\alpha\Delta x$ over as wide a bandwidth as possible. However, from the standpoint of wave propagation, it is also important to ensure that the group velocity of the finite difference scheme is a good approximation of that of the original equation. It will be shown later that the group velocity of a finite difference scheme is determined by $d\bar{\alpha}/d\alpha$. $d\bar{\alpha}/d\alpha$, the slope of the $\alpha\Delta x - \bar{\alpha}\Delta x$ curve, should be equal to 1 if the scheme is to reproduce the same group velocity or speed of propagation of the original partial differential equation.

Figure 2.2 shows the $d\bar{\alpha}/d\alpha$ curve of the optimized 7-point stencil scheme as a function of $\alpha\Delta x$. It is seen that for $0.4 < \alpha\Delta x < 1.15$, $d\bar{\alpha}/d\alpha$ is larger than 1. This will lead to numerical dispersion, because some wave components would propagate faster than others and faster than the wave speed of the original partial differential equation. Around $\alpha\Delta x = 0.8$ to 1.0, the slope is greater than 1.02. Thus, the wave component in this wave number range will propagate at a speed 2 percent faster (assuming time is computed exactly). That is to say, after propagating over 100 mesh spacings, the group of fast waves has reached the 102 mesh point. This is too much dispersion for long-range propagation.

One way to reduce numerical dispersion is to reduce the range of optimization. Instead of the range of $\alpha\Delta x = 0$ to $\pi/2$, one may use the range of 0 to η. That is, the integrated error is

$$E = \int_0^\eta |\alpha\Delta x - \bar{\alpha}\Delta x|^2\, d(\alpha\Delta x).$$

By setting η to a specified set of values, the coefficients a_1, a_2, and a_3 of the 7-point central difference stencil may be found as before. Upon examining the corresponding $d\bar{\alpha}/d\alpha$ curve, it is recommended that the case $\eta = 1.1$ has the best overall wave

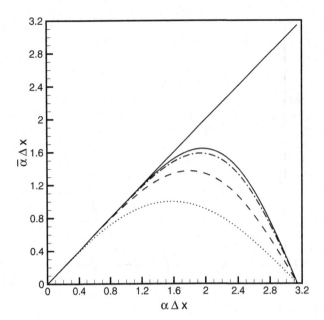

Figure 2.3. $\bar{\alpha}\Delta x$ versus $\alpha\Delta x$. ———, optimized fourth-order scheme; —·—·—, standard sixth-order scheme; – – –, standard fourth-order scheme; ·······, standard second-order scheme.

propagation characteristics. The stencil coefficients are

$$a_0 = 0 \qquad\qquad a_1 = -a_{-1} = 0.77088238051822552$$
$$a_2 = -a_{-2} = -0.166705904414580469 \quad a_3 = -a_{-3} = 0.02084314277031176.$$

Figure 2.3 shows the $\bar{\alpha}\,\Delta x$ versus $\alpha\Delta x$ relations for the standard sixth-order, fourth-order, and second-order central difference scheme and the optimized scheme ($\eta = 1.1$). The values of $\alpha\Delta x$ at approximately the point of deviation from the $\bar{\alpha}\,\Delta x - \alpha\Delta x$ straight-line relation for these schemes, are

Scheme	$\alpha_c\Delta x$	Resolution, $\frac{\lambda}{\Delta x}$
Sixth-order	0.85	7.4
Fourth-order	0.60	10.5
Second-order	0.30	21.0
Optimized	0.95	6.6

where λ is the wavelength. Thus, a high-order finite difference scheme can better resolve small-scale features.

Figure 2.4 is a comparison of the $\frac{d(\bar{\alpha}\Delta x)}{d(\alpha\Delta x)}$ curves. Figure 2.5 is the same plot but with an enlarged vertical scale. The increase in group velocity near $\alpha\Delta x = 0.7$ is around 0.3 percent. This means that the scheme has very small numerical dispersion. The resolved bandwidth for $|d\bar{\alpha}/d\alpha - 1.0| < 0.003$ is about 15 percent wider than that of the standard sixth-order scheme. This improvement is achieved at no cost because both schemes have a 7-point stencil.

EXAMPLE. Consider the initial value problem governed by the one-dimensional convective wave equation. Dimensionless variables with Δx as the length scale,

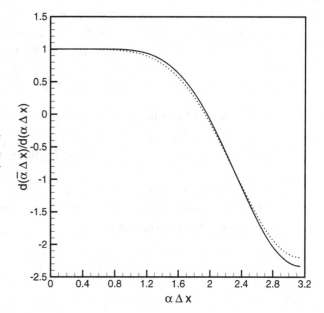

Figure 2.4. $\frac{d(\bar{\alpha}\Delta x)}{d(\alpha\Delta x)}$ as a function of $\alpha\Delta x$ for the ——, optimized scheme ($\eta = 1.1$) and the ······, standard sixth-order scheme.

$\Delta x/c$ ($c =$ sound speed) as the timescale will be used. The mathematical problem is

$$\frac{\partial u}{\partial t} + \frac{\partial u}{\partial x} = 0, \qquad -\infty < x < \infty. \tag{2.15}$$

$$t = 0 \qquad u = f(x). \tag{2.16}$$

The exact solution is

$$u = f(x - t).$$

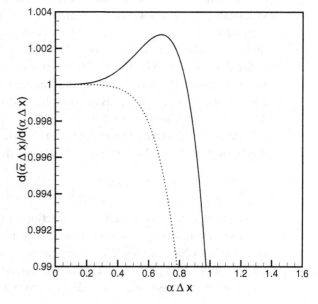

Figure 2.5. Enlarged $\frac{d(\bar{\alpha}\Delta x)}{d(\alpha\Delta x)}$ versus $\alpha\Delta x$ curves. ——, optimized scheme and the ······, standard sixth-order central difference scheme.

In other words, the solution consists of the initial disturbance propagating to the right at a dimensionless speed of 1 without any distortion of the waveform. In particular, if the initial disturbance is in the form of a Gaussian function,

$$f(x) = 0.5 \exp\left[-\ln 2 \left(\frac{x}{3}\right)^2\right], \tag{2.17}$$

with a half-width of three mesh spacings, then the exact solution is

$$u = 0.5 \exp\left[-\ln 2 \left(\frac{x-t}{3}\right)^2\right]. \tag{2.18}$$

On using a $(2N + 1)$-point stencil finite difference approximation to $\partial/\partial x$ in Eq. (2.15), the original initial value problem is converted to the following system of ordinary differential equations:

$$\frac{du_\ell}{dt} + \sum_{j=-N}^{N} a_j u_{\ell+j} = 0, \qquad \ell = \text{integer} \tag{2.19}$$

$$t = 0 \qquad u_\ell = f(\ell). \tag{2.20}$$

ℓ is the spatial index.

For the Gaussian pulse, the initial condition is

$$t = 0 \qquad u_\ell = 0.5 \exp\left[-\ln 2 \left(\frac{\ell}{3}\right)^2\right]. \tag{2.21}$$

The system of ordinary differential equations (2.19) and initial condition (2.21) can be integrated in time numerically by using the Runge-Kutta or a multistep method. The solutions using the standard central second-order ($N = 1$), fourth-order ($N = 2$), sixth-order ($N = 3$), and the optimized 7-point ($N = 3$, $\eta = 1.1$) finite difference schemes at $t = 400$ are shown in Figure 2.6. The exact solution in the form of a Gaussian pulse is shown as the dotted curves. The second-order solution is in the form of an oscillatory wave train. There is no resemblance to the exact solution. The numerical solution is totally dispersed. The fourth-order solution is better. The sixth-order scheme is even better, but there are still some dispersive waves trailing behind the main pulse. The wavelength of the trailing wave is about 7, which corresponds to a wave number of 0.9, which is beyond the resolved range of the sixth-order scheme. The optimized scheme gives very acceptable numerical results.

To understand why the different finite difference scheme performs in this way, it is to be noted that the Fourier transform of the initial disturbance is

$$\tilde{f} = \frac{3}{4(\pi \ln 2)^{1/2}} \exp\left[-\frac{9\alpha^2}{4 \ln 2}\right]. \tag{2.22}$$

It is easy to verify that a significant fraction of the initial wave spectrum lies in the short-wave range ($\alpha \Delta x > \alpha_c \Delta x$) of the $\bar{\alpha} \Delta x$ versus $\alpha \Delta x$ curve of the standard second- and fourth-order finite difference schemes. (Note: $\Delta x = 1$ in this problem.) These wave components propagate at a slower speed (smaller than the group velocity). This causes the pulse solution to disperse in space and exhibits a wavy tail.

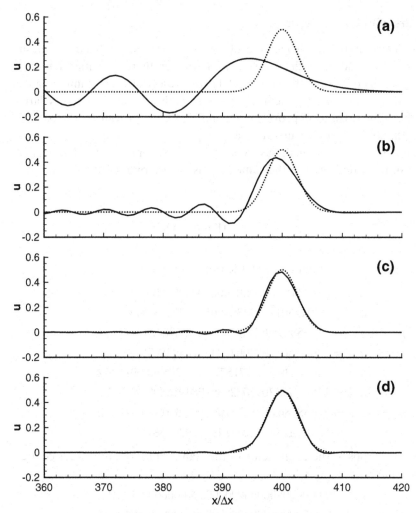

Figure 2.6. Comparison between the computed and the exact solution of the simple one-dimensional convective wave equation. (a) second-order, (b) fourth-order, and (c) sixth-order standard central difference scheme, (d) 7-point optimized scheme. ———, numerical solution; · · · · · · ·, exact solution.

Another way to understand the dispersion effect is to superimpose the $\tilde{f}(\alpha)$ curve on the $d\bar{\alpha}/d\alpha$ versus α curve. The standard sixth-order scheme maintains a wave speed nearly equal to 1 ($d\bar{\alpha}/d\alpha \cong 1.0$) up to $\alpha\Delta x = 0.85$. Beyond this value the wave speed decreases. The optimized scheme maintains an accurate wave speed up to $\alpha\Delta x = 1.0$. Now, for the given initial condition, there is a small quantity of wave energy around $\alpha\Delta x = 0.9$. This component, having a slower wave speed, appears as trailing waves in the solution of the standard sixth-order scheme. The wavelength of the trailing wave is $\lambda = (2\pi/\alpha) \simeq 7.0$. This is consistent with the numerical result shown in Figure 2.6. For the optimized scheme, there are trailing waves with wave number $\simeq 1.0$ or $\lambda \simeq 6.3$, but the amplitude is much smaller and is, therefore, not so easily detectable. Basically, none of these schemes can resolve waves with wavelengths less than six mesh spacings really well.

2.4 Schemes with Large Stencils

By now, it is clear that, if one wishes to resolve short waves using a fixed size mesh, it is necessary to use schemes with large stencils. However, there is limitation to this. Beyond the optimized 7-point stencil ($\eta = 1.1$) with resolved wave number range up to $\alpha \Delta x = 0.95$ (seven mesh spacings per wavelength), it will take a very large increase in the stencil size to improve the resolution to five or four mesh spacings per wavelength. So the cost goes up quite fast.

Consider a very large stencil, say $N = 7$ or a 15-point stencil. The stencil coefficients can be determined in the same way as before by minimizing the integrated error E.

$$E = \int_{-\eta}^{\eta} |\alpha \Delta x - \overline{\alpha} \Delta x|^2 \, d(\alpha \Delta x).$$

For $\eta = 1.8$ (a good overall compromise), the coefficients are

$$
\begin{aligned}
a_0 &= 0.000000000000000000000000000000D + 0 \\
a_1 = -a_{-1} &= 9.194250111034304505927772288 5D - 1 \\
a_2 = -a_{-2} &= -3.558295992683526875566764240 1D - 1 \\
a_3 = -a_{-3} &= 1.525150160840649246910492867 9D - 1 \\
a_4 = -a_{-4} &= -5.946304082971577266682859689 9D - 2 \\
a_5 = -a_{-5} &= 1.901075270950829865984916798 8D - 2 \\
a_6 = -a_{-6} &= -4.380864929733648185113700090 7D - 3 \\
a_7 = -a_{-7} &= 5.389612186862338465969295587 8D - 4
\end{aligned}
$$

For comparison purposes, the coefficients of the standard fourteenth-order central difference scheme are

$$
\begin{aligned}
a_0 &= 0.000000000000000000000000000000D + 0 \\
a_1 = -a_{-1} &= 8.750000000000000000000000000000D - 1 \\
a_2 = -a_{-2} &= -2.916666666666666666666666667D - 1 \\
a_3 = -a_{-3} &= 9.722222222222222222222222222D - 2 \\
a_4 = -a_{-4} &= -2.651515151515151515151515101D - 2 \\
a_5 = -a_{-5} &= 5.303030303030303030303030303030D - 3 \\
a_6 = -a_{-6} &= -6.798756798756798756798756798 7D - 4 \\
a_7 = -a_{-7} &= 4.162504162504162504162504162 5D - 5
\end{aligned}
$$

The $\overline{\alpha} \Delta x$ versus $\alpha \Delta x$ and the $d\overline{\alpha}/d\alpha$ versus $\alpha \Delta x$ curves are shown in Figures 2.7 and 2.8, respectively. The choice of η is based on the enlarged $d\overline{\alpha}/d\alpha$ curve. For very large stencils, these curves exhibit small-amplitude oscillations. The $\eta = 1.8$ case gives a good balance between the resolved band width and constancy of group velocity. This stencil can resolve waves as short as $3.5\Delta x$.

To demonstrate the effectiveness of the use of a large computation stencil to resolve short waves, again consider the initial value problem of Eqs. (2.15) and (2.16). The initial disturbance is again taken to be a Gaussian

$$f(x) = h e^{-\ln 2 (\frac{x}{b})^2}, \tag{2.23}$$

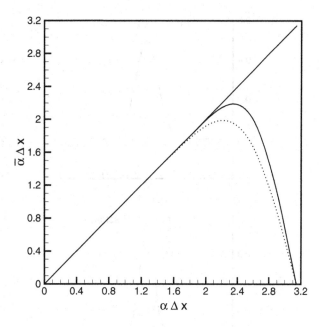

Figure 2.7. $\bar{\alpha}\Delta x$ versus $\alpha\Delta x$ curve for the optimized 15-point stencil ($\eta = 1.8$) and the standard fourteenth-order central difference scheme. ———, optimized scheme; $\cdots\cdots\cdots$, standard fourteenth-order scheme.

where h is the initial maximum amplitude of the pulse and b is the half-width. The Fourier transform of (2.23) is

$$\tilde{f} = \frac{hb}{2(\pi \ln 2)^{\frac{1}{2}}} e^{-\frac{\alpha^2 b^2}{4 \ln 2}}. \tag{2.24}$$

Notice that the smaller the value of b, the sharper is the pulse waveform in physical space. Also, the corresponding wave number spectrum of the pulse is wider. This makes it more difficult to produce an accurate numerical solution. The numerical results correspond to $h = 0.5$ and $b = 3, 2$, and 1 using the standard fourteenth-order

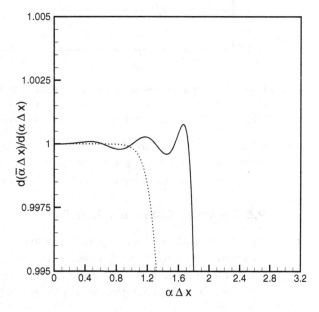

Figure 2.8. $\frac{d(\bar{\alpha}\Delta x)}{d(\alpha\Delta x)}$ versus $\alpha\Delta x$ for the —— 15-point optimized scheme ($\eta = 1.8$) and the $\cdots\cdots$ standard fourteenth-order central difference scheme.

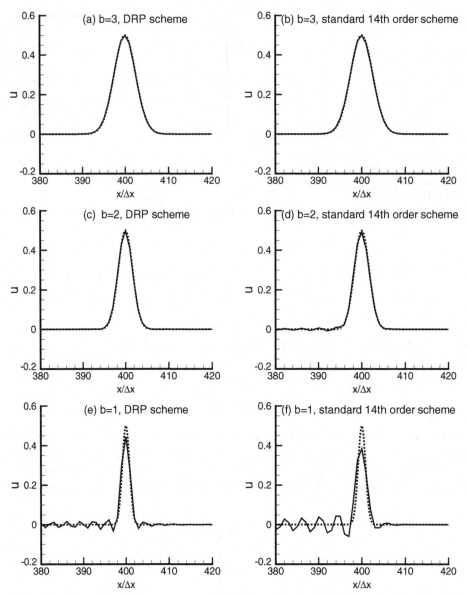

Figure 2.9. Comparisons between the computed and the exact solutions of the convective wave equation. $t = 400$, $h = 0.5$. Dispersion-relation-preserving scheme is the optimized scheme, $\cdots\cdots\cdots$ exact solution, ———— computed solution.

scheme and the optimized scheme ($\eta = 1.8$) are given in Figure 2.9. With $b \geq 2$, the optimized scheme gives good waveform. However, for an extremely narrow pulse, $b = 1$, none of these schemes can provide a solution free of spurious waves.

2.5 Backward Difference Stencils

Near the boundary of a computation domain, a symmetric stencil cannot be used; backward difference stencils are needed. Figure 2.10 shows the various 7-point backward difference stencils. Unlike symmetric stencil, the wave number of a backward difference stencil is complex. As will be shown later, a stencil with a complex wave

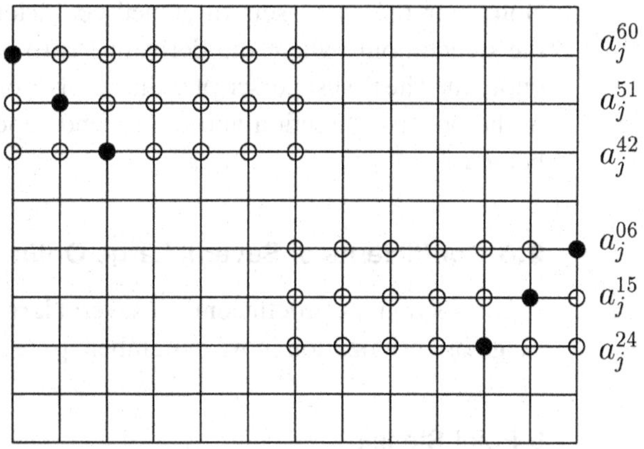

Figure 2.10. 7-point backward difference stencils for mesh points shown as black circles.

number would lead to either damping or amplification depending on the direction of wave propagation. For numerical stability, artificial damping must be added. The subject of artificial damping is discussed in Chapter 7. The following is a set of stencil coefficients that has been obtained through optimization taking into consideration both good resolution and minimal damping/amplification requirements. These backward difference stencils are also useful near a wall or surface of discontinuity.

$$a_0^{60} = -a_0^{06} = -2.19228033900$$
$$a_1^{60} = -a_{-1}^{06} = 4.74861140100$$
$$a_2^{60} = -a_{-2}^{06} = -5.10885191500$$
$$a_3^{60} = -a_{-3}^{06} = 4.46156710400$$
$$a_4^{60} = -a_{-4}^{06} = -2.83349874100$$
$$a_5^{60} = -a_{-5}^{06} = 1.12832886100$$
$$a_6^{60} = -a_{-6}^{06} = -0.20387637100$$

$$a_{-1}^{51} = -a_1^{15} = -0.20933762200$$
$$a_0^{51} = -a_0^{15} = -1.08487567600$$
$$a_1^{51} = -a_{-1}^{15} = 2.14777605000$$
$$a_2^{51} = -a_{-2}^{15} = -1.38892832200$$
$$a_3^{51} = -a_{-3}^{15} = 0.76894976600$$
$$a_4^{51} = -a_{-4}^{15} = -0.28181465000$$
$$a_5^{51} = -a_{-5}^{15} = 0.48230454000E - 1$$

$$a_{-2}^{42} = -a_2^{24} = 0.49041958000E - 1$$
$$a_{-1}^{42} = -a_1^{24} = -0.46884035700$$
$$a_0^{42} = -a_0^{24} = -0.47476091400$$
$$a_1^{42} = -a_{-1}^{24} = 1.27327473700$$
$$a_2^{42} = -a_{-2}^{24} = -0.51848452600$$
$$a_3^{42} = -a_{-3}^{24} = 0.16613853300$$
$$a_4^{42} = -a_{-4}^{24} = -0.26369431000E - 1$$

Note: For the three sets of stencil coefficients a_j^{60}, a_j^{51}, and a_j^{42} subscript $j = 0$ is the stencil point where the derivative is to be approximated by a finite difference quotient. The first superscript is the number of mesh points ahead of the stencil point in the positive direction and the second superscript is the number of mesh points behind.

2.6 Coefficients of Several Large Optimized Stencils

In this section, the coefficients of several large optimized stencils are provided. The range of optimization in wave number space is $\alpha \Delta x = \eta$.

9-Point Stencil

$$\eta = 1.28, \quad a_0 = 0.0$$

$$a_1 = -a_{-1} = 0.8301178834769906875382633360472085$$
$$a_2 = -a_{-2} = -0.23175333877690181900845126210 9655756$$
$$a_3 = -a_{-3} = 0.052872050204836964235921565029 01203$$
$$a_4 = -a_{-4} = -6.3068146383663000192506972352 82424\text{E-}3$$

11-Point Stencil

$$\eta = 1.45, a_0 = 0.0$$

$$a_1 = -a_{-1} = 0.8691451973307874495001324546317289$$
$$a_2 = -a_{-2} = -0.28182159562075193452800172953580692$$
$$a_3 = -a_{-3} = 0.08707110821545964532454798738037454$$
$$a_4 = -a_{-4} = -0.01951085872803834759459471291 4321635$$
$$a_5 = -a_{-5} = 2.2656208352981747921211787912 09576\text{E-}3$$

13-Point Stencil

$$\eta = 1.63, \quad a_0 = 0.0$$

$$a_1 = -a_{-1} = 0.8978538704842304955720852088 41816$$
$$a_2 = -a_{-2} = -0.32269821467978701682118324993041657$$
$$a_3 = -a_{-3} = 0.12096287073505875036103692011036235$$
$$a_4 = -a_{-4} = -0.03798910219344821097001738425 1701696$$
$$a_5 = -a_{-5} = 8.52610760898908781618447085963 4284\text{E-}3$$
$$a_6 = -a_{-6} = -1.0033637668308847022803811005724351\text{E-}3$$

The $\bar{\alpha} \Delta x$ versus $\alpha \Delta x$ curves of these stencils are shown in Figure 2.11. Figure 2.12 shows the variation of the slope $\frac{d(\bar{\alpha} \Delta x)}{d(\alpha \Delta x)}$ versus $\alpha \Delta x$. Figure 2.13 shows an enlarged portion of Figure 2.11 near the maxima of the curves.

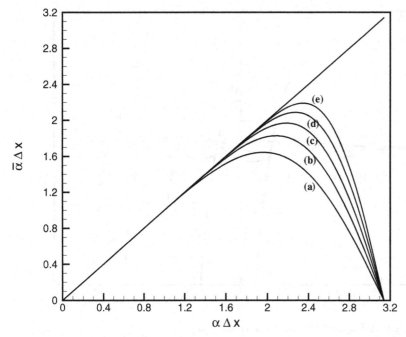

Figure 2.11. The $\bar{\alpha}\Delta x$ versus $\alpha\Delta x$ curves. (a) 7-point stencil, (b) 9-point stencil, (c) 11-point stencil, (d) 13-point stencil, (e) 15-point stencil.

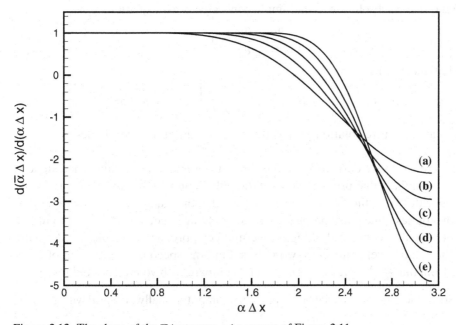

Figure 2.12. The slope of the $\bar{\alpha}\Delta x$ versus $\alpha\Delta x$ curves of Figure 2.11.

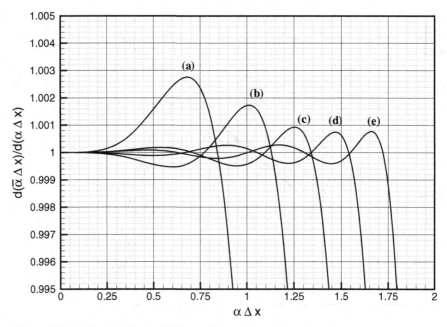

Figure 2.13. Enlarged Figure 2.12 near the maxima of the curves.

EXERCISES

2.1. Dimensionless variables with respect to length scale $= \Delta x$, velocity scale $= c$ (speed of sound), and timescale $= \Delta x/a$ are to be used.

Find the exact solution of the simple convective wave equation

$$\frac{\partial u}{\partial t} + \frac{\partial u}{\partial x} = 0$$

with initial condition

$$t = 0, \qquad u = \exp\left[-(\ln 2)\left(\frac{x}{12}\right)^2\right](1.0 + \cos(2.5x)).$$

Note that the entire solution propagates to the right at a dimensionless speed of unity.

Discretize the x derivative by the 7-point optimized finite difference stencil. Solve the initial value problem computationally. You may use any time marching scheme with a small time step Δt. Demonstrate that the numerical solution separates into two pulses. One pulse propagates to the right at a speed nearly equal to unity. The other pulse, composed of high wave numbers, propagates to the left. Explain why the solution separates into two pulses. Find the speed of propagation of each pulse from the $\frac{d(\bar{\alpha}\Delta x)}{d(\alpha\Delta x)}$ versus $\alpha\Delta x$ curve and compare with your computed results.

2.2. Solve the following initial value problem computationally on a grid with $\Delta x = 1$

$$\frac{\partial u}{\partial t} + \frac{\partial u}{\partial x} = 0.$$

$$t = 0, \qquad u(x, 0) = e^{-\sigma x^2}\cos(\mu x)$$

Use the standard sixth-order central difference scheme.

Take $\sigma = 0.035$ and consider three cases:

1. $\mu = 0.8$
2. $\mu = 1.6$
3. $\mu = 2.1$

For each case, plot $u(x, t)$ at $t = 300$ and compare with the exact solution. Explain your results.

2.3. Solve the initial value problem:

$$\frac{\partial u}{\partial t} + \frac{\partial u}{\partial x} = 0 \qquad \text{at } t = 0, \quad u = H(x + 50) - H(x - 50),$$

computationally on a grid with $\Delta x = 1$, where $H(x)$ is the unit step function,

Plot $u(x, t)$ at $t = 300$. Determine from your numerical solution

(a) the most negative wave velocity (waves propagating to the left)
(b) the wave number of the wave with zero group velocity Compare the wave speed and wave number of (a) and (b) with the values from the $\frac{d\bar{\alpha}}{d\alpha}$ curve.

Do the above using

(1) the standard sixth-order central difference scheme and
(2) the optimized scheme (7-point stencil optimized over $-1.1 \leq \alpha\Delta x \leq 1.1$).

3 Time Discretization

There is a fundamental difference between space and time. Space extends from negative infinity to positive infinity, whereas time only increases in one direction. This inherent difference requires finite difference approximation for time derivatives to be different from spatial derivatives.

In general, there are two ways to form a discretized approximation of a time derivative. They are the single time step method and the multilevel time discretization method. In this chapter, the single time step method will be discussed briefly. The multilevel discretization method will be discussed in greater detail. One major difference between the single time step method and multilevel methods is that, for wave propagation problems, the latter, when properly implemented, would lead to dispersion-relation-preserving (DRP) schemes (see Tam and Webb, 1993). DRP property is very desirable for computing wave propagation problems. DRP schemes will be discussed in detail in the next two chapters.

3.1 Single Time Step Method: Runge-Kutta Scheme

One of the most popular time marching methods is the Runge-Kutta (RK) scheme. The RK scheme is based on Taylor series truncation. A widely used RK scheme is the fourth-order method. To illustrate the basic approach of the RK method, let the time axis be divided into increments of Δt. Suppose the solution of a differential equation,

$$\frac{d\mathbf{u}}{dt} = \mathbf{F}(\mathbf{u}), \tag{3.1}$$

where $\mathbf{u}$ and $\mathbf{F}$ are vectors, and time level n is known. To find the solution at the next time level $(n + 1)$, four evaluations of the derivative function $\mathbf{F}$ are performed (fourth-order RK). These intermediate evaluations provide indirectly the high-order derivatives of $\mathbf{u}$ so that a matching of high-order terms in Δt in a Taylor series expansion becomes possible. The following is a very general form of the fourth-order RK scheme:

$$\mathbf{u}^{(n+1)} = \mathbf{u}^{(n)} + \sum_{j=1}^{4} w_j \mathbf{k}_j \tag{3.2}$$

$$\mathbf{k}_1 = \mathbf{F}\left(\mathbf{u}^{(n)}\right)\Delta t$$
$$\mathbf{k}_2 = \mathbf{F}\left(\mathbf{u}^{(n)} + \beta_2 \mathbf{k}_1\right)\Delta t$$
$$\mathbf{k}_3 = \mathbf{F}\left(\mathbf{u}^{(n)} + \beta_3 \mathbf{k}_2\right)\Delta t$$
$$\mathbf{k}_4 = \mathbf{F}\left(\mathbf{u}^{(n)} + \beta_4 \mathbf{k}_3\right)\Delta t.$$

Superscript (n) indicates the time level. The constants β_2, β_3, β_4, w_1, w_2, w_3, and w_4 are chosen so that, when the left and right sides of Eq. (3.2) are expanded in Taylor series for small Δt, they are matched to order $(\Delta t)^4$. For the standard fourth-order scheme, the constants are assigned the following values:

$$\beta_2 = \beta_3 = \frac{1}{2}, \qquad \beta_4 = 1.0, \qquad w_1 = w_4 = \frac{1}{6}, \qquad w_2 = w_3 = \frac{1}{3}. \qquad (3.3)$$

It is known that this choice is not unique. Other choices are possible within the requirement that the Taylor series expansions of Eq. (3.2) are matched to terms of order $(\Delta t)^4$. An alternative choice proposed by Hu *et al.* (1996) called the low-dissipation low-dispersion Runge-Kutta (LDDRK) scheme is widely used in computational aeroacoustics (CAA). In formulating the LDDRK scheme, attention is focused on discretizing the time derivative of the convective wave equation,

$$\frac{\partial u}{\partial t} + c\frac{\partial u}{\partial x} = 0, \qquad (3.4)$$

rather than Eq. (3.1). Suppose the spatial derivative of Eq. (3.4) is approximated by a high-order finite difference scheme. The Fourier transform of the finite difference equation is

$$\frac{d\tilde{u}}{dt} = -ic\bar{\alpha}\tilde{u}, \qquad (3.5)$$

where $\tilde{u}$ is the Fourier transform of u and $\bar{\alpha}(\alpha)$ is the wave number of the spatial finite difference approximation of $\partial/\partial x$.

Now, if the fourth-order RK scheme (3.2) is applied to Eq. (3.5), it is easy to find, after some algebra, that the time marching scheme becomes

$$u^{(n+1)} = u^{(n)}\left[1 + \sum_{j=1}^{4} c_j\left(-ic\bar{\alpha}\Delta t\right)^j\right], \qquad (3.6)$$

where

$$c_1 = \sum_{j=1}^{4} w_j, \qquad c_2 = \sum_{j=2}^{4} w_j\beta_j$$
$$c_3 = w_3\beta_3\beta_2 + w_4\beta_4\beta_3, \qquad c_4 = w_4\beta_4\beta_3\beta_2. \qquad (3.7)$$

It is a simple matter to show that, for the standard fourth-order RK scheme, $c_j = \frac{1}{j!}$.

According to LDDRK, the c_j constants are assigned the following values:

$$c_1 = 1.0, \qquad c_2 = 0.5, \qquad c_3 = 0.162997, \qquad c_4 = 0.0407574. \qquad (3.8)$$

This choice is motivated by numerical stability consideration and the desire to use the largest Δt permissible. It turns out that the choice also keeps the numerical dispersion low, as pointed out by Tam (2004). That LDDRK is a low dispersion scheme will be discussed in the next chapter.

To implement the LDDRK scheme, it is necessary to find a set of numerical values of β_j and w_j's when the c_j's are given by Eq. (3.8). A simple way to keep the scheme close to the standard fourth-order scheme is to keep the values of β_j the same as the standard scheme and to determine the values of w_j by solving Eq. (3.7) as a linear system of equations. The values of w_j's are as follows:

$$w_1 = w_4 = 0.1630296, \quad w_2 = 0.348012, \quad w_3 = 0.3259288.$$

3.2 Optimized Multilevel Time Discretization

Suppose $\mathbf{u}(t)$ is the solution of Eq. (3.1). In a multilevel time discretization method, the time axis is divided into a uniform grid with time step Δt. It will be assumed that the values of $\mathbf{u}$ and $d\mathbf{u}/dt$ are known at time level $n, n-1, n-2, n-3$. (Note: In CAA, $d\mathbf{u}/dt$ is given by the governing equations of motion.) To advance to the next time level, the following four-level finite difference approximation is used:

$$\mathbf{u}^{(n+1)} = \mathbf{u}^{(n)} + \Delta t \sum_{j=0}^{3} b_j \left(\frac{\partial \mathbf{u}}{\partial t} \right)^{(n-j)}. \tag{3.9}$$

The last term on the right side of Eq. (3.9) may be regarded as a weighted average of the time derivatives at the last 4 mesh points. There are four constants, namely, b_0, b_1, b_2, and b_3, that are to be selected. To ensure that the scheme is consistent, it is suggested to choose three of the four coefficients, say b_j $(j = 1, 2, 3)$, so that when the terms in Eq. (3.9) are expanded in a Taylor series in Δt they match to the order $(\Delta t)^2$. This leaves one free parameter, b_0. The relation between the other coefficients and b_0 are as follows:

$$b_1 = -3b_0 + \frac{53}{12}, \quad b_2 = 3b_0 - \frac{16}{3}, \quad b_3 = -b_0 + \frac{23}{12}. \tag{3.10}$$

The remaining coefficient b_0 will now be determined by requiring the Laplace transform of the finite difference scheme (3.9) to be a good approximation of that of the partial derivative.

The Laplace transform and its inverse of a function $f(t)$ are related by (see Appendix A) the following:

$$\tilde{f}(\omega) = \frac{1}{2\pi} \int_0^\infty f(t) e^{i\omega t} \, dt, \quad f(t) = \int_\Gamma \tilde{f}(\omega) e^{-i\omega t} \, d\omega. \tag{3.11}$$

The inverse contour Γ is a line in the upper half ω plane parallel to the real ω-axis above all poles and singularities. Before Laplace transform can be applied to Eq. (3.9), it is necessary to generalize the equation to one with a continuous variable. This can easily be done. The result is as follows:

$$\mathbf{u}(t + \Delta t) = \mathbf{u}(t) + \Delta t \sum_{j=0}^{3} b_j \frac{\partial \mathbf{u}(t - j\Delta t)}{\partial t}. \tag{3.12}$$

Eq. (3.12) reduces to Eq. (3.9) by setting $t = n\Delta t$. It will be assumed, for the time being, that $\mathbf{u}(t)$ satisfies the following trivial initial conditions (the case with general initial conditions will be discussed later); i.e.,

$$\mathbf{u}(t) = 0, \quad t < \Delta t. \tag{3.13}$$

In Appendix A, the following shifting theorem for Laplace transform is proven:

$$\text{Transform of } f(t + \Delta) = e^{-i\omega\Delta}\tilde{f}; \quad \tilde{} = \text{Laplace transform} \quad (3.14)$$

On applying the shifting theorem to Eq. (3.12), it is easy to find

$$\tilde{u}e^{-i\omega\Delta t} = \tilde{u} + \Delta t \left(\sum_{j=0}^{3} b_j e^{i\omega j\Delta t} \right) \left(\text{transform of } \frac{\partial u}{\partial t} \right).$$

Thus,

$$-i\frac{i(e^{-i\omega\Delta t} - 1)}{\Delta t \sum_{j=0}^{3} b_j e^{ij\omega\Delta t}} \tilde{u} = \left(\text{transform of } \frac{\partial u}{\partial t} \right). \quad (3.15)$$

The Laplace transform of the time derivative of u is $-i\omega\tilde{u}$. Thus, by comparing the two sides of Eq. (3.15), the quantity

$$\overline{\omega} = \frac{i(e^{-i\omega\Delta t} - 1)}{\Delta t \sum_{j=0}^{3} b_j e^{ij\omega\Delta t}} \quad (3.16)$$

is the effective angular frequency of the time marching scheme (3.9). The weighted error E_1 incurred by using $\overline{\omega}$ to approximate ω will be defined as

$$E_1 = \int_{-\zeta}^{\zeta} \{\sigma[\text{Re}(\overline{\omega}\Delta t - \omega\Delta t)]^2 + (1 - \sigma)[\text{Im}(\overline{\omega}\Delta t - \omega\Delta t)]^2\}d(\omega\Delta t), \quad (3.17)$$

where $\text{Re}(\ )$ and $\text{Im}(\ )$ are the real and imaginary part of $(\ )$. σ is the weight and ζ is the frequency range. One expects $\overline{\omega}$ to be a good approximation of ω. On substituting $\overline{\omega}$ from Eqs. (3.16) and (3.10) into Eq. (3.17), E_1 becomes a function of b_0 alone. b_0 will be chosen so that E_1 is a minimum. Therefore, b_0 is given by the root of

$$\frac{dE_1}{db_0} = 0. \quad (3.18)$$

A range of possible values of σ and ζ has been used in Eq. (3.18) to determine b_0 and, hence, b_1, b_2, and b_3. After considering the effect on the range of useful frequency and numerical damping rate, the values $\sigma = 0.36$ and $\zeta = 0.5$ were selected. For this value of σ and ζ, the values of the coefficients b_j ($j = 0, 1, 2, 3$) are

$$
\begin{aligned}
b_0 &= 2.3025580888383, \quad b_1 = -2.4910075998482 \\
b_2 &= 1.5743409331815, \quad b_3 = -0.3858914221716.
\end{aligned}
\quad (3.19)
$$

3.3 Stability Diagram

It is easily seen from Eq. (3.16) that the relationship between $\overline{\omega}\Delta t$ and $\omega\Delta t$ is not one to one. This means that there are spurious numerical solutions. These spurious solutions could cause numerical instability, so it is necessary to find out about their characteristics.

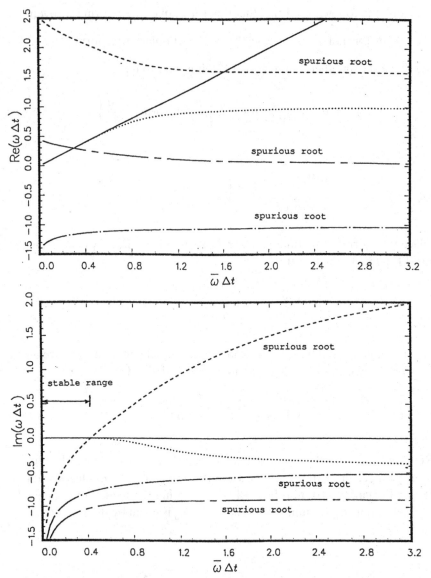

Figure 3.1. The real and imaginary parts of the four roots of $\omega\Delta t$ as functions of $\overline{\omega}\Delta t$.

Eq. (3.16) may be rewritten in the form as follows:

$$b_3 z^4 + b_2 z^3 + b_1 z^2 + \left(b_0 + \frac{i}{\overline{\omega}\Delta t}\right) z - \frac{i}{\overline{\omega}\Delta t} = 0, \qquad (3.20)$$

where $z = e^{i\omega\Delta t}$. Thus, given a $\overline{\omega}\Delta t$, there are four roots of $e^{i\omega\Delta t}$. In other words, there are four values of $\omega\Delta t$ that would give the same $\overline{\omega}\Delta t$ when substituted into Eq. (3.16). For the values of b_j ($j = 0, 1, 2, 3$) given by formula (3.19), the values of the four roots of $\omega\Delta t$ as functions of $\overline{\omega}\Delta t$ have been found. The real and imaginary part of these roots over the range $0 \leq \overline{\omega}\Delta t \leq \pi$ are plotted in Figure 3.1. In the range $\overline{\omega}\Delta t < 0.41$, the imaginary parts of all the four roots are negative. Recall that the solution ω, given by the inverse transform of Eq. (3.11), is a superposition of elemental solutions having time dependence of the form $e^{-i\omega t}$. If $\text{Im}(\omega)$ is negative, the solution is damped in time. However, for $\overline{\omega}\Delta t > 0.41$,

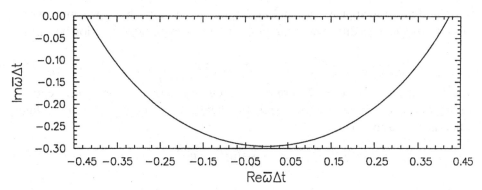

Figure 3.2. Region of stability in the complex $\overline{\omega}\Delta t$ plane.

one spurious root has a positive imaginary part. The solution corresponding to this root will grow, in time overwhelming the physical solution, and lead to numerical instability.

On examining the real parts of the four roots, it is seen that one of the roots gives $\overline{\omega}\Delta t \cong \omega\Delta t$ over the range $\overline{\omega}\Delta t < 0.41$. This is the physical root. The others are spurious. The physical root has a very small imaginary part up to $\overline{\omega}\Delta t = 0.41$. That is, there is very little numerical damping. A more detailed investigation reveals that, for this root, Im $(\omega\Delta t) = -0.118 \times 10^{-4}$ for $\overline{\omega}\Delta t = 0.19$. This is reasonably small. It is recommended that Δt be chosen such that $\overline{\omega}\Delta t < 0.19$. This would automatically guarantee numerical stability and negligible numerical damping.

In most wave propagation problems, to be discussed later, $\overline{\omega}\Delta t$ is real. In this case, this stability consideration is sufficient. However, if the wave is damped by some physical process, or artificial damping is added to the computation scheme for the purpose of removing spurious waves from the numerical solution, then $\overline{\omega}\Delta t$ is complex. To include this more general case, it is necessary to consider the four solutions of Eq. (3.20) for $\overline{\omega}\Delta t$ in the complex plane. It is straightforward to find that the stable region (all the four roots of $\omega\Delta t$ have a negative imaginary part) is confined to a small nearly semicircular region in the lower $\overline{\omega}\Delta t$ plane. This is shown in Figure 3.2. This stability diagram plays an important role in the choice of time step Δt in time marching computation. Numerical stability will be discussed in Chapter 5.

EXERCISES

3.1. Construct an optimized two-level time marching scheme for the differential equation

$$\frac{du}{dt} = F(u).$$

Find the stability region in the complex $\overline{\omega}\Delta t$ plane.

3.2. Consider the solution of the initial value problem

$$\frac{\partial u}{\partial t} + c\frac{\partial u}{\partial x} = 0 \tag{A}$$

$$t = 0, \qquad u = e^{-(\ln 2)\left(\frac{x}{3\Delta x}\right)^2} \tag{B}$$

by finite difference approximation. Let the spatial derivative be approximated by a second-order central difference equation. The convective wave equation becomes

$$\frac{du_\ell}{dt} = -\frac{c}{2\Delta x}(u_{\ell+1} - u_{\ell-1}), \tag{C}$$

where ℓ is the spatial index. Difference-differential equation (C) may be integrated by the four-level optimized time marching scheme. On writing out in full, the finite difference equations are

$$E_\ell^{(n)} = -\frac{c}{2\Delta x}\left(u_{\ell+1}^{(n)} - u_{\ell-1}^{(n)}\right) \tag{D}$$

$$u_\ell^{(n+1)} = u_\ell^{(n)} + \Delta t \sum_{j=0}^{3} b_j E_\ell^{(n-j)}. \tag{E}$$

Consider a Fourier-Laplace transform component of the solution

$$u_\ell^{(n)} = A e^{i(\ell\alpha\Delta x - n\omega\Delta t)}. \tag{F}$$

Perform a Von Neumann stability analysis of the finite difference scheme by substituting Eq. (F) into (D) and (E). Show that this yields the dispersion relation as follows:

$$\overline{\omega}(\omega) = c\overline{\alpha}(\alpha), \tag{G}$$

where $\overline{\alpha}$ is the stencil wave number of the second-order central difference scheme and $\overline{\omega}$ is the angular frequency of the four-level time marching scheme.

(a) Show that the scheme is stable if

$$\Delta t \le \frac{0.41\Delta x}{c}. \tag{H}$$

(b) Implement the time marching scheme on a computer. Take Δt to be slightly larger than that given by Eq. (H). Compute the solution in time up to $t = 400$ with (B) as the initial condition to see whether the solution is unstable (i.e., grows exponentially in time).

(c) Verify computationally that, if Δt is chosen according to condition (H), the numerical scheme is stable. Furthermore, if Δt is taken to be less than or equal to $0.19\Delta x/c$, an excellent numerical solution can be obtained.

4 Finite Difference Scheme as Dispersive Waves

Acoustics are governed by the compressible Navier-Stokes equations. In most cases, molecular viscosity is unimportant, so that the use of Euler equations is sufficient. In solving the Navier-Stokes or Euler equations computationally, the first step is to perform a discretized approximation to the spatial and temporal derivatives. This converts the partial differential equations to a set of partial difference equations. However, one must recognize that the solutions of the discretized equations are not the same as those of the original partial differential equations. A central effort of computational aeroacoustics (CAA) is to understand mathematically the behavior of the solution of the discretized equations and to quantify and minimize the error. Here, error is referred to as the difference between the solution of the original partial differential equations and the partial difference system.

Invariably, finite difference approximation of the governing equations of acoustics will result in a dispersive wave system (Vichnevetsky and Bowles, 1982; Trefethen, 1982; Tam and Webb, 1993; Tam 1995), even though the waves supported by the original partial differential equations are nondispersive. This is an extremely important point and should be clearly understood by all CAA investigators and users.

4.1 Dispersive Waves of Physical Systems

Many physical systems support dispersive waves. Examples of commonly encountered dispersive waves are small-amplitude water waves, waves in stratified fluid, elastic waves, and magnetohydrodynamic waves. A common feature of these dispersive waves is that they are governed by linear partial differential equations with constant coefficients. The following are some of these equations.

a. *Klein-Gordon equation*

$$\frac{\partial^2 \phi}{\partial t^2} - c^2 \frac{\partial^2 \phi}{\partial x^2} + \mu^2 \phi = 0.$$

b. *Beam equation*

$$\frac{\partial^2 \phi}{\partial t^2} + \mu^2 \frac{\partial^4 \phi}{\partial x^4} = 0.$$

c. Linearized Korteweg de Vries equation for water waves

$$\frac{\partial \phi}{\partial t} + \mu \frac{\partial^3 \phi}{\partial x^3} = 0.$$

d. Linearized Boussinesq equation for water and elastic waves

$$\frac{\partial^2 \phi}{\partial t^2} - a^2 \frac{\partial^2 \phi}{\partial x^2} - \mu^2 \frac{\partial^4 \phi}{\partial x^2 \partial t^2} = 0.$$

Because these equations are linear with constant coefficients, they can be readily solved by Fourier-Laplace transform. The Fourier-Laplace transforms of these equations are

(a) $(\omega^2 - c^2 \alpha^2 - \mu^2)\tilde{\phi} = H_1(\alpha, \omega).$

(b) $(\omega^2 - \mu^2 \alpha^4)\tilde{\phi} = H_2(\alpha, \omega).$

(c) $(\omega + \mu \alpha^3)\tilde{\phi} = H_3(\alpha).$

(d) $(\omega^2 - a^2 \alpha^2 + \mu^2 \alpha^2 \omega^2)\tilde{\phi} = H_4(\alpha, \omega).$

The right-hand side of each of these equations represents some arbitrary initial conditions. The bracket multiplying the transform of the unknown on the left side of each equation is called the dispersion function, which will be denoted by $D(\alpha, \omega)$. The dispersion relation of the dispersive waves is given by the zeros of the dispersion function; i.e.,

$$D(\alpha, \omega) = 0 \quad \Rightarrow \quad \omega = \omega(\alpha). \tag{4.1}$$

Eq. (4.1) is a relationship between wave number α and angular frequency ω.

The solution of a dispersive wave system in wave number space, in general, may be written in the following form:

$$\tilde{u} = \frac{H(\alpha, \omega)}{D(\alpha, \omega)}. \tag{4.2}$$

The corresponding solution in physical space is found by inverting the transforms of Eq. (4.2) as follows:

$$u(x, t) = \frac{1}{(2\pi)^2} \int_{-\infty}^{\infty} \int_{\Gamma} \frac{H(\alpha, \omega)}{D(\alpha, \omega)} e^{i(\alpha x - \omega t)} d\omega \, d\alpha. \tag{4.3}$$

Suppose, for real α, the dispersion relation has a simple real zero. That is, $\omega(\alpha)$ is a solution of

$$D(\alpha, \omega(\alpha)) = 0. \tag{4.4}$$

Thus, for the ω-integral of Eq. (4.3), there is a pole lying on real ω-axis. The contribution of this pole can be found by evaluating the ω-integral by the Residue Theorem. This gives, $t > 0$

$$u(x, t) = -\frac{i}{2\pi} \int_{-\infty}^{\infty} \frac{H(\alpha, \omega(\alpha))}{\left(\frac{\partial D}{\partial \omega}\right)_{\omega = \omega(\alpha)}} e^{i(\alpha x - \omega(\alpha) t)} d\alpha$$

$$= -\frac{i}{2\pi} \int_{-\infty}^{\infty} \frac{H(\alpha, \omega(\alpha))}{\left(\frac{\partial D}{\partial \omega}\right)_{\omega(\alpha)}} e^{i\Phi(\alpha, (x/t)) t} d\alpha, \tag{4.5}$$

where $\Phi = \alpha(x/t) - \omega(\alpha)$ is the phase function of the integral. For large t, Eq. (4.5) may be evaluated asymptotically by the method of stationary phase (see Appendix B). The stationary phase point α_s is given by

$$\frac{\partial \Phi}{\partial \alpha} = 0 \quad \Rightarrow \quad \frac{d\omega}{d\alpha}\bigg|_{\alpha=\alpha_s} = \frac{x}{t}. \tag{4.6}$$

This yields α_s as a function of x/t, i.e., $\alpha_s = \alpha_s(x/t)$. $d\omega/d\alpha$ will be referred to as the group velocity. The integral is equal to

$$u(x,t) \underset{t\to\infty}{\sim} -\frac{iH(\alpha_s,\omega_s)}{2\pi\left(\frac{\partial D}{\partial \omega}\right)_s} \left(\frac{2\pi}{|\omega''(\alpha_s)|t}\right)^{\frac{1}{2}} e^{i\Phi(\alpha_s,(x/t))t+(\pi i/4)\mathrm{sgn}\Phi_s''}. \tag{4.7a}$$

Subscript s indicates the evaluation at $\alpha = \alpha_s(x/t)$. In the following, the subscript s is dropped with the understanding that α is $\alpha_s(x/t)$ and ω is $\omega(\alpha_s)$. The asymptotic formula may be rewritten in the compact form as follows:

$$u(x,t) \sim A(x,t)e^{i\theta(x,t)}, \tag{4.7b}$$

where θ is the phase function and A is amplitude.

$$\theta = x\alpha(x/t) - t\omega(x/t) \tag{4.8}$$

$$A = -\frac{iH(\alpha,\omega)}{(2\pi)^{\frac{1}{2}}\left(\frac{\partial D}{\partial \omega}\right)} \left(\frac{1}{|\omega''(\alpha)|t}\right)^{\frac{1}{2}} e^{i(\pi/4)\mathrm{sgn}\Phi''}. \tag{4.9}$$

4.2 Group Velocity and Dispersion

Note that by Eq. (4.8), for large t,

$$\frac{\partial\theta(x,t)}{\partial x} = \alpha + [x - \omega'(\alpha)t]\frac{\partial\alpha}{\partial x} \to \alpha$$

$$\frac{\partial\theta(x,t)}{\partial t} = -\omega + [x - \omega'(\alpha)t]\frac{\partial\alpha}{\partial t} \to -\omega,$$

where $\omega'(\alpha) = (d\omega/d\alpha)$. Thus, locally, θ_x and $-\theta_t$ (subscript denotes partial derivative) may be taken as the wave number and angular frequency as $t \to \infty$. To find the phase velocity, one may keep θ constant in Eq. (4.8) and differentiate with respect to t to obtain

$$[x - \omega'(\alpha)t]\frac{\partial\alpha}{\partial t} + \alpha\frac{dx}{dt} - \omega = 0.$$

For large t, by the above equation, the phase velocity is found to be

$$\frac{dx}{dt}\bigg|_{\theta=\mathrm{constant}} = \frac{\omega}{\alpha}. \tag{4.10}$$

Now the asymptotic solution (4.6)–(4.9) may be given the following interpretation.

1. Each Fourier component with wave number α is constant along a ray in the x–t plane and propagates with group velocity $\omega'(\alpha)$ as shown in Figure 4.1. This is because $x = \omega'(\alpha)t$ so that α is a constant along a given line (ray) $x/t = \mathrm{constant}$.

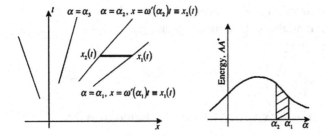

Figure 4.1. (a) Rays of $x = \omega'(\alpha)t$ for constant α in the x–t diagram. (b) Initial energy spectrum.

Furthermore, $dx/dt = \omega'(\alpha)$, suggesting that $\omega'(\alpha)$, the group velocity, is the speed of propagation of wave number α.

2. Far away, the waves may be regarded as initiated locally near $x = 0$. After a period of time, the waves spread out in space because a different Fourier component with different wave number α propagates with different speed ($\omega''(\alpha) \neq 0$), see Figure 4.1a.

A quantity of interest in wave propagation is the wave energy that is $|A|^2 = AA^*$ (* denotes the complex conjugate). The wave energy, $Q(t)$, between two rays $x = x_1(t) = \omega'(\alpha_1)\,t$ and $x = x_2(t) = \omega'(\alpha_2)\,t$ (see Figure 4.1a) is

$$Q(t) = \int_{x_2(t)}^{x_1(t)} AA^* dx = \int_{x_2(t)}^{x_1(t)} \frac{1}{2\pi} \left| \frac{H(\alpha, \omega)}{\frac{\partial D}{\partial \omega}} \right|^2 \frac{1}{|\omega''| t} dx.$$

With t fixed, a change of integration variable from x to α gives

$$dx = t\omega''(\alpha)d\alpha.$$

Hence,

$$Q(t) = \frac{1}{2\pi} \int_{\alpha_2}^{\alpha_1} \left| \frac{H(\alpha, \omega)}{\frac{\partial D}{\partial \omega}} \right|^2 \frac{\omega''(\alpha)}{|\omega''(\alpha)|} d\alpha = \text{constant (independent of } t).$$

Therefore, energy is constant between group lines. In other words, the wave energy of the initial energy spectrum (Figure 4.1b) propagates with group velocity. The group lines diverge with a separation that increases with t. Hence, the wave amplitude $|A|$ decreases as $t^{-(1/2)}$.

It is important to point out that this analysis clearly shows that propagation speed and dispersion have nothing to do with phase velocity. Dispersion arises because of the variation of group velocity with wave number. For a more detailed treatment of dispersive waves, a reading of the book by Whitham (1974) is highly recommended.

4.3 Origin of Numerical Dispersion

Let us consider the solution of the simple convective wave equation:

$$\frac{\partial u}{\partial t} + c\frac{\partial u}{\partial x} = 0 \tag{4.11}$$

and initial condition

$$t = 0, \qquad u(x, 0) = \Phi(x).$$

The Fourier-Laplace transform of (4.11) leads to

$$(\omega - c\alpha)\tilde{u} = \frac{i\hat{\Phi}(\alpha)}{2\pi}, \tag{4.12}$$

where $\tilde{u}$ is the Fourier-Laplace transform of u. $\hat{\Phi}(\alpha)$ is the Fourier transform of initial condition $u(x, 0)$. Thus, $\tilde{u}$ is given by

$$\tilde{u} = \frac{i\hat{\Phi}(\alpha)}{2\pi(\omega - c\alpha)}. \tag{4.13}$$

The poles of Eq. (4.13) are found by setting the denominator to zero. This yields

$$\omega = c\alpha, \tag{4.14}$$

which is the dispersion relation. The group velocity or wave speed is given by $d\omega/d\alpha$. In the case of Eq. (4.11), by using dispersion relation (4.14), the group velocity is

$$\frac{d\omega}{d\alpha} = c, \tag{4.15}$$

so that all wave components propagate with the same speed. That is to say, the waves supported by the convective wave equation are nondispersive.

Now Eq. (4.11) may be fully discretized by first rewriting it as a system of two equations, each involving only either a temporal or spatial derivative, as follows:

$$\frac{\partial u}{\partial t} = K$$

$$K = -c\frac{\partial u}{\partial x}.$$

The derivative of each of these equations may be discretized according to Chapter 2 for spatial derivatives and Chapter 3 for time derivatives. The fully discretized form of Eq. (4.11) is

$$K_\ell^{(n)} = -\frac{c}{\Delta x} \sum_{j=-N}^{N} a_j u_{\ell+j}^{(n)} \tag{4.16a}$$

$$u_\ell^{(n+1)} = u_\ell^{(n)} + \Delta t \sum_{j=0}^{3} b_j K_\ell^{(n-j)}. \tag{4.16b}$$

The initial condition for the four-level marching scheme, following Tam and Webb (1993), will be taken as follows:

$$u_\ell^{(0)} = \Phi(\ell\Delta x), \quad u_\ell^{(n)} = 0 \text{ for negative } n, \tag{4.16c}$$

This set of finite difference equations (Eq. 4.16) with discrete variables ℓ and n may be generalized into a set of finite difference equations with continuous variables x

and t. They are as follows:

$$K(x, t) = -\frac{c}{\Delta x} \sum_{j=-N}^{N} a_j u(x + j\Delta x, t) \tag{4.17a}$$

$$u(x, t + \Delta t) = u(x, t) + \Delta t \sum_{j=0}^{3} b_j K(x, t - j\Delta t), \tag{4.17b}$$

with initial condition

$$u(x, t) = \begin{cases} \Phi(x), & 0 \le t < \Delta t \\ 0, & t < 0. \end{cases} \tag{4.17c}$$

By taking the Fourier-Laplace transform of problem (4.17) and using the Shifting Theorem (see Appendix A), it is straightforward to find, after some algebra, that the solution is

$$\tilde{u} = \frac{i\hat{\Phi}(\alpha)}{2\pi (\bar{\omega} - c\bar{\alpha})} \left(\frac{\bar{\omega}}{\omega}\right), \tag{4.18}$$

where $\tilde{u}$ is the Fourier-Laplace transform of the solution of the finite difference equation (4.17). Eq. (4.18) is identical to the Fourier-Laplace transform of the solution of the original convective wave equation; i.e., Eq. (4.13), if $\bar{\omega}$ and $\bar{\alpha}$ in Eq. (4.18) are replaced by ω and α.

The exact solution of the finite difference solution of (4.11) is given by the inverse transform of Eq. (4.18) as follows:

$$u(x, t) = \frac{i}{2\pi} \int_{-\infty}^{\infty} \int_{\Gamma} \frac{\hat{\Phi}(\alpha)}{(\bar{\omega} - c\bar{\alpha})} \frac{\bar{\omega}}{\omega} e^{i(\alpha x - \omega t)} d\omega \, d\alpha. \tag{4.19}$$

The double integral may be evaluated by first calculating the contributions of the poles in the ω plane. The poles are given by the zeros of the denominator of the integrand, that is, the roots of

$$\bar{\omega}(\omega) = c\bar{\alpha}(\alpha). \tag{4.20}$$

Eq. (4.20) is the dispersion relation. This is formally the same as the dispersion relation of the original partial differential equation; i.e., Eq. (4.14), provided $\bar{\omega}$ is replaced by ω and $\bar{\alpha}$ is replaced by α. For this reason, the finite difference scheme is referred to as dispersion-relation-preserving (DRP).

For a four-level time marching scheme, there would be four poles or Eq. (4.20) has four roots. Let the poles be $\omega = \omega_j(\alpha)$, $j = 1, 2, 3, 4$. One of the poles has $\mathrm{Im}(\omega)$ nearly equal to zero. This is the pole that corresponds to the physical solution. The other poles lead to solutions that are heavily damped.

Numerical dispersion can arise from spatial as well as temporal discretization. Numerical dispersion due to spatial discretization alone will be considered first. For this purpose, the limit $\Delta t \to 0$ is applied, that is, time is treated exactly. This is equivalent to replacing $\bar{\omega}$ by ω in Eq. (4.19) so that the expression for the exact solution of the spatially discretized equation (4.11) is

$$u(x,t) = \frac{i}{2\pi} \int_{-\infty}^{\infty} \int_{\Gamma} \frac{\hat{\Phi}(\alpha)}{\omega - c\bar{\alpha}(\alpha)} e^{i(\alpha x - \omega t)} d\omega \, d\alpha. \qquad (4.21)$$

Dispersion relation (4.20) now simplifies to

$$\omega = c\bar{\alpha}(\alpha). \qquad (4.22)$$

The integrand has a pole at $\omega = c\bar{\alpha}(\alpha)$. On deforming the inverse contour Γ to pick up this pole in the ω plane, it is easy to find through the use of the Residue Theorem

$$u(x,t) = \int_{-\infty}^{\infty} \hat{\Phi}(\alpha) e^{i(\alpha \frac{x}{t} - \omega)t} \Bigg|_{\omega = c\bar{\alpha}(\alpha)} d\alpha. \qquad (4.23)$$

Now for large t (with x/t fixed), the α-integral of Eq. (4.23) may be evaluated by the method of stationary phase (see Appendix A). The stationary phase point $\alpha = \alpha_s$ is given by the zero of the derivative of the phase function $\Psi = (\alpha x/t - c\bar{\alpha}(\alpha))$ with respect to α. This leads to

$$\frac{d\omega}{d\alpha} = c\frac{d\bar{\alpha}(\alpha)}{d\alpha}\Bigg|_{\alpha = \alpha_s} = \frac{x}{t}, \qquad (4.24)$$

which yields α_s as a function of x/t. The asymptotic solution is

$$u(x,t) \underset{t \to \infty}{\sim} \left(\frac{2\pi}{|c\bar{\alpha}''(\alpha_s)| t}\right)^{\frac{1}{2}} \hat{\Phi}(\alpha_s) e^{i[\alpha_s \frac{x}{t} - c\bar{\alpha}(\alpha_s)]t + i\frac{\pi}{4}\text{sgn}(-c\bar{\alpha}'')}, \qquad (4.25)$$

where $\text{sgn}(z)$ is the sign of z.

Now, ω and α are related by Eq. (4.22). In general, $d\bar{\alpha}/d\alpha$ is not a constant equal to 1, so that the group velocity varies with wave number. Different wave numbers of the initial disturbances will propagate with a slightly different speed leading to numerical dispersion.

In the more general case of solution (4.19), when both spatial and temporal discretization are used, the group velocity may be calculated by implicit differentiation of Eq. (4.20). This gives

$$\frac{d\omega}{d\alpha} = \frac{c\dfrac{d\bar{\alpha}}{d\alpha}}{\dfrac{d\bar{\omega}}{d\omega}}. \qquad (4.26)$$

Eq. (4.26) indicates that numerical dispersion can be the result of imperfect spatial discretization, i.e., $d\bar{\alpha}/d\alpha \neq 1$, or imperfect temporal discretization, i.e., $d\bar{\omega}/d\omega \neq 1$. Consideration of imperfect time discretization will be deferred to later because it is quite involved. The effect of dispersion due to spatial discretization will now be illustrated by a numerical example.

For the convective wave equation (4.11), Eq. (4.24) reveals that the variation in group velocity is caused by the variation of the slope of the $\bar{\alpha}\Delta x$ versus $\alpha \Delta x$ curve. Figure 4.2 shows the $d\bar{\alpha}/d\alpha$ versus $\alpha \Delta x$ curves for the 7-point and 15-point

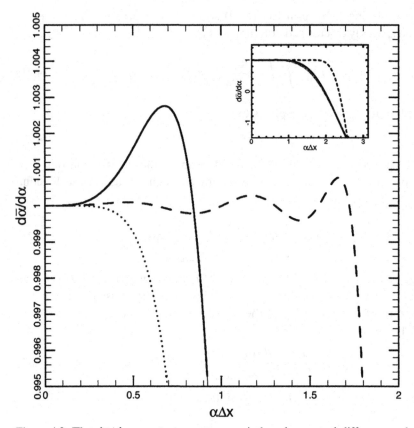

Figure 4.2. The $d\bar{\alpha}/d\alpha$ vs. $\alpha\Delta x$ curves. ·······, sixth-order central difference scheme; ———, 7-point DRP scheme – – – –, 15-point DRP scheme.

stencil DRP finite difference schemes. Notice that the $d\bar{\alpha}/d\alpha$ curve for the 7-point stencil DRP scheme has a peak at $\alpha\Delta x = 0.67$ at which $d\bar{\alpha}/d\alpha = 1.00276$. That is, the wave component with $\alpha\Delta x = 0.67$ (wave with 9.38 mesh points per wavelength) will propagate faster than the exact wave speed $d\bar{\alpha}/d\alpha = 1.0$ by about 0.3 percent. Suppose, in a computation, the waves propagate over a distance of 400 mesh points. In this case, when the main part of the wave packet reaches $x = 400$, the wave component with $\alpha\Delta x = 0.67$ will reach $x = 401.1$. Under these circumstances, numerical dispersion would just be noticeable as wave dispersion exceeds one mesh point. On the other hand, if the standard sixth-order scheme is used instead, severe numerical dispersion will result when the main wave reaches $x = 400$. Waves in the wave number range of $\alpha\Delta x > 0.6$, having a group velocity less than 1.0, would form trailing waves.

A relevant question to ask is how to reduce numerical dispersion due to spatial discretization. This can be done by using a scheme with a larger stencil. Figure 4.2 shows the $d\bar{\alpha}/d\alpha$ curve for the 15-point stencil DRP scheme. For $\alpha\Delta x < 1.73$, the group velocity differs from 1.0 by no more than 0.1 percent. Thus, if the wave packet of the solution contains only waves with wave number $\alpha\Delta x < 1.73$, then there will be no observable numerical dispersion even after the wave packet propagates over a distance of up to 1000 mesh points. As a concrete example, Figure 4.3 shows the

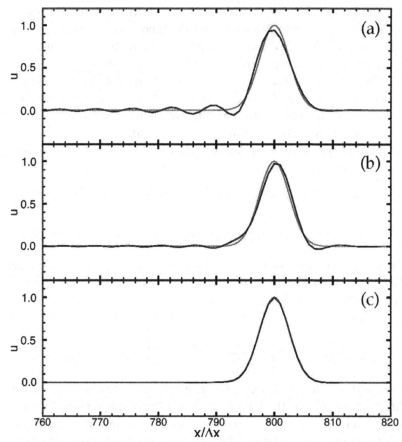

Figure 4.3. Comparisons between computed solutions and exact solution for the one-dimensional convective wave equation. (a) sixth-order central difference scheme; (b) 7-point DRP scheme; (c) 15-point DRP scheme.

computed solution of the convective wave equation (4.11) with initial condition in the form of a Gaussian with a half-width of 3 mesh spacings,

$$u(x, 0) = e^{-(\ln 2)\left(\frac{x}{3\Delta x}\right)^2},$$

using the standard sixth-order scheme, the 7-point stencil, and the 15-point stencil DRP scheme. It is easy to see that there is significant numerical dispersion (with extensive trailing waves) when computed by the standard sixth-order scheme. The 7-point stencil DRP scheme is better. Both schemes, having the same stencil size, require the same amount of computation. The solution by the 15-point stencil is even better. For all intents and purposes, the solution is identical numerically to the exact solution.

4.4 Numerical Dispersion Arising from Temporal Discretization

As pointed out in Section 4.3, computation schemes that have a DRP property will exhibit numerical dispersion due to temporal discretization if $d\overline{\omega}/d\omega$ of the angular frequency relationship $\overline{\omega}(\omega)$ is not equal to 1.0 (see Tam, 2004). Most single time step schemes, such as the Runge-Kutta method, do not have a DRP property. However, they do lead to numerical dispersion. For schemes of this type, there is no general way to derive the dispersion relation. To illustrate this point, consider again the numerical solution of the convective wave equation (4.11) by finite difference. Suppose the 15-point stencil DRP scheme is used to approximate the spatial derivative. This spatial scheme can accurately resolve waves with as few as 3.6 mesh points per wavelength so that, unless an acoustic pulse has extremely narrow half-width, there is negligible dispersion due to spatial discretization. This leaves any observed numerical dispersion to come primarily from temporal discretization. The Fourier transform of the discretized form of Eq. (4.11) is

$$\frac{d\hat{u}}{dt} = f(\hat{u}), \tag{4.27a}$$

$$\text{where } f(\hat{u}) = -ic\overline{\alpha}(\alpha)\hat{u}. \tag{4.27b}$$

Now suppose the time derivative of Eq. (4.27a) is discretized by the fourth-order Runge-Kutta scheme with time step Δt (see Eq. (3.2)). It is straightforward to find that the time marching finite difference equation for $\tilde{u}^{(n)}$ is

$$\hat{u}^{(n+1)} = \hat{u}^{(n)} \left[1 + \sum_{j=1}^{4} c_j \left(-ic\overline{\alpha}\Delta t \right)^j \right]. \tag{4.28}$$

Eq. (4.28) is a first-order finite difference equation in n. It can be readily generalized into a finite difference equation with a continuous variable. By applying the Laplace transform to the difference equation, it is easy to determine that the dispersion relation of the finite difference algorithm of the convective wave equation discretized temporally by a fourth-order Runge-Kutta scheme is

$$e^{-i\omega\Delta t} = \left[1 + \sum_{j=1}^{4} c_j \left(-ic\overline{\alpha}\Delta t \right)^j \right]. \tag{4.29}$$

By differentiating Eq. (4.29) with respect to α, the following formula for the group velocity is derived:

$$\frac{d\omega}{d\alpha} = \frac{c \sum_{j=1}^{4} j c_j \left(-ic\overline{\alpha}\Delta t \right)^{j-1}}{\left[1 + \sum_{j=1}^{4} c_j \left(-ic\overline{\alpha}\Delta t \right)^j \right]} \frac{d\overline{\alpha}}{d\alpha}. \tag{4.30}$$

Since $d\omega/d\alpha$ is not a constant equal to c even if $d\overline{\alpha}/d\alpha = 1.0$, numerical dispersion is introduced by the use of the Runge-Kutta scheme. A plot of $\text{Re}(d\omega/d\alpha)$ as a function of $c\alpha\Delta t$ ($\overline{\alpha} = \alpha$ assumed) for the low-dissipation low-dispersion Runge-Kutta (LDDRK) scheme is given as one of the curves in Figure 4.5. Notice that the group velocity drops below c for $c\alpha\Delta t > 0.5$. The group velocity is less than the exact wave speed by at least 0.2 percent for $c\alpha\Delta t > 0.65$. Waves in this range will form trailing waves.

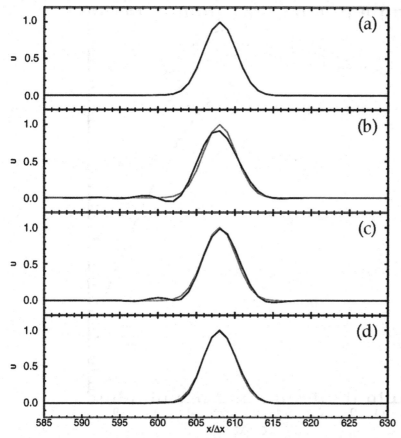

Figure 4.4. Wave form of Gaussian pulse with half-width $= 2.5\Delta x$ after propagating a distance of 608 mesh spacings. (a) $\Delta t/\Delta x = 0.25$, standard fourth-order Runge-Kutta scheme; (b) $\Delta t/\Delta x = 0.76$, standard fourth-order Runge-Kutta scheme; (c) $\Delta t/\Delta x = 0.76$, LDDRK scheme of Hu *et al.* (1996); (d) $\Delta t/\Delta x = 0.19$, four-level DRP scheme.

To illustrate numerical dispersion due to temporal discretization, consider computing the solution of Eq. (4.11) with initial condition,

$$t = 0, \qquad u\,(x, 0) = \exp\left[-(\ln 2)\left(\frac{x}{2.5\Delta x}\right)^2\right], \qquad (4.31)$$

using the 15-point stencil DRP scheme for spatial discretization and the standard fourth-order Runge-Kutta scheme for temporal discretization. For convenience, the wave speed c is set equal to unity. The computed results are shown in Figure 4.4. Figure 4.4a shows the computed waveform after the pulse has propagated a distance of 608 mesh spacings using a very small time step $\Delta t = 0.25\Delta x$. The computed result is indistinguishable from the exact solution. Figure 4.4b shows the computed result when a larger Δt, $\Delta t = 0.76\Delta x$, is used. There is no question that the pulse is slightly dispersed. The cause is temporal discretization. For comparison purposes, Figure 4.4c shows the computed waveform using the LDDRK scheme of Hu *et al.* (1996) with $\Delta t = 0.76\Delta x$. There is definitely an improvement with the use of this scheme. However, there are still some trailing waves, a manifestation of numerical dispersion. For the initial condition (4.31), a small fraction of the initial pulse has

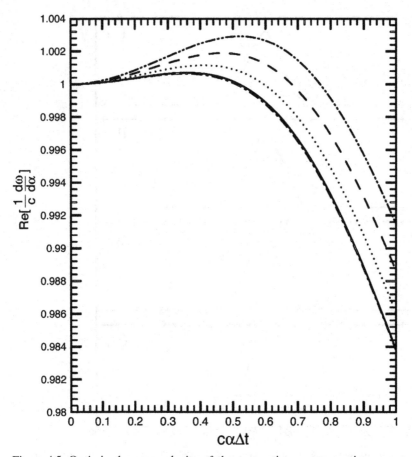

Figure 4.5. Optimized group velocity of the convective wave equation. — — — —, $\eta = 0.6$; ...., $\eta = 0.7$; – – – – – –, $\eta = 0.8$; ----------, $\eta = 0.9$; ————, Hu *et al.* (1996) LDDRK scheme.

a wave number larger than $\alpha \Delta x = 0.855$ corresponding to $\alpha \Delta t = 0.65$. This part of the initial spectrum will propagate slower according to the group velocity curve of Figure 4.5. They show up as trailing waves. Finally, Figure 4.4d shows the computed result of the four-level time marching DRP scheme. The four-level scheme computes the derivative function once per step. The single step Runge-Kutta scheme computes the derivative function four times per step. Thus, to compare the results based on nearly the same workload, the time step of the DRP scheme is taken to be a quarter of that of the Runge-Kutta scheme, i.e., $\Delta t = 0.19 \Delta x$. As can be seen, the results of Figures 4.4b, 4.4c, and 4.4d seem to indicate that the DRP scheme has less dispersion due to temporal discretization.

    For certain wave propagation problems, minimizing dispersion error could be an important consideration. To do so, one might choose the free parameters or constants of the Runge-Kutta scheme to minimize the difference between the group velocity and the exact velocity of propagation of the governing wave equation. For the simple convective wave equation, the group velocity is given by Eq. (4.30). Suppose in (4.28) c_1 and c_2 are given the values of 1.0 and 0.5, but allow c_3 and c_4 to be free parameters. These parameters are then chosen to minimize the square of

Table 4.1. *Values of parameter c_3 and c_4*

η	c_3	c_4
0.6	0.163168	4.07351 E-02
0.7	0.161940	4.04090 E-02
0.8	0.160545	4.00390 E-02
0.9	0.158992	3.96280 E-02
Hu *et al.* (1996)	0.162997	4.07574 E-02
Standard fourth-order RK	0.166667	4.16667 E-02

the difference between the group velocity and the exact wave speed $c = 1$ over a band of wave number. It will be assumed that the spatial discretization is exact; i.e., $\bar{\alpha} = \alpha$. Thus, c_3 and c_4 are chosen to minimize the integral,

$$E = \int_0^\eta \left| \frac{1}{c} \frac{d\omega}{d\alpha} - 1 \right|^2 d(c\alpha\Delta t). \tag{4.32}$$

Table 4.1 gives the values of c_3 and c_4 for different choices of η. The group velocity $\mathrm{Re}(d\omega/d\alpha)$ as a function of $c\alpha\Delta t$ corresponding to these choices of c_3 and c_4 is plotted in Figure 4.5. It is clear from this figure that by using a large η the minimization procedure forces a large band of wave number to have a group velocity closer to c. However, it is a feature of mean squared minimization that this also leads to a larger overshoot. So one might select a compromise between a larger bandwidth with a normalized group velocity closer to 1.0 and a smaller overshoot. Plotted also in Figure 4.5 is the group velocity curve of the LDDRK scheme. It turns out that it is almost identical to that of the choice $\eta = 0.6$. This is strictly a coincidence! But this appears to offer an explanation of why the LDDRK scheme does not incur excessive numerical dispersion error arising from temporal discretization.

4.5 Origin of Numerical Dissipation

Discretization of a partial differential equation into a finite difference equation, in general, leads to numerical dissipation in addition to numerical dispersion. To illustrate this, again consider solving the convective wave equation (4.11) using a large-stencil finite difference scheme. Suppose the spatial derivative is approximated by a finite difference quotient and solve time exactly. The exact solution of the semidiscretized problem is given by Eq. (4.21) and, upon evaluating the ω-integral, the solution is given by Eq. (4.23).

$$u(x,t) = \int_{-\infty}^{\infty} \hat{\Phi}(\alpha) e^{i(\alpha x - c\bar{\alpha}(\alpha)t)} d\alpha. \tag{4.33}$$

Now, whether the numerical solution is damped or not depends critically on whether a central difference stencil or an unsymmetric difference stencil is used. If a central difference stencil is used, then $\bar{\alpha}(\alpha)$ is a real function for real α. In this case, Eq. (4.33) represents a dispersive wave packet without being damped in time. If an unsymmetric stencil is used, then $\bar{\alpha}(\alpha)$ is complex for real α. In this case, the solution is damped in time provided that $\mathrm{Im}(c\bar{\alpha})$ is negative for all α. The time rate of damping for the wave component with wave number α is given by $c\,\mathrm{Im}[\bar{\alpha}(\alpha)]$. For many popular numerical schemes $c\,\mathrm{Im}(\bar{\alpha})$ is negative. Such schemes have built-in numerical damping. The damping rate can be calculated precisely once the stencil size and stencil coefficients are known.

It is worthwhile to point out that it is possible that c and $\mathrm{Im}(\bar{\alpha})$ have the same sign. In this case, the numerical solution will grow exponentially in time leading to numerical instability. Therefore, a necessary condition for numerical stability is

$$c\,\mathrm{Im}\left[\bar{\alpha}(\alpha)\right] < 0, \qquad -\pi \le \alpha\Delta x \le \pi. \tag{4.34}$$

Eq. (4.34) is the upwinding requirement. It is easy to show numerically that only upwind unsymmetric stencils could satisfy condition (4.34). The upwinding requirement is extremely important when solving the full Euler equations. Euler equations support several modes of waves (acoustic, vorticity, and entropy waves). The upwinding condition must be satisfied by each wave mode if the numerical solution is to remain stable.

Recently, a number of large-stencil upwind schemes were proposed by Lockard *et al.* (1995), Li (1997), Zhuang *et al.* (1998, 2002), Zingg (2000), and others for use in CAA. Each of these authors offered examples to demonstrate the numerical damping of his/her upwind scheme that could, indeed, damp out spurious waves to produce an acceptable solution. However, strictly speaking, the spatial derivative terms of the Euler equations are wave propagation terms. They are not damping terms. To use the discretized form of the derivative terms to perform two functions, namely, to support wave propagation and to damp out spurious short waves at the same time, might be imposing too many constraints on the stencil coefficients. This would make it less likely that the stencil design is optimal. A more natural strategy is to use a central difference scheme and incorporate needed numerical damping by the addition of artificial selective damping terms.

Temporal discretization also introduces numerical damping. The damping rate can be determined quantitatively from the dispersion relation. Because $\bar{\omega}(\omega)$ is complex, the root $\omega = \omega(\alpha)$ of $\bar{\omega}(\omega) = c\bar{\alpha}(\alpha)$ has an imaginary part even when $\bar{\alpha}(\alpha)$ is real for real α. For the convective wave equation, the exact solution of the fully discretized system is given by Eq. (4.19). The integrals can be evaluated by the Residue Theorem. This leads to the replacement of ω by $\omega = \omega(\alpha)$, the physical root of $\bar{\omega}(\omega) = c\bar{\alpha}(\alpha)$. Since $\omega(\alpha)$ is now complex, the solution is damped in time. The time rate of damping for wave component with wave number α is $\mathrm{Im}[\omega(\alpha)]$.

A simple way to reduce numerical damping due to temporal discretization is to use a smaller Δt. It is not too difficult to see why reducing time step size would reduce numerical damping. Suppose a spatial discretization scheme is chosen so that the problem would be solved accurately by using N mesh points per wavelength. In other words, if λ is the shortest wavelength, the scheme is designed to resolve, then $\lambda/\Delta x > N$. This yields, in terms of wave number, $2\pi/N > \bar{\alpha}_{\mathrm{largest}}\Delta x$. Now $\bar{\omega}$ is related to $\bar{\alpha}$ by the dispersion relation (4.20), thus,

$$\frac{2\pi}{N}c\left(\frac{\Delta t}{\Delta x}\right) > \bar{\omega}_{\mathrm{largest}}\Delta t.$$

By choosing $\Delta t / \Delta x$ small, $\bar{\omega}_{\mathrm{largest}} \Delta t$ would be confined to a range that is closer to zero. For most temporal marching schemes, including the Runge-Kutta method, the DRP scheme, the $\bar{\omega}\Delta t$ versus $\omega \Delta t$ relation is such that $\mathrm{Im}(\bar{\omega}\Delta t)$ and, hence, $\mathrm{Im}(\omega \Delta t)$ become smaller when $\bar{\omega}\Delta t$ is closer to zero. It follows that the root $\omega(\alpha)$ of Eq. (4.20) would be nearly real when Δt is reduced. This means that the temporal damping rate $\mathrm{Im}(\omega(\alpha))$ would be reduced.

4.6 Multidimensional Waves

The analysis of the previous sections is primary for one-dimensional waves. However, it can easily be generalized to waves in two or three dimensions.

Suppose the multidimensional dispersion relation of a wave system is

$$D(\alpha_i, \omega) = 0, \quad i = 1, 2, 3.$$

Here, for convenience, Cartesian tensor subscript notation will be used. A straightforward generalization of Eq. (4.2) for three-dimensional waves is

$$\tilde{u}(\alpha_i, \omega) = \frac{H(\alpha_i, \omega)}{D(\alpha_i, \omega)}. \tag{4.35}$$

The inverse transform yields

$$u(x_i, t) = \frac{1}{(2\pi)^4} \int\limits_{-\infty}^{\infty} \int \int\limits_{\Gamma} \int \frac{H(\alpha_i, \omega)}{D(\alpha_i, \omega)} e^{i(\alpha_j x_j - \omega t)} d\omega \, d\alpha_1 \, d\alpha_2 \, d\alpha_3. \tag{4.36}$$

Again, assuming there is a simple pole on the real ω-axis given by $\omega(\alpha_i)$, it is straightforward to obtain by the Residue Theorem that the contribution of this pole to the solution is

$$u(x_i, t) = -\frac{i}{(2\pi)^3} \int\limits_{-\infty}^{\infty} \int \int \frac{H(\alpha_i, \omega)}{\left(\frac{\partial D}{\partial \omega}\right)_{\omega = \omega(\alpha_i)}} e^{i(\alpha_j x_j - \omega(\alpha_i)t)} d\alpha_1 \, d\alpha_2 \, d\alpha_3. \tag{4.37}$$

The triple integral may be evaluated asymptotically for large t by the method of stationary phase in multidimensions. Thus,

$$u(x_i, t) \underset{t \to \infty}{\longrightarrow} \frac{iH(\alpha_i, \omega)}{\left(\frac{\partial D}{\partial \omega}\right)} \frac{1}{(2\pi t)^{\frac{3}{2}}} \frac{e^{i(\alpha_j x_j - \omega(\alpha_i)t) + i\varsigma}}{\left(\det \left|\frac{\partial^2 \omega}{\partial x_i \partial x_j}\right|\right)^{\frac{1}{2}}}, \tag{4.38}$$

where

$$\frac{x_i}{t} = \frac{\partial \omega}{\partial \alpha_i}, \quad i = 1, 2, 3. \tag{4.39}$$

In Eq. (4.38), det is the determinant of and ς depends on the number of factors of $\pi/4$ arising from path rotation. Now Eq. (4.39) gives the group velocity, V_i, in three dimensions as follows:

$$V_i = \frac{\partial \omega}{\partial \alpha_i}, \quad i = 1, 2, 3. \tag{4.40}$$

It is easy to see, for a multidimensional finite difference system, that the same analysis applies. Therefore, the conclusions concerning numerical dispersion and dissipation obtained for a one-dimensional system are also true for multidimensional finite difference approximations.

EXERCISES

4.1. Suppose the initial value problem,

$$\frac{\partial u}{\partial t} + c\frac{\partial u}{\partial x} = 0, \qquad u(x,0) = f(x),$$

is computed by approximating the $\partial u/\partial x$ term by a finite difference quotient, show

  (i) If the standard sixth-order central difference approximation is used, the numerical solution can only have trailing waves.
  (ii) If the 7-point stencil DRP scheme is used, the numerical solution may have both leading and trailing waves. Estimate the wavelength of the leading waves.

4.2. In solving the convective wave equation,

$$\frac{\partial u}{\partial t} + c\frac{\partial u}{\partial x} = 0,$$

by approximating the spatial derivative by finite difference quotient and marching the solution in time by the fourth-order RK scheme, it has been shown that the solution, in wave number space, is governed by the equation

$$\hat{u}^{(n+1)} = \hat{u}^{(n)}\left[1 + \sum_{j=1}^{4} c_j\, (-ic\bar{\alpha}\Delta t)^j\,\right]$$

where $\bar{\alpha}(\alpha)$ is the wave number of the finite difference scheme and n is the time level. One may consider this as a first-order finite difference equation in n (see Chapter 1). Find the exact solution of this finite difference equation. To avoid numerical instability, the absolute value of the characteristic root must be smaller than 1.0. Suppose $\bar{\alpha}$ is the wave number of the standard sixth-order central difference scheme and c_j's are the coefficients of the standard fourth-order RK scheme. Determine the largest time step Δt that may be used without encountering numerical instability.

5 Finite Difference Solution of the Linearized Euler Equations

5.1 Dispersion Relations and Asymptotic Solutions of the Linearized Euler Equations

Consider small-amplitude disturbances superimposed on a uniform mean flow of density ρ_0, pressure p_0, and velocity u_0 in the x direction (see Figure 5.1). The linearized Euler equations for two-dimensional disturbances are as follows:

$$\frac{\partial \mathbf{U}}{\partial t} + \frac{\partial \mathbf{E}}{\partial x} + \frac{\partial \mathbf{F}}{\partial y} = \mathbf{H}, \tag{5.1}$$

where

$$\mathbf{U} = \begin{bmatrix} \rho \\ u \\ v \\ p \end{bmatrix}, \qquad \mathbf{E} = \begin{bmatrix} \rho_0 u + \rho u_0 \\ u_0 u + \frac{p}{\rho_0} \\ u_0 v \\ u_0 p + \gamma p_0 u \end{bmatrix}, \qquad \mathbf{F} = \begin{bmatrix} \rho_0 v \\ 0 \\ \frac{p}{\rho_0} \\ \gamma p_0 v \end{bmatrix}.$$

The nonhomogeneous term $\mathbf{H}$ on the right side of Eq. (5.1) represents distributed time-dependent sources.

The Fourier-Laplace transform $\tilde{f}(\alpha, \beta, \omega)$ of a function $f(x, y, t)$ is related to the function by

$$\tilde{f}(\alpha, \beta, \omega) = \frac{1}{(2\pi)^3} \int_0^\infty \int_{-\infty}^\infty \int_{-\infty}^\infty f(x, y, t) e^{-i(\alpha x + \beta y - \omega t)} dx \, dy \, dt$$

$$f(x, y, t) = \int_\Gamma \int_{-\infty}^\infty \int_{-\infty}^\infty \tilde{f}(\alpha, \beta, \omega) e^{i(\alpha x + \beta y - \omega t)} d\alpha \, d\beta \, d\omega.$$

In these equations, the contour Γ is a line parallel to the real axis in the complex ω plane above all poles and singularities of the integrand (see Appendix A).

Now, consider the general initial value problem governed by the linearized Euler equation (5.1). The initial condition at $t = 0$ is

$$\mathbf{u} = \mathbf{u}_{\text{initial}}(x, y). \tag{5.2}$$

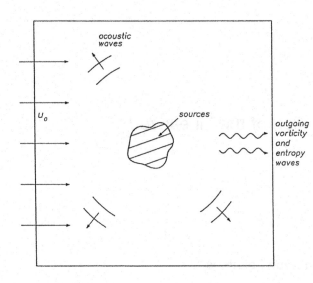

Figure 5.1. Sources of acoustic, vorticity, and entropy waves in a uniform flow.

The general initial value problem for arbitrary nonhomogeneous source term **H** can be solved formally by Fourier-Laplace transforms. As shown in Appendix A, the Laplace transform of a time derivative is

$$\frac{1}{2\pi} \int_0^\infty \frac{\partial \mathbf{u}}{\partial t} e^{i\omega t}\, dt = -\frac{1}{2\pi}\mathbf{u}_{\text{initial}} - i\omega\tilde{\mathbf{u}}. \tag{5.3}$$

The Fourier-Laplace transform of Eq. (5.1) is a system of linear algebraic equations that may be written in the following form:

$$\mathbf{A}\tilde{\mathbf{U}} = \tilde{\mathbf{G}}, \tag{5.4}$$

where

$$\mathbf{A} = \begin{bmatrix} (\omega - \alpha u_0) & -\rho_0\alpha & -\rho_0\beta & 0 \\ 0 & (\omega - \alpha u_0) & 0 & -\frac{\alpha}{\rho_0} \\ 0 & 0 & (\omega - \alpha u_0) & -\frac{\beta}{\rho_0} \\ 0 & -\gamma p_0\alpha & -\gamma p_0\beta & (\omega - \alpha u_0) \end{bmatrix} \tag{5.4a}$$

and $\tilde{\mathbf{G}} = i[\tilde{\mathbf{H}} + (\tilde{\mathbf{U}}_{\text{initial}}/2\pi)]$ represents the sum of the transforms of the source term and the initial condition.

It is easy to show that the eigenvalues λ and eigenvectors $\mathbf{X}_j$ ($j = 1, 2, 3, 4$) of matrix **A** are

$$\lambda_1 = \lambda_2 = (\omega - \alpha u_0) \tag{5.5}$$

$$\lambda_3 = (\omega - \alpha u_0) + a_0(\alpha^2 + \beta^2)^{\frac{1}{2}} \tag{5.6}$$

$$\lambda_4 = (\omega - \alpha u_0) - a_0(\alpha^2 + \beta^2)^{\frac{1}{2}} \tag{5.7}$$

$$\mathbf{X}_1 = \begin{bmatrix} 1 \\ 0 \\ 0 \\ 0 \end{bmatrix}, \quad \mathbf{X}_2 = \begin{bmatrix} 0 \\ \beta \\ -\alpha \\ 0 \end{bmatrix}, \quad \mathbf{X}_3 = \begin{bmatrix} \frac{1}{a_0^2} \\ \frac{-\alpha}{\rho_0 a_0 (\alpha^2+\beta^2)^{1/2}} \\ \frac{-\beta}{\rho_0 a_0 (\alpha^2+\beta^2)^{1/2}} \\ 1 \end{bmatrix}, \quad \mathbf{X}_4 = \begin{bmatrix} \frac{1}{a_0^2} \\ \frac{\alpha}{\rho_0 a_0 (\alpha^2+\beta^2)^{1/2}} \\ \frac{\beta}{\rho_0 a_0 (\alpha^2+\beta^2)^{1/2}} \\ 1 \end{bmatrix}, \quad (5.8)$$

where $a_0 = (\gamma p_0/\rho_0)^{1/2}$ is the speed of sound. The solution of Eq. (5.4) may be expressed as a linear combination of the eigenvectors as follows:

$$\tilde{\mathbf{U}} = \frac{C_1}{\lambda_1}\mathbf{X}_1 + \frac{C_2}{\lambda_2}\mathbf{X}_2 + \frac{C_3}{\lambda_3}\mathbf{X}_3 + \frac{C_4}{\lambda_4}\mathbf{X}_4. \quad (5.9)$$

The coefficient vector $\mathbf{C}$, having elements C_j ($j = 1$ to 4), is given by

$$\mathbf{C} = \mathbf{X}^{-1}\tilde{\mathbf{G}}. \quad (5.10)$$

$\mathbf{X}^{-1}$ is the inverse of the fundamental matrix:

$$\mathbf{X}^{-1} = \begin{bmatrix} 1 & 0 & 0 & -\frac{1}{a_0^2} \\ 0 & \frac{\beta}{(\alpha^2+\beta^2)} & -\frac{\alpha}{(\alpha^2+\beta^2)} & 0 \\ 0 & -\frac{1}{2}\frac{\rho_0 a_0 \alpha}{(\alpha^2+\beta^2)^{1/2}} & -\frac{1}{2}\frac{\rho_0 a_0 \beta}{(\alpha^2+\beta^2)^{1/2}} & \frac{1}{2} \\ 0 & \frac{1}{2}\frac{\rho_0 a_0 \alpha}{(\alpha^2+\beta^2)^{1/2}} & \frac{1}{2}\frac{\rho_0 a_0 \beta}{(\alpha^2+\beta^2)^{1/2}} & \frac{1}{2} \end{bmatrix}. \quad (5.11)$$

Eq. (5.9) represents the decomposition of the solution into the entropy wave, $\mathbf{X}_1$, the vorticity wave, $\mathbf{X}_2$, and the two modes of acoustic waves, $\mathbf{X}_3$ and $\mathbf{X}_4$. Now, each wave solution will be considered individually.

5.1.1 The Entropy Wave

The entropy wave consists of density fluctuations alone; i.e., $u = v = p = 0$. By inverting the Fourier-Laplace transform, it is straightforward to find that ρ is given by

$$\rho(x, y, t) = \int_\Gamma \int\int_{-\infty}^{\infty} \frac{C_1}{(\omega - \alpha u_0)} e^{i(\alpha x + \beta y - \omega t)} \, d\alpha \, d\beta \, d\omega. \quad (5.12)$$

The zero of the denominator gives rise to a pole of the integrand. The relationship between ω, α, and β arising from this zero is the dispersion relation. In this case, it is

$$\lambda_1 = (\omega - \alpha u_0) = 0. \quad (5.13)$$

C_1 may have poles in the ω plane. This is due to periodic forcing. If the wave is excited by initial condition alone, then C_1 is independent of ω.

In the α plane, the zero of the integrand of Eq. (5.13) is given by

$$\alpha = \frac{\omega}{u_0} \qquad (\text{Im}\,(\omega) > 0 \text{ when } \omega \text{ is on } \Gamma).$$

The α-integral of Eq. (5.12) may now be evaluated by means of the Residue Theorem. For $x > 0$, Jordan's Lemma requires the completion of the contour on the top half

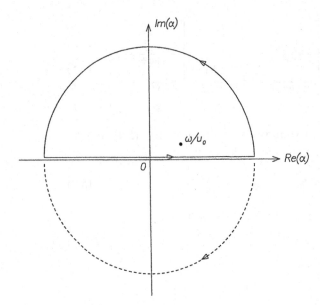

Figure 5.2. Inversion contours and pole in the complex α plane.

α plane as shown in Figure 5.2. For $x < 0$, the contour is completed in the lower half-plane. Thus,

$$\rho(x, y, t) = \begin{cases} -2\pi i \displaystyle\int_\Gamma \int_\beta \dfrac{C_1 e^{i[(x/u_0)-t]\omega+i\beta y}}{u_0}\, d\omega\, d\beta, & x \to \infty \\ 0, & x \to -\infty \end{cases}$$

or

$$\rho(x, y, t) = \begin{cases} \chi(x - u_0 t, y) & x \to \infty \\ 0, & x \to -\infty \end{cases}. \qquad (5.14)$$

5.1.2 The Vorticity Wave

The vorticity wave consists of velocity fluctuations alone. That is, there are no pressure and density fluctuations associated with this wave mode ($p = \rho = 0$). By inverting the Fourier-Laplace transforms, it is found that

$$\begin{bmatrix} u \\ v \end{bmatrix} = \int_\Gamma \int \int_{-\infty}^{\infty} \begin{bmatrix} \beta \\ -\alpha \end{bmatrix} \frac{C_2(\alpha, \beta)}{(\omega - \alpha u_0)} e^{i(\alpha x + \beta y - \omega t)}\, d\alpha\, d\beta\, d\omega. \qquad (5.15)$$

The dispersion relation is

$$\lambda_2 = (\omega - \alpha u_0) = 0.$$

Let

$$\psi(x, y, t) = \int_\Gamma \int \int_{-\infty}^{\infty} \frac{-iC_2(\alpha, \beta)}{(\omega - \alpha u_0)} e^{i(\alpha x + \beta y - \omega t)}\, d\alpha\, d\beta\, d\omega, \qquad (5.16)$$

then clearly

$$u = \frac{\partial \psi}{\partial y} \quad \text{and} \quad v = -\frac{\partial \psi}{\partial x}. \qquad (5.16a)$$

The integral in Eq. (5.16) may be evaluated in the same way as that in Eq. (5.12). This leads to

$$
\psi = \begin{cases} \chi(x - u_0 t, y) & x \to \infty \\ 0, & x \to -\infty \end{cases}.
\tag{5.16b}
$$

Just as for entropy waves, the vorticity wave is convected downstream as a frozen pattern at the speed of the mean flow.

5.1.3 The Acoustic Wave

The acoustic waves involve fluctuations in all the physical variables. The dispersion relation is given by

$$
\lambda_3 \lambda_4 = (\omega - \alpha u_0)^2 - a_0^2(\alpha^2 + \beta^2) = 0.
\tag{5.17}
$$

The formal solution may be found by inverting the Fourier-Laplace transforms. After some algebra, the formal solution may be written as

$$
\begin{bmatrix} \rho \\ p \end{bmatrix} = \int_\Gamma \int_{-\infty}^{\infty} \int \frac{\rho_0 a_0^2(\alpha \tilde{G}_2 + \beta \tilde{G}_3) + (\omega - \alpha u_0)\tilde{G}_4}{(\omega - \alpha u_0)^2 - a_0^2(\alpha^2 + \beta^2)} \begin{bmatrix} \frac{1}{a_0^2} \\ 1 \end{bmatrix} e^{i(\alpha x + \beta y - \omega t)} \, d\alpha \, d\beta \, d\omega.
\tag{5.18}
$$

$$
\begin{bmatrix} u \\ v \end{bmatrix} = \int_\Gamma \int_{-\infty}^{\infty} \int \frac{[(\omega - \alpha u_0)(\alpha \tilde{G}_2 + \beta \tilde{G}_3)/(\alpha^2 + \beta^2)] + (\tilde{G}_4/\rho_0)}{(\omega - \alpha u_0)^2 - a_0^2(\alpha^2 + \beta^2)} \begin{bmatrix} \alpha \\ \beta \end{bmatrix} e^{i(\alpha x + \beta y - \omega t)} \, d\alpha \, d\beta \, d\omega.
\tag{5.19}
$$

Now, the integrals of Eq. (5.18) will be evaluated in the limit $(x^2 + y^2) \to \infty$. The β-integral will be evaluated first. The poles in the β plane are at $\beta_\pm$ which are given by

$$
\beta_\pm = \pm i \left[\alpha^2 - \frac{(\omega - \alpha u_0)^2}{a_0^2} \right]^{\frac{1}{2}}.
\tag{5.20}
$$

The branch cuts of this square root function in the α plane are taken to be

$$
-\frac{\pi}{2} \le \arg \left[\alpha^2 - \frac{(\omega - \alpha u_0)^2}{a_0^2} \right]^{\frac{1}{2}} \le \frac{\pi}{2},
\tag{5.21}
$$

where the left (right) equality sign is to be used when ω is real and positive (negative). The branch cut configuration and the position of the inverse α-contour are shown in Figure 5.3. This configuration is valid regardless whether ω is real and positive or negative. In the β plane the integrand of Eq. (5.18) has a simple pole at β_+ in the upper half-plane and a simple pole at β_- in the lower half-plane (see Figure 5.4). For $y > 0$ ($y < 0$), the inversion contour may be closed in the upper (lower) half-plane by adding a large semicircle as shown (by Jordan's lemma). By invoking the

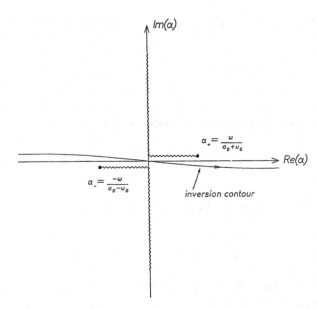

Figure 5.3. The branch cut configuration and inversion contour in the complex α plane.

Residue Theorem, the expressions for ρ and p simplify to

$$\begin{bmatrix} \rho \\ p \end{bmatrix} = \int_\Gamma \int_{-\infty}^{\infty} \frac{-\pi i}{a_0^2 \beta_+} [\rho_0 a_0^2 (\alpha \tilde{G}_2 + \beta_+ \tilde{G}_3) + (\omega - \alpha u_0) \tilde{G}_4] \begin{bmatrix} \frac{1}{a_0^2} \\ 1 \end{bmatrix} e^{i\Phi(\alpha,\theta)r - i\omega t} \, d\alpha \, d\omega.$$

(5.22)

In Eq. (5.22), r and θ are the polar coordinates and $\Phi = \alpha \cos\theta + i[\alpha^2 - (\omega - \alpha u_0)^2/a_0^2]^{1/2} \sin\theta$.

In the far field, where $r \to \infty$, the α-integral of Eq. (5.22) can be evaluated by the method of stationary phase (see Appendix B). A straightforward application of this method gives

$$\begin{bmatrix} \rho \\ p \end{bmatrix} = \int_\Gamma \frac{-\pi i}{a_0^2 \beta_s} [\rho_0 a_0^2 (\alpha_s \tilde{G}_2 + \beta_s \tilde{G}_3) + (\omega - \alpha_s u_0) \tilde{G}_4] \begin{bmatrix} \frac{1}{a_0^2} \\ 1 \end{bmatrix}$$

$$\cdot \left(\frac{2\pi}{r|\Phi''|} \right)^{\frac{1}{2}} e^{i[(r/V(\theta)) - t]\omega + (i\pi/4)\mathrm{sgn}(\Phi'')} \, d\omega,$$

(5.23)

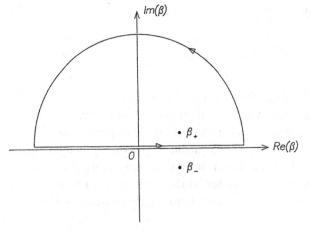

Figure 5.4. The poles and inversion contour in the complex β plane.

where α_s is the stationary phase point, $\beta_s = \beta_+(\alpha_s)$, $\Phi''(\alpha_s) = d^2\Phi/d\alpha^2|_{\alpha=\alpha_s}$, and $V(\theta) = u_0\cos\theta + a_0(1 - u_0^2\sin^2\theta/a_0^2)^{1/2}$. Eq. (5.23) may be rewritten in the more convenient form:

$$\begin{bmatrix} \rho \\ p \end{bmatrix} = \frac{F\left(\frac{r}{V(\theta)} - t, \theta\right)}{r^{1/2}} \begin{bmatrix} \frac{1}{a_0^2} \\ 1 \end{bmatrix} + O\left(r^{-\frac{3}{2}}\right). \tag{5.24}$$

Similarly, the integrals for u and v can be evaluated. The complete asymptotic solution has the following form:

$$\begin{bmatrix} \rho \\ u \\ v \\ p \end{bmatrix} = \frac{F\left(\frac{r}{V(\theta)} - t, \theta\right)}{r^{1/2}} \begin{bmatrix} \frac{1}{a_0^2} \\ \frac{\hat{u}(\theta)}{\rho_0 a_0} \\ \frac{\hat{v}(\theta)}{\rho_0 a_0} \\ 1 \end{bmatrix} + O\left(r^{-\frac{3}{2}}\right), \tag{5.25}$$

where

$$V(\theta) = u_0\cos\theta + a_0(1 - M^2\sin^2\theta)^{\frac{1}{2}}; \quad M = u_0/a_0$$

$$\hat{u}(\theta) = \frac{\cos\theta - M(1 - M^2\sin^2\theta)^{1/2}}{(1 - M^2\sin^2\theta)^{1/2} - M\cos\theta}$$

$$\hat{v}(\theta) = \sin\theta\left[(1 - M^2\sin^2\theta)^{\frac{1}{2}} + M\cos\theta\right].$$

$V(\theta)$ is the effective velocity of propagation in the θ-direction. This asymptotic solution and that of the entropy and vorticity waves will become useful later.

5.2 Dispersion-Relation-Preserving (DRP) Scheme

Now, consider solving the linearized Euler equation (5.1) computationally. It would be advantageous to first rewrite the equation in the following form:

$$\mathbf{K} = -\frac{\partial\mathbf{E}}{\partial x} - \frac{\partial\mathbf{F}}{\partial y} + \mathbf{H} \tag{5.26}$$

$$\frac{\partial\mathbf{U}}{\partial t} = \mathbf{K}, \tag{5.27}$$

so that the time derivative is by itself. Eq. (5.26) may be discretized by using the 7-point optimized finite difference scheme and Eq. (5.27) by using the four-level optimized time marching scheme. The fully discretized equations are as follows:

$$\mathbf{K}_{\ell,m}^{(n)} = -\frac{1}{\Delta x}\sum_{j=-3}^{3} a_j \mathbf{E}_{\ell+j,m}^{(n)} - \frac{1}{\Delta y}\sum_{j=-3}^{3} a_j \mathbf{F}_{\ell,m+j}^{(n)} + \mathbf{H}_{\ell,m}^{(n)} \tag{5.28}$$

$$\mathbf{U}_{\ell,m}^{(n+1)} = \mathbf{U}_{\ell,m}^{(n)} + \Delta t\sum_{j=0}^{3} b_j \mathbf{K}_{\ell,m}^{(n-j)}, \tag{5.29}$$

where ℓ, m, and n are the x, y, and t indices and Δx, Δy, and Δt are the mesh sizes. Eqs. (5.28) and (5.29) form a self-contained time marching scheme. If the values of $\mathbf{U}$ are known at time level n, $n-1$, $n-2$, and $n-3$, then $\mathbf{K}_{\ell,m}^{(i)}$ ($i = n, n-1, n-2, n-3$) may be calculated according to Eq. (5.28). By Eq. (5.29) the values of $\mathbf{U}_{\ell,m}^{(n+1)}$

may be determined. With $\mathbf{U}_{\ell,m}^{(n+1)}$ known, $\mathbf{K}_{\ell,m}^{(n+1)}$ may be calculated by Eq. (5.28) and marched to the next time level, $n + 2$, by using Eq. (5.29). The process is repeated until the desired time level is reached.

The Euler equations are first-order in time, so the initial conditions are as follows:

$$t = 0 \qquad \mathbf{U}(x, y, 0) = \mathbf{U}_{\text{initial}}(x, y). \tag{5.30}$$

This initial condition is sufficient for the single time step method. However, it will be shown that the appropriate starting condition for a multilevel time marching scheme is

$$\mathbf{U}_{\ell,m}^{(0)} = \mathbf{U}_{\text{initial}}; \qquad \mathbf{U}_{\ell,m}^{(n)} = 0 \quad \text{for negative } n. \tag{5.31}$$

At this time, the two pertinent questions to ask are, "How good is the computation scheme (5.28), (5.29), and (5.31)?"; "Do they provide entropy, vorticity, and acoustic wave solutions that have the same wave propagation characteristics as those of the original linearized Euler equations?" The answer to the second question is that the marching scheme does guarantee that the long waves ($\lambda > 7\Delta x$) of the numerical solution would have almost identical wave propagation properties as the corresponding waves of the linearized Euler equations. The answer to the first question is that the solution would be numerically accurate. This is because it will be shown that the finite difference algorithm has formally identical dispersion relations as the original linearized Euler equations. For this reason, the algorithm is referred to as the DRP scheme (Tam and Webb, 1993).

To find the dispersion relations of the time marching scheme, it is necessary first to generalize the discrete finite difference equations to finite difference equations with continuous variables. The complete initial value problem consists of the following equations and initial conditions.

$$\mathbf{K}(x, y, t) = -\frac{1}{\Delta x} \sum_{j=-3}^{3} a_j \mathbf{E}(x + j\Delta x, y, t)$$

$$- \frac{1}{\Delta y} \sum_{j=-3}^{3} a_j \mathbf{F}(x, y + j\Delta y, t) + \mathbf{H}(x, y, t) \tag{5.32}$$

$$\mathbf{U}(x, y, t + \Delta t) = \mathbf{U}(x, y, t) + \Delta t \sum_{j=0}^{3} b_j \mathbf{K}(x, y, t - j\Delta t) \tag{5.33}$$

$$\mathbf{U}(x, y, t) = \begin{cases} \mathbf{U}_{\text{initial}}(x, y), & 0 \le t < \Delta t \\ 0, & t < 0 \end{cases}. \tag{5.34}$$

The following Shifting Theorems for Laplace transforms (see Appendix A) are useful.

$$\Delta > 0$$

$$\frac{1}{2\pi} \int_0^\infty f(t + \Delta) e^{i\omega t}\, dt = e^{-i\omega\Delta} \tilde{f}(\omega) - \left(\frac{1}{2\pi} \int_0^\Delta f(t) e^{i\omega t}\, dt \right) e^{-i\omega\Delta}$$

$$\frac{1}{2\pi} \int_0^\infty f(t - \Delta) e^{i\omega t}\, dt = e^{i\omega\Delta} \tilde{f}(\omega) + \left(\frac{1}{2\pi} \int_{-\Delta}^0 f(t) e^{i\omega t}\, dt \right) e^{i\omega\Delta}. \tag{5.35}$$

The Fourier-Laplace transforms of Eqs. (5.32) and (5.33) with initial condition (5.34) are

$$\tilde{\mathbf{K}} = -i\,\overline{\alpha}\tilde{\mathbf{E}} - i\,\overline{\beta}\tilde{\mathbf{F}} + \tilde{\mathbf{H}} \tag{5.36}$$

$$-i\,\overline{\omega}\tilde{\mathbf{U}} = \tilde{\mathbf{K}} + \frac{\overline{\omega}}{2\pi\omega}\tilde{\mathbf{U}}_{\text{initial}}. \tag{5.37}$$

Upon eliminating $\tilde{\mathbf{K}}$, the two equations may be combined to yield the following:

$$\overline{\mathbf{A}}\tilde{\mathbf{U}} = \overline{\mathbf{G}}, \tag{5.38}$$

where $\overline{\mathbf{A}}$ is the same as $\mathbf{A}$ of Eq. (5.4) provided α, β, and ω are replaced by $\overline{\alpha}$, $\overline{\beta}$, $\overline{\omega}$, and $\overline{\mathbf{G}} = i[\tilde{\mathbf{H}} + (\overline{\omega}\tilde{\mathbf{U}}_{\text{initial}}/\omega 2\pi)]$.

A closer examination of Eq. (5.38) indicates that it is the same as Eq. (5.4) except with the replacement of α, β, ω by $\overline{\alpha}$, $\overline{\beta}$, $\overline{\omega}$. In the long-wave range, it was shown that

$$\overline{\alpha} \simeq \alpha, \qquad \overline{\beta} \simeq \beta, \quad \text{and} \quad \overline{\omega} \simeq \omega.$$

This means that the Fourier-Laplace transform of the DRP scheme is formally the same as the Fourier-Laplace transform of the original partial differential equations. This guarantees that the two systems would have formally the same dispersion relations. The solutions would automatically have nearly the same properties and the same asymptotic solutions. In other words, for long waves, the numerical solution is identical to that of the solution of the partial differential equations to the extent $\overline{\alpha} \simeq \alpha$, $\overline{\beta} \simeq \beta$, and $\overline{\omega} \simeq \omega$.

5.3 Numerical Stability

All numerical schemes are liable to be numerically unstable. One can usually avoid numerical instability by taking an appropriately small Δt. The numerical stability requirement of the DRP scheme will now be investigated. Recall that the Fourier-Laplace transform of the DRP scheme is formally the same as that of the original Euler equations (with α, β, ω replaced by $\overline{\alpha}$, $\overline{\beta}$, $\overline{\omega}$, respectively). Thus, by modifying Eq. (5.18) accordingly, the pressure associated with the acoustic waves computed by the DRP scheme is given by

$$p(x, y, t) = \int_{\Gamma} \int_{-\infty}^{\infty} \int \frac{[\rho_0 a_0^2(\overline{\alpha}\tilde{G}_2 + \overline{\beta}\,\tilde{G}_3) + (\overline{\omega} - \overline{\alpha}u_0)\tilde{G}_4]}{(\overline{\omega} - \overline{\alpha}u_0)^2 - a_0^2(\overline{\alpha}^2 + \overline{\beta}^2)} e^{i(\alpha x + \beta y - \omega t)} \, d\alpha \, d\beta \, d\omega,$$

$$\tag{5.39}$$

where $\overline{\alpha}(\alpha)$, $\overline{\beta}(\beta)$, and $\overline{\omega}(\omega)$ and its inverse $\omega = \omega(\overline{\omega})$ are given by Eqs. (2.9) and (3.16). From the theory of Laplace transform, it is well known that for large t the dominant contribution to the ω-integral comes from the residue of the pole of the integrand that has the largest imaginary part in the complex ω plane. In Eq. (5.39), the pole corresponding to the outgoing acoustic wave is given implicitly by

$$\overline{\omega}(\omega) = \overline{\alpha}(\alpha)u_0 + a_0(\overline{\alpha}^2(\alpha) + \overline{\beta}^2(\beta))^{\frac{1}{2}}. \tag{5.40}$$

As discussed in Chapter 3, for a given $\overline{\omega}$, there are four values of ω. Thus, there are four wave solutions associated with the $\overline{\omega}$ pole of dispersion relation (5.40);

three of which are spurious. Let the roots be denoted by ω_k ($k = 1, 2, 3, 4$) with ω_1 corresponding to the physical acoustic wave. On completing the contour in the lower half ω plane by a large semicircle, by the Residue Theorem, the pressure field given by Eq. (5.39) becomes

$$p(x, y, t) = \sum_{k=1}^{4} - (2\pi i)$$

$$\times \left[\int_{-\infty}^{\infty} \int_{-\infty}^{\infty} \frac{[\rho_0 a_0(\bar{\alpha}\tilde{G}_2 + \bar{\beta}\,\tilde{G}_3) + (\bar{\omega} - \bar{\alpha}u_0)\tilde{G}_4]}{2(\bar{\omega} - \bar{\alpha}u_0)\frac{d\bar{\omega}}{d\omega}} \bigg|_{\omega=\omega_k} e^{i(\alpha x + \beta y)} d\alpha\, d\beta \right] e^{-i\omega_k t}$$

$$(5.41)$$

To obtain numerical stability, a sufficient condition is

$$\mathrm{Im}(\omega_k) \le 0, \qquad k = 1, 2, 3, 4. \tag{5.42}$$

Note, from Figure 2.3, that for arbitrary values of α and β the inequalities

$$\bar{\alpha}\Delta x < 1.75, \qquad \bar{\beta}\Delta y < 1.75 \tag{5.43}$$

hold true. Substitution of inequality (5.43) into Eq. (5.40), and upon multiplying by Δt, it is found that

$$\bar{\omega}\Delta t \le \frac{1.75 a_0}{\Delta x} \left[M + \left(1 + \left(\frac{\Delta x}{\Delta y}\right)^2 \right)^{\frac{1}{2}} \right] \Delta t, \tag{5.44}$$

where M is the Mach number. From Figure 3.1 it is clear that if $|\bar{\omega}\Delta t|$ is less than 0.41, then all the roots of ω_k, especially the spurious roots, are damped. Therefore, to ensure numerical stability, it is sufficient by Eq. (5.44) to restrict Δt to less than $\Delta t_{\max}$, where $\Delta t_{\max}$ is given by

$$\Delta t_{\max} = \frac{0.41}{1.75 \left[M + \left(1 + \left(\frac{\Delta x}{\Delta y}\right)^2 \right)^{1/2} \right]} \frac{\Delta x}{a_0}. \tag{5.45}$$

Similar analysis for the entropy and the vorticity wave modes of the finite difference scheme yields the following criterion for numerical stability:

$$\Delta t < \frac{0.4}{1.75 M} \frac{\Delta x}{a_0}. \tag{5.46}$$

Formula (5.45) is a more stringent condition than formula (5.46). Therefore, for $\Delta t < \Delta t_{\max}$, the DRP scheme would be numerically stable.

5.4 Group Velocity for Finite Difference Schemes

Suppose $\omega = \omega(\alpha, \beta)$ is the dispersion relation for a certain wave mode in two dimensions, then, as discussed in Section 4.6, the group velocity is

$$\mathbf{v}_{\mathrm{group}} = \frac{\partial \omega}{\partial \alpha}\mathbf{e_x} + \frac{\partial \omega}{\partial \beta}\mathbf{e_y},$$

where $\mathbf{e_x}$ and $\mathbf{e_y}$ are the unit vectors in the x and y directions. For instance, the dispersion relation for both the entropy and the vorticity wave is Eq. (5.13),

$$\omega = u_0\alpha, \tag{5.47}$$

so that

$$\frac{\partial \omega}{\partial \alpha} = u_0, \qquad \frac{\partial \omega}{\partial \beta} = 0.$$

Hence,

$$\mathbf{v}_{\text{group}} = u_0\mathbf{e_x}. \tag{5.48}$$

That is, the wave is convected downstream at the speed of the mean flow.

The dispersion relations for acoustic waves in a uniform mean flow are given by Eq. (5.17) as follows:

$$\omega = u_0\alpha \pm a_0(\alpha^2 + \beta^2)^{1/2}. \tag{5.49}$$

The group velocity is

$$\mathbf{v}_{\text{group}} = \left[u_0 \pm \frac{\alpha a_0}{(\alpha^2 + \beta^2)^{1/2}} \right] \mathbf{e_x} \pm \frac{\beta a_0}{(\alpha^2 + \beta^2)^{1/2}} \mathbf{e_y}. \tag{5.50}$$

For waves propagating in the x direction, β is equal to zero. For these waves,

$$\mathbf{v}_{\text{group}} = u_0 \pm a_0.$$

Now, for the DRP scheme, the dispersion relation of the corresponding wave is given by

$$\overline{\omega}(\omega) = \overline{\omega}(\overline{\alpha}(\alpha), \overline{\beta}(\beta)). \tag{5.51}$$

By implicit differentiation, the group velocity of the wave is

$$\mathbf{v}_{\text{group,DRP}} = \frac{\frac{\partial \overline{\omega}}{\partial \overline{\alpha}} \frac{d\overline{\alpha}}{d\alpha}}{\frac{d\overline{\omega}}{d\omega}} \mathbf{e_x} + \frac{\frac{\partial \overline{\omega}}{\partial \overline{\beta}} \frac{d\overline{\beta}}{d\beta}}{\frac{d\overline{\omega}}{d\omega}} \mathbf{e_y}. \tag{5.52}$$

The dispersion relations for acoustic waves, by Eq. (5.40), are

$$\overline{\omega} = u_0\overline{\alpha} \pm a_0(\overline{\alpha}^2 + \overline{\beta}^2)^{\frac{1}{2}}. \tag{5.53}$$

By Eq. (5.52), it is found that

$$\mathbf{v}_{\text{group,DRP}} = \frac{\left[u_0 \pm \frac{a_0\overline{\alpha}}{(\overline{\alpha}^2 + \overline{\beta}^2)^{1/2}} \right] \frac{d\overline{\alpha}}{d\alpha}}{\frac{d\overline{\omega}}{d\omega}} \mathbf{e_x} + \frac{\left[\pm \frac{a_0\overline{\beta}}{(\overline{\alpha}^2 + \overline{\beta}^2)^{1/2}} \right] \frac{d\overline{\beta}}{d\beta}}{\frac{d\overline{\omega}}{d\omega}} \mathbf{e_y}. \tag{5.54}$$

For waves propagating in the x direction alone with $\beta = \overline{\beta} = 0$, the group velocity is

$$\mathbf{v}_{\text{group,DRP}} = \frac{(u_0 \pm a_0)\frac{d\overline{\alpha}}{d\alpha}}{\frac{d\overline{\omega}}{d\omega}} \hat{\mathbf{e}}_\mathbf{x}. \tag{5.55}$$

For small Δt, $d\overline{\omega}/d\omega \simeq 1$. In this case, the group velocity is proportional to $d\overline{\alpha}/d\alpha$. It is interesting to point out that for grid-to-grid oscillations (i.e., short waves with $\alpha\Delta x \simeq \pi$) $d\overline{\alpha}/d\alpha \simeq -2.3$ from Figure 2.4. Thus, the grid-to-grid oscillations propagate at 2.3 times the speed of sound. The waves travel supersonically in the opposite direction.

5.5 Time Step Δt: Accuracy Consideration

In Section 5.3, formulas to determine the largest time step, Δt, permitted by numerical stability requirement for waves governed by a given dispersion relation were developed. Here, the choice of Δt is revisited from the point of view of numerical accuracy. The Δt required by accuracy consideration may not be the same as that required by numerical stability. It is recommended that the smaller of the two Δt's be used in actual computation.

From Figure 2.5, it is easy to find that, for the 7-point stencil DRP scheme, if the permissible numerical dispersion is limited to the range $d\bar{\alpha}/d\alpha \leq 0.003$, the wave number $\bar{\alpha}\Delta x$ of the numerical solution must not be larger than $\alpha^* \Delta x$ ($\alpha^* \Delta x = 0.9$ from the figure). Also, from an enlarged Figure 3.1, it is easy to see that, if the range of $\bar{\omega}\Delta t$ to be used is restricted, then it is possible to make both the numerical damping rate small (say, less than 1.2×10^{-5} or $-\text{Im}(\omega\Delta t) \leq 1.2 \times 10^{-5}$) and that $\bar{\omega}\Delta t$ and $\text{Re}(\omega\Delta t)$ differ by no more than 0.0002. This is accomplished if $\bar{\omega}\Delta t$ is limited to less than $\omega^* \Delta t$ ($\omega^* \Delta t = 0.19$ from the figure). To keep the computation to within this accuracy in the wave number as well as the angular frequency space, the size of Δt must be chosen appropriately. From dispersion relation (5.40), it is easily found that

$$\bar{\omega}\Delta t = \left\{ M\bar{\alpha}\Delta x + \left[(\bar{\alpha}\Delta x)^2 + (\bar{\beta}\Delta y)^2 \left(\frac{\Delta x}{\Delta y} \right)^2 \right]^{\frac{1}{2}} \right\} \frac{a_0 \Delta t}{\Delta x}. \tag{5.56}$$

Now, to maintain numerical accuracy to $d\bar{\alpha}/d\alpha \leq 0.003$, it is necessary to keep $\bar{\alpha}\Delta x$ and $\bar{\beta}\Delta y$ smaller than $\alpha^*\Delta x = 0.9$. By Eq. (5.56), this yields

$$\bar{\omega}\Delta t \leq \left\{ M + \left[1 + \left(\frac{\Delta x}{\Delta y} \right)^2 \right]^{\frac{1}{2}} \right\} 0.9 \frac{a_0 \Delta t}{\Delta x}. \tag{5.57}$$

The largest value of $\bar{\omega}\Delta t$ that complies with the this accuracy is $\bar{\omega}\Delta t = 0.19$. Therefore, by Eq. (5.57), Δt must be such that

$$\left\{ M + \left[1 + \left(\frac{\Delta x}{\Delta y} \right)^2 \right]^{\frac{1}{2}} \right\} 0.9 \frac{a_0 \Delta t}{\Delta x} \leq 0.19$$

or

$$\frac{a_0 \Delta t}{\Delta x} \leq \frac{0.211}{M + \left[1 + \left(\frac{\Delta x}{\Delta y} \right)^2 \right]^{1/2}}. \tag{5.58}$$

Eq. (5.58) is the formula for Δt if this accuracy is to be guaranteed in the computation. The requirement on Δt for numerical stability, from Eq. (5.45), is

$$\frac{a_0 \Delta t}{\Delta x} \leq \frac{0.228}{M + \left[1 + \left(\frac{\Delta x}{\Delta y} \right)^2 \right]^{1/2}}. \tag{5.59}$$

The right side of Eq. (5.59) is slightly larger than that of Eq. (5.58). In other words, the requirement for numerical accuracy is slightly more stringent than that for numerical stability. Eq. (5.58) should be used for determining the step size Δt in a numerical simulation. Note that the choice of Eq. (5.58) ensures that the dispersion relation matches exactly the wave number and angular frequency range of acceptable accuracy.

5.6 DRP Scheme in Curvilinear Coordinates

Most practical computations are often not carried out in Cartesian coordinates. Instead, the actual computations usually are done on a set of body-fitted curvilinear coordinates. Thus, the space of the curvilinear coordinate system forms the computation domain. In other words, the governing equations, written in curvilinear coordinates, will be discretized and computed. It will be shown that, if the optimized spatial and temporal discretization method developed in Chapters 2 and 3 are used, then the discretized equations retain the DRP property.

The three-dimensional compressible Euler equations for a perfect gas in Cartesian coordinates may be written as

$$\frac{\partial \mathbf{u}}{\partial t} + \mathbf{A}\frac{\partial \mathbf{u}}{\partial x} + \mathbf{B}\frac{\partial \mathbf{u}}{\partial y} + \mathbf{C}\frac{\partial \mathbf{u}}{\partial z} = 0, \qquad (5.60)$$

where

$$\mathbf{u} = \begin{bmatrix} \rho \\ u \\ v \\ w \\ p \end{bmatrix} \quad \mathbf{A} = \begin{bmatrix} u & \rho & 0 & 0 & 0 \\ 0 & u & 0 & 0 & \frac{1}{\rho} \\ 0 & 0 & u & 0 & 0 \\ 0 & 0 & 0 & u & 0 \\ 0 & \gamma p & 0 & 0 & u \end{bmatrix} \quad \mathbf{B} = \begin{bmatrix} v & 0 & \rho & 0 & 0 \\ 0 & v & 0 & 0 & 0 \\ 0 & 0 & v & 0 & \frac{1}{\rho} \\ 0 & 0 & 0 & v & 0 \\ 0 & 0 & \gamma p & 0 & v \end{bmatrix} \quad \mathbf{C} = \begin{bmatrix} w & 0 & 0 & \rho & 0 \\ 0 & w & 0 & 0 & 0 \\ 0 & 0 & w & 0 & 0 \\ 0 & 0 & 0 & w & \frac{1}{\rho} \\ 0 & 0 & 0 & \gamma p & w \end{bmatrix}.$$

Now, let $\xi(x, y, z)$, $\eta(x, y, z)$, and $\zeta(x, y, z)$ be a set of curvilinear coordinates. This set of coordinates will be used for discretizing the Euler equations and for computing the solution. A simple change of variables by means of the chain rule transforms Eq. (5.60) into an equation with (ξ, η, ζ) as independent variables in the following form:

$$\frac{\partial \mathbf{u}}{\partial t} + \overline{\mathbf{A}}\frac{\partial \mathbf{u}}{\partial \xi} + \overline{\mathbf{B}}\frac{\partial \mathbf{u}}{\partial \eta} + \overline{\mathbf{C}}\frac{\partial \mathbf{u}}{\partial \zeta} = 0, \qquad (5.61)$$

where

$$\overline{\mathbf{A}} = \frac{\partial \xi}{\partial x}\mathbf{A} + \frac{\partial \xi}{\partial y}\mathbf{B} + \frac{\partial \xi}{\partial z}\mathbf{C}$$

$$\overline{\mathbf{B}} = \frac{\partial \eta}{\partial x}\mathbf{A} + \frac{\partial \eta}{\partial y}\mathbf{B} + \frac{\partial \eta}{\partial z}\mathbf{C}$$

$$\overline{\mathbf{C}} = \frac{\partial \zeta}{\partial x}\mathbf{A} + \frac{\partial \zeta}{\partial y}\mathbf{B} + \frac{\partial \zeta}{\partial z}\mathbf{C}.$$

Suppose the Fourier-Laplace transform of a function $f(\xi, \eta, \zeta, t)$ is $\tilde{f}(\alpha, \beta, \gamma, \omega)$ and that of another function $g(\xi, \eta, \zeta, t)$ is $\tilde{g}(\alpha, \beta, \gamma, \omega)$, that is,

$$\begin{bmatrix} \tilde{f}(\alpha, \beta, \gamma, \omega) \\ \tilde{g}(\alpha, \beta, \gamma, \omega) \end{bmatrix} = \frac{1}{(2\pi)^4} \int_0^\infty \int\int\int_{-\infty}^\infty \begin{bmatrix} f(\xi, \eta, \zeta, t) \\ g(\xi, \eta, \zeta, t) \end{bmatrix} e^{i(\alpha\xi + \beta\eta + \gamma\zeta - \omega t)} d\xi\, d\eta\, d\zeta\, dt.$$

It can easily be proven that the Fourier-Laplace transform of the product $f(\xi, \eta, \zeta, t)\, g(\xi, \eta, \zeta, t)$ is given by the convolution integral of $\tilde{f}(\alpha, \beta, \gamma, \omega)$ and $\tilde{g}(\alpha, \beta, \gamma, \omega)$, namely,

$$\frac{1}{(2\pi)^4} \int_0^\infty \int\int\int_{-\infty}^\infty f(\xi, \eta, \zeta, t) g(\xi, \eta, \zeta, t)\, e^{i(\alpha\xi + \beta\eta + \gamma\zeta - \omega t)} d\xi\, d\eta\, d\zeta\, dt$$

$$= \int\int\int_{-\infty}^\infty \int \tilde{f}(\alpha - \alpha_1, \beta - \beta_1, \gamma - \gamma_1, \omega - \omega_1)$$

$$\times\, \tilde{g}(\alpha_1, \beta_1, \gamma_1, \omega_1)\, d\alpha_1\, d\beta_1\, d\gamma_1\, d\omega_1. \tag{5.62}$$

By means of convolution integral (5.62), the Fourier-Laplace transform of the Euler equation (5.61) may be written as follows:

$$-i\omega\tilde{u} = \int\int\int_{-\infty}^\infty \int \left[i\alpha_1 \tilde{\overline{A}}(\alpha - \alpha_1, \beta - \beta_1, \gamma - \gamma_1, \omega - \omega_1) \right.$$

$$+ i\beta_1 \tilde{\overline{B}}(\alpha - \alpha_1, \beta - \beta_1, \gamma - \gamma_1, \omega - \omega_1)$$

$$+ \left. i\gamma_1 \tilde{\overline{C}}(\alpha - \alpha_1, \beta - \beta_1, \gamma - \gamma_1, \omega - \omega_1) \right]$$

$$\times\, \tilde{u}(\alpha_1, \beta_1, \gamma_1, \omega_1)\, d\alpha_1\, d\beta_1\, d\gamma_1\, d\omega_1. \tag{5.63}$$

Now, for the purpose of computation, let Eq. (5.61) be discretized in the (ξ, η, ζ) spacing using the 7-point stencil optimized scheme and the four-level time marching algorithm developed in Chapters 2 and 3. The discretized equations are as follows:

$$\mathbf{u}_{\ell,m,n}^{(k+1)} = \mathbf{u}_{\ell,m,n}^{(k)} + \Delta t \sum_{j=0}^3 b_j \mathbf{E}_{\ell,m,n}^{k+j} \tag{5.64}$$

$$\mathbf{E}_{\ell,m,n}^{(k)} = -\overline{\mathbf{A}}_{\ell,m,n}^{(k)} \frac{1}{\Delta\xi} \left(\sum_{j=-3}^3 a_j \mathbf{u}_{\ell+j,m,n}^{(k)} \right) - \overline{\mathbf{B}}_{\ell,m,n}^{(k)} \frac{1}{\Delta\eta} \left(\sum_{j=-3}^3 a_j \mathbf{u}_{\ell,m+j,n}^{(k)} \right)$$

$$- \overline{\mathbf{C}}_{\ell,m,n}^{(k)} \frac{1}{\Delta\zeta} \left(\sum_{j=-3}^3 a_j \mathbf{u}_{\ell,m,n+j}^{(k)} \right), \tag{5.65}$$

where subscripts (ℓ, m, n) are the spatial indices and superscript k is the time level.

The Fourier-Laplace transforms of the discretized Eqs. (5.64) and (5.65) are

$$-i\,\bar{\omega}(\omega)\tilde{\mathbf{u}} = \tilde{\mathbf{E}} \tag{5.66}$$

$$\tilde{\mathbf{E}} = \int\int\int_{-\infty}^{\infty}\int \left[i\,\bar{\alpha}(\alpha_1)\tilde{\mathbf{A}}(\alpha - \alpha_1, \beta - \beta_1, \gamma - \gamma_1, \omega - \omega_1)\right.$$

$$+ i\,\bar{\beta}(\beta_1)\tilde{\mathbf{B}}(\alpha - \alpha_1, \beta - \beta_1, \gamma - \gamma_1, \omega - \omega_1)$$

$$\left. + i\,\bar{\gamma}(\gamma_1)\tilde{\mathbf{C}}(\alpha - \alpha_1, \beta - \beta_1, \gamma - \gamma_1, \omega - \omega_1)\right]$$

$$\times \tilde{\mathbf{u}}(\alpha_1, \beta_1, \gamma_1, \omega_1)\,d\alpha_1\,d\beta_1\,d\gamma_1\,d\omega_1. \tag{5.67}$$

Upon eliminating $\tilde{\mathbf{E}}$ from Eqs. (5.66) and (5.67), the Fourier-Laplace transform of the discretized solution is given by

$$-i\,\bar{\omega}(\omega)\tilde{\mathbf{u}} = \int\int\int_{-\infty}^{\infty}\int \left[i\,\bar{\alpha}(\alpha_1)\tilde{\mathbf{A}}(\alpha - \alpha_1, \beta - \beta_1, \gamma - \gamma_1, \omega - \omega_1)\right.$$

$$+ i\,\bar{\beta}(\beta_1)\tilde{\mathbf{B}}(\alpha - \alpha_1, \beta - \beta_1, \gamma - \gamma_1, \omega - \omega_1).$$

$$\left. + i\,\bar{\gamma}(\gamma_1)\tilde{\mathbf{C}}(\alpha - \alpha_1, \beta - \beta_1, \gamma - \gamma_1, \omega - \omega_1)\right]$$

$$\times \tilde{\mathbf{u}}(\alpha_1, \beta_1, \gamma_1, \omega_1)\,d\alpha_1\,d\beta_1\,d\gamma_1\,d\omega_1. \tag{5.68}$$

By comparing Eqs. (5.63) and (5.68), it becomes clear that, if the discretized solution is restricted to the range for which $\bar{\alpha}(\alpha) = \alpha$, $\bar{\beta}(\beta) = \beta$, $\bar{\gamma}(\gamma) = \gamma$, and $\bar{\omega}(\omega) = \omega$, the equations are identical. Thus, the discretized Eqs. (5.64) and (5.65) are DRP in the sense that their Fourier-Laplace transforms are formally the same as the exact solution of Euler equations.

EXERCISES

5.1. The propagation of electromagnetic waves is governed by the Maxwell equations:

$$\frac{\mu\varepsilon}{c}\frac{\partial\mathbf{E}}{\partial t} - \nabla \times \mathbf{B} = 0 \tag{1}$$

$$\frac{1}{c}\frac{\partial\mathbf{B}}{\partial t} + \nabla \times \mathbf{E} = 0 \tag{2}$$

$$\nabla \cdot \mathbf{E} = 0 \tag{3}$$

$$\nabla \cdot \mathbf{B} = 0, \tag{4}$$

where $\mathbf{E}$ and $\mathbf{B}$ are the electric and magnetic fields and μ and ε are material constants. Now, look for plane wave solutions in the following form:

$$\mathbf{E} = \hat{\mathbf{e}}E_0 e^{i(\mathbf{k}\cdot\mathbf{x}-\omega t)}, \tag{5}$$

where $\hat{\mathbf{e}}$ is a constant unit vector characterizing the direction of the electric field. E_0 is the amplitude and $\mathbf{k} = \alpha\hat{\mathbf{e}}_x + \beta\hat{\mathbf{e}}_y + \gamma\hat{\mathbf{e}}_z$, where $\hat{\mathbf{e}}_x, \hat{\mathbf{e}}_y, \hat{\mathbf{e}}_z$ are the unit vectors in the

x, y, and z directions, is the wave vector and ω is the angular frequency. Eqs. (2) and (4) are satisfied automatically if the magnetic field is given by

$$\mathbf{B} = \frac{c}{\omega}(\mathbf{k} \times \mathbf{E}).\tag{6}$$

Substitution of (5) and (6) into (1) and (3) leads to

$$\frac{\omega\mu\varepsilon}{c}\hat{\mathbf{e}} = \frac{c}{\omega}\mathbf{k} \times (\mathbf{k} \times \hat{\mathbf{e}})\tag{7}$$

$$\hat{\mathbf{e}} \cdot \mathbf{k} = 0.\tag{8}$$

(Note: Eqs. (7) and (8) can also be obtained by taking the Fourier-Laplace transform of Eqs. (1) to (4). In this case, the transform variables are $(\alpha, \beta, \gamma, \omega)$.)

If $\hat{\mathbf{e}} = \xi\hat{\mathbf{e}}_{\mathbf{x}+} + \eta\hat{\mathbf{e}}_{\mathbf{y}} + \zeta\hat{\mathbf{e}}_{\mathbf{z}}$ is the solution of Eq. (7), then Eq. (8) is automatically satisfied. By writing Eq. (7) in a matrix form for the unknowns (ξ, η, ζ), the dispersion relations of electromagnetic waves are given by setting the determinant of the coefficient matrix to zero.

Discretize the Maxwell equations according to the DRP scheme. Show that the discretized equation formally preserves the dispersion relation of the electromagnetic waves; i.e., $\bar{\omega}$ replaces ω, $(\bar{\alpha}, \bar{\beta}, \bar{\gamma})$ replaces (α, β, γ), respectively.

5.2. Consider solving the convective equation

$$\frac{\partial u}{\partial t} + c\frac{\partial u}{\partial x} = 0$$

computationally by using central difference approximation. Suppose we use different size stencils; e.g., $N = 7$, 9, 11 etc. Assume that the same time marching scheme is used. Show that the maximum time step Δt allowed by numerical stability requirement is smaller when a larger size stencil is used.

(Note: This result is not restricted to the convective equation. It is true for the solution of the Euler equations and the Maxwell equations.)

5.3. Consider solving numerically the following initial value problem:

$$\frac{\partial u}{\partial t} + \frac{\partial u}{\partial x} = 0, \quad t = 0, \quad u = f(x).$$

In standard numerical analysis books, one finds that the spatial derivative may be discretized in three ways.

(a) Forward difference

$$\left(\frac{\partial u}{\partial x}\right)_\ell = \frac{-u_{\ell+2} + 4u_{\ell+1} - 3u_\ell}{2\Delta x} + O(\Delta x^2)$$

(b) Central difference

$$\left(\frac{\partial u}{\partial x}\right)_\ell = \frac{u_{\ell+1} - u_{\ell-1}}{2\Delta x} + O(\Delta x^2)$$

(c) Backward difference

$$\left(\frac{\partial u}{\partial x}\right)_\ell = \frac{3u_\ell - 4u_{\ell-1} + 3u_{\ell-2}}{2\Delta x} + O(\Delta x^2)$$

1. Determine which of these finite difference schemes would lead to stable numerical solution, assuming that the time derivative is computed exactly.
2. Now apply the same discretization to the linearized Euler equations in one dimension.

$$\frac{\partial u}{\partial t} + \frac{\partial p}{\partial x} = 0, \qquad \frac{\partial p}{\partial t} + \frac{\partial u}{\partial x} = 0.$$

Determine which of these schemes are suitable for solving this system of equations.

3. If the central difference scheme is used for the x derivative and the four-level time marching DRP scheme is used to advance the solution in time, determine the largest Δt one may use for solving the one-dimensional Euler equations.

5.4. Suppose an upwind difference

$$\left(\frac{\partial u}{\partial x}\right)_\ell = \frac{3u_\ell - 4u_{\ell-1} + u_{\ell-2}}{2\Delta x} + O((\Delta x)^2)$$

is used to approximate the x derivative of the equation

$$\frac{\partial u}{\partial t} + c\frac{\partial u}{\partial x} = 0$$

in the solution of the initial value problem

$$u(x, 0) = e^{-(\ell n2)\left[\frac{x}{10\Delta x}\right]^2}.$$

By using $\mathrm{Im}(\bar{\alpha}\Delta x)$ or otherwise estimate the decrease in the area of the pulse after it has propagated over a distance of 400 mesh points. Estimate the shape of the pulse. Compare your estimate with your numerical solution.

5.5. Solve the dimensionless spherical wave problem computationally

$$\frac{\partial u}{\partial t} + \frac{u}{x} + \frac{\partial u}{\partial x} = 0$$

over the domain $5 \le x \le 5$ with initial condition $t = 0, u = 0$. The boundary condition at $x = 5$ is

$$x = 5, \qquad u = \sin\left(\frac{\pi t}{4}\right).$$

Here, x is nondimensionalized by Δx, u by a (the speed of propagation), and t by $\Delta x/a$. Give the numerical solution at $t = 50$ and 200 over the interval $150 \le x \le 250$. Plot the result in graphical form. Compare your numerical solution with the exact solution as follows:

$$u(x, t) = \begin{cases} \frac{5}{x}\sin\left[\frac{\pi}{4}(t - x + 5)\right], & x \le t + 5 \\ 0, & x > t + 5, \quad t - x + 5 < 0 \end{cases}$$

5.6. Consider simulating the propagation and reflection of acoustic disturbance in a semiinfinite long tube with a closed end at $x = 0$ as shown. The governing equations are the linearized Euler equations in one dimension. On using Δx (the mesh size) as the length scale, a_0 (the speed of sound) as the velocity scale, $\Delta x/a_0$ as the time scale, $\rho_0 a_0^2$ as the pressure scale, where ρ_0 is the gas density inside the tube, the dimensionless governing equations may be written as

$$\frac{\partial u}{\partial t} + \frac{\partial p}{\partial x} = 0$$

$$\frac{\partial p}{\partial t} + \frac{\partial u}{\partial x} = 0$$

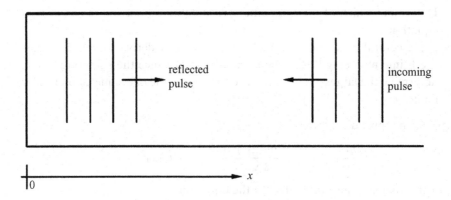

The gas is set into motion by a pressure pulse at $t = 0$; i.e.,

$$t = 0, \qquad u = 0, \qquad p = 10^{-5} \exp\left[-\ln 2 \left(\frac{x - 25}{5} \right)^2 \right].$$

Use a computational domain $0 \le x \le 100$; calculate the pressure and velocity distribution at $t = 40$ by a time marching scheme. Compare your solution with the exact solution as follows:

$$p = 0.5 \times 10^{-5} \left[e^{-\ln 2 \left(\frac{x-t-25}{5} \right)^2} + e^{-\ln 2 \left(\frac{x+t-25}{5} \right)^2} + e^{-\ln 2 \left(\frac{x-t+25}{5} \right)^2} \right].$$

5.7. To compute the sound field radiated by two-dimensional unsteady sources in the vicinity of two walls intersecting at an angle as shown in the figure below, a natural

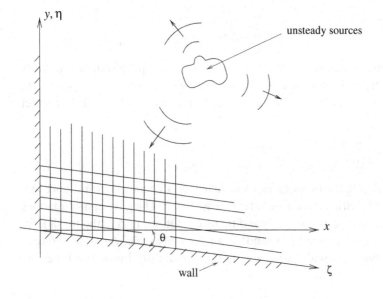

choice is to use oblique Cartesian coordinates (ξ, η). (ξ, η) are related to Cartesian coordinates (x, y) by

$$\xi = \frac{x}{\cos \theta}, \qquad \eta = y + x \sin \theta.$$

a. Transform the governing equations (linearized Euler equations) from Cartesian to oblique Cartesian coordinates.
b. Discretize the equations in oblique Cartesian coordinates by the use of the 7-point stencil DRP scheme.
c. Show that in the computation domain (the $\xi - \eta$ plane) the discretized equations are dispersion-relation preserving.

6 Radiation, Outflow, and Wall Boundary Conditions

A computational domain is inevitably finite. For open domain problems, this automatically creates a set of artificial exterior boundaries. There are two basic reasons why exterior boundary conditions are needed at the artificial boundaries. First, exterior boundary conditions must reproduce the effects the outside world exerts on the flow inside the computation domain. Since only the computed results inside the computational domain are known, in general, it is not possible to know the outside influence. For this reason, a computation domain is often taken as large as possible so that all sources are inside. This will minimize any external influence that is unaccounted for.

Another reason for imposing exterior boundary conditions is to avoid the reflection of outgoing disturbances back into the computational domain and thus contaminate the computed solution. One way to avoid reflection is to construct boundary conditions in such a way to allow the smooth exit of all disturbances.

In this chapter, it will be assumed that there is little external influence and there are no incoming disturbances. Issues of external influence and incoming entropy, vorticity, or acoustic waves will be considered in later chapters.

As discussed previously, the linearized Euler equations support three types of waves. Two types of these waves, namely, the entropy and vorticity waves, are convected downstream by the mean flow. The acoustic waves, however, propagate and radiate out in all directions if the mean flow is subsonic. Thus, at a subsonic inflow region, the outgoing waves consist of acoustic waves alone. In the outflow region, the outgoing disturbances now comprise of all three types of waves. As a result, it is prudent to develop separate inflow and outflow boundary conditions.

Many aeroacoustics problems involve solid surfaces. To specify the presence of these surfaces, the no-slip boundary conditions are imposed if the fluid is viscous. On the other hand, if the fluid is inviscid, the no-through flow boundary condition is to be used. How to impose these physical boundary conditions on a high-order finite difference computation without creating excessive spurious short waves is one of the main topics of this chapter.

Now, consider an exterior problem involving a uniform flow of velocity u_0 and sound speed a_0 past some arbitrary acoustic, vorticity, and entropy sources as shown in Figure 5.1. It will be assumed that the boundaries of the computation domain are quite far from the sources. From the analysis of Chapter 5, it is clear that outflow boundary conditions are needed on the right side of the computational domain with outgoing disturbances formed by a superposition of acoustic, entropy, and vorticity waves. On the other hand, at the top and bottom boundaries, as well as the left inflow boundary of the computational domain, the outgoing disturbances are acoustic waves only. Now, the exterior boundaries are far from the sources, the outgoing waves in the regions close to these boundaries are, therefore, given by the asymptotic solutions of the discretized form of the governing partial differential equations. It will now be assumed that the DRP scheme is used for time-marching computation. However, in the resolved wave number range, the DRP scheme and the original partial differential equations have (almost) identical dispersion relations and asymptotic solutions. These solutions are given by Eqs. (5.14), (5.16a,b), and (5.25). A set of radiation and outflow boundary conditions will now be constructed based on these asymptotic solutions.

6.1 Radiation Boundary Conditions

At boundaries where there are only outgoing acoustic waves, the solution is given by an asymptotic solution (5.25) as follows:

$$
\begin{bmatrix} \rho \\ u \\ v \\ p \end{bmatrix} \equiv \begin{bmatrix} \rho_a \\ u_a \\ v_a \\ p_a \end{bmatrix} = \frac{F\left(\frac{r}{V(\theta)} - t, \theta\right)}{r^{1/2}} \begin{bmatrix} \frac{1}{a_0^2} \\ \frac{\hat{u}(\theta)}{\rho_0 a_0} \\ \frac{\hat{v}(\theta)}{\rho_0 a_0} \\ 1 \end{bmatrix} + O\left(r^{\frac{-3}{2}}\right),
\tag{6.1}
$$

where $V(\theta) = a_0[M \cos \theta + (1 - M^2 \sin^2 \theta)^{1/2}]$. The subscript 'a' in (ρ_a, u_a, v_a, p_a) in Eq. (6.1) indicates that the disturbances are associated with the acoustic waves alone. By taking the time t and r derivatives of formula (6.1), it is straightforward to find that for arbitrary function F the acoustic disturbances satisfy the following equations:

$$
\left(\frac{1}{V(\theta)} \frac{\partial}{\partial t} + \frac{\partial}{\partial r} + \frac{1}{2r}\right) \begin{bmatrix} \rho \\ u \\ v \\ p \end{bmatrix} = 0 + O\left(r^{\frac{-5}{2}}\right).
\tag{6.2}
$$

Eq. (6.2) provides a set of radiation boundary conditions.

6.2 Outflow Boundary Conditions

At the outflow region, the outgoing disturbances, in general, consist of a combination of acoustic, entropy, and vorticity waves. Thus, by means of the asymptotic solutions obtained before, the density, velocity, and pressure fluctuations are given by

$$
\begin{bmatrix} \rho \\ u \\ v \\ p \end{bmatrix} = \begin{bmatrix} \chi(x - u_0 t, y) + \rho_a \\ \frac{\partial \psi}{\partial y}(x - u_0 t, y) + u_a \\ -\frac{\partial \psi}{\partial x}(x - u_0 t, y) + v_a \\ p_a \end{bmatrix} + \cdots,
\tag{6.3}
$$

where the explicit form of (ρ_a, u_a, v_a, p_a) may be found in Eq. (6.1). The functions χ, ψ, and F are entirely arbitrary. It is observed that the total pressure fluctuation comes only from the acoustic component of the outgoing disturbances. Thus, the appropriate outflow boundary condition for p is the same as that of Eq. (6.2). On writing out Cartesian coordinates in two dimensions, it is

$$\frac{1}{V(\theta)}\frac{\partial p}{\partial t} + \cos\theta\frac{\partial p}{\partial x} + \sin\theta\frac{\partial p}{\partial y} + \frac{p}{2r} = 0, \tag{6.4}$$

where θ is the angular coordinate of the boundary point.

By differentiating the expression for ρ in Eq. (6.3) with respect to t and x, the following equation is found:

$$\frac{\partial\rho}{\partial t} + u_0\frac{\partial\rho}{\partial x} = \frac{\partial\rho_a}{\partial t} + u_0\frac{\partial\rho_a}{\partial x}. \tag{6.5}$$

But $\rho_a = p_a/a_0^2 = p/a_0^2$ and p is known from Eq. (6.4). By eliminating ρ_a, the outflow boundary condition for ρ becomes

$$\frac{\partial\rho}{\partial t} + u_0\frac{\partial\rho}{\partial x} = \frac{1}{a_0^2}\left(\frac{\partial p}{\partial t} + u_0\frac{\partial p}{\partial x}\right). \tag{6.6}$$

Similarly, by differentiating the expressions of u and v in Eq. (6.3), it is easy to find that

$$\frac{\partial u}{\partial t} + u_0\frac{\partial u}{\partial x} = \frac{\partial u_a}{\partial t} + u_0\frac{\partial u_a}{\partial x}. \tag{6.7}$$

$$\frac{\partial v}{\partial t} + u_0\frac{\partial v}{\partial x} = \frac{\partial v_a}{\partial t} + u_0\frac{\partial v_a}{\partial x}. \tag{6.8}$$

However, the acoustic component satisfies the momentum equations of the (linearized) Euler equations; i.e.,

$$\frac{\partial u_a}{\partial t} + u_0\frac{\partial u_a}{\partial x} = -\frac{1}{\rho_0}\frac{\partial p_a}{\partial x} = -\frac{1}{\rho_0}\frac{\partial p}{\partial x} \tag{6.9}$$

$$\frac{\partial v_a}{\partial t} + u_0\frac{\partial v_a}{\partial x} = -\frac{1}{\rho_0}\frac{\partial p_a}{\partial y} = -\frac{1}{\rho_0}\frac{\partial p}{\partial y}. \tag{6.10}$$

Upon eliminating u_a and v_a from Eqs. (6.7) and (6.8) by Eqs. (6.9) and (6.10), the outflow boundary conditions for the velocity components may be written as

$$\frac{\partial u}{\partial t} + u_0\frac{\partial u}{\partial x} = -\frac{1}{\rho_0}\frac{\partial p}{\partial x} \tag{6.11}$$

$$\frac{\partial v}{\partial t} + u_0\frac{\partial v}{\partial x} = -\frac{1}{\rho_0}\frac{\partial p}{\partial y}. \tag{6.12}$$

In summary, the outflow boundary conditions are as follows:

$$\frac{\partial\rho}{\partial t} + u_0\frac{\partial\rho}{\partial x} = \frac{1}{a_0^2}\left(\frac{\partial p}{\partial t} + u_0\frac{\partial p}{\partial x}\right)$$

$$\frac{\partial u}{\partial t} + u_0\frac{\partial u}{\partial x} = -\frac{1}{\rho_0}\frac{\partial p}{\partial x}$$

$$\frac{\partial v}{\partial t} + u_0\frac{\partial v}{\partial x} = -\frac{1}{\rho_0}\frac{\partial p}{\partial y}$$

$$\frac{1}{V(\theta)}\frac{\partial p}{\partial t} + \cos\theta\frac{\partial p}{\partial x} + \sin\theta\frac{\partial p}{\partial y} + \frac{p}{2r} = 0. \tag{6.13}$$

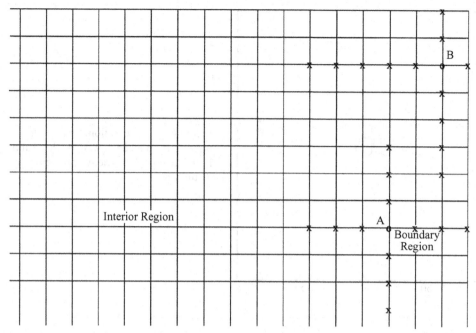

Figure 6.1. The interior and boundary regions of a computational domain. Also shown are typical stencils for interior point A and boundary point B.

6.3 Implementation of Radiation and Outflow Boundary Conditions

The optimized central difference scheme for approximating spatial derivatives introduced in Chapter 2 will give rise to ghost points (points outside the computational domain) at the boundaries of the computational domain. For a 7-point stencil, 3 ghost points are created. To advance the entire computation to the next time level, a way to calculate the unknowns at the ghost points must be specified. Here, it is suggested that this is done using the radiation or outflow boundary conditions.

Figure 6.1 shows the upper right-hand corner of a computation domain. Three columns or rows of ghost points are added to form a boundary region surrounding the original computation domain. For grid points inside and on the boundary of the interior region, Eqs. (5.28) and (5.29) discretized according to the DRP scheme are used to advance the calculation of the unknowns to the next time level. In the boundary region it will be assumed that the solution is made up of outgoing disturbances satisfying the radiation or outflow boundary conditions of Eq. (6.2) or Eq. (6.13). The time derivatives of these equations are discretized in the same manner as described in Chapter 3. These equations are to be used to advance the solution in time. Now, all the variables in the boundary region as well as in the interior regions can be advanced in time simultaneously. To approximate spatial derivatives, symmetric spatial stencils are not always possible for points in the boundary region. The optimized backward difference stencils of Chapter 2 are to be used whenever necessary. An example of such backward difference stencils is illustrated by that of the corner point B in Figure 6.1. In the boundary region, the domain of dependence of the outgoing waves is consistent with that of the backward difference approximations.

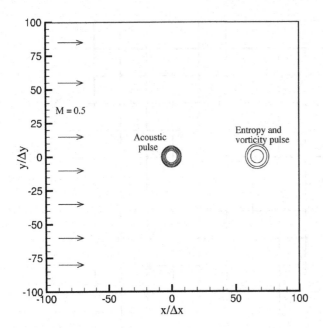

Figure 6.2. Computational plane showing a uniform mean flow of Mach 0.5 and initial acoustic, vorticity, and entropy disturbances.

Unlike the use of unsymmetric stencils at interior points where waves may propagate in any direction, no numerical instability would be created unless there are strong reflections of waves back into the interior.

6.4 Numerical Simulation: An Example

To show the effectiveness of the DRP schemes and the radiation and outflow boundary conditions, the results of one special initial value problem involving all three types of disturbances will be presented below in some detail. The case under consideration consists of an acoustic pulse generated by an initial Gaussian pressure distribution at the center of the computational domain as shown in Figure 6.2. The mean flow Mach number is 0.5. Downstream of the pressure pulse, at a distance equal to one-third of the length of the computational domain, a vorticity pulse and an entropy pulse (also with Gaussian distribution) are also generated at the initial time. Since the acoustic pulse travels three times faster than the vorticity and entropy pulses in the downstream direction, this arrangement ensures that all the three pulses reach the outflow boundary simultaneously. In this way, it is possible to obtain a critical test of the effectiveness of the radiation as well as the outflow boundary conditions in a single simulation.

In the simulation, the variables are nondimensionalized by the following scales:

$$\text{length scale} = \Delta x = \Delta y$$
$$\text{velocity scale} = a_0$$
$$\text{time scale} = \frac{\Delta x}{a_0}$$
$$\text{density scale} = \rho_0$$
$$\text{pressure scale} = \rho_0 a_0^2.$$

The computational domain is divided into a 200×200 mesh. The initial conditions imposed at time $t = 0$ are as follows:

$$\left. \begin{array}{l} p = \varepsilon_1 \exp[-\alpha_1(x^2 + y^2)] \\ \rho = \varepsilon_1 \exp[-\alpha_1(x^2 + y^2)] \end{array} \right\} + \varepsilon_2 \exp[-\alpha_2((x - 67)^2 + y^2)] \quad \begin{array}{l} \text{(acoustic pulse)} \\ \text{(entropy pulse)} \end{array}$$

$$\left. \begin{array}{l} u = \varepsilon_3 y \exp[-\alpha_3((x - 67)^2 + y^2)] \\ v = -\varepsilon_3(x - 67) \exp[-\alpha_3((x - 67)^2 + y^2)] \end{array} \right\} \quad \text{(vorticity wave)},$$

where $\varepsilon_1 = 0.01$, $\varepsilon_2 = 0.001$, $\varepsilon_3 = 0.0004$, $\alpha_1 = (\ln 2)/9$, $\alpha_2 = \alpha_3 = (\ln 2)/25$.
    The exact solution (see Tam and Webb, 1993) is.

6.4.1 Acoustic Waves

$$u(x, y, t) = \frac{\varepsilon_1(x - Mt)}{2\alpha_1 \eta} \int_0^\infty e^{-\xi^2/4\alpha_1} \sin(\xi t) J_1(\xi \eta) \xi \, d\xi$$

$$v(x, y, t) = \frac{\varepsilon_1 y}{2\alpha_1 \eta} \int_0^\infty e^{-\xi^2/4\alpha_1} \sin(\xi t) J_1(\xi \eta) \xi \, d\xi$$

$$p(x, y, t) = \rho = \frac{\varepsilon_1}{2\alpha_1} \int_0^\infty e^{-\xi^2/4\alpha_1} \cos(\xi t) J_0(\xi \eta) \xi \, d\xi,$$

where $\eta = [(x - Mt)^2 + y^2]^{1/2}$, J_1 and J_0 are the Bessel functions of order 1 and 0. M is the Mach number.

6.4.2 Entropy Waves

$$p = u = v = 0, \qquad \rho = \varepsilon_2 e^{-\alpha_2[(x-67-Mt)^2 + y^2]}.$$

6.4.3 Vorticity Waves

$$p = \rho = 0$$
$$u(x, y, t) = \varepsilon_3 y e^{-\alpha_3[(x-67-Mt)^2 + y^2]}$$
$$v(x, y, t) = -\varepsilon_3(x - 67 - Mt) e^{-\alpha_3[(x-67-Mt)^2 + y^2]}.$$

The results of the simulation will now be presented. Figure 6.3a shows the density contours of the acoustic pulse at the center of the computation domain and the entropy pulse downstream at time $t = 0$. The vorticity pulse has no density fluctuations and, therefore, cannot be seen in this figure. Figure 6.3b shows the computed density contours after 500 time steps ($\Delta t = 0.0569$). At this time, the radius of the acoustic pulse has expanded considerably, while that of the entropy pulse remains the same. The centers of the two pulses have moved downstream by an equal

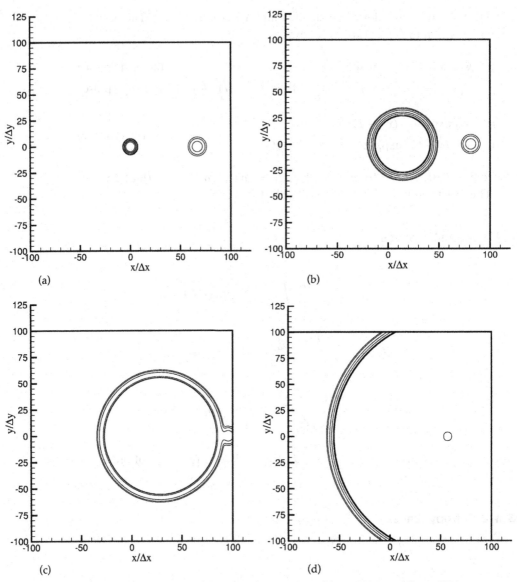

Figure 6.3. Density contours. (a) $t = 0$, (b) $t = 500\ \Delta t$, (c) $t = 1000\ \Delta t$, (d) $t = 2000\ \Delta t$.

distance. Figure 6.3c shows the locations of the two pulses at 1000 time steps. The
acoustic pulse has now caught up with the entropy pulse. At a slightly later time, the
density contours of the two pulses merge and exit the outflow boundary together.
Based on the 1 percent contour plot, no noticeable reflections have been observed,
indicating that the outflow boundary condition is transparent. At a still later time,
the acoustic pulse reaches and leaves the computation domain through the top and
bottom boundaries. Again, few or no reflections can be found (to 1 percent of the
incident wave amplitude). This is shown in Figure 6.3d. Finally, at 3200 time steps,
the acoustic wave front reaches the inflow boundary on the left. The pulse exits the
computation domain again with few observable reflections. Contours of the exact

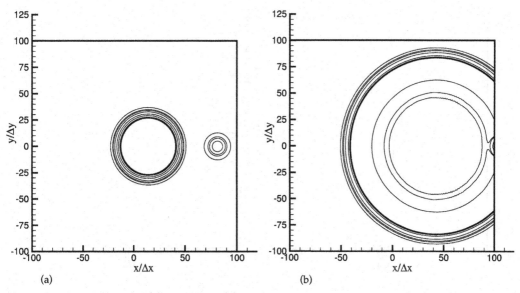

Figure 6.4. Speed contours. (a) $t = 500\ \Delta t$, (b) $t = 1500\ \Delta t$.

solution are also plotted in these figures. However, the difference between the exact and numerical solution is so small, they cannot be easily seen.

Figure 6.4a shows the contours of the magnitude of the velocity fluctuations (or speed) at 500 time steps. The center circles are those of the acoustic pulse. The circles to the right are contours associated with the vorticity pulse. The contours are constructed by interpolation between mesh points. To the accuracy allowed by this procedure, the expanding circles are, indeed, circular so that the computed acoustic wave speed is the same in all directions. Figure 6.4b gives the speed contours at 1500 time steps. At this time, the main part of the acoustic pulse has already left the outflow boundary. The vorticity pulse, having a slower velocity, has, however, not completely left the right boundary. A small piece of it can still be seen just at the outflow boundary. In this figure, the 1 percent contour exhibits some minor wiggles. A closer examination of the computed data indicates that they are generated by the graphic program and not by the simulation.

The computed density waveform along the x-axis at 500 time steps is given in Figure 6.5a. The exact solution is shown by the dotted line. The exact and computed waveforms are clearly almost identical. Figure 6.5b provides both the computed and the exact waveforms at 1000 time steps when the acoustic pulse has caught up with the entropy pulse. Again, there is good agreement between the two waveforms. Extensive comparisons between the computed density, pressure, and velocity waveforms with the exact solution have been carried out at different directions of propagation up to 4500 time steps when nearly all the disturbances have exited the computation domain. Good agreements are found regardless of the direction of wave propagation. Such good agreements are maintained in time up to the termination of the simulation.

In addition to the simulation just described in detail, several series of simulations using more than one acoustic pulse generated at various locations of the computation

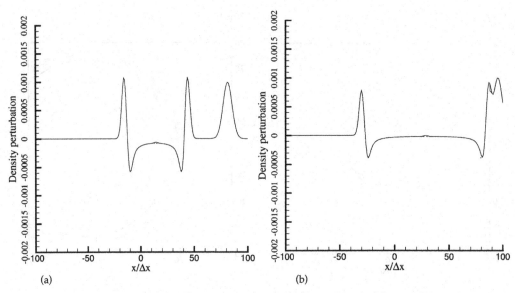

Figure 6.5. Density waveform along the x axis. (a) $t = 500 \, \Delta t$, (b) $t = 1000 \, \Delta t$. Full line is numerical solution; dotted line is the exact solution.

domain have been carried out. Good agreements are again found when compared with the exact solutions of the linearized Euler equations. This is true as long as the predominant part of the wave spectrum has wave numbers α and β, such that $\alpha \Delta x$ and $\beta \Delta y$ are both less than 1.1. Overall, the results of all the simulations strongly suggest that the DRP scheme can be relied on to yield accurate results when used to simulate isotropic, nondispersive, and nondissipative acoustic, vorticity, and entropy waves. Furthermore, the scheme can be expected to reproduce the wave speeds correctly.

It is important to point out that the radiation boundary conditions (6.2) and the outflow boundary conditions (6.13) depend on the angle θ. If the boundaries are far from the source, then the exact location of the source is not important. But, if the source is close to a boundary, the effectiveness of these boundary conditions would deteriorate because the direction of wave propagation is in error. An extensive series of tests involving an acoustic source put closer and closer to the boundaries have been carried out. The radiation boundary conditions appear to perform quite well even when the center of the acoustic pulse is 20 mesh points away from the boundary. The reflected wave amplitude is generally less than 2 percent of the incident wave amplitude. The outflow boundary conditions, on the other hand, have been observed to cause a moderate level of reflection; 15 percent for acoustic pulse initiated 20 mesh points away. This is true with or without mean flow. Recalling that the radiation and outflow boundary conditions were developed from the asymptotic solutions, the degradation of the effectiveness of the boundary conditions for sources located close to a boundary should, therefore, be expected.

6.5 Generalized Radiation and Outflow Boundary Conditions

Radiation boundary condition (6.2) and outflow boundary condition (6.13) were derived for a uniform mean flow in the x direction. In many problems, the mean flow is not in the x direction. However, if the mean flow is not in the x direction, and even has slow spatial variation, Eqs. (6.2) and (6.13) may be extended to account for a general direction and for a slightly nonuniform mean flow. Let the nonuniform mean flow in the boundary region of the computation domain be $(\bar{\rho}, \bar{u}, \bar{v}, \bar{p})$, then a generalization of radiation boundary condition (6.2) (see Tam and Dong, 1996) is

$$\frac{1}{V(r, \theta)} \frac{\partial}{\partial t} \begin{bmatrix} \rho \\ u \\ v \\ p \end{bmatrix} + \frac{\partial}{\partial r} \begin{bmatrix} \rho - \bar{\rho} \\ u - \bar{u} \\ v - \bar{v} \\ p - \bar{p} \end{bmatrix} + \frac{1}{2r} \begin{bmatrix} \rho - \bar{\rho} \\ u - \bar{u} \\ v - \bar{v} \\ p - \bar{p} \end{bmatrix} = 0, \qquad (6.14)$$

where $V(r, \theta) = \bar{u} \cos \theta + \bar{v} \sin \theta + [\bar{a}^2 - (\bar{v} \cos \theta - \bar{u} \sin \theta)^2]^{1/2}$ and $\bar{a}$ is the local speed of sound. Note: The variables in Eq. (6.2) are the perturbation quantities, whereas (ρ, u, v, p) are the full variables.

The generalized outflow boundary conditions are as follows:

$$\frac{\partial \rho}{\partial t} + \mathbf{v} \cdot \nabla (\rho - \bar{\rho}) = \frac{1}{\bar{a}^2} \frac{\partial p}{\partial t} + \mathbf{v} \cdot \nabla \left(\frac{p - \bar{p}}{\bar{a}^2} \right)$$

$$\frac{\partial u}{\partial t} + \mathbf{v} \cdot \nabla (u - \bar{u}) = -\frac{1}{\bar{\rho}} \frac{\partial}{\partial x} (p - \bar{p})$$

$$\frac{\partial v}{\partial t} + \mathbf{v} \cdot \nabla (v - \bar{v}) = -\frac{1}{\bar{\rho}} \frac{\partial}{\partial y} (p - \bar{p})$$

$$\frac{1}{V(r, \theta)} \frac{\partial p}{\partial t} + \frac{\partial}{\partial r} (p - \bar{p}) + \frac{1}{2r} (p - \bar{p}) = 0, \qquad (6.15)$$

where $\mathbf{v} = (\bar{u}, \bar{v})$.

It is worthwhile to point out that radiation boundary condition (6.14) allows an automatic adjustment of the mean flow. For time-independent solution, this equation has a solution in the following form:

$$(v - \bar{v}) = \frac{A}{r^{1/2}},$$

and similarly for the other variables. Thus, this set of boundary conditions permits a steady entrainment of ambient fluid when the computed solution requires.

6.6 The Ghost Point Method for Wall Boundary Conditions

Large-stencil finite difference methods are used for CAA problems because they are generally less dispersive and less dissipative, and they tend to be more isotropic. In addition, they yield numerical wave speeds that are nearly equal to the wave speeds of the original partial differential equations. On the other hand, large-stencil or high-order methods support spurious numerical waves. These waves are contaminants or pollutants of the numerical solution. Some of these spurious waves have very short wavelengths (grid-to-grid oscillations) and propagate with ultrafast speeds. They have been referred to as parasite waves. Other spurious waves are spatially damped. Their presence is, therefore, confined locally in space. To ensure a high-quality computational aeroacoustics solution, it is important that these spurious waves are not excited in the computation.

For inviscid flows, the well-known boundary condition at a solid wall is that the velocity component normal to the wall is zero. This condition is sufficient for the determination of a unique solution to the Euler equations. When a large-stencil finite difference scheme is used, the order of the difference equations is higher than that of the Euler equations. Thus, the zero normal velocity boundary condition is insufficient to define a unique solution. Extraneous conditions must be imposed. Unfortunately, these extraneous conditions would often lead to the generation of spurious numerical waves as mentioned before. The net result is the degradation of the quality of the numerical solution. The degradation may be divided into three types. First, if an acoustic wave is incident on a solid wall, perfect reflection (e.g., reflected wave amplitude equal to incident wave amplitude) may not be obtained. Second, parasite waves may be reflected off the solid surface. Third, a numerical boundary layer can form adjacent to the solid wall surface by the spatially damped spurious numerical waves.

6.6.1 Concept of Ghost Points and Ghost Values

Consider two-dimensional small-amplitude disturbances superimposed on a uniform mean flow of density ρ_0, pressure p_0, and velocity u_0 in the x direction. The linearized Navier-Stokes equations may be written in the following form:

$$\frac{\partial \mathbf{U}}{\partial t} + \frac{\partial \mathbf{E}}{\partial x} + \frac{\partial \mathbf{F}}{\partial y} = \frac{\partial \mathbf{G}}{\partial x} + \frac{\partial \mathbf{H}}{\partial y} + \mathbf{Q}, \tag{6.16}$$

where

$$\mathbf{U} = \begin{bmatrix} \rho \\ u \\ v \\ p \end{bmatrix}, \quad \mathbf{E} = \begin{bmatrix} \rho_0 u + \rho u_0 \\ u_0 u + \frac{p}{\rho_0} \\ u_0 v \\ u_0 p + \gamma p_0 u \end{bmatrix}, \quad \mathbf{F} = \begin{bmatrix} \rho_0 v \\ 0 \\ \frac{p}{\rho_0} \\ \gamma p_0 v \end{bmatrix}, \quad \mathbf{G} = \frac{1}{\rho_0} \begin{bmatrix} 0 \\ \tau_{xx} \\ \tau_{yx} \\ 0 \end{bmatrix}, \quad \mathbf{H} = \frac{1}{\rho_0} \begin{bmatrix} 0 \\ \tau_{xy} \\ \tau_{yy} \\ 0 \end{bmatrix}$$

$$\tau_{xx} = 2\mu \frac{\partial u}{\partial x}, \quad \tau_{xy} = \tau_{yx} = \mu \left(\frac{\partial u}{\partial y} + \frac{\partial v}{\partial x} \right), \quad \tau_{yy} = 2\mu \frac{\partial v}{\partial y}.$$

The first two terms on the right side of Eq. (6.16) are viscous stress terms. They are to be discarded for inviscid fluids. The last term $\mathbf{Q}$ represents possible acoustic sources. In addition, viscous dissipation terms are omitted in the energy equation. They are unimportant in most acoustics problems.

Let the $x - y$ plane be divided into a mesh of spacings Δx and Δy in the x and y directions, respectively. The discretized form of Eq. (6.16) according to the 7-point stencil DRP scheme is

$$\mathbf{K}_{\ell,m}^{(n)} = -\frac{1}{\Delta x} \sum_{j=-3}^{3} a_j \left(\mathbf{E}_{\ell+j,m}^{(n)} - \mathbf{G}_{\ell+j,m}^{(n)} \right)$$

$$- \frac{1}{\Delta y} \sum_{j=-3}^{3} a_j \left(\mathbf{F}_{\ell,m+j}^{(n)} - \mathbf{H}_{\ell,m+j}^{(n)} \right) + \mathbf{Q}_{\ell,m}^{(n)} \tag{6.17}$$

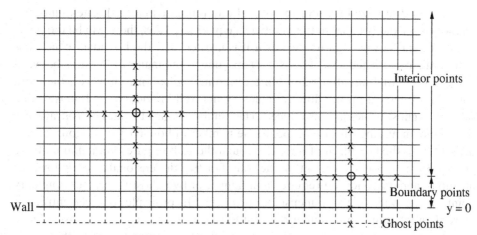

Figure 6.6. Computational mesh adjacent to a wall $y = 0$ showing an interior point with 7-point computation stencil. Also shown are boundary points and ghost points.

$$(\tau_{xx})_{\ell,m}^{(n)} = \frac{2\mu}{\Delta x} \sum_{j=-3}^{3} a_j u_{\ell+j,m}^{(n)},$$

$$(\tau_{xy})_{\ell,m}^{(n)} = (\tau_{yx})_{\ell,m}^{(n)} = \mu \sum_{j=-3}^{3} a_j \left(\frac{1}{\Delta y} u_{\ell,m+j}^{(n)} + \frac{1}{\Delta x} v_{\ell+j,m}^{(n)} \right),$$
(6.18)

$$(\tau_{yy})_{\ell,m}^{(n)} = \frac{2\mu}{\Delta y} \sum_{j=-3}^{3} a_j v_{\ell,m+j}^{(n)},$$

$$\mathbf{U}_{\ell,m}^{(n+1)} = \mathbf{U}_{\ell,m}^{(n)} + \Delta t \sum_{j=0}^{3} b_j \mathbf{K}_{\ell,m}^{(n-j)},$$
(6.19)

where ℓ, m are the indices of the mesh points and the superscript n is the time level. If $\mathbf{U} = \mathbf{U}_{\text{initial}}$ at $t = 0$ is the initial condition for the Navier-Stokes or Euler equations, the appropriate initial conditions for the DRP scheme (6.19) are

$$\mathbf{U}_{\ell,m}^{(0)} = \mathbf{U}_{\text{initial}}, \qquad \mathbf{U}_{\ell,m}^{(n)} = 0 \quad \text{for negative } n.$$
(6.20)

Suppose the fluid is inviscid and a solid wall lies at $y = 0$ as shown in Figure 6.6. For points lying three rows or more away from the wall, the computation stencil for them lies entirely inside the physical domain. They are referred to as interior points. For the first three rows of points adjacent to the wall, their 7-point stencil will extend outside the physical domain. They are referred to as boundary points. The points outside the computation domain have no physical meaning. They are called ghost points.

Although ghost points are fictitious points with no physical meaning, they are necessary for high-order finite difference schemes. To see this, recall that the solution of the Euler or Navier-Stokes equations satisfies the partial differential equations at every interior or boundary point. In addition, at a point on the wall, the solution also satisfies the appropriate boundary conditions. Now, the discretized governing equations are no more than a set of algebraic equations. In the discretized system, each flow variable at either an interior or boundary point is governed by an algebraic equation (a discretized form of the partial differential equations). The number of unknowns is exactly equal to the number of equations. Thus, there will be too many equations and not enough unknowns if it is insisted that the boundary conditions at the wall are satisfied also. This is, perhaps, one of the major differences between partial differential equations and finite difference equations. However, the extra conditions imposed on the flow variables by the wall boundary conditions can be satisfied if ghost values are introduced. The number of ghost values is arbitrary, but the minimum number must be equal to the number of boundary conditions. For an inviscid flow, the condition of no flux through the wall requires a minimum of one ghost value per boundary point on the wall. For two-dimensional viscous flow, the no-slip boundary condition requires a minimum of two ghost values per boundary point on the wall for two-dimensional problems. In principle, one can introduce more than the minimum number of ghost values. But for each extra ghost value, a condition must be imposed so that it can be uniquely determined. Clearly, the quality of the solid wall boundary conditions in terms of acoustic wave reflection, acoustic boundary layer characteristics, etc., would be affected by these extra ghost values. Since there is no compelling advantage to introduce extra wall boundary conditions, in the present approach, only the minimum number of ghost values is used.

To fix ideas, let us first consider an inviscid fluid adjacent to a solid wall at $y = 0$ (see Figure 6.6). In this case, the wall boundary condition is $v = 0$ at $y = 0$, where (u, v) are the velocity components in the x and y direction, respectively. Since there is one boundary condition, one ghost value is needed for each boundary point at the wall. Physically, the wall exerts a pressure on the fluid with a magnitude just enough to make $v = 0$ at its surface. This suggests that a ghost value in p (pressure) at the ghost point immediately below the wall should be used to simulate the pressure exerted by the wall. Because only one ghost value is adopted, the difference stencil for the y derivative in the boundary region needs to be modified. Here, it is proposed that the y derivatives are to be computed according to the backward difference stencils shown in Figure 6.7. According to this scheme, the quantities $\partial p/\partial y$, $\partial u/\partial y$, and $\partial v/\partial y$ are all calculated using values of the variables lying inside the physical domain. For the y derivative of p the stencils extend to the ghost point below the wall. Now, the ghost value of p at the ghost point $(\ell, -1)$ or $p_{\ell,-1}^{(n)}$ is to be chosen so that $v_{\ell,0}^{(n)}$ is zero for all n. This can be accomplished through the discretized form of the y-momentum equation (the third equation of Eq. (6.16)), which is used to calculate the value of v at the next time level. Upon incorporating the backward difference stencils of Figure 6.7, the y-momentum equation at a wall point $(1, 0)$ is

$$v_{\ell,0}^{(n+1)} = v_{\ell,0}^{(n)} + \Delta t \sum_{j=0}^{3} b_j K_{\ell,0}^{(n-j)} \qquad (6.21)$$

$$K_{\ell,0}^{(n-j)} = -\frac{u_0}{\Delta x} \sum_{i=-3}^{3} a_i v_{\ell+i,0}^{(n-j)} - \frac{1}{\rho_0 \Delta y} \sum_{i=-1}^{5} a_i^{51} p_{\ell,i}^{(n-j)}, \qquad (6.22)$$

where a_i^{51} are the optimized coefficients of the backward difference stencil. Equations (6.21) and (6.22) are to be used to find $v_{\ell,0}^{(n+1)}$ after all the physical quantities except $p_{\ell,-1}^n$, the ghost value, are found at the end of the nth time level computation. To ensure that the wall boundary condition $v_{\ell,0}^{(n+1)} = 0$ is satisfied, the ghost value $p_{\ell,-1}^{(n)}$

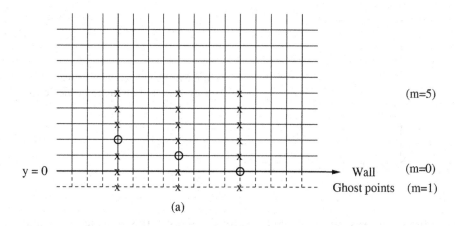

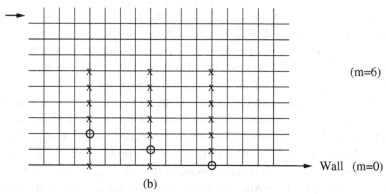

Figure 6.7. Seven-point backward difference stencils to be used to compute (a) y derivative of p, (b) y derivatives of ρ, u, and v in the boundary region near a wall at $y = 0$.

may be found by setting $v_{\ell,0}^{(n+1)} = 0$ in Eqs. (6.21) and (6.22) and then solve for $p_{\ell,-1}^{(n)}$. This gives

$$p_{\ell,-1}^{(n)} = -\frac{1}{a_{-1}^{51}} \sum_{i=0}^{5} a_i^{51} p_{\ell,i}^{(n)}. \tag{6.23}$$

Eq. (6.23) is tantamount to setting the ghost value such that $\partial p / \partial y = 0$ at the wall. It is worthwhile to point out that if no ghost value is introduced, and the boundary condition $v_{\ell,0}^{(n)}$ is imposed, Eq. (6.21) or its equivalent will, in general, not be satisfied. This means that $\partial p / \partial y$ of the solution so computed would not necessarily be equal to zero at the wall.

Now attention is turned to a viscous compressible fluid. In this case, the viscous stress terms of Eqs. (6.16) and (6.17) are to be retained. The stresses are physical quantities and can be computed from the velocity field by the same 7-point difference stencil at an interior point or by the backward difference stencils of Figure 6.7 at a boundary point. For example, at an interior point (ℓ, m), the shear stress τ_{xy} at the nth time level is given by

$$(\tau_{xy})_{\ell,m}^{(n)} = \mu \sum_{j=-3}^{3} a_j \left(\frac{1}{\Delta y} u_{\ell,m+j}^{(n)} + \frac{1}{\Delta x} v_{\ell+j,m}^{(n)} \right), \tag{6.24}$$

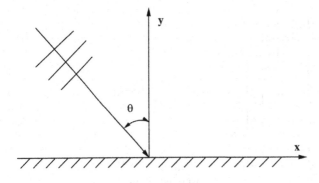

Figure 6.8. Plane acoustic waves incident on a wall at an angle of incidence of θ.

where μ is the shear viscosity coefficient. (Note: The stress terms are calculated and stored at each time level as a part of the solution. The derivatives of the stresses are computed by the first derivative stencil applied directly on the stresses. No second derivatives need be computed.) At the wall, the no-slip boundary condition $u = v = 0$ must be enforced. Two ghost values are, therefore, needed per each boundary point at the wall. On following the inviscid case, the ghost value $p_{\ell,-1}^{(n)}$ will again be used to ensure that $v_{\ell,0}^{(n)} = 0$ for all n. Physically, in addition to applying a pressure on the fluid normal to the wall, the wall also exerts a shear stress τ_{xy} on the fluid to reduce the velocity component u to zero at its surface. This suggests that a ghost value $(\tau_{xy})_{\ell,-1}^{(n)}$ be included in the computation. To enforce boundary condition $u_{\ell,0}^{(n)} = 0$, it is noted that u is determined by advancing the x-momentum equation in time. By following the treatment of $v = 0$, it is recommended that the y derivative of τ_{xy} in the x-momentum equation be approximated by the same backward finite difference stencils as those for p (see Figure 6.7). It is easy to see that the ghost value $(\tau_{xy})_{\ell,-1}^{(n)}$ can be found (in exactly the same way by which the ghost value $p_{\ell,-1}^{(n)}$ is determined) by imposing the no-slip boundary condition $u_{\ell,0}^{(n+1)} = 0$ on the x-momentum equation at the boundary point $(1, 0)$ at the wall. In this way, a unique solution satisfying both the discretized form of the Navier-Stokes equations and the no-slip boundary conditions can be calculated.

6.6.2 Reflection of Acoustic Waves by a Plane Wall

A detailed analytical analysis of the reflection of acoustic waves from a wall (see Figure 6.8) using the solid wall boundary conditions described above has been carried out by Tam and Dong (1994). Their analysis clearly shows that, in addition to the reflected acoustic wave, spurious waves (parasite waves) are reflected off the wall. Furthermore, spatially damped numerical waves of the computation scheme are also excited at the wall boundary. These waves form a numerical boundary layer. Space limitations would not allow a discussion of the analytical part of their work. However, their numerical results are useful to provide an assessment of the accuracy and quality of solution in adopting the ghost point method to enforce wall boundary conditions.

Suppose numerical boundary layer thickness, δ_z, is defined to be the distance between the wall and where the spurious numerical wave solution drops to z times the magnitude of the reflected acoustic wave amplitude, e.g., $\delta_{0.005}$ denotes the boundary

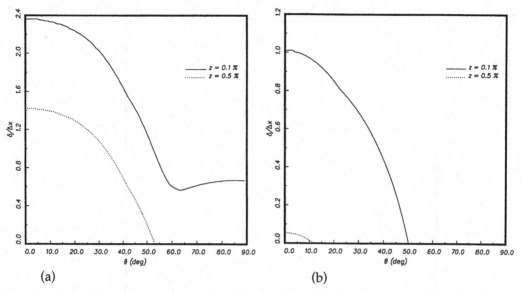

Figure 6.9. Thickness of numerical boundary layer. (a) $\lambda/\Delta x = 6$, (b) $\lambda/\Delta x = 10$.

layer thickness based on 0.5 percent of the reflected wave amplitude. Figure 6.9a shows the calculated numerical boundary layer thickness as a function of the angle of incidence when a spatial resolution of $\lambda/\Delta x = 6$ ($\Delta y = \Delta x$) is used for computing the acoustic waves; λ is the wavelength. Two thicknesses are shown, one corresponds to $z = 0.005$ (0.5 percent), the other $z = 0.001$ (0.1 percent). This figure indicates that the numerical boundary layer is thickest for normal incidence. In this case $\delta_{0.001}$ is almost equal to $2.5\Delta x$. Figure 6.9b shows the corresponding numerical boundary layer thickness if a spatial resolution of $\lambda/\Delta x = 10$ is used in the computation. It is clear that with better spatial resolution the numerical boundary layer thickness decreases.

Figure 6.10 shows the dependence of the magnitude of the reflected parasite wave (grid-to-grid oscillations), $|p_{\text{parasite}}/p_{\text{incident}}|$, on the angle of incidence for different spatial resolution. At $\lambda/\Delta x = 6$, the reflected parasite wave can be as large as more than 1 percent of the incident wave amplitude at normal incidence. At glazing incidence, i.e., beyond $\theta = 50°$, there is generally very little parasite wave reflected off the wall. The magnitude of the reflected parasite wave would be greatly reduced if the spatial resolution in the computation is increased. By using 10 or more mesh points per acoustic wavelength in the computation, the reflected parasite wave is so weak that they may be ignored entirely.

Figure 6.11 is a plot of the deviation from perfect acoustic reflection. It is a measure of the quality of the numerical solid wall boundary condition. For perfect reflection, the ratio of the reflected wave amplitude to that of the incident wave, $|p_{\text{reflected}}/p_{\text{incident}}|$, is unity. Figure 6.11 shows that, by using the proposed numerical solid wall boundary condition, the deviation from perfect reflection is small. It is of the order of 1 percent even when only 6 mesh points per acoustic wavelength are used in the computation. When acoustic waves impinge on a wall, the pressure at the wall is twice the incident sound pressure. This phenomenon is usually referred to as pressure doubling. Figure 6.12 shows that the computed wall pressure would,

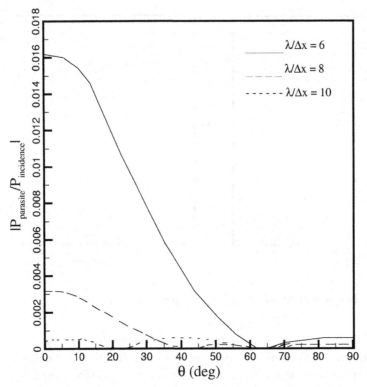

Figure 6.10. Magnitude of reflected parasite waves as a function of θ.

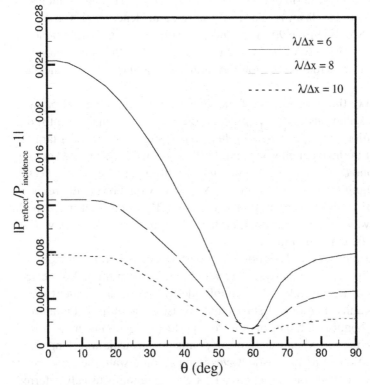

Figure 6.11. Magnitude of reflected wave.

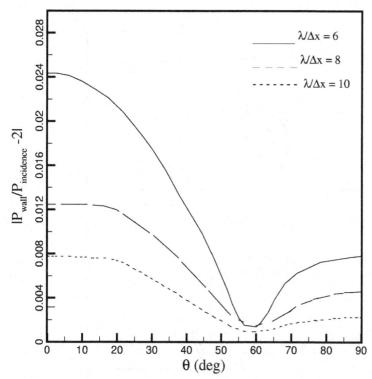

Figure 6.12. Dependence of wall pressure on θ.

indeed, be close to twice the incident sound pressure level for $\lambda/\Delta x = 6$. Again, when a better spatial resolution is used in the computation, e.g., $\lambda/\Delta x = 8$, the computed results would better reproduce the pressure-doubling phenomenon.

6.6.3 Numerical Examples

To assess the effectiveness of the ghost point solid wall boundary conditions, a number of direct numerical simulations have been carried out. Comparisons between numerical and analytical solutions are made. This permits an evaluation of the fidelity of both the proposed inviscid and the viscous boundary conditions. In all the examples discussed below, dimensionless variables are used. The characteristic scales are as follows:

$$\text{length scale} = \Delta x = \Delta y \text{ (mesh size)}$$

$$\text{velocity scale} = a_0 \text{ (speed of sound)}$$

$$\text{time scale} = \frac{\Delta x}{a_0}$$

$$\text{density scale} = \rho_0 \text{ (ambient density)}$$

$$\text{pressure scale} = \rho_0 a_0^2$$

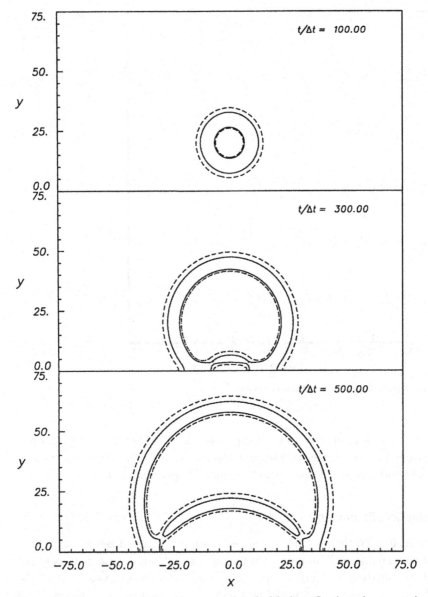

Figure 6.13. Pressure contour patterns associated with the reflection of an acoustic pulse by a wall.

6.6.3.1 Reflection of a Transient Acoustic Pulse by a Wall

Consider the reflection of a two-dimensional acoustic pulse by a plane wall as shown in Figure 6.13. The fluid is inviscid and is at rest at time $t = 0$. An acoustic pulse is generated by an initial pressure disturbance with a Gaussian spatial distribution centered at $(0, 20)$. The wall is located at $y = 0$. The initial conditions are as follows:

$$p = 0.01 \exp \left\{ \frac{-\ln 2[x^2 + (y - 20)^2]}{9} \right\}$$

$$\rho = p, \qquad u = v = 0.$$

In the numerical simulation, the 7-point DRP scheme is used, and the time step Δt is set equal to 0.07677. This value of Δt satisfies numerical stability requirement and ensures that the amount of numerical damping due to time discretization is insignificant.

Figure 6.13 shows the calculated pressure contour patterns associated with the acoustic pulse at 100, 300, and 500 time steps. The corresponding contours of the exact solution are also plotted in this figure. To the accuracy given by the thickness of the contour lines, the two sets of contours are almost indistinguishable. At 100 time steps, the pulse has not reached the wall, so the pressure contours are circular. At 300 time steps, the front part of the pulse reaches the wall. It is immediately reflected back. At 500 time steps, the entire pulse has effectively been reflected off the wall, creating a double-pulse pattern: one from the original source, and the other from the image source below the wall.

Figure 6.14 shows the computed pressure waveforms along the line $x = y$. The distance measured along this line from the origin is denoted by s. The computed waveforms at 400, 700, and 1000 time steps are shown together with the exact solution. As can be seen, there is excellent agreement between the exact and computed results. At 400 time steps, the pulse has just been reflected off the wall. At 700 time steps, the double-pulse characteristic waveform is fully formed. Both pulses propagate away from the wall with essentially the same waveform. The amplitude, however, decreases at a rate inversely proportional to the square root of the distance.

6.6.3.2 Effect of Mean Flow

To test the efficacy of the inviscid solid wall boundary condition in the presence of a mean flow, this acoustic pulse problem is modified to include a uniform stream of Mach number 0.5 flowing parallel to the wall. In this case, the reflection process is modified by the effect of mean flow convection. For this problem, a computational domain $-100 \le x \le 100$, $0 \le y \le 200$ is used. The wall is at $y = 0$. The linearized Euler equation in two dimensions is as follows:

$$
\frac{\partial}{\partial t}
\begin{bmatrix} \rho \\ u \\ v \\ p \end{bmatrix}
+ \frac{\partial}{\partial x}
\begin{bmatrix} M\rho + u \\ Mu + p \\ Mv \\ Mp + u \end{bmatrix}
+ \frac{\partial}{\partial y}
\begin{bmatrix} v \\ 0 \\ p \\ v \end{bmatrix}
= 0,
$$

where $M = 0.5$. The initial conditions are

$$ t = 0, \qquad u = v = 0 $$

$$ p = \rho = 0.01 \exp\left\{ -(\ln 2) \left[\frac{x^2 + (y-25)^2}{25} \right] \right\}. $$

The exact solution of this problem is available. Let

$$ \alpha = (\ln 2)/25, \qquad \eta = [(x - Mt)^2 + (y - 25)^2]^{\frac{1}{2}}, \qquad \varsigma = [(x - Mt)^2 + (y + 25)^2]^{\frac{1}{2}}. $$

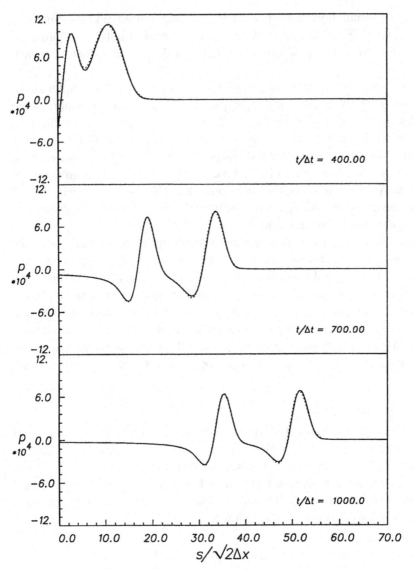

Figure 6.14. Pressure waveforms of an acoustic pulse reflected off a solid wall along the line $x = y$.

The solution is

$$u = 0.01\frac{(x - Mt)}{2\alpha\eta}\int_0^\infty e^{-\frac{\xi^2}{4\alpha}}\sin(\xi t)J_1(\xi\eta)\xi\,d\xi + 0.01\frac{(x - Mt)}{2\alpha\varsigma}\int_0^\infty e^{-\frac{\xi^2}{4\alpha}}\sin(\xi t)J_1(\xi\varsigma)\xi\,d\xi$$

$$v = 0.01\frac{(y - 25)}{2\alpha\eta}\int_0^\infty e^{-\frac{\xi^2}{4\alpha}}\sin(\xi t)J_1(\xi\eta)\xi\,d\xi + 0.01\frac{(y - 25)}{2\alpha\varsigma}\int_0^\infty e^{-\frac{\xi^2}{4\alpha}}\sin(\xi t)J_1(\xi\varsigma)\xi\,d\xi$$

$$p = \rho = \frac{0.01}{2\alpha}\int_0^\infty e^{-\frac{\xi^2}{4\alpha}}\cos(\xi t)[J_0(\xi\eta) + J_0(\xi\varsigma)]\xi\,d\xi,$$

where $J_0(z), J_1(z)$ are Bessel functions of order 0 and 1.

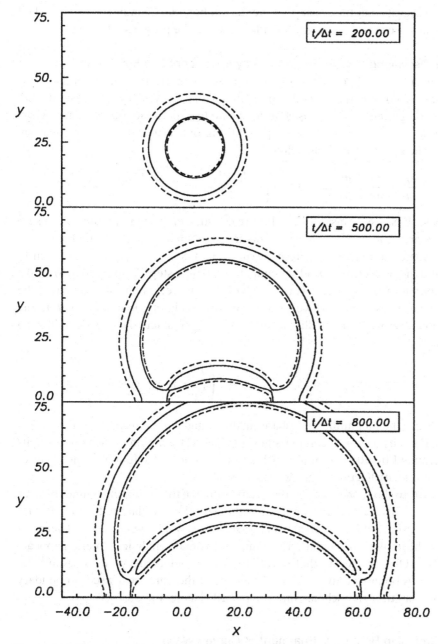

Figure 6.15. Pressure contour patterns associated with the reflection of an acoustic pulse by a solid wall in the presence of a Mach 0.5 uniform mean flow.

Figure 6.15 shows the computed pressure contour patterns of the acoustic pulse, with the use of the 7-point DRP scheme, at 200, 500, and 800 time steps. Again, the computed pressure contours are indistinguishable from those of the exact solution. Geometrically, the pressure contour patterns are identical to those without mean flow (see Figure 6.13). The uniform mean flow translates the entire pulse downstream at a Mach number of 0.5. The good agreement between the computed and the exact solutions suggests that the ghost point numerical boundary conditions can be relied upon to simulate correctly a solid boundary even in the presence of a flow.

6.6.3.3 Oscillating Viscous Boundary Layer Adjacent to a Solid Wall

To demonstrate the effectiveness of the proposed numerical viscous boundary conditions for a solid wall, the case of an oscillating viscous boundary layer is simulated. The oscillating boundary layer is generated by a time periodic source in the energy equation at a very low frequency. Upon writing out in full, the energy equation, including the source term, is as follows:

$$\frac{\partial p}{\partial t} + \frac{\partial u}{\partial x} + \frac{\partial v}{\partial y} = 0.01 \exp\left\{-\frac{(\ln 2)[x^2 + (y - 60)^2]}{9}\right\} \cos(\omega t).$$

All the viscous stress terms of Eq. (6.16) are included in the computation. To ensure that there are at least 7 to 8 mesh points in the boundary layer (so that it can be resolved computationally) the mesh Reynolds number, $R = \rho_0 a_0 \Delta x / \mu$ is taken to be 5.0, and the oscillation period set equal to 1250 time steps (at $\Delta t = 0.07677$). An exact solution of this problem is not available. However, far from the source, the boundary layer resembles the well-known Stokes oscillatory boundary layer. In the Stokes boundary layer the oscillatory velocity component u is given by the following formula:

$$u = \varepsilon \left[\cos\left(\omega t + \delta\right) - e^{-\left(\frac{\omega}{2\mu}\right)^{1/2} y} \cos\left(\left(\frac{\omega}{2\mu}\right)^{\frac{1}{2}} y - \omega t - \delta\right) \right] \tag{6.25}$$

In Eq. (6.25), the amplitude and phase factors ε and δ are constants of the Stokes solution. For the purpose of comparison with numerical solution, these two constants are determined by fitting solution (6.25) to the numerical solution at the point of maximum velocity fluctuation at two instants of time.

Figure 6.16 shows the computed velocity profile of the oscillating boundary layer along the wall at $x = 0$. In this figure, T is the period of oscillation. Shown also are the velocity profile of Eq. (6.25) at every quarter period. The agreement between the results of the numerical simulation and the approximate analytical solution appears to be good. On considering that only 7 to 8 mesh points are used to resolve the entire viscous boundary layer, the performance of the numerical viscous boundary conditions and the DRP scheme must be regarded as very good.

6.6.4 Cartesian Boundary Treatment of Curved Walls

The ghost point method has been extended to treat curved walls in two dimensions by Kurbatskii and Tam (1997). In general, depending on the geometry of the boundary curve and the Cartesian mesh, ghost points can be classified into many types. One major difference between the ghost values associated with curved walls and those of a straight wall aligned with a mesh line is that the ghost values are coupled together. They form a linear matrix system that has to be solved at the end of each time step. Another difference is that there is a stronger tendency for a curved wall to generate high wave number (grid-to-grid oscillations) spurious numerical waves. To ensure stability, a large amount of artificial selective damping is necessary (artificial

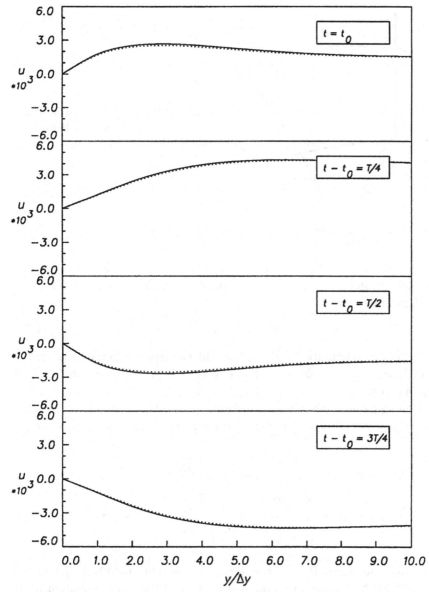

Figure 6.16. Velocity profile of an oscillatory viscous boundary layer at each quarter cycle. Full line is the numerical solution. Dotted line is the Stokes solution.

selective damping will be discussed Chapter 7). This tends to degrade slightly the quality of the computed solution. A detailed discussion of the method is beyond the scope of this book. Interested persons should consult the original work.

6.7 Enforcing Wall Boundary Conditions on Curved Surfaces

When curved surfaces are involved in a computation, very often a body-fitted grid is used. It will now be assumed that the computation grid is on a set of orthogonal

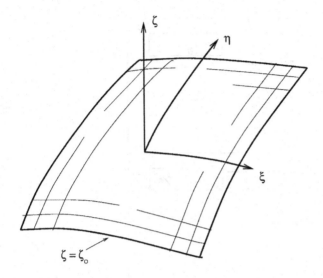

Figure 6.17. Orthogonal curvilinear co-ordinates (ξ, η, ζ) forming a body-fitted curved surface.

curvilinear coordinates $\xi(x, y, z)$, $\eta(x, y, z)$, and $\zeta(x, y, z)$. The coordinate system is orthogonal, meaning that

$$\nabla\xi \cdot \nabla\eta = \nabla\eta \cdot \nabla\zeta = \nabla\zeta \cdot \nabla\xi = 0. \tag{6.26}$$

Without the loss of generality, the curved boundary surface is taken to be on the $\zeta = \zeta_0$ coordinate surface as shown in Figure 6.17. The normal to the surface is in the direction of $\nabla\zeta$.

The Euler equation in curvilinear coordinates (ξ, η, ζ) is given by Eq. (5.61). Let $U = \mathbf{v} \cdot \nabla\xi$, $V = \mathbf{v} \cdot \nabla\eta$, $W = \mathbf{v} \cdot \nabla\zeta$ (the contravariant velocity components), i.e.,

$$U = u\frac{\partial\xi}{\partial x} + v\frac{\partial\xi}{\partial y} + w\frac{\partial\xi}{\partial z}$$

$$V = u\frac{\partial\eta}{\partial x} + v\frac{\partial\eta}{\partial y} + w\frac{\partial\eta}{\partial z}$$

$$W = u\frac{\partial\zeta}{\partial x} + v\frac{\partial\zeta}{\partial y} + w\frac{\partial\zeta}{\partial z}. \tag{6.27}$$

The momentum equation in the ζ direction may be found by multiplying the second equation of (5.61) by ζ_x, the third equation by ζ_y, and the fourth equation by ζ_z (subscript denotes partial derivatives), then summing up the three equations. On being written out in full, the equation is as follows:

$$\frac{\partial W}{\partial t} + U\left(\frac{\partial u}{\partial \xi}\zeta_x + \frac{\partial v}{\partial \xi}\zeta_y + \frac{\partial w}{\partial \xi}\zeta_z\right) + V\left(\frac{\partial u}{\partial \eta}\zeta_x + \frac{\partial v}{\partial \eta}\zeta_y + \frac{\partial w}{\partial \eta}\zeta_z\right)$$

$$+ W\left(\frac{\partial u}{\partial \zeta}\zeta_x + \frac{\partial v}{\partial \zeta}\zeta_y + \frac{\partial w}{\partial \zeta}\zeta_z\right) = -\frac{1}{\rho}(\zeta_x^2 + \zeta_y^2 + \zeta_z^2)\frac{\partial p}{\partial \zeta}. \tag{6.28}$$

For inviscid flow, the wall boundary condition is $W = 0$ on $\zeta = \zeta_0$ (the solid surface). To enforce the wall boundary condition, it is recommended to include a set of ghost points on the surface $\zeta = \zeta_0 - \Delta\zeta$ as shown in Figure 6.18. At each ghost point, a ghost value of p is included in the computation. The computation is carried out in the (ξ, η, ζ) space. It is similar to computation in a Cartesian coordinate system.

Figure 6.18. Computational grid and ghost points near a solid wall in a computational domain.

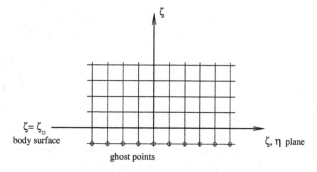

For all mesh points inside the computation domain, the discretized form of Eq. (5.61) is used. For all flow variables, with the exception of p, backward difference stencils for $\partial/\partial\zeta$ are implemented for points in the boundary region. This is to keep the stencil points inside the computational domain. For $\partial p/\partial\zeta$, the stencils are extended to the ghost points. For mesh points lying on the solid surface $\zeta = \zeta_0$, instead of the second, third, and fourth equation of (5.61), Eq. (6.28) and two of the three equations are used. Let (ℓ, m, n) be the spatial indices in the ξ, η, and ζ directions, respectively, and $\zeta = \zeta_0$ corresponds to $n = N$, the discretized form of Eq. (6.28) at $n = N$, taking into account that the boundary condition $W = 0$ at $\zeta = \zeta_0$, is as follows:

$$
U^{(k)}_{\ell,m,N} \left[(\zeta_x)_{\ell,m,N} \left(\frac{1}{\Delta\xi} \sum_{j=-3}^{3} a_j u^{(k)}_{\ell+j,m,N} \right) + (\zeta_y)_{\ell,m,N} \left(\frac{1}{\Delta\xi} \sum_{j=-3}^{3} a_j v^{(k)}_{\ell+j,m,N} \right) \right.
$$
$$
\left. + (\zeta_z)_{\ell,m,N} \left(\frac{1}{\Delta\xi} \sum_{j=-3}^{3} a_j w^{(k)}_{\ell+j,m,N} \right) \right] + V^{(k)}_{\ell,m,N} \left[(\zeta_x)_{\ell,m,N} \left(\frac{1}{\Delta\eta} \sum_{j=-3}^{3} a_j u^{(k)}_{\ell,m+j,N} \right) \right.
$$
$$
\left. + (\zeta_y)_{\ell,m,N} \left(\frac{1}{\Delta\eta} \sum_{j=-3}^{3} a_j v^{(k)}_{\ell,m+j,N} \right) + (\zeta_z)_{\ell,m,N} \left(\frac{1}{\Delta\eta} \sum_{j=-3}^{3} a_j w^{(k)}_{\ell,m+j,N} \right) \right]
$$
$$
= -\left(\frac{\zeta_x^2 + \zeta_y^2 + \zeta_z^2}{\rho} \right)_{\ell,m,N} \left[\frac{1}{\Delta\zeta} \sum_{j=-1}^{5} a_j^{51} p^{(k)}_{\ell,m,N+j} \right], \tag{6.29}
$$

where superscript k is the time level.

The ghost value of pressure, $p^{(k)}_{\ell,m,N-1}$, may now be calculated by Eq. (6.29). The explicit formula is as follows:

$$
p^{(k)}_{\ell,m,N-1} = -\frac{1}{a^{51}_{-1}} \left\{ \sum_{j=0}^{5} a_j^{51} p^{(k)}_{\ell,m,N+j} + \left(\frac{\rho}{\zeta_x^2 + \zeta_y^2 + \zeta_z^2} \right)_{\ell,m,N} \Delta\zeta \right.
$$
$$
\cdot \left\{ U^{(k)}_{\ell,m,N} \left[(\zeta_x)_{\ell,m,N} \left(\frac{1}{\Delta\xi} \sum_{j=-3}^{3} a_j u^{(k)}_{\ell+j,m,N} \right) \right. \right.
$$
$$
\left. + (\zeta_y)_{\ell,m,N} \left(\frac{1}{\Delta\xi} \sum_{j=-3}^{3} a_j v^{(k)}_{\ell+j,m,N} \right) + (\zeta_z)_{\ell,m,N} \left(\frac{1}{\Delta\xi} \sum_{j=-3}^{3} a_j w^{(k)}_{\ell+j,k,N} \right) \right]
$$

$$+ V_{\ell,m,N}^{(k)}\left[(\zeta_x)_{\ell,m,N}\left(\frac{1}{\Delta\eta}\sum_{j=-3}^{3}a_j u_{\ell,m+j,N}^{(k)}\right)+(\zeta_y)_{\ell,m,N}\left(\frac{1}{\Delta\eta}\sum_{j=-3}^{3}a_j v_{\ell,m+j,N}^{(k)}\right)\right.$$

$$\left.+(\zeta_z)_{\ell,m,N}\left(\frac{1}{\Delta\eta}\sum_{j=-3}^{3}a_j w_{\ell,m+j,N}^{(k)}\right)\right]\bigg\}. \tag{6.30}$$

With the ghost value of p found, the computation can proceed as in the case of a flat wall.

For viscous flows, the full Navier-Stokes equation must be used in the computation. The boundary conditions on the curved solid surface are the no-slip conditions. Because the stress terms are fairly complicated in curvilinear coordinates, the use of ghost values of shear stresses to enforce the no-slip boundary conditions would require a good deal of effort. A less complicated approach is to set (u, v, w) to zero on $\zeta = \zeta_0$ $(n = N)$. To avoid the computation becoming an overdetermined system, the three momentum equations will not be used for mesh points right at the solid surface. For density ρ and pressure p on the solid surface, they are to be computed by means of the continuity and energy equations. Although this approach is a bit less accurate, this way of treating the curved wall boundary condition has been found to be quite efficient, easy to implement, and offers reasonably accurate results.

EXERCISES

6.1. Consider solving the one-dimensional wave equation

$$\frac{\partial^2 u}{\partial t^2} = c^2\frac{\partial^2 u}{\partial x^2}$$

in a finite domain. The equation has two solutions:

$$u = F(x - ct) + G(x + ct),$$

where F and G are arbitrary functions. The solution $F(x - ct)$ represents a wave propagating to the right, and the solution $G(x + ct)$ represents a wave propagating to the left.

Develop a radiation boundary condition for the right and the left boundary of the computation domain.

6.2. For problems with spherical symmetry, it is advantageous to use spherical coordinates (R, θ, ϕ). The linearized Euler equations in spherical coordinates for problems with spherical symmetry are $(\partial/\partial\theta = 0, \partial/\partial\phi = 0)$ as follows:

$$\frac{\partial v}{\partial t} + \frac{1}{\rho_0}\frac{\partial p}{\partial R} = 0 \tag{1}$$

$$\frac{\partial p}{\partial t} + \gamma p_0\frac{1}{R^2}\frac{\partial}{\partial R}(R^2 v) = 0 \tag{2}$$

By eliminating v, it is easy to obtain

$$\frac{\partial^2(Rp)}{\partial t^2} - a_0^2\frac{\partial^2}{\partial R^2}(Rp) = 0, \tag{3}$$

where $a_0^2 = \gamma p_0/\rho_0$. The general solution of (3) is

$$p = \frac{F(R - a_0 t)}{R} + \frac{G(R + a_0 t)}{R}. \tag{4}$$

F and G are arbitrary functions. Derive a radiation boundary condition for p at large R. In solving the system of equations (1) and (2) a boundary condition for v is also needed. Derive a boundary condition for v.

6.3. Use dimensionless variables with respect to the following scales:

$$\Delta x = \text{length scale}$$

$$a_\infty \text{ (ambient sound speed)} = \text{velocity scale}$$

$$\frac{\Delta x}{a_\infty} = \text{time scale}$$

$$\rho_\infty = \text{density scale}$$

$$\rho_\infty a_\infty^2 = \text{pressure scale}$$

The linearized two-dimensional Euler equations on a uniform mean flow are to be solved

$$\frac{\partial \mathbf{U}}{\partial t} + \frac{\partial \mathbf{E}}{\partial x} + \frac{\partial \mathbf{F}}{\partial y} = 0,$$

where

$$\mathbf{U} = \begin{bmatrix} \rho \\ u \\ v \\ p \end{bmatrix}, \qquad \mathbf{E} = \begin{bmatrix} M_x \rho + u \\ M_x u + p \\ M_x v \\ M_x p + u \end{bmatrix}, \qquad \mathbf{F} = \begin{bmatrix} M_y \rho + v \\ M_y u \\ M_y v + p \\ M_y p + v \end{bmatrix}.$$

M_x and M_y are constant mean flow Mach number in the x and y directions, respectively. Use a computation domain $-100 \le x \le 100$, $-100 \le y \le 100$ embedded in free space.

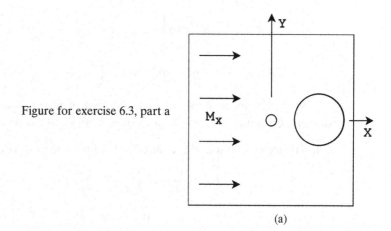

Figure for exercise 6.3, part a

(a)

(a) Let $M_x = 0.5$, $M_y = 0$. Solve the initial value problem, $t = 0$.

$$p = \exp\left[-(\ln 2)\left(\frac{x^2 + y^2}{9}\right)\right]$$

$$\rho = \exp\left[-(\ln 2)\left(\frac{x^2 + y^2}{9}\right)\right] + 0.1\exp\left[-(\ln 2)\left(\frac{(x-67)^2 + y^2}{25}\right)\right]$$

$$u = 0.04y\exp\left[-(\ln 2)\left(\frac{(x-67)^2 + y^2}{25}\right)\right]$$

$$v = -0.04(x-67)\exp\left[-(\ln 2)\left(\frac{(x-67)^2 + y^2}{25}\right)\right]$$

Compute the distributions of p, ρ, and u at $t = 30, 60, 80, 100$, and 200.

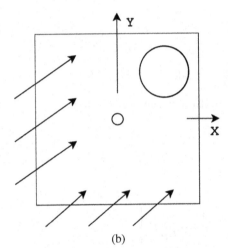

(b)

Figure for exercise 6.3, part b

(b) Let $M_x = M_y = 0.5\cos(\pi/4)$. Solve the initial value problem, $t = 0$.

$$p = \exp\left[-(\ln 2)\left(\frac{x^2 + y^2}{9}\right)\right]$$

$$\rho = \exp\left[-(\ln 2)\left(\frac{x^2 + y^2}{9}\right)\right] + 0.1\exp\left[-(\ln 2)\left(\frac{(x-67)^2 + (y-67)^2}{25}\right)\right]$$

$$u = 0.04(y-67)\exp\left[-(\ln 2)\left(\frac{(x-67)^2 + (y-67)^2}{25}\right)\right]$$

$$v = -0.04(x-67)\exp\left[-(\ln 2)\left(\frac{(x-67)^2 + (y-67)^2}{25}\right)\right]$$

Note: The mean flow is in the direction of the diagonal of the computational domain. Compute the distributions of p, ρ, and u at $t = 60, 80, 100, 200$, and 600.

6.4. The exact solution of the initial value problem of the diffusion equation,

$$\frac{\partial u}{\partial t} = \kappa\left(\frac{\partial^2 u}{\partial x^2} + \frac{\partial^2 u}{\partial y^2}\right)$$

$$t = 0, \qquad u = e^{-(\ln 2)\frac{(x^2 + y^2)}{b^2}}$$

is

$$u(x, y, t) = \frac{1}{1 + \frac{4(\ln 2)\kappa t}{b^2}} e^{\frac{-(x^2+y^2)}{(b^2/(\ln 2)+4\kappa t)}}.$$

Suppose one is interested to find the solution computationally by using the DRP scheme. Let $\Delta x = \Delta y$ be the length scale and $(\Delta x)^2/\kappa$ be the time scale, so that the nondimensional diffusion equation becomes

$$\frac{\partial u}{\partial t} = \frac{\partial^2 u}{\partial x^2} + \frac{\partial^2 u}{\partial y^2}$$

$$t = 0$$

$$u = e^{-(\ln 2)\frac{(x^2+y^2)}{b^2}}.$$

To solve this problem computationally in a finite domain requires numerical boundary conditions. For diffusion equations, the asymptotic solution is not useful for constructing a numerical boundary condition.

It is observed, because of the diffusive nature of the solution, that the solution at a point depends mainly on the solution on the side closer to the source. For this reason, it is recommended that one use backward difference stencils at mesh points close to the exterior boundaries of the computation domain. No other specific boundary conditions at the outer boundaries of the computation domain is needed.

Compute the solution on a 100×100 mesh with $b = 3$. Compare your numerical solution with the exact solution by plotting $u(x, 0, t)$ at $t = 20, 40, 80, 160, 300$. See Appendix D for helpful suggestions.

6.5. Consider one-dimensional flow and acoustics in a nozzle attached to a long duct of uniform cross section as shown. For calculating the flow and acoustic waves in the nozzle, it is decided to use a computation domain encompassing the variable area part of the nozzle. The right boundary of the computation domain is in the uniform duct. A small-amplitude sound wave of angular frequency ω propagates from the uniform duct into the nozzle region. Part of this wave is transmitted through the nozzle throat and a part of it is reflected back into the uniform duct.

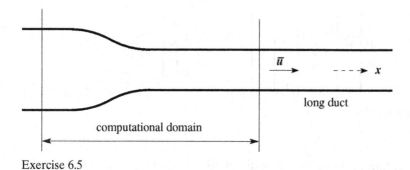

Exercise 6.5

Let $\bar{u}$, $\bar{\rho}$, and $\bar{p}$ be the mean flow in the uniform area duct. The linearized governing equations in the duct are the continuity, momentum, and energy equations.

$$\frac{\partial \rho}{\partial t} + \bar{\rho}\frac{\partial u}{\partial x} + \bar{u}\frac{\partial \rho}{\partial x} = 0 \tag{1}$$

$$\frac{\partial u}{\partial t} + \bar{u}\frac{\partial u}{\partial x} + \frac{1}{\bar{\rho}}\frac{\partial p}{\partial x} = 0 \tag{2}$$

$$\frac{\partial p}{\partial t} + \bar{u}\frac{\partial p}{\partial x} + \gamma\bar{p}\frac{\partial u}{\partial x} = 0. \tag{3}$$

The upstream propagating acoustic wave in the uniform duct is given by

$$\begin{bmatrix} \rho \\ u \\ p \end{bmatrix} = \begin{bmatrix} \frac{1}{\bar{a}^2} \\ -\frac{1}{\bar{\rho}\bar{a}} \\ 1 \end{bmatrix} F\left(\frac{x}{\bar{a}-\bar{u}} + t\right), \tag{4}$$

where $\bar{a} = (\gamma\bar{p}/\bar{\rho})^{1/2}$ is the sound speed. F is a known function.

Develop a set of outflow boundary conditions for use on the right side of the computation domain.

If the left side of the nozzle is attached to another long uniform duct, hot spots in the form of entropy waves are convected downstream by the mean flow into the nozzle. The entropy wave solution may be written as

$$\begin{bmatrix} \rho \\ u \\ p \end{bmatrix} = \begin{bmatrix} 1 \\ 0 \\ 0 \end{bmatrix} H(x - \tilde{u}t),$$

where $\tilde{u}$ is the local mean flow velocity of the uniform duct and H is a known function characterizing the density (temperature) distribution.

Develop a set of inflow boundary conditions for use on the left side of the computation domain.

6.6. Consider acoustic radiation from an oscillating circular piston in a wall as shown below.

Let the radius of piston be R and the velocity of piston be $u = \varepsilon \sin(\pi t/5)$. Use a computational domain $0 \le x \le 100$, $0 \le r \le 100$. The wall and the piston are at $x = 0$. The cylindrical coordinate system is centered at the center of the piston. With axisymmetry, the dimensionless linearized Euler equation is

$$\frac{\partial}{\partial t}\begin{bmatrix} \rho \\ u \\ v \\ p \end{bmatrix} + \frac{\partial}{\partial r}\begin{bmatrix} v \\ 0 \\ p \\ v \end{bmatrix} + \begin{bmatrix} \frac{v}{r} \\ 0 \\ 0 \\ \frac{v}{r} \end{bmatrix} + \frac{\partial}{\partial x}\begin{bmatrix} u \\ p \\ 0 \\ u \end{bmatrix} = 0.$$

The initial conditions are

$$t = 0, \qquad \rho = u = v = p = 0.$$

Compute the time harmonic pressure distribution at the beginning, 1/4, 1/2, and 3/4 of a period of piston oscillation.

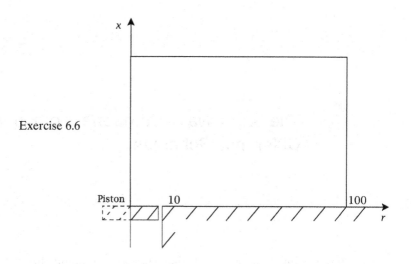

Exercise 6.6

Let $\varepsilon = 10^{-4}$, $R = 10$, $\omega = \pi/5$. The exact solutions is

$$p = \mathrm{Re}\left[\varepsilon R \omega \int_0^\infty \frac{J_1(\xi R)}{(\xi^2 - \omega^2)^{\frac{1}{2}}} J_0(\xi r)\, e^{-(\xi^2 - \omega^2)^{\frac{1}{2}} x - i\omega t}\, d\xi\right]$$

$$u = \mathrm{Im}\left[-\varepsilon R \int_0^\infty J_1(\xi R) J_0(\xi r)\, e^{-(\xi^2 - \omega^2)^{\frac{1}{2}} x - i\omega t}\, d\xi\right]$$

$$v = \mathrm{Re}\left[-\varepsilon R \int_0^\infty \frac{J_1(\xi R)}{(\xi^2 - \omega^2)^{\frac{1}{2}}} J_1(\xi r)\, e^{-(\xi^2 - \omega^2)^{\frac{1}{2}} x - i\omega t}\, d\xi\right],$$

where Re[] is the real part of and Im[] is the imaginary part of.

Note: The branch cut for the square root function is chosen so that $(\xi^2 - \omega^2)^{\frac{1}{2}} = -i|\xi^2 - \omega^2|^{\frac{1}{2}}$ for $\xi < \omega$.

6.7. A spherically symmetric acoustic pulse is initiated at time $t = 0$ by a small-amplitude pressure pulse of the following form:

$$p = \rho = \varepsilon e^{-(\ell n 2)[(x^2 + y^2 + z^2)/b^2]}$$

in a static environment. Here, the length scale is $\Delta x = \Delta y = \Delta z$, the velocity scale is a_0 (the speed of sound), the time scale is $\Delta x / a_0$, the density scale is ρ_0 (the mean gas density), and the pressure scale is $\rho_0 a_0^2$. Compute the solution by solving the three-dimensional Euler equations in Cartesian coordinates. Take $\varepsilon = 10^{-4}$ and $b = 3$.

Let $R = (x^2 + y^2 + z^2)^{1/2}$ be the radial distance. The exact linearized solution is

$$p = \varepsilon\left[\frac{R - t}{2R} e^{-(\ell n 2)[(R-t)/b]^2} + \frac{R + t}{2R} e^{-(\ell n 2)[(R+t)/b]^2}\right]$$

Note: A computer code for this problem is provided in Appendix G.

7 The Short Wave Component of Finite Difference Schemes

It was pointed out previously that when a central finite difference scheme was used to approximate the spatial derivative, the wave number of the finite difference scheme $\bar{\alpha}$ was related to the actual wave number α. Figure 2.1 shows a typical relationship between $\bar{\alpha}\Delta x$ and $\alpha\Delta x$. Earlier, it was also suggested that the wave spectrum might be divided into two parts; the long waves (waves for which $\bar{\alpha} \simeq \alpha$) and the short waves (waves for which $\bar{\alpha}$ is very different from α). The long waves behave like the corresponding wave component of the exact solution. In this chapter, attention is focused on the short waves.

7.1 The Short Waves

To fix ideas, consider the initial value problem associated with the linearized Euler equations in one space dimension without mean flow. The same dimensionless variable as in Section 6.4 is used. The dimensionless linearized momentum and energy equations are as follows:

$$\frac{\partial u}{\partial t} + \frac{\partial p}{\partial x} = 0 \tag{7.1}$$

$$\frac{\partial p}{\partial t} + \frac{\partial u}{\partial x} = 0. \tag{7.2}$$

For simplicity, consider the following initial conditions,

$$t = 0, \qquad u = 0 \tag{7.3}$$

$$p = f(x). \tag{7.4}$$

It is easy to verify that the exact solution of (7.1) to (7.4) is

$$p = \frac{1}{2}[f(x-t) + f(x+t)]. \tag{7.5}$$

This solution suggests that half the initial pressure pulse propagates to the left and the other half propagates to the right at the speed of sound (unity in dimensionless units).

Now consider a "boxcar" initial condition as follows:

$$f(x) = H(x+M) - H(x-M), \tag{7.6}$$

where M is a large positive number and $H(x)$ is the unit step function. The wave number spectrum of Eq. (7.6) extends well beyond the range $-\pi < \alpha < \pi$. A considerable fraction of the spectrum falls in the short wave range. This offers an excellent example on the wave propagation characteristics of the short waves of finite difference schemes. From Eq. (7.5), the exact solution is

$$p(x, t) = \frac{1}{2}[H(x - t + M) - H(x - t - M)] + \frac{1}{2}[H(x + t + M) - H(x + t - M)].$$
$$(7.7)$$

It is of interest to find out what happens when the problem is solved by the 7-point stencil dispersion-relation-preserving (DRP) scheme. On discretizing Eqs. (7.1) and (7.2) according to the DRP scheme, the resulting finite difference equations are

$$E_\ell^{(n)} = -\sum_{j=-3}^{3} a_j p_{\ell+j}^{(n)}$$

$$F_\ell^{(n)} = -\sum_{j=-3}^{3} a_j u_{\ell+j}^{(n)}$$

$$u_\ell^{(n+1)} = u_\ell^{(n)} + \Delta t \sum_{j=0}^{3} b_j E_\ell^{(n-j)}$$

$$p_\ell^{(n+1)} = p_\ell^{(n+1)} + \Delta t \sum_{j=0}^{3} b_j F_\ell^{(n-j)}.$$
$$(7.8)$$

The initial conditions corresponding to Eqs. (7.3) and (7.6) are

$$u_\ell^{(n)} = 0, \qquad\qquad\qquad n \le 0$$
$$p_\ell^{(n)} = \begin{cases} H(l + M) - H(l - M), & n = 0 \\ 0, & n < 0 \end{cases}.$$
$$(7.9)$$

Figure 7.1 shows the numerical solution with the boxcar initial condition ($M = 50$). As can be seen, the solution is badly contaminated by short waves. The lead waves are the grid-to-grid oscillatory waves. The envelope of the amplitude of the short waves oscillates spatially (giving the appearance of lumps of waves). The longer dispersive waves are trailing the main pulse solution.

This example clearly shows that short waves are pollutants of numerical solutions. They can be generated by discontinuous initial or boundary conditions. To render numerical solutions acceptable, a way must be found to suppress or eliminate the short waves without interfering with the long waves.

7.2 Artificial Selective Damping

The short waves are numerical contaminants of a computed solution. To improve the quality of a numerical solution, it is imperative that the short waves be automatically removed from the computation as soon as they are generated. A way to remove short waves is to add artificial selective damping terms to the finite difference equations.

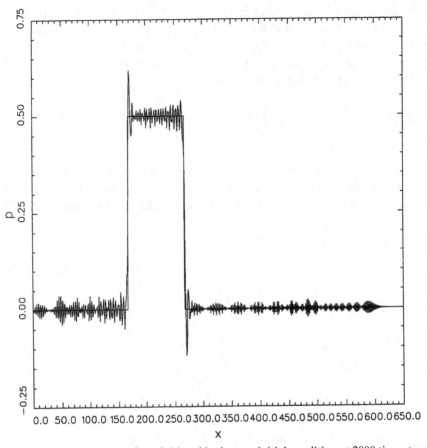

Figure 7.1. Pressure waveform initiated by boxcar initial condition at 2000 time steps.

For this method to be acceptable, the damping terms must selectively damp out the short waves and have minimal effect on the long waves.

7.2.1 Basic Concept

Suppose a damping term, $D(x)$, is added to the right side of the momentum equation of the linearized Euler equations in dimensional form as follows:

$$\frac{\partial u}{\partial t} + \frac{1}{\rho_0} \frac{\partial p}{\partial x} = D(x). \tag{7.10}$$

Let the spatial derivative be approximated by the 7-point stencil DRP scheme. The discretized form of Eq. (7.10) is

$$\frac{du_\ell}{dt} + \frac{1}{\rho_0 \Delta x} \sum_{j=-3}^{3} a_j p_{\ell+j} = D_\ell. \tag{7.11}$$

It will now be assumed that D_ℓ is proportional to the values of u_ℓ within the stencil. Let d_j be the weight coefficients. Eq. (7.11) may be written as

$$\frac{du_\ell}{dt} + \frac{1}{\rho_0 \Delta x} \sum_{j=-3}^{3} a_j p_{\ell+j} = -\frac{\nu}{(\Delta x)^2} \sum_{j=-3}^{3} d_j u_{\ell+j}, \tag{7.12}$$

where ν is the artificial kinematic viscosity. $\nu/(\Delta x)^2$ has the dimension of $(\text{time})^{-1}$, so that d_j's are pure numbers. Now d_j's are to be chosen so that the artificial damping would be effective mainly for high wave number or short waves.

The Fourier transform of the generalized form of Eq. (7.12) is,

$$\frac{d\tilde{u}}{dt} + \cdots = -\frac{\nu}{(\Delta x)^2}\tilde{D}(\alpha\Delta x)\tilde{u} \tag{7.13}$$

where

$$\tilde{D}(\alpha\Delta x) = \sum_{j=-3}^{3} d_j e^{ij\alpha\Delta x}. \tag{7.14}$$

On ignoring the terms not shown in Eq. (7.13), the solution is,

$$\tilde{u} \sim e^{-\frac{\nu}{(\Delta x)^2}\tilde{D}(\alpha\Delta x)t}. \tag{7.15}$$

Since $\tilde{D}(\alpha\Delta x)$ depends on the wave number, the damping will vary with wave number. It is desirable for $\tilde{D}$ to be zero, or small for small $\alpha\Delta x$, but large for large $\alpha\Delta x$. This can be done by choosing d_j's appropriately. But one must be careful to make sure that the damping term would not cause undamping or numerical instability. The purposes of the following three conditions that are to be imposed on $\tilde{D}$ are self-evident.

1. $\tilde{D}(\alpha\Delta x)$ should be a positive even function of $\alpha\Delta x$. The even function condition is ensured by setting $d_{-j} = d_j$. Thus,

$$\tilde{D}(a\Delta x) = d_0 + 2\sum_{j=-3}^{3} d_j \cos(ja\Delta x). \tag{7.16}$$

2. There should be no damping for long waves. That is, $\tilde{D}(\alpha\Delta x) \to 0$ as $\alpha\Delta x \to 0$. This requires

$$d_0 + 2\sum_{j=1}^{3} d_j = 0. \tag{7.17}$$

3. For convenience, $\tilde{D}(\alpha\Delta x)$ may be normalized so that

$$\tilde{D}(\pi) = 1. \tag{7.18}$$

The right side of Eq. (7.14) is a truncated Fourier cosine series. To ensure that $\tilde{D}(\alpha\Delta x)$ is small when $\alpha\Delta x$ is small but becomes large when $\alpha\Delta x \to \pi$, one may use a Gaussian function centered at $\alpha\Delta x = \pi$ with half-width σ as a template for choosing the Fourier coefficients d_j. This is done by determining d_j such that the integral

$$\int_{0}^{\beta} \left[\tilde{D}(\alpha\Delta x) - e^{-\ln 2\left(\frac{\alpha\Delta x-\pi}{\sigma}\right)^2}\right]^2 d(\alpha\Delta x)$$

is a minimum subjected to conditions (7.17) and (7.18). The upper limit of the integral β is a parameter that may be adjusted to yield the most desirable properties for $\tilde{D}$.

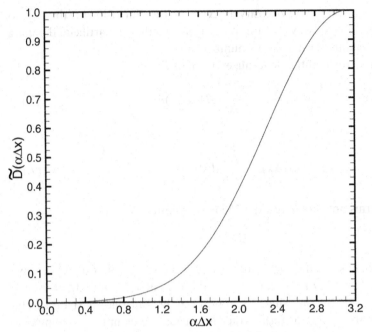

Figure 7.2. Damping curve. 7-point stencil, $\sigma = 0.3\,\pi$.

For example, the coefficients of a very useful damping curve for problems involving discontinuous solution are ($\sigma = 0.3\pi$, $\beta = 0.65\pi$) as follows

$$d_0 = 0.3217949913$$
$$d_1 = d_{-1} = -0.2328759104$$
$$d_2 = d_{-2} = 0.08910250435$$
$$d_3 = d_{-3} = -0.01712408960$$

A plot of $\tilde{D}$ versus $\alpha \Delta x$ for this choice is shown in Figure 7.2.

7.2.2 Numerical Example

To illustrate the effectiveness of artificial selective damping, consider again the simple wave problem with discontinuous initial condition. On incorporating artificial selective damping to the DRP scheme, the finite difference equations are

$$u_\ell^{(n+1)} = u_\ell^{(n)} + \Delta t \sum_{j=0}^{3} b_j E_\ell^{(n-j)}$$

$$p_\ell^{(n+1)} = p_\ell^{(n)} + \Delta t \sum_{j=0}^{3} b_j F_\ell^{(n-j)}$$

$$E_\ell^{(n)} = -\sum_{j=-3}^{3} a_j p_{\ell+j}^{(n)} - \frac{1}{R_\Delta} \sum_{j=-3}^{3} d_j u_{\ell+j}^{(n)}$$

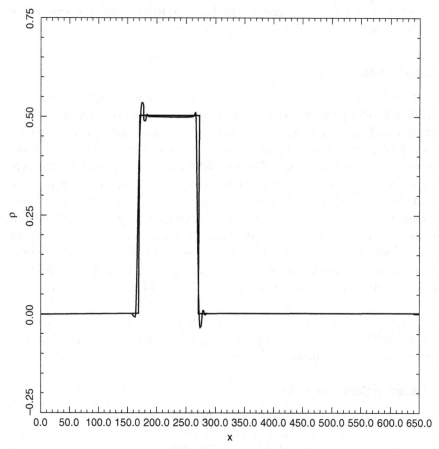

Figure 7.3. Pressure waveform initiated by boxcar initial condition computed with artificial selective damping. $t = 2000$ time steps.

$$F_\ell^{(n)} = -\sum_{j=-3}^{3} a_j u_{\ell+j}^{(n)} - \frac{1}{R_\Delta} \sum_{j=-3}^{3} d_j p_{\ell+j}^{(n)}$$

$$u_\ell^{(n)} = 0 (n \leq 0)$$

$$p_\ell^{(n)} = \begin{cases} [H(\ell + M) - H(\ell - M)], & n = 0 \\ 0, & n < 0 \end{cases} \qquad (7.19)$$

where $R_\Delta = a_0 \Delta x/\nu$ is the artificial mesh Reynolds number. In the numerical simulation below R_Δ^{-1} is given the value of 0.3.

In the course of computing the numerical solution of Eq. (7.19), it is noticed that the short waves corresponding to grid-to-grid oscillations that are subjected to most intense damping are damped out almost immediately after they are generated. The damping for the spurious waves with longer wavelengths is smaller. Their presence in the numerical solution can still be seen after 200 time steps. At $t = 500\Delta t$, they, too, are almost completely eliminated. Figure 7.3 shows the calculated pressure waveform at $t = 2000\Delta t$. Essentially all the spurious waves, except for two spikes, have been damped out. The computed solution is now a reasonably good approximation of the exact boxcar solution. Of course, the finite difference solution is not perfect. It does not faithfully reproduce the sharp discontinuities. Each discontinuity is spread over a distance of about 5 to 6 mesh spacings. This is closed to the minimum wavelength that the scheme can resolve.

7.2.3 Other Damping Stencils

The $\sigma = 0.3\,\pi$ 7-point damping stencil with damping function shown in Figure 7.2 is primarily designed for problems with discontinuities or shocks. Because of the steep gradients involved, strong damping over a wide band of wave numbers is necessary. But even in problems for which the solutions are shock and discontinuity free, short spurious waves are often generated. Wall boundaries, mesh size change interfaces and regions with steep gradients are potential sources of short waves. For this reason, it is a good practice to add a small amount of background artificial selective damping to a computation. Also because of the need to use backward difference stencils at the boundary region of a computation domain, the numerical solution is subjected to weak numerical instability. These instabilities usually manifest themselves in the form of grid-to-grid oscillations with very gradual increase in amplitude. This type of weak instability can easily be suppressed by the addition of artificial selective damping.

For general background damping for which only short waves need to be removed, the following $\sigma = 0.2\,\pi$ 7-point damping stencil is recommended. The stencil coefficients are as follows:

7-Point Damping Stencil ($\sigma = 0.2\pi$)

$$\begin{aligned}
d_0 &= 0.2873928425 \\
d_1 = d_{-1} &= -0.2261469518 \\
d_2 = d_{-2} &= 0.1063035788 \\
d_3 = d_{-3} &= -0.0238530482
\end{aligned}$$

The damping function of this damping stencil is shown in Figure 7.4.

When very high resolution is required, a 15-point stencil DRP scheme may be used. The following is a set of damping coefficients for such a stencil.

15-Point Damping Stencil

$$\begin{aligned}
d_0 &= 0.2042241813072920 \\
d_1 = d_{-1} &= -0.1799016298200503 \\
d_2 = d_{-2} &= 0.1224349282118140 \\
d_3 = d_{-3} &= -6.34562798275548 90\text{E} - 02 \\
d_4 = d_{-4} &= 2.4341225689340974\text{E} - 02 \\
d_5 = d_{-5} &= -6.5519987489327603\text{E} - 03 \\
d_6 = d_{-6} &= 1.1117554451990776\text{E} - 03 \\
d_7 = d_{-7} &= -9.0091603462069583\text{E} - 05
\end{aligned}$$

The damping curve is shown in Figure 7.4.

7.3 Excessive Damping

It was shown in previous sections that, by adding artificial selective damping to a finite difference equation, the short waves are eliminated. However, if too much

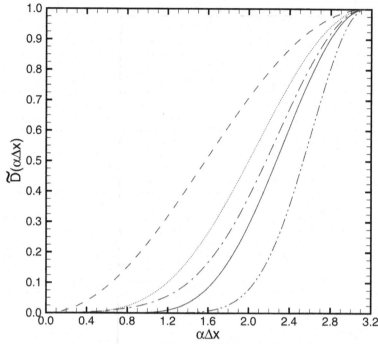

Figure 7.4. Damping functions $\tilde{D}(\alpha\Delta x)$. ——, 7-point stencil ($\sigma = 0.2\,\pi$); — - - - - ,7-point stencil ($\sigma = 0.3\,\pi$); $\cdots\cdots$, 5-point stencil; – – – –, 3-point stencil; – - – - – - –, 15-point stencil.

damping is used, there are also negative effects. Excessive selective damping can basically result in "artificial viscous diffusion" and "numerical instability." Here, the phenomenon of "artificial viscous diffusion" will be discussed first.

7.3.1 Artificial Viscous Diffusion

Consider again the one-dimensional convective wave equation in dimensionless form as follows:

$$\frac{\partial u}{\partial t} + \frac{\partial u}{\partial x} = 0 \quad -\infty < x < \infty. \tag{7.20}$$

Suppose Eq. (7.20) is discretized according to the DRP scheme with the addition of artificial selective damping terms. The discretized equations are

$$u_\ell^{(n+1)} = u_\ell^{(n)} + \Delta t \sum_{j=0}^{3} b_j K_\ell^{(n-j)}$$

$$K_\ell^{(n)} = -\sum_{j=-3}^{3} a_j u_{\ell+j}^{(n)} - \frac{1}{R_\Delta} \sum_{j=-3}^{3} d_j u_{\ell+j}^{(n)} \tag{7.21}$$

For the stated purpose, R_Δ^{-1} is taken to be equal to 5. This is about 15 times more damping than what was used in the previous example. Suppose the initial condition is

$$u(x, 0) = 0.5 e^{-\ln 2 \left(\frac{x}{3}\right)^2}. \tag{7.22}$$

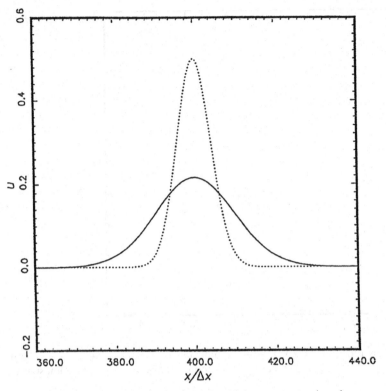

Figure 7.5. Effects of excessive artificial damping on an acoustic pulse. ...., waveform without damping; _____, waveform with large damping $R_{\Delta}^{-1} = 5, t = 400$.

In Chapter 2, it was demonstrated that the 7-point stencil DRP scheme can provide an almost error-free solution for this initial value problem when there is no artificial damping. Figure 7.5 shows the computed result at $t = 400$ by using a damping stencil with $\sigma = 0.3\,\pi$ and $\Delta t = 0.02$. By comparing the computed waveform and the exact solution, it is clear that there is an overall reduction in the wave amplitude. However, at points farther away from the center of the pulse, the wave amplitude has actually increased, as if the pulse has diffused outward on both sides. This is a very unexpected phenomenon.

The cause of this apparent diffusion effect is subtle but can be understood by viewing the damping process in wave number space. Since the Fourier transform of a Gaussian function is also a Gaussian function, the initial pulse is also a concentrated pulse around $\alpha\Delta x = 0$ in the wave number space. Now in time, the artificial selective damping terms gradually reduce the amplitude of the pulse in the wave number space, but not evenly. By design, there is no reduction at zero wave number. The reduction increases with wave number. Because of this, the half-width of the pulse in wave number space is reduced over time. The maximum height that occurs at $\alpha\Delta x = 0$, nevertheless, remains the same. The waveform in the physical space is the Fourier inverse transform of the waveform in the wave number space. With a narrower pulse, as time increases, the physical waveform spreads out according to the Inverse Spreading Theorem of Fourier transform. Thus, artificial selective damping has the unintended side effect of inducing artificial viscous diffusion.

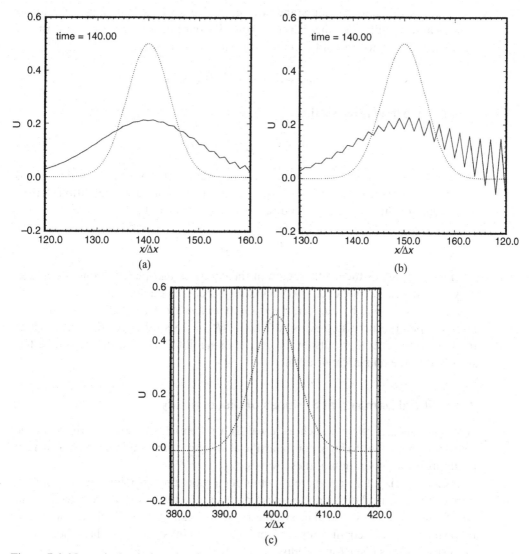

Figure 7.6. Numerical solution showing the onset of instability due to excessive damping. $R_\Delta^{-1} = 15$ ········, waveform without damping; ———, waveform with damping. (a) $t = 140$, (b) $t = 150$, (c) $t = 400$.

7.3.2 Damping-Induced Numerical Instability

Now the phenomenon of "numerical instability" caused by excessive artificial damping is investigated. Suppose the solution of Eq. (7.21) is again computed with the initial condition (7.22), but with even larger damping, say $R_\Delta^{-1} = 15$. Figure 7.6 shows the computed result as time increases. At $t = 140$, Figure 7.6a shows the onset of numerical instability. The instability wave has a wave length of about two mesh spacings or $\alpha \Delta x = \pi$. This is grid-to-grid oscillation. At $t = 150$, Figure 7.6b shows that the numerical instability has grown in amplitude rapidly. At $t = 400$, Figure 7.6c shows that the instability completely overwhelms the numerical solution.

It is not difficult to understand why excessive artificial damping can lead to numerical instability. For this purpose, consider the Fourier-Laplace transform of Eq. (7.21). The transform of the finite difference equation is

$$-i\overline{\omega}\tilde{u} + i\overline{\alpha}\tilde{u} = -\frac{1}{R_\Delta}\tilde{D}(\alpha\Delta x)\tilde{u}. \tag{7.23}$$

Thus, the dispersion relation is

$$\overline{\omega} = \overline{\alpha} - \frac{i}{R_\Delta}\tilde{D}(\alpha\Delta x). \tag{7.24}$$

Consider grid-to-grid oscillation instability wave with $\alpha\Delta x = \pi$. It is to be noted that $\overline{\alpha}(\pi) = 0.0$ (see Figure 2.3) and $\tilde{D}(\pi) = 1.0$ (see normalization condition (7.18)). Therefore, for this wave, dispersion relation (7.24) reduces to

$$\overline{\omega}\Delta t = -i\frac{\Delta t}{R_\Delta}. \tag{7.25}$$

Figure 3.2 gives the stable region in the complex $\overline{\omega}\Delta t$-plane. Along the imaginary axis, the four-time level DRP marching scheme is stable if $\text{Im}(\overline{\omega}\Delta t) > -0.29$. Therefore, there will be numerical instability unless $\Delta t/R_\Delta = 0.29$. In the numerical example, the time step Δt is 0.02 and $1/R_\Delta = 15$ so that $\Delta t/R_\Delta = 0.3$. Thus, the numerical scheme is outside the region of stability, and one should not be too surprised to see numerical instability.

7.4 Artificial Damping at Surfaces of Discontinuity

A surface of discontinuity such as a wall is a potential source of spurious waves. Thus, it is important to incorporate artificial selective damping near such a surface to suppress the generation of these waves.

Near a horizontal wall, there is not enough room for a 7-point central damping stencil in the vertical direction for mesh points in the two rows adjacent to the wall (see Figure 7.7). Only a smaller symmetric stencil can fit in the limited space. The following are stencil coefficients for 5-point and 3-point stencils that have been used extensively with satisfactory results.

5-Point Damping Stencil

$$d_0 = 0.375$$
$$d_1 = d_{-1} = -0.25$$
$$d_2 = d_{-2} = 0.0625.$$

3-Point Damping Stencil

$$d_0 = 0.5$$
$$d_1 = d_{-1} = -0.25.$$

The damping curves for these stencils are shown in Figure 7.4. These smaller damping stencils impose a nonnegligible damping on the long waves. They should not be used as general damping stencils over the entire computational domain.

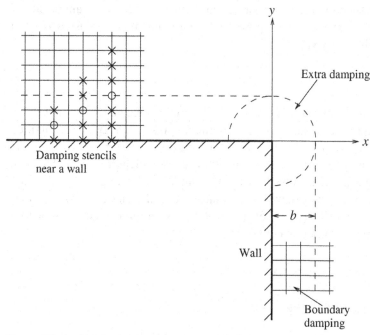

Figure 7.7 Artificial selective damping and damping stencils near a wall and at a sharp corner.

Experience indicates that, when used only near a wall or at the outside boundary of a computational domain, they do not cause excessive damping to the physical solution.

For general background damping, the use of an inverse mesh Reynolds number, R_Δ^{-1}, of 0.05 or slightly less has been found to be satisfactory. The amount of damping is sufficient to remove spurious short waves in most problems. For viscous fluid, additional damping near a wall, especially at a sharp corner (see Figure 7.7), is often necessary. To impose extra damping, a simple but effective way is to use a Gaussian distribution of R_Δ^{-1} in the direction normal to the wall. The peak value of the Gaussian is at the wall. A good choice of the half-width of the Gaussian is three or four mesh spacings. For instance, along the wall $y = 0$ in Figure 7.7, an additional inverse mesh Reynolds number distribution of the following form,

$$R_\Delta^{-1} = 0.15\, e^{-(\ln 2)\left(\frac{y}{4\Delta y}\right)^2} H(y), \tag{7.26}$$

where $H(y)$ is the unit step function, may be incorporated into the artificial selective damping terms of the computation scheme. The highest value of R_Δ^{-1} at the wall may be larger than 0.1 as indicated above. A similar distribution of inverse mesh Reynolds number distribution may also be imposed adjacent to the wall at $x = 0$.

At a sharp corner, as shown in Figure 7.7, large amount of artificial selective damping is required to stabilize the computation. This may be implemented by using an inverse mesh Reynolds number distribution at the corner point in the form of a multidimensional Gaussian function,

$$R_\Delta^{-1} = 0.3 e^{-\ln 2\left[\frac{x^2+y^2}{16\Delta x^2}\right]}. \tag{7.27}$$

At the present time, there are no formal guidelines to setting the amplitude of the Gaussian function at the corner point. This value has to be adjusted in each problem until a satisfactory value is selected.

7.5 Aliasing

The fundamental wave number range of a finite difference scheme is $-\pi \leq \alpha \Delta x < \pi$ (see Figure 2.1). The $\bar{\alpha}\Delta x$ versus $\alpha \Delta x$ curve in the range $-\pi \leq \alpha \Delta x \leq 0$ is obtained by an antisymmetric extension of that in the range $0 \leq \alpha \Delta x \leq \pi$. Wave numbers that fall outside this range are underresolved. Their wavelength is less than two mesh spacings. They are aliased back inside the fundamental range. To determine the relationship between the original wave number and the aliased wave number, consider the initial condition (7.4). Let $f(x) = \Phi(x)$, where Φ is a Gaussian with a concentration of wave number around α_0 ; i.e.,

$$t = 0, \quad \Phi(x) = e^{-(\ln 2)(\frac{x}{b})^2 + i\alpha_0 x}. \tag{7.28}$$

The Fourier transform of Eq. (7.28) is

$$t = 0, \quad \hat{\Phi}(\alpha) = \frac{b}{2(\pi \ln 2)^{\frac{1}{2}}} e^{-(\ln 2)\left[\frac{(\alpha - \alpha_0)b}{2\ln 2}\right]^2}. \tag{7.29}$$

If b, the half-width of the Gaussian, is large, then Eq. (7.29) confirms that there is an essential concentration of wave number around $\alpha = \alpha_0$.

Let Δx be the mesh size used in a computation. ℓ is the spatial index so that $x = \ell \Delta x$. As far as the computation on the mesh is concerned, Eq. (7.28), in the discretized form, is

$$\Phi_\ell = e^{-(\ln 2)(\frac{\ell \Delta x}{b})^2 + i\alpha_0 \ell \Delta x}. \tag{7.30}$$

It will be assumed that $\alpha_0 \Delta x > \pi$ but less than 2π. Since $\alpha_0 \Delta x$ is larger than π, let

$$\alpha_0 \Delta x = \pi + \delta. \tag{7.31}$$

But

$$e^{i\alpha_0 \Delta x \ell} = e^{i[2\pi + (\delta - \pi)]\ell} = e^{i(\delta - \pi)\ell} = e^{i(\frac{\delta - \pi}{\Delta x})x} = e^{i(\alpha_0 - \frac{2\pi}{\Delta x})x}.$$

Therefore, the initial condition is the same as

$$t = 0 \quad \Phi(x) = e^{-(\ln 2)(\frac{x}{b})^2 + i(\alpha_0 - \frac{2\pi}{\Delta x})x} \tag{7.32}$$

for the computation. In other words, the effective wave number is

$$\alpha = \alpha_0 - \frac{2\pi}{\Delta x} \tag{7.33}$$

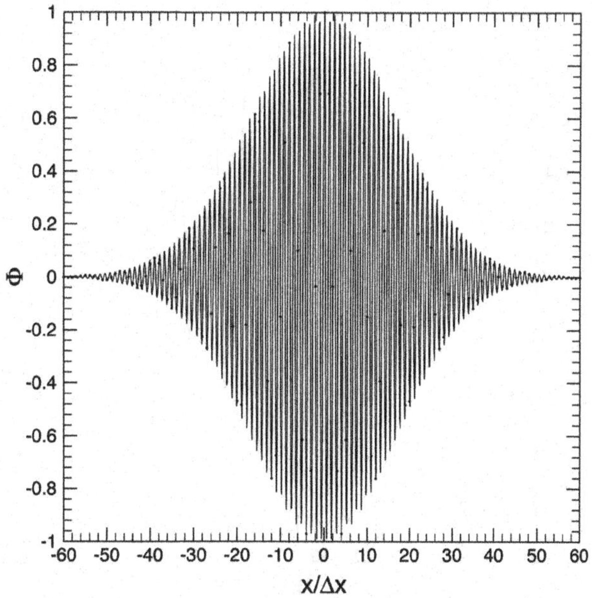

Figure 7.8. The function $\exp[-(\ln 2)(0.05x/\Delta x)^2]\cos(5.48x/\Delta x)$. Black dots are the sampling points at $x/\Delta x$ equal to integers.

or

$$\alpha \Delta x = \alpha_0 \Delta x - 2\pi. \tag{7.34}$$

Hence, although $\alpha_0 \Delta x$ is larger than π, it is aliased back into the fundamental range by a shift of 2π.

As an example, consider the function

$$\Phi(x) = e^{-(\ln 2)\left(\frac{0.05x}{\Delta x}\right)^2} \cos\left(\frac{5.48x}{\Delta x}\right). \tag{7.35}$$

The wave number of this function concentrates around $\alpha_0 \Delta x = \pm 5.48$. This is outside the fundamental range. For computation on the mesh, the effective wave number, according to Eq. (7.34), is $\alpha \Delta x = \pm 0.803$. Figure 7.8 shows the original function (7.35), with wave number concentrates around $\alpha_0 \Delta x = \pm 5.48$. The black dots in the figure are sampling points. From the finite difference computation point of view, only the sampling points matter. Figure 7.9 shows the function representing the sampling points. This function is the same as

$$\Phi(x) = e^{-(\ln 2)\left(\frac{0.05\Delta x}{x}\right)^2} \cos\left(\frac{0.803x}{\Delta x}\right).$$

This function has a low wave number of $\alpha \Delta x = \pm 0.803$ and lies inside the fundamental range of wave numbers.

7.6 Coefficients of Several Large Damping Stencils

When large computational stencils are used for increased spatial resolution, it is a good practice to use large artificial selective damping stencils as well. The coefficients of several large damping stencils are provided below.

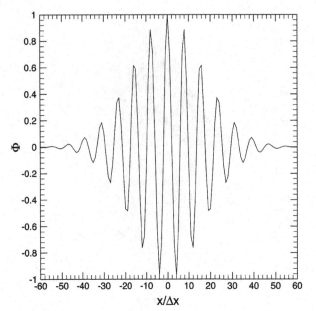

Figure 7.9. The aliased function of Figure 7.8.

9-Point Stencil

$$\sigma = 0.1975\pi, \quad \beta = 1.28$$
$$d_0 = 0.24771185403726$$
$$d_1 = d_{-1} = -0.20711771710355$$
$$d_2 = d_{-2} = 0.11853742369381$$
$$d_3 = d_{-3} = -0.04288228289644$$
$$d_4 = d_{-4} = 7.606649287550E\text{-}3.$$

11-Point Stencil

$$\sigma = 0.195\pi, \quad \beta = 1.45$$
$$d_0 = 0.22311349696678$$
$$d_1 = d_{-1} = -0.19226925045875$$
$$d_2 = d_{-2} = 0.12201457655441$$
$$d_3 = d_{-3} = -0.05522898418946$$
$$d_4 = d_{-4} = 0.01642867496219$$
$$d_5 = d_{-5} = -2.5017653517767E\text{-}3.$$

13-Point Stencil

$$\sigma = 0.1925\pi, \quad \beta = 1.63$$
$$d_0 = 0.20975644405137$$
$$d_1 = d_{-1} = -0.18327489344470$$
$$d_2 = d_{-2} = 0.12192070829492$$

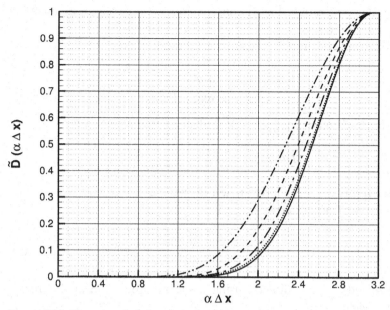

Figure 7.10. Damping functions corresponding to several large damping stencils. $- \cdots -$, 7-point stencil; $----$, 9-point stencil; $- \cdot -$, 11-point stencil; $\cdots\cdots$, 13-point stencil; ———, 15-point stencil.

$$d_3 = d_{-3} = -0.06111106887629$$
$$d_4 = d_{-4} = 0.02244708528388$$
$$d_5 = d_{-5} = -5.6140376790013E\text{-}3$$
$$d_6 = d_{-6} = 7.539843954985E\text{-}4$$

A plot of these damping functions is given in Figure 7.10.

EXERCISES

7.1. Let us reconsider the initial value problem

$$\frac{\partial u}{\partial t} + \frac{\partial u}{\partial x} = 0$$

$t = 0$, $u = H(x + 50) - H(x - 50)$, where $H(x)$ is the unit step function.

Compute the solution using the 7-point DRP scheme with artificial selective damping. Use a Δt that satisfies the numerical stability requirement.

1. Use $R_\Delta^{-1} = 0.3$, where R_Δ is the artificial mesh Reynolds number.
2. Use $R_\Delta^{-1} = 3$.
3. Use $R_\Delta^{-1} = 0.3/\Delta t$.

Plot $u(x, t)$ at $t = 300$ for each case.

7.2. Real fluids have molecular viscosity that tends to damp out long and short wave disturbances. This problem focuses on how best to compute viscous terms to simulate real viscous effects. At the same time, it provides an example illustrating why artificial selective damping rather than viscous damping is preferred for eliminating short spurious numerical waves.

Consider the linearized Burger's equation as follows:

$$\frac{\partial u}{\partial t} + c\frac{\partial u}{\partial x} = v\frac{\partial^2 u}{\partial x^2}. \tag{1}$$

The exact damping function on the right side of the equation is

$$D(\alpha\Delta x) = (\alpha\Delta x)^2. \tag{2}$$

Now, the viscous terms can be approximated in two ways: directly, as a second derivative, or as two first-order derivatives; i.e.,

$$\frac{\partial^2 u}{\partial x^2} = \frac{\partial v}{\partial x}, \qquad v = \frac{\partial u}{\partial x}. \tag{3}$$

(a) Discretize $v(\partial^2 u/\partial x^2)$ directly using a second-order central difference scheme to yield

$$v\left(\frac{\partial^2 u}{\partial x^2}\right)_\ell = \frac{v}{(\Delta x)^2}(u_{\ell+1} - 2u_\ell + u_{\ell+1}).$$

Find the damping function in wave number space.

(b) Discretize (3) using second-order central difference; i.e.,

$$v_\ell = \frac{u_{\ell+1} - u_{\ell-1}}{2\Delta x}, \qquad v\left(\frac{\partial^2 u}{\partial x^2}\right)_\ell = \frac{v}{(\Delta x)^2}\left(\frac{v_{\ell+1} - v_{\ell-1}}{2}\Delta x\right).$$

Find the damping function in wave number space.

(c) Which approximation is more accurate in terms of reproducing the effect of physical viscosity over the resolved wave range?

(d) Which approximation would help to damp out spurious short waves?

(e) If the objective is to damp out spurious short waves alone with little damping on the long waves, is artificial selective damping better, or are physical viscous terms better?

7.3. Develop an artificial selective damping stencil for a high-order scheme involving 17 mesh points.

7.4. In some applications, a symmetric stencil with regular size mesh may not be the most appropriate. Consider the seven-point stencil as shown,

Develop a set of damping coefficients for the stencil.

7.5. In two-dimensional problems artificial selective damping is added both in the x and y directions. For instance, the continuity equation with damping terms added may be written as follows:

$$\frac{d\rho_{\ell,m}}{dt} + \left(\frac{\partial \rho u}{\partial x}\right)_{\ell,m} + \left(\frac{\partial \rho v}{\partial y}\right)_{\ell,m} = v_a \sum_{j=-3}^{3} d_j \left(\frac{\rho_{\ell+j,m}}{\Delta x^2} + \frac{\rho_{\ell,m+j}}{\Delta y^2}\right).$$

(a) Plot contours of constant damping rate in the $\alpha \Delta x - \beta \Delta y$ plane for $-\pi \leq \alpha \Delta x, \beta \Delta y \leq \pi$ (assume $\Delta x = \Delta y$).

(b) What is the damping rate for a wave propagating at an angle of 30 degrees to the positive x-axis; the total wave number $k\Delta x = 2.5$, where $k = (\alpha^2 + \beta^2)^{1/2}$.

7.6. Compute the solution of the convective wave equation

$$\frac{\partial u}{\partial t} + \frac{\partial u}{\partial x} = 0$$

on a uniform mesh with $\Delta x = 1$ and the following initial condition,

$$t = 0, \qquad u = [2 + \cos(\alpha x)]\exp[-(\ln 2)(x/10)^2]$$

Consider two cases.

(i) $\alpha = 1.7$
(ii) $\alpha = 4.6$

Results to be reported are the spatial distributions of u at $t = 400$ and $t = 800$.
    The exact solution to (i) is

$$u(x, t) = [2 + \cos\alpha(x - t)]\exp\left[-(\ln 2)\left(\frac{x - t}{10}\right)^2\right].$$

The exact (aliased) solution to (ii) is

$$u(x, t) = [2 + \cos[(2\pi - \alpha)(x - t)]]\exp\left[-(\ln 2)\left(\frac{x - t}{10}\right)^2\right].$$

8 Computation of Nonlinear Acoustic Waves

8.1 Nonlinear Simple Waves

One-dimensional acoustic waves are governed by the continuity, momentum (inviscid), and energy equations. They can be written in the following form:

$$\frac{\partial \rho}{\partial t} + u\frac{\partial \rho}{\partial x} + \rho\frac{\partial u}{\partial x} = 0 \tag{8.1}$$

$$\frac{\partial u}{\partial t} + u\frac{\partial u}{\partial x} + \frac{1}{\rho}\frac{\partial p}{\partial x} = 0 \tag{8.2}$$

$$\frac{\partial p}{\partial t} + u\frac{\partial p}{\partial x} + \gamma p\frac{\partial u}{\partial x} = 0. \tag{8.3}$$

It is well-known that the above first-order system has three characteristics: the P, the C_+, and the C_- characteristics (see Appendix C for derivation).

8.1.1 The P Characteristic

Along the P characteristic

$$\frac{dx}{dt} = u, \tag{8.4}$$

the dependent variables are related by

$$dp - \frac{\gamma p}{\rho}d\rho = 0 \quad \text{or} \quad \frac{p}{\rho^{\gamma}} = \text{constant.} \tag{8.5}$$

That is, the flow is isentropic following a fluid element.

8.1.2 The C_+ and C_- Characteristic

Along these characteristics,

$$\frac{dx}{dt} = u \pm a, \tag{8.6}$$

where $a = (\gamma p/\rho)^{\frac{1}{2}}$ is the speed of sound, the dependent variables are related by

$$\frac{dp}{\rho a} \pm du = 0. \tag{8.7}$$

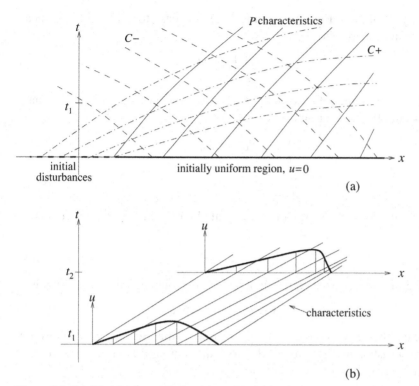

Figure 8.1. (a) The P, C_+, and C_- characteristics in the $x - t$ plane. (b) Characteristics of equation (8.13) and nonlinear steepening of the wave front.

Now consider the problem of acoustic waves propagating into a region initially at rest; i.e., $t = 0$, $u = 0$, $\rho = \rho_0$, $a = a_0$. Without loss of generality, it will be assumed that the waves propagate in the positive x direction. Since this region is covered by the P characteristics that start out from the same initially undisturbed region (see Figure 8.1), the isentropic relation

$$\frac{p}{\rho^\gamma} = \frac{p_0}{\rho_0^\gamma} \quad \text{or} \quad \frac{dp}{\rho a} = a\frac{d\rho}{\rho} = \frac{2}{\gamma - 1}da \tag{8.8}$$

holds throughout the region. The same is true with the C_- characteristics. By eliminating $dp/\rho a$ from Eqs. (8.7) and (8.8), it is found that

$$\frac{2}{\gamma - 1}da - du = 0. \tag{8.9}$$

Upon integrating Eq. (8.9) and noting that the constant of integration is the same for all points in the region, the following integral relation is obtained:

$$\frac{2a}{\gamma - 1} - u = \frac{2a_0}{\gamma - 1}. \tag{8.10}$$

Now, in the region of the $x - t$ plane covered by both the P and C_- characteristics starting from the initially uniform region, Eq. (8.2) may be cast into the following form, by means of the relation (8.7) of the C_- characteristics:

$$\frac{\partial u}{\partial t} + (u + a)\frac{\partial u}{\partial x} = 0. \tag{8.11}$$

Finally, by combining Eqs. (8.10) and (8.11), it is easy to derive the nonlinear simple wave equation as follows:

$$\frac{\partial u}{\partial t} + \left(a_0 + \frac{\gamma+1}{2}u\right)\frac{\partial u}{\partial x} = 0. \tag{8.12}$$

Eq. (8.12) is a nonlinear first-order partial differential equation for u. The characteristic system of this equation is

$$\frac{dx}{dt} = a_0 + \frac{\gamma+1}{2}u$$

$$\frac{du}{dt} = 0. \tag{8.13}$$

The two equations of (8.13) can be integrated easily. Suppose the initial condition is

$$t = 0, \qquad u = \Phi(x). \tag{8.14}$$

The integral of the second equation of (8.13) satisfying the initial condition is

$$t = 0, \qquad x = s, \qquad u = \Phi(s). \tag{8.15}$$

This means that u is a constant along a characteristic. Because u is a constant along a characteristic, the right side of the first equation of (8.13) is a constant. The integral of this equation is

$$x = s + \left(a_0 + \frac{\gamma+1}{2}\Phi(s)\right)t. \tag{8.16}$$

The solution of the nonlinear simple wave equation (8.12) and initial condition (8.14) may now be found by eliminating the parameter s from Eqs. (8.15) and (8.16). Eq. (8.16) indicates that the characteristics are straight lines in the $x - t$ plane. The slopes of the characteristics, however, depend on $\Phi(x)$. If $\Phi(x)$ is not a constant, the slope of the characteristics will vary from characteristic to characteristic (see Figure 8.1b). Note that adjacent straight lines that are not parallel but convergent will eventually intersect each other. Therefore, over a sufficiently long period of time, the characteristics of Eq. (8.16) will overlap each other. When this happens, the solution will become multivalued. To render the solution single-valued, a discontinuity or a shock must be inserted into the solution. In this chapter, shock-capturing computation using the dispersion-relation-preserving (DRP) scheme is considered.

8.2 Spurious Oscillations: Origin and Characteristics

Consider the computation of the solution of the nonlinear simple wave equation (8.12) by a high-order finite difference scheme. For this purpose, it is more convenient to use dimensionless variables with mesh size Δx as the length scale, speed of sound a_0 as the velocity scale, and $\Delta x/a_0$ as the time scale. On recasting (8.12) into a conservation form, the initial value problem is

$$\frac{\partial u}{\partial t} + \frac{\partial u}{\partial x} + \frac{\gamma+1}{4}\frac{\partial u^2}{\partial x} = 0 \tag{8.17}$$

$$t = 0 \qquad u = f(x). \tag{8.18}$$

The discretized form of (8.17) according to the 7-point stencil DRP scheme is

$$K_\ell^{(n)} = -\sum_{j=-3}^{3} a_j u_{j+\ell}^{(n)} - \frac{\gamma+1}{4} \sum_{j=-3}^{3} a_j \left(u_{j+\ell}^{(n)}\right)^2$$

$$u_\ell^{(n+1)} = u_\ell^{(n)} + \Delta t \sum_{j=0}^{3} b_j K_\ell^{(n-j)}. \tag{8.19}$$

Consider the initial condition in the form of a Gaussian function as follows:

$$f(x) = h \exp\left[-\ln 2 \left(\frac{x}{b}\right)^2\right]. \tag{8.20}$$

The discretized form of Eq. (8.20) is

$$\left.\begin{array}{l} u_\ell^{(0)} = h \exp\left[-\ln 2 \left(\frac{\ell}{b}\right)^2\right] \\[2mm] u_\ell^{(n)} = 0; \quad n < 0 \end{array}\right\} \tag{8.21}$$

The solution of Eq. (8.19) with initial condition (8.21) can be computed easily. Figure 8.2 shows the computed results for the case $h = 0.5$ and $b = 5.0$ at $t = 6$ and $t = 16$. At $t = 6$, the waveform, unlike the initial shape, is no longer symmetric about its maximum. It has been substantially distorted by nonlinear steepening effects. The waveform is in good agreement with the exact solution. $t = 6$ is very near to the wave breaking time when a shock will form.

At $t = 16$, a shock has formed in the exact solution. Clearly, the high-order finite difference scheme cannot reproduce a sharp discontinuity. The shock is replaced by a steep gradient. However, following the steep gradient are spurious spatial oscillations. The presence of these spurious oscillations renders the numerical solution totally unacceptable.

The spurious oscillations are not related to the Gibbs phenomenon as many investigators had speculated. What causes the spurious spatial oscillations then? It turns out that a good deal of the understanding of the origin of the spurious oscillations can be found by studying and comparing the time evolution of the exact and finite difference solutions in wave number space.

The Fourier transform of Eq. (8.17) and initial condition (8.20) is

$$\frac{d\tilde{u}(\alpha,t)}{dt} + i\alpha \left[\tilde{u}(\alpha,t) + \frac{\gamma+1}{4} \int_{-\infty}^{\infty} \tilde{u}(k,t)\tilde{u}(\alpha-k,t)\,dk\right] = 0 \tag{8.22}$$

$$t = 0, \qquad \tilde{u}(\alpha,t) = \frac{1}{2}\left(\frac{1}{\pi \ln 2}\right)^{\frac{1}{2}} hb \exp\left[-\frac{(\alpha b)^2}{4 \ln 2}\right]. \tag{8.23}$$

Now the spatial discretized form of Eq. (8.17) is

$$\frac{du_\ell}{dt} + \left[\sum_{j=-3}^{3} a_j u_{j+\ell} + \frac{\gamma+1}{4} \sum_{j=-3}^{3} a_j (u_{j+\ell})^2\right] = 0. \tag{8.24}$$

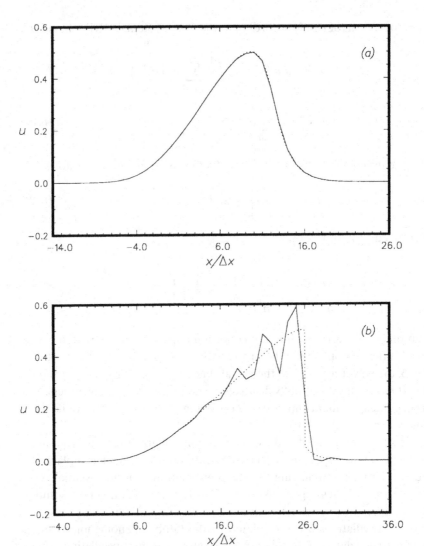

Figure 8.2. Comparisons between computed and exact solutions of the nonlinear simple wave equation. $\cdots\cdots\cdots$ exact, _____ DRP scheme. (a) $t = 6.0$, (b) $t = 16.0$.

This is a special case of the following finite difference differential equation in which x is a continuous variable.

$$\frac{\partial u(x,t)}{\partial t} + \sum_{j=-3}^{3} a_j u(x + j\Delta x, t) + \frac{\gamma+1}{4} \sum_{j=-3}^{3} a_j u^2(x + j\Delta x, t) = 0. \quad (8.25)$$

The Fourier transform of Eq. (8.25) is

$$\frac{d\hat{u}(\alpha,t)}{dt} + i\bar{\alpha}\left[\hat{\tilde{u}}(\alpha,t) + \frac{\gamma+1}{4}\int_{-\infty}^{\infty}\hat{u}(k,t)\hat{\tilde{u}}(\alpha-k,t)dk\right] = 0, \quad (8.26)$$

where $\bar{\alpha}(\alpha)$ is the wave number of the finite difference scheme.

Both Eqs. (8.22) and (8.26) with initial condition (8.23) describe the time evolution of a nonlinear acoustic pulse in wave number space. Eq. (8.22) is for the exact solution. Eq. (8.26) is for the finite difference equation. These nonlinear integral-differential equations can be solved numerically by first discretizing the wave number axis into a fine mesh of spacing $\Delta\alpha$. The integrals are then converted to sums. This reduces the equation to a system of $(2M + 1)$ first-order ordinary differential equations in time, where M is a large integer.

Eqs. (8.22) and (8.26) become, upon discretizating the α-axis,

$$\frac{d\tilde{u}}{dt} + in\Delta\alpha \left[\tilde{u}_n + \frac{\gamma+1}{4} \sum_{j=-M}^{M} \tilde{u}_j \tilde{u}_{n-j} \Delta\alpha \right] = 0 \tag{8.27}$$

$$\frac{d\hat{\tilde{u}}_n}{dt} + i\bar{\alpha}(n\Delta\alpha) \left[\hat{\tilde{u}}_n + \frac{\gamma+1}{4} \sum_{j=-M}^{M} \hat{\tilde{u}}_j \hat{\tilde{u}}_{n-j} \Delta\alpha \right] = 0, \tag{8.28}$$

where $n = -M, -M+1, \ldots, M$, and, $\tilde{u}_j = \hat{\tilde{u}}_j = 0$ if $|j| > M$. The initial condition is

$$\tilde{u}_n(0) = \hat{\tilde{u}}_n(0) = \frac{1}{2}\left(\frac{1}{\pi \ln 2}\right)^{\frac{1}{2}} hbe^{-\frac{(nb\Delta\alpha)^2}{4\ln 2}}. \tag{8.29}$$

The systems of Eqs. (8.27) and (8.28) with initial condition (8.29) can be integrated in time numerically by the Runge-Kutta method. The computed results are shown in Figure 8.3 for the case $h = 0.5$ and $b = 5.0$. The time evolution of the pulse in physical space is given in Figure 8.2. At $t = 6$, the wave number spectrum is smooth and nearly Gaussian as shown in Figure 8.3a. There is very little difference between the exact and the finite difference solution. At $t = 11.67$, the shock formation time, the wave number spectra are shown in Figure 8.3b. The computed and the exact wave number spectra are no longer identical. A small spurious peak appears at $\alpha\Delta x = 1.8$ for the finite difference solution. This peak grows in height as time progresses. At time $t = 16$, a significant peak has developed as can be seen in Figure 8.3c.

In the physical space, Figure 8.2b, the dominant wavelength of the spurious spatial oscillations is about $3.5\Delta x$. The corresponding wave number is $\alpha\Delta x = \frac{2\pi \Delta x}{3.5\Delta x} \simeq 1.8$. Clearly, the accumulation of wave energy around the spurious peak in the wave number space is the cause of the appearance of the extraneous spatial oscillations in the waveform in physical space. Let us now determine the factors that could significantly influence the wave number at the maximum of the spurious peak, namely, α_{peak}. A list of the likely factors is as follows:

1. The initial waveform of the acoustic pulse.
2. Whether the equation is cast in conservation form or nonconservation form.
3. The size of the finite difference stencil, N.

To test the effect of the initial waveform, the following families of initial profile shapes will be considered.

(a) Gaussian Profile

$$u(x, 0) = h \exp\left[-\ln 2\left(\frac{x}{b}\right)^2\right]. \tag{8.30}$$

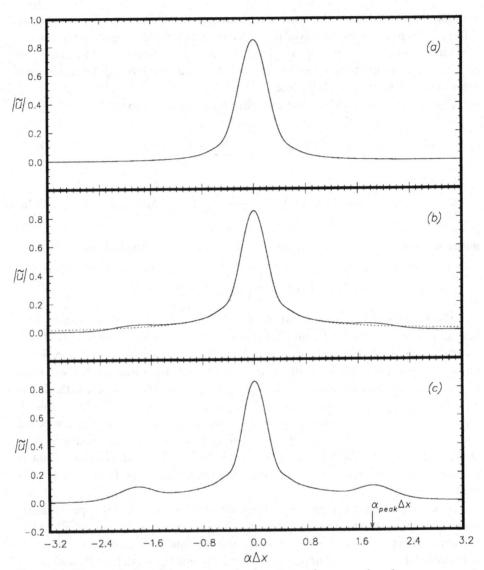

Figure 8.3. Time evolution of the wave number spectrum of an acoustic pulse $\cdots\cdots$ exact solution, _____ DRP scheme. (a) $t = 6.0$, (b) $t = 11.67$, (c) $t = 16.0$.

(b) The Lorentz Line Shape

$$u(x, 0) = \frac{h}{1 + \left(\frac{x}{b}\right)^2}. \tag{8.31}$$

(c) The Half-Period Cosine Function

$$u(x, 0) = \begin{cases} h \cos\left(\dfrac{\pi x}{3b}\right); & -\dfrac{3b}{2} \le x \le \dfrac{3b}{2} \\ 0; & \text{otherwise} \end{cases} \tag{8.32}$$

Each of these waveforms is characterized by two parameters, namely, h (the maximum amplitude) and b (the half-width). For each initial waveform, the systems of Eqs. (8.27), (8.28), and the counterpart of (8.28) in nonconservation form; i.e.,

$$\frac{d\hat{\bar{u}}}{dt} + i\left[\bar{\alpha}\hat{\bar{u}} + \frac{\gamma+1}{2}\int_{-\infty}^{\infty}\bar{k}(k)\hat{u}(k,t)\hat{\bar{u}}(\alpha-k,t)dk\right] = 0$$

or in discretized form,

$$\frac{d\hat{\bar{u}}_n}{dt} + i\left[\bar{\alpha}(n\Delta\alpha)\hat{\bar{u}} + \frac{\gamma+1}{2}\sum_{j=-M}^{M}\bar{k}(j\Delta\alpha)\hat{\bar{u}}_j\,\hat{\bar{u}}_{n-j}\Delta\alpha\right] = 0, \tag{8.33}$$

where $\bar{k}(k)$ is $\bar{\alpha}(\alpha)$ with $\alpha \to k$, is integrated in time. The location $\alpha_{peak}\Delta x$ in wave number space using stencil size $N = 3, 5$, and 7 are determined computationally. The results are given in Figures 8.4a, 8.4b, and 8.4c. These results suggest the following conclusions.

1. Whether the governing partial differential equation is cast in conservation form or not has only a minor effect on the location of $\alpha_{peak}\Delta x$.
2. The initial profile shape of the waveform has a stronger influence than whether the equation is in conservation form or not.
3. The stencil size has the dominant influence. The value of $\alpha_{peak}\Delta x$ shifts to higher values with increase in stencil size.

A different presentation of this result is given in Figure 8.5. It is found that $\alpha_{peak}\Delta x$ invariably lies in the range marked AB in the $\bar{\alpha}\Delta x$ versus $\alpha\Delta x$ plot. Two other important points are observed. (a) Point A is nearly equal to $\alpha_c\Delta x$ so that the wave number of the spurious spectrum peak is definitely in the short wave range. (b) $\alpha_{peak}\Delta x$ lies close to where $\bar{\alpha}\,\Delta x$ is largest. It is not difficult to understand these results. The term $u(\partial u/\partial x)$ or $\frac{1}{2}(\partial u^2/\partial x)$ in the partial differential equation is the nonlinear steepening term. It causes the solution to cascade to finer and finer scales or higher wave numbers. For the finite difference scheme, the nonlinear term in wave number space is

$$\bar{\alpha}\int_{-\infty}^{\infty}\tilde{u}(k,t)\tilde{u}(\alpha-k,t)dk$$

This is weighted by $\bar{\alpha}$ so that it favors the transfer of energy to wave numbers for which $\bar{\alpha}$ is the largest.

Once energy cascades into this wave number range, further cascading is prevented by the dispersive nature of the numerical waves. Dispersive waves propagate with different wave speeds. This reduces the effectiveness of nonlinear interaction and hence the cascading process. This is why the wave energy piles up around wave number $\alpha_{peak}\Delta x$ without cascading to even higher wave numbers.

8.3 Variable Artificial Selective Damping

Now the origin of the spurious spatial oscillations around a computed shock solution is known; what is needed to obtain an acceptable nonlinear solution is to eliminate the short waves around the peak at $\alpha\,\Delta x = \alpha_{peak}\Delta x$. One way to do this is to use artificial selective damping, which was discussed in Chapter 7. For a nonlinear wave

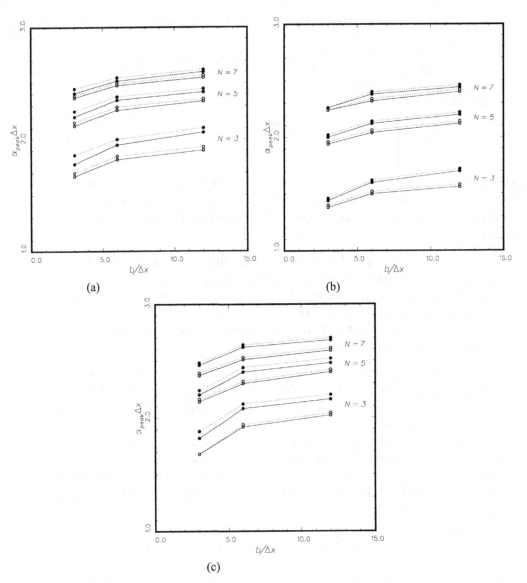

Figure 8.4. Variation of $\alpha_{peak}\Delta x$ on initial pulse profile and other parameters. _____ equation in conservation form, $\cdots\cdots$ equation in nonconservation form. $\circ$ $h = 0.5$, $\bullet$ $h = 1.0$. (a) Gaussian profile, (b) Lorentz line shape, (c) half-cosine function.

propagation problem, artificial selective damping is needed only in the limited region of space where large gradients occur. The use of a constant artificial viscosity, which introduces uniform damping everywhere, is not a good strategy. For a high-quality numerical solution, the use of a variable artificial viscosity or mesh Reynolds number is desirable.

To establish a variable damping algorithm, a way to measure the shock strength or velocity gradient is needed. A shock is expected to be smeared out over a few mesh points, so the relevant measure of length is Δx (the mesh size). As a measure of velocity, one may use $u_{\text{stencil}} = |u_{\text{max}} - u_{\text{min}}|$; the difference between the maximum velocity, u_{max}, and the minimum velocity, u_{min}, within the stencil. Note: if a very large stencil is used in the finite difference scheme, one might consider restricting

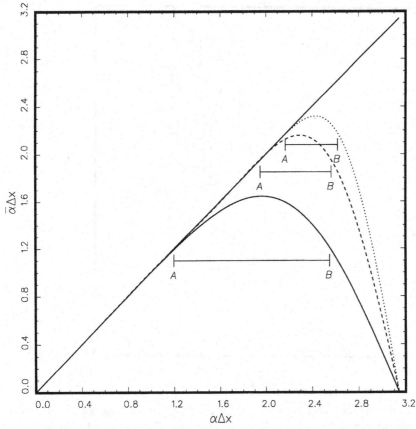

Figure 8.5. $\bar{\alpha}\Delta x$ versus $\alpha\Delta x$ relation for the DRP scheme. ———, $N = 3$; – – – –, $N = 5$, $\cdots\cdots\cdot N = 7$.

the determination of $u_{\max}$ and $u_{\min}$ to a smaller substencil in forming u_{stencil}. It is useful to define a stencil Reynolds number, R_{stencil}, by

$$R_{\text{stencil}} = \frac{u_{\text{stencil}}\Delta x}{\nu}. \tag{8.34}$$

On adding artificial selective damping terms to the discretized nonlinear simple wave equation (8.12), it is straightforward to find (in dimensional form) the following:

$$\frac{du_\ell}{dt} + \left(a_0 + \frac{\gamma+1}{2}u_\ell\right)\frac{1}{\Delta x}\sum_{j=-3}^{3}a_j u_{\ell+j} = -\frac{\gamma+1}{2}\frac{u_{\text{stencil}}}{\Delta x R_{\text{stencil}}}\sum_{j=-3}^{3}d_j u_{\ell+j}. \tag{8.35}$$

In Eq. (8.35), the ratio of the magnitude of the nonlinear steepening term to the damping term is (Note: velocity scale is u_{stencil}) as follows:

$$\frac{\text{magnitude of nonlinear steepening term}}{\text{magnitude of damping terms}} \cong \frac{\frac{u_{\text{stencil}}^2}{\Delta x}}{\frac{u_{\text{stencil}}^2}{(\Delta x R_{\text{stencil}})}} = R_{\text{stencil}}.$$

So R_{stencil} may be interpreted as the rate of generation of spurious short waves around $\alpha\Delta x = \alpha_{\text{peak}}\Delta x$ to the rate of removal of the short waves by artificial selective damping. What one would like to see is that, as soon as the spurious waves are

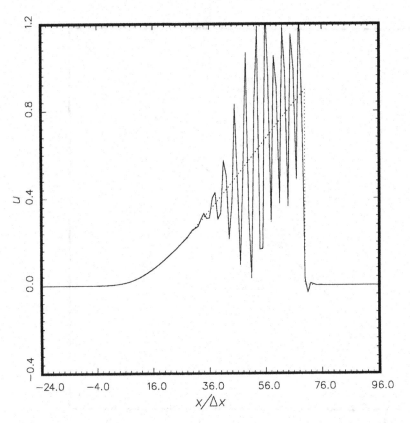

Figure 8.6. Computation of nonlinear acoustic pulse without artificial damping. $t = 36$. Gaussian initial waveform with $h = 1.0$, $b = 12\Delta x$. ———, computed, $\cdots\cdots\cdots$ exact solution.

generated, they are eliminated by damping. In other words, the generation and damping processes are matched or balanced. This means that R_{stencil} is a universal constant. Here, a universal constant means that it is independent of the nonlinear problem to be solved. However, the value of the constant obviously depends on the damping curve and the finite difference scheme being used.

To find R_{stencil} for the 7-point stencil DRP scheme and a damping curve of half-width 0.3π, extensive numerical experiments have been carried out. It is found that by taking $R_{\text{stencil}} \simeq 0.06$ to 0.1 most of the preshock and postshock spurious spatial oscillations are removed from the numerical solutions for all the initial pulse profiles that have been considered.

Two numerical examples are provided here. Figure 8.6 shows the computed waveform at $t = 36$ (in $\Delta x/a_0$ units) when no artificial selective damping terms are included. The initial pulse profile is a Gaussian with $h = 1.0$ and $b = 12$. The waveform is clearly dominated by spurious spatial oscillations bearing little resemblance to the exact solution. Figure 8.7 is the computed solution with artificial selective damping terms. R_{stencil} is set at 0.1 (for general application, $R_{\text{stencil}} = 0.06$ gives the best overall results). The numerical solution compares favorably with the exact solution with a fitted shock according to the Whitham (1974) method. Figure 8.8 shows the case with $h = 0.5$ and $b = 3$. At $t = 72$ the numerical solution without artificial selective damping consists of random spikes. The computed results seemingly have no relationship to

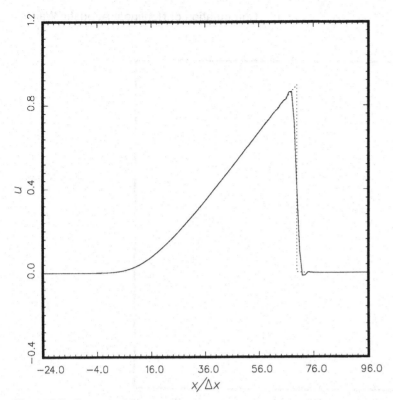

Figure 8.7. Computation of nonlinear acoustic pulse with variable artificial damping. $R_{\text{stencil}} = 0.1$, $t = 36$. Gaussian initial waveform with $h = 1.0$, $b = 12\Delta x$. ——, computed, $\cdots\cdots$ exact solution.

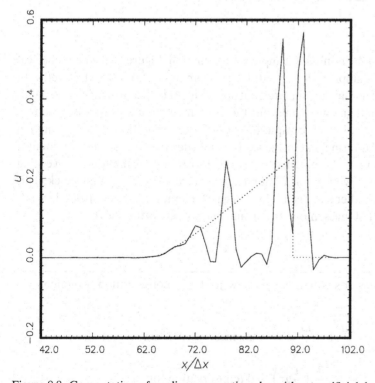

Figure 8.8. Computation of nonlinear acoustic pulse without artificial damping. $t = 72$. Gaussian initial waveform with $h = 0.5$, $b = 3\Delta x$. ——, computed, $\cdots\cdots$ exact solution.

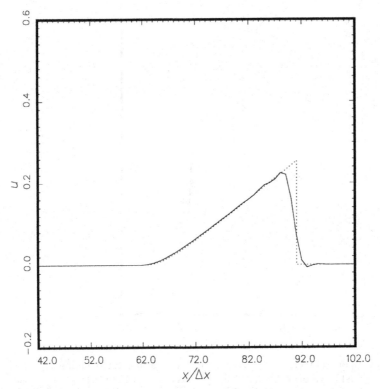

Figure 8.9. Computation of nonlinear acoustic pulse with variable artificial damping. $R_{\text{stencil}} = 0.1, t = 72$. Gaussian initial waveform with $h = 0.5, b = 3\Delta x$. ——, computed, $\cdots\cdots$ exact solution.

the exact solution of the nonlinear simple wave equation. Figure 8.9 is the numerical result with variable damping ($R_{\text{stencil}} = 0.1$). The computed shock matches well with the exact solution. The shock is smeared out approximately over five mesh spacings.

Concerning whether one should put the governing equations of gas dynamics in a conservation form before computing the solution using the variable damping method for shock capturing, extensive testing indicates that this is a requirement if an accurate solution is to be obtained. Test results reveal that the use of governing equations in nonconservation form often leads to a slower propagation shock. This is, perhaps, not altogether surprising since in gas dynamics the shock speed depends on the conservation of mass, momentum, and energy across the shock.

EXERCISES

8.1. The inviscid Burger's equation can be written in a conservation or nonconservation form

$$\frac{\partial u}{\partial t} + u \frac{\partial u}{\partial x} = 0 \ \text{(nonconservation form)}$$

$$\frac{\partial u}{\partial x} + \frac{\partial}{\partial x}\left(\frac{1}{2}u^2\right) = 0 \ \text{(conservation form)}$$

Solve these equations computationally using the variable damping method with the initial condition,

$$t = 0 \quad u = 0.5e^{-(\ln 2)\left(\frac{x}{5}\right)^2}$$

where $\Delta x = 1$.

Assess the accuracy of the two solutions, especially the location of the shock as a function of time by comparing the solutions with the shock-fitted solution following Whitham's equal area rule (see Whitham (1974) section 2.8).

8.2. Solve the simple nonlinear wave equation,

$$\frac{\partial u}{\partial t} + \left(1 + \frac{\gamma + 1}{2}u\right)\frac{\partial u}{\partial x} = 0,$$

with initial condition, $t = 0$

$$u = 0.5\exp\left[-(\ln 2)\left(\frac{x}{10\Delta x}\right)^2\right]$$

by adding variable damping to the numerical scheme. Compare the solutions obtained using a 7-point stencil and a 15-point stencil (using the same size damping stencil as the computation algorithm). Compare the time history of the shock with that obtained by Witham's equal area rule.

8.3. In dimensionless form with respect to the following scales:

$$\text{length scale} = \Delta x$$
$$\text{velocity scale} = a_{\text{ref}}$$
$$\text{time scale} = \Delta x/a_{\text{ref}}$$
$$\text{density scale} = \rho_{\text{ref}}$$
$$\text{pressure scale} = \rho_{\text{ref}}a_{\text{ref}}^2.$$

the Euler equations in one dimension are

$$\frac{\partial \rho}{\partial t} + \frac{\partial \rho u}{\partial x} = 0$$

$$\rho\left(\frac{\partial u}{\partial t} + u\frac{\partial u}{\partial x}\right) = -\frac{\partial p}{\partial x}$$

$$\frac{\partial \rho E}{\partial \tau} + \frac{\partial}{\partial x}\left[(\rho E + p)u\right] = 0,$$

where $E = \frac{p}{(\gamma-1)\rho} + \frac{1}{2}u^2$.

Solve the shock tube problem using the following initial conditions and a computation domain $-100 \leq x \leq 100$, $t = 0$, $u = 0$:

$$p = \begin{cases} 4.4 & , \quad x < -2 \\ 2.7 + 1.7\cos\left(\frac{(x+2)\pi}{4}\right), & -2 \leq x \leq 2 \\ 1 & , \quad x > 2 \end{cases}$$

$$\rho = (\gamma p)^{1/\gamma}, \qquad \gamma = 1.4.$$

Find the spatial distribution of p, ρ, and u at $t = 40, 50, 60,$ and 70. Compare your numerical solution with the standard shock tube solution.

9 Advanced Numerical Boundary Treatments

High-quality boundary conditions are an essential part of computational aeroacoustics (CAA). Because a computational domain is necessarily finite in size, numerical boundary conditions play several diverse roles in numerical simulation. First and foremost, they must assist any outgoing disturbances to leave the computational domain with little or no reflection. The alternative is to use a perfectly absorbing layer as a numerical boundary treatment. Such a layer absorbs all outgoing disturbances without reflection as in the case of an anechoic chamber. In addition, if the problem to be simulated involves incoming disturbances, then these disturbances must be generated by the boundary conditions prescribed at the outer boundary of the computational domain. Furthermore, if there are flows that are originated from outside the computational domain, they must be reproduced by the boundary treatment as well. In this chapter, methods to construct numerical boundary conditions that perform these various functions are discussed.

9.1 Boundaries with Incoming Disturbances

In many aeroacoustics problems, there are disturbances that enter the computational domain through the outer boundary. For example, in computing the scattering characteristics of an object, acoustic waves must be allowed to pass through the boundary of a computational domain in a specified direction and intensity. The scattering phenomenon produces scattered waves. These waves are radiated in all directions. They propagate to the far field as outgoing waves through the boundary of the computational domain. In this case, the boundary condition of the computational domain must take on the responsibility of generating the incoming sound and, at the same time, they also serve as radiation boundary conditions for the scattered waves. To handle the dual role, a split-variable method has been developed. The essence of this method is discussed below.

The equations governing the propagation of small-amplitude disturbances are the linearized Euler equations. Both the incoming disturbances and the outgoing scattered waves are solutions of these equations. However, in a scattering problem, incoming disturbances are known or specified, whereas the scattered outgoing waves are unknown. To illustrate the basic concept of the split-variable method, consider a two-dimensional problem with a uniform mean flow velocity u_0 in the x direction.

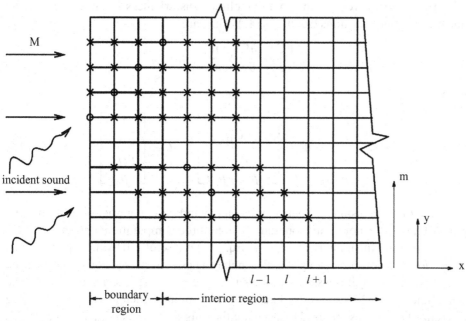

Figure 9.1. Computational mesh showing the boundary and the interior regions and the 7-point stencils for the DRP scheme. Incident sound waves enter the computational domain on the left.

It will be assumed that the incoming sound waves enter the computational domain from the left boundary as shown in Figure 9.1. The variables associated with the incoming waves are labeled by a subscript "in" and variables associated with the outgoing scattered waves are labeled by a subscript "out." In the left boundary region of the computational domain, the total variables are the direct sum of the incoming and outgoing disturbances; i.e.,

$$
\begin{bmatrix} \rho \\ u \\ v \\ p \end{bmatrix} = \begin{bmatrix} \rho_{in} \\ u_{in} \\ v_{in} \\ p_{in} \end{bmatrix} + \begin{bmatrix} \rho_{out} \\ u_{out} \\ v_{out} \\ p_{out} \end{bmatrix}. \tag{9.1}
$$

In Eq. (9.1) the variables associated with the incoming disturbances are known. They are prescribed by the physical problem. The aim is to compute the outgoing scattered waves. In the boundary region (see Figure 9.1), the outgoing acoustic waves are governed by the radiation boundary condition or Eq. (6.2). That is,

$$
\left[\frac{1}{V(\theta)} \frac{\partial}{\partial t} + \cos\theta \frac{\partial}{\partial x} + \sin\theta \frac{\partial}{\partial y} + \frac{1}{2(x^2 + y^2)^{\frac{1}{2}}} \right] \begin{bmatrix} \rho_{out} \\ u_{out} \\ v_{out} \\ p_{out} \end{bmatrix} = 0 + 0\left(r^{-\frac{5}{2}}\right), \tag{9.2}
$$

where $V(\theta) = [u_0 \cos\theta + (a_0^2 - u_0^2 \sin^2\theta)^{1/2}]$ and a_0 is the speed of sound.

In the interior region, the small-amplitude disturbances are governed by the linearized Euler equations (see Eq. (5.1)). They are

$$\frac{\partial \mathbf{U}}{\partial t} + \frac{\partial \mathbf{E}}{\partial x} + \frac{\partial \mathbf{F}}{\partial y} = 0,$$

where

$$\mathbf{U} = \begin{bmatrix} \rho \\ u \\ v \\ p \end{bmatrix}, \quad \mathbf{E} = \begin{bmatrix} \rho_0 u + \rho u_0 \\ u_0 u + p/\rho_0 \\ u_0 v \\ u_0 p + \gamma p_0 u \end{bmatrix}, \quad \mathbf{F} = \begin{bmatrix} \rho_0 v \\ 0 \\ p/\rho_0 \\ \gamma p_0 v \end{bmatrix}. \tag{9.3}$$

Note that the distinction between incoming and outgoing waves is lost once the point of interest is not in the boundary region of the computational mesh.

To compute the numerical solution, Eq. (9.3) is discretized by the 7-point stencil dispersion-relation-preserving (DRP) scheme. These discretized equations are used to determine the full physical variables (ρ, u, v, p) in the interior region. At the same time Eq. (9.2) is discretized again by the 7-point stencil DRP scheme. These discretized equations are used to calculate the outgoing waves (ρ_{out}, u_{out}, v_{out}, p_{out}). They are applied only in the boundary region. Before the computation can be implemented, however, one must recognize that once the 7-point stencil DRP scheme is applied to Eq. (9.2), the finite difference stencil in the x direction would extend outside the boundary region into the interior region as shown in Figure 9.1. In the interior region, only the full variables (ρ, u, v, p) are computed. But (ρ_{out}, u_{out}, v_{out}, p_{out}) are needed at the extended stencil points to support the computation. To find these quantities, one may use Eq. (9.1). Since the incoming disturbances are known, the outgoing wave variables can be found by subtracting the incoming disturbance variables from the full variables at mesh points near the outer boundary of the interior region.

Now, in computing the full variables (ρ, u, v, p) in the interior region, one would encounter a similar problem. For mesh points in the three columns closest to the boundary region, the 7-point stencil will extend into the boundary region where only the outgoing disturbances are computed. To find the values of the full variables in the boundary region, one may again use Eq. (9.1) by adding the known incoming disturbances to the computed outgoing waves. In this way, the variables in the entire computational domain can be marched forward in time. It is straightforward to see that the split-variable method essentially converts the boundary region into a source of the incoming disturbances. At the same time, it acts as radiation boundary condition for the outgoing scattered waves.

9.2 Entrainment Flow

In performing numerical simulation of practical problems, sometimes one is forced to use a relatively small computation domain. This may be because of computer memory constraint or the need to reduce computation time. In these cases, flow into or out of the scaled down computational domain could be due to the influence of sources or forces located outside as well as inside the computational domain.

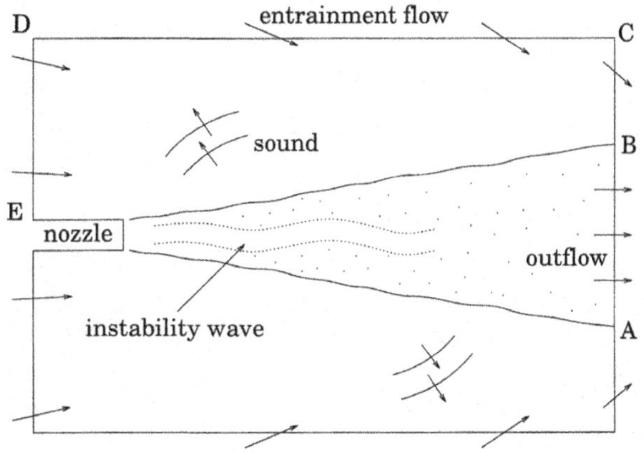

Figure 9.2. Computational domain for numerical simulation of jet noise generation.

To maintain an accurate simulation, it is necessary to develop special boundary conditions to account for these effects on the flow inside the computational domain.

As an example, consider simulating a jet flow and noise radiation as shown in Figure 9.2. For practical reasons, the size of the computational domain is typically thirty to forty jet diameters in the axial direction and twenty to thirty diameters in the radial direction or smaller. These dimensions are somewhat smaller than those of a typical anechoic chamber used in physical experiments. Because of the proximity of the computation boundary to the jet flow, the boundary conditions along boundary $BCDE$ are burdened with multiple tasks. Obviously, the boundary conditions must be transparent to the outgoing acoustic waves radiating from the jet. In addition, the boundary conditions must impose the ambient conditions on the numerical simulation. In other words, they specify the static conditions far away from the jet. Furthermore, the jet entrains a large volume of ambient gas. The entrainment flow velocity at the computational boundary, although small, is not entirely negligible. For high-quality numerical simulation, the boundary condition must, therefore, allow an as yet unknown entrainment flow to enter or leave the computational domain smoothly as well.

In Section 6.5, a set of radiation boundary conditions for nonuniform mean flow is provided. Let $\bar{p}$, $\bar{u}$, $\bar{v}$, and $\bar{p}$ be the weakly nonuniform mean flow at the boundary region of the computational domain. The radiation boundary condition in three-dimensional cylindrical coordinates may be written as

$$\frac{1}{V(\theta, r)} \frac{\partial}{\partial t} \begin{bmatrix} \rho \\ u \\ v \\ w \\ p \end{bmatrix} + \left[\sin\theta \frac{\partial}{\partial r} + \cos\theta \frac{\partial}{\partial x} + \frac{1}{(r^2 + x^2)^{\frac{1}{2}}} \right] \begin{bmatrix} \rho - \bar{\rho} \\ u - \bar{u} \\ v - \bar{v} \\ w \\ p - \bar{p} \end{bmatrix} = 0, \quad (9.4)$$

where (r, ϕ, x) are the cylindrical coordinates, θ is the polar angle in spherical coordinates with the x-axis (flow direction of the jet) as the polar axis, and (u, v, w) are the velocity components in the axial (x), radial (r), and azimuthal (ϕ) directions. $V(\theta, r) = \bar{u}\cos\theta + \bar{v}\sin\theta + [\bar{a}^2 - (\bar{v}\cos\theta - \bar{u}\sin\theta)^2]^{1/2}$, and $\bar{a}$ is the speed of sound.

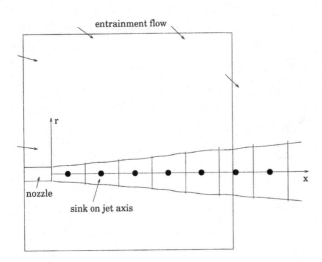

Figure 9.3. Determination of the entrainment flow of a jet by the point-sink approximation.

Note that the entrainment flow at the boundary region of the computation domain would be influenced by the jet flow outside the computational domain as well as inside. To develop an asymptotic entrainment flow solution, a simple way is to divide the jet into many slices as shown in Figure 9.3. The energetic portion of a jet may extend, depending on jet Mach number, beyond the computation domain to about forty diameters downstream. The mass fluxes across the boundaries of each jet slice may be found from available empirical jet flow data. The difference in mass fluxes at the two ends of each slice of the jet gives the amount of entrainment flow for the particular slice. This entrainment may be simulated by a point sink located at the center of the slice. The asymptotic solutions for a point sink located on the x-axis at x_s in a compressible fluid is given by (a subscript "e" is used to indicate entrainment flow)

$$\frac{\rho_e}{\rho_\infty} = 1 - \frac{Q^2}{32\pi^2[(x - x_s)^2 + r^2]^2} + \cdots$$

$$\frac{u_e}{a_\infty} = -\frac{Q(x - x_s)}{4\pi[(x - x_s)^2 + r^2]^{\frac{3}{2}}} + \cdots$$

$$\frac{v_e}{a_\infty} = -\frac{Qr}{4\pi[(x - x_s)^2 + r^2]^{\frac{3}{2}}} + \cdots$$

$$\frac{p_e}{\rho_\infty a_\infty^2} = \frac{1}{\gamma} - \frac{Q^2}{32\pi^2[(x - x_s)^2 + r^2]^2} + \cdots, \qquad (9.5)$$

where ρ_∞, a_∞, and γ are the ambient gas density, the sound speed, and the ratio of specific heats, and Q is the strength of the sink. It has dimensions of $\rho_\infty a_\infty D^2$ where D is the jet diameter. On replacing $(\overline{\rho}, \overline{u}, \overline{v}, \overline{p})$ of Eq. (9.4) by (ρ_e, u_e, v_e, p_e) of Eq. (9.5) and upon summing over the contributions from all the sinks to approximately $60D$ downstream from the nozzle exit, the desired radiation-entrainment flow boundary conditions are obtained.

Figure 9.4 shows the entrainment flow streamlines of a Mach 1.13 cold jet from a convergent nozzle computed using the radiation-entrainment flow boundary condition. It is worthwhile to point out that along the right-hand boundary BC, the mean

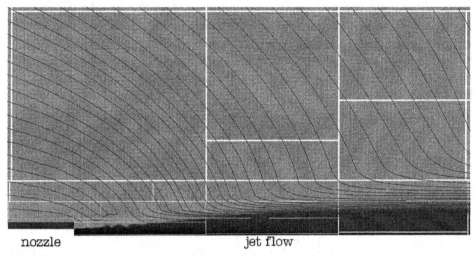

nozzle   jet flow

Figure 9.4 Computed streamlines of the entrainment flow around a supersonic screeching jet at Mach 1.13. White lines separate different computation blocks.

flow actually flows out of the computational domain, exactly as observed in free jet experiments. The streamline pattern would be vastly different had the entrainment flow outside the computational domain not been included in the sink flow computation. If the sinks outside the computational domain are excluded in forming ($\bar{\rho}$, $\bar{u}$, $\bar{v}$, $\bar{p}$) then a recirculation pattern would emerge. This, however, is inconsistent with experimental observations.

In this jet flow and noise radiation example, the jet may be subsonic and supersonic. For a subsonic jet, the nozzle exit pressure is nearly equal to ambient pressure, but for supersonic jets operating at off-design conditions, the pressure at the nozzle exit is quite different from that at ambient pressure. In this case, it is vital that the boundary conditions can communicate to the computation what the ambient pressure, density, and temperature are. Radiation boundary conditions (9.4) allow the specification of an ambient condition. The time-independent solution of (9.4) is $u = \bar{u}$, $v = \bar{v}$, $\rho = \bar{\rho}$, and $p = \bar{p}$. Many other radiation boundary conditions, designed with different objectives in mind, do not allow the imposition of an independent ambient condition. If these boundary conditions are used for computing an imperfectly expanded jet, the jet will not expand properly in the simulation.

9.3 Outflow Boundary Conditions: Further Consideration

In many CAA simulations, the boundary of the computation domain cuts across an outflow. An example is the case of simulating jet noise generation as shown in Figure 9.2. Since numerical boundary conditions imposed at the outflow boundary must reproduce all the external effects on the flow field inside the computational domain, the outflow boundary is usually placed sufficiently far downstream of the nozzle exit to avoid significant upstream influence of the jet flow left outside. In the jet noise simulation problem, as in many other problems, the mean flow is not known a priori. The numerical boundary conditions are required to capture the mean flow as well as to be transparent to the outgoing acoustic, vorticity, and entropy waves.

At present, many CAA boundary conditions are formulated under the assumption that the mean flow is known. Their main concern is to avoid any reflection of the outgoing unsteady disturbances. Dong (1997) recognized the need for CAA outflow boundary conditions to be capable of capturing the mean flow correctly as well. He proposed a set of asymptotic outflow boundary conditions based on sink/source flows. In two dimensions, these conditions are as follows:

$$\frac{1}{a_\infty}\frac{\partial \rho}{\partial t} + \frac{\partial \rho}{\partial r} + \frac{2}{r}(\rho - \rho_\infty) = 0$$

$$\frac{1}{a_\infty}\frac{\partial u}{\partial t} + \frac{\partial u}{\partial r} + \frac{u}{r} = 0$$

$$\frac{1}{a_\infty}\frac{\partial v}{\partial t} + \frac{\partial v}{\partial r} + \frac{v}{r} = 0$$

$$\frac{1}{a_\infty}\frac{\partial p}{\partial t} + \frac{\partial p}{\partial r} + \frac{2}{r}(p - p_\infty) = 0, \tag{9.6}$$

where r is the radial distance in polar coordinates, and p_∞, ρ_∞, and a_∞ are the ambient pressure, density, and sound speed, respectively. Dong derived this set of outflow boundary conditions by noting that the asymptotic solution of a time-independent localized source in two dimensions, denoted by an overbar, is

$$\overline{\rho} = \rho_\infty - \frac{Q^2}{8\pi^2 \gamma p_\infty r^2} + O\left(\frac{1}{r^4}\right)$$

$$\overline{v}_r = \frac{Q}{2\pi^2 \rho_\infty r} + O\left(\frac{1}{r^3}\right)$$

$$\overline{p} = p_\infty - \frac{Q^2}{8\pi^2 \rho_\infty r^2} + O\left(\frac{1}{r^4}\right), \tag{9.7}$$

where Q is the source strength, γ is the ratio of specific heats, and v_r is the radical velocity component. It is easy to verify that Eq. (9.7) is a solution of Eq. (9.6) to order $O(r^{-4})$. On the other hand, the asymptotic solution of outgoing acoustic waves in two dimensions without mean flow (see Eq. (6.1)) is

$$\begin{bmatrix} \rho \\ u \\ v \\ p \end{bmatrix} = \frac{F\left(\frac{r}{a_\infty} - t, \theta\right)}{r^{\frac{1}{2}}} \begin{bmatrix} \frac{1}{a_\infty^2} \\ \frac{\hat{u}(\theta)}{\rho_\infty a_\infty} \\ \frac{\hat{v}(\theta)}{\rho_\infty a_\infty} \\ 1 \end{bmatrix} + O\left(r^{\frac{3}{2}}\right). \tag{9.8}$$

By substitution of asymptotic solution (9.8) into Eq. (9.6), it is easy to find that Eq. (9.6) is satisfied to order $O(r^{-3/2})$. It has been pointed out that Eq. (9.6) is very similar to the asymptotic boundary conditions of Chapter 6, Eq. (6.14). The major difference is in the numerical coefficients of the $1/r$ terms. It is straightforward to see that the asymptotic solution of the mean flow of a localized source and that of the outgoing acoustic waves have different dependences on r. Eq. (9.6) as an outflow boundary condition is a reasonable compromise.

Outflow boundary condition (9.6) may be generalized to three dimensions. The corresponding equations in spherical polar coordinates (R, θ, ϕ) are as follows:

$$\frac{1}{a_\infty}\frac{\partial \rho}{\partial t} + \frac{\partial \rho}{\partial R} + \frac{3(\rho - \rho_\infty)}{R} = 0$$

$$\frac{1}{a_\infty}\frac{\partial u}{\partial t} + \frac{\partial u}{\partial R} + \frac{2u}{R} = 0$$

$$\frac{1}{a_\infty}\frac{\partial v}{\partial t} + \frac{\partial v}{\partial R} + \frac{2v}{R} = 0$$

$$\frac{1}{a_\infty}\frac{\partial w}{\partial t} + \frac{\partial w}{\partial R} + \frac{2w}{R} = 0$$

$$\frac{1}{a_\infty}\frac{\partial p}{\partial t} + \frac{\partial p}{\partial R} + \frac{3(p - p_\infty)}{R} = 0. \tag{9.9}$$

Jet flows are not source flows. Boundary layer arguments as well as experimental measurements suggest that the static pressure across a jet far downstream is practically constant and is equal to the ambient pressure. In this case, $\bar{p} = p_\infty$ is a good mean flow pressure boundary condition. This is the steady-state solution of the jet outflow boundary condition. Now, if the outflow boundary conditions (6.13) are nonlinearized, i.e., replacing the linear convection term $u_0(\partial u/\partial x)$ by the full convection terms etc., it is easy to arrive at the following set of nonlinear outflow boundary conditions. Written in cylindrical coordinates (r, ϕ, x), they are

$$\frac{\partial \rho}{\partial t} + u\frac{\partial \rho}{\partial x} + v\frac{\partial \rho}{\partial r} + \frac{w}{r}\frac{\partial \rho}{\partial \phi} = \frac{1}{a^2}\left(\frac{\partial p}{\partial t} + u\frac{\partial p}{\partial x} + v\frac{\partial p}{\partial r} + \frac{w}{r}\frac{\partial p}{\partial \phi}\right) \tag{9.10a}$$

$$\frac{\partial u}{\partial t} + u\frac{\partial u}{\partial x} + v\frac{\partial u}{\partial r} + \frac{w}{r}\frac{\partial u}{\partial \phi} = -\frac{1}{\rho}\frac{\partial p}{\partial x} \tag{9.10b}$$

$$\frac{\partial v}{\partial t} + u\frac{\partial v}{\partial x} + v\frac{\partial v}{\partial r} + \frac{w}{r}\frac{\partial v}{\partial \phi} - \frac{w^2}{r} = -\frac{1}{\rho}\frac{\partial p}{\partial r} \tag{9.10c}$$

$$\frac{\partial w}{\partial t} + u\frac{\partial w}{\partial x} + v\frac{\partial w}{\partial r} + \frac{w}{r}\frac{\partial w}{\partial \phi} + \frac{vw}{r} = -\frac{1}{\rho}\frac{1}{r}\frac{\partial w}{\partial \phi} \tag{9.10d}$$

$$\frac{1}{V(\theta)}\frac{\partial p}{\partial t} + \frac{\partial p}{\partial R} + \frac{p - p_\infty}{R} = 0, \tag{9.10e}$$

where $V(\theta) = u\cos\theta + a(1 - \sin^2\theta)^{1/2}$ and $M = u/a$; $a = (\gamma p/\rho)^{1/2}$ is the speed of sound; and (R, θ, ϕ) are the spherical polar coordinates with the polar axis coinciding with the centerline of the jet. For the mean flow, $p = p_\infty$ is a solution for Eq. (9.10e). Eqs. (9.10b), (9.10c), and (9.10d) are identical to the momentum equations of the Euler equations. They should yield the correct velocity field. For a time-independent solution, Eq. (9.10a) reduces to

$$u\frac{\partial \rho}{\partial x} + v\frac{\partial \rho}{\partial r} + \frac{w}{r}\frac{\partial \rho}{\partial \phi} = 0. \tag{9.11}$$

Eq. (9.11) is the same as $\rho = $ constant along a streamline. This should be reasonably good for jet flow far downstream where there is no intense mixing. Thus, outflow boundary conditions (Eq. (9.10)), although originally designed for outgoing disturbances alone, are also capable of capturing the mean flow of a jet.

Now, outflow boundary conditions (9.9) and (9.10) are very different. Each is designed specifically for a class of outflows. If the mean flow is source-like, then use Eq. (9.9). On the other hand, if the mean flow is jet-like, then Eq. (9.10) would be more appropriate.

9.4 Axis Boundary Treatment

There are many aeroacoustics problems for which the use of a cylindrical coordinate system is the most natural choice for computing the solution. A simple example is the case of computing the noise from a circular jet. Another example is the numerical calculation of the propagation of acoustic waves inside a circular duct. The governing equations are the Navier-Stokes equations. When written in cylindrical coordinates, some terms of these equations with $1/r$ coefficient apparently could blow up at the axis. In other words, there is an apparent singularity (although not a true one) at the axis ($r = 0$) of the coordinate system. In this section, a way to treat such apparent singularity is discussed.

9.4.1 Linear Problem Involving a Single Azimuthal Fourier Component

Acoustic problems are often linear. If the problem is linear and involves only a single azimuthal Fourier component, say the nth mode, then usually it is advantageous to separate out the azimuthal dependence by assuming that all variables have the dependence of $e^{in\phi}$. On factoring out $e^{in\phi}$, the reduced set of equations effectively becomes two-dimensional. Let the mean flow be axisymmetric. Near the x-axis, the mean flow may be assumed to be locally uniform; i.e., $u = \overline{u}$, $\overline{v} = \overline{w} = 0$, $\rho = \overline{\rho}$, $p = \overline{p}$, and $T = \overline{T}$. The linearized equations of motion are

$$\frac{\partial \rho}{\partial t} + \overline{u}\frac{\partial \rho}{\partial x} + \overline{\rho}\nabla \cdot \mathbf{v} = 0 \tag{9.12}$$

$$\frac{\partial \mathbf{v}}{\partial t} + \overline{u}\frac{\partial \mathbf{v}}{\partial x} = -\frac{1}{\overline{\rho}}\nabla p + v_t \nabla^2 \mathbf{v} \tag{9.13}$$

$$c_v \overline{\rho}\left(\frac{\partial T}{\partial t} + \overline{u}\frac{\partial T}{\partial x}\right) + \overline{p}\nabla \cdot \mathbf{v} = k_t \nabla^2 T \tag{9.14}$$

$$p = \overline{\rho}RT + \rho R\overline{T}, \tag{9.15}$$

where v_t and k_t are the turbulent kinematic viscosity and thermal conductivity, if the flow is turbulent; otherwise, the molecular values should be used.

It can easily be shown that when Eqs. (9.12) to (9.15) are written in cylindrical coordinates (r, ϕ, x), it is an eighth-order differential system in r. Four of the solutions are bounded at $r = 0$. The other four are unbounded. Here, interest is confined to the bounded solutions alone.

In general, the velocity vector may be represented by a scalar potential Φ and vector potential $\mathbf{A}$, that is,

$$\mathbf{v} = \nabla \Phi + \nabla \times \mathbf{A}. \tag{9.16}$$

For Eqs. (9.12) to (9.15), the solutions for the scalar and vector potentials can be obtained separately. It turns out that the viscous solutions are related only to the vector potential. For convenience, the vector potential solutions will be referred to as viscous solutions.

Substituting Eq. (9.16) into Eqs. (9.12) to (9.15), it is easy to find that the governing equations for the scalar and vector potentials are

$$\frac{\partial \rho}{\partial t} + \bar{u}\frac{\partial \rho}{\partial x} + \bar{\rho}\nabla^2 \Phi = 0 \tag{9.17}$$

$$\frac{\partial \Phi}{\partial t} + \bar{u}\frac{\partial \Phi}{\partial x} = -\frac{1}{\rho}p + v_t \nabla^2 \Phi \tag{9.18}$$

$$\bar{\rho}c_v\left(\frac{\partial T}{\partial t} + \bar{u}\frac{\partial T}{\partial x}\right) + \bar{p}\nabla^2 \Phi = k_t \nabla^2 T \tag{9.19}$$

$$p = \bar{\rho}RT + \rho R\bar{T} \tag{9.20}$$

and

$$\left(\frac{\partial \mathbf{A}}{\partial t} + \bar{u}\frac{\partial \mathbf{A}}{\partial x}\right) = v_t \nabla^2 \mathbf{A}. \tag{9.21}$$

The viscous solutions are given by $\rho = p = T = 0, \mathbf{v} = \nabla \times \mathbf{A}$.

9.4.1.1 Scalar Potential Solutions

General solutions of the scalar potential, Φ, may be found by applying Fourier-Laplace transforms to x and t and expanding ϕ dependence in a Fourier series, that is,

$$\Phi(r, \phi, x, t) = \sum_{n=-\infty}^{\infty} \int_{-\infty}^{\infty} \int_{\Gamma} \tilde{\Phi}_n(r, k, \omega) e^{i(kx-\omega t+n\phi)} d\omega \, dk. \tag{9.22}$$

This reduces Eqs. (9.17) to (9.20) to a fourth-order differential system in r. The two solutions bounded at $r = 0$ may be expressed in terms of Bessel functions of order n. The complete solutions when written out in full are as follows:

$$\tilde{u}_n = ikJ_n(\lambda_\pm r), \quad \tilde{v}_n = \frac{d}{dr}J_n(\lambda_\pm r)$$

$$\tilde{w}_n = \frac{in}{r}J_n(\lambda_\pm r), \quad \tilde{\rho}_n = \frac{i\tilde{\rho}(\lambda_\pm^2 + k^2)}{(\omega - \bar{u}k)}J_n(\lambda_\pm r)$$

$$\tilde{p}_n = \bar{\rho}\left[i(\omega - \bar{u}k) - v_t(\lambda_\pm^2 + k^2)\right]J_n(\lambda_\pm r)$$

$$\tilde{T}_n = \left[\frac{i(\omega - \bar{u}k)}{R} - (\lambda_\pm^2 + k^2)\left(\frac{i\bar{T}}{\omega - \bar{u}k} + \frac{v_t}{R}\right)\right]J_n(\lambda_\pm r) \tag{9.23}$$

where

$$\lambda_\pm = \left[\frac{\alpha \pm (\alpha^2 + 4\beta)^{\frac{1}{2}}}{2} - k^2\right]^{\frac{1}{2}}$$

$$\alpha = \frac{ik_t(\omega - \bar{u}k)/R - \bar{p} - c_v\bar{\rho}\bar{T} + ic_v\bar{\rho}v_t[(\omega - \bar{u}k)/R]}{k_t[v_t/R + i\bar{T}/(\omega - \bar{u}k)]}$$

$$\beta = \frac{-c_v\bar{\rho}(\omega - \bar{u}k)^2}{Rk_t[v_t/R + i\bar{T}/(\omega - \bar{u}k)]}.$$

9.4.1.2 Viscous Solutions

There are two sets of viscous solutions. The first set is found by letting

$$\mathbf{A} = \psi \mathbf{e_x}, \qquad p = \rho = T = 0, \tag{9.24}$$

where $\mathbf{e_x}$ is the unit vectors in the x direction. Again, by the use of Fourier-Laplace transforms and Fourier expansion, the solution that is bounded at $r = 0$ is

$$\tilde{p}_n = \hat{\rho}_n = \tilde{T}_n = \tilde{u}_n = 0$$

$$\tilde{v}_n = \frac{in}{r} J_n \left\{ i \left[k^2 - \frac{i(\omega - \bar{u}k)}{\nu_t} \right]^{\frac{1}{2}} r \right\}$$

$$\tilde{w}_n = \frac{d}{dr} J_n \left\{ i \left[k^2 - \frac{i(\omega - \bar{u}k)}{\nu_t} \right]^{\frac{1}{2}} r \right\}. \tag{9.25}$$

The second set may be found by letting

$$\mathbf{A} = \chi \mathbf{e_r} - i\chi \mathbf{e_\phi}, \qquad p = \rho = T = 0. \tag{9.26}$$

On following these steps, the bounded solution, after some algebra, is found to be

$$\tilde{p}_n = \tilde{\rho}_n = \tilde{T}_n = 0$$

$$\tilde{u}_n = -i \left(\frac{d}{dr} J_{n+1} \left\{ i \left[k^2 - \frac{i(\omega - \bar{u}k)}{\nu_t} \right]^{\frac{1}{2}} r \right\} + \frac{n+1}{r} J_{n+1} \left\{ i \left[k^2 - \frac{i(\omega - \bar{u}k)}{\nu_t} \right]^{\frac{1}{2}} r \right\} \right)$$

$$\tilde{v}_n = -k J_{n+1} \left\{ i \left[k^2 - \frac{i(\omega - \bar{u}k)}{\nu_t} \right]^{\frac{1}{2}} r \right\}$$

$$\tilde{w}_n = ik J_{n+1} \left\{ i \left[k^2 - \frac{i(\omega - \bar{u}k)}{\nu_t} \right]^{\frac{1}{2}} r \right\}. \tag{9.27}$$

9.4.1.3 Analytic Continuation into the $r < 0$ Region

The general solution is a linear combination of the scalar and vector potential solutions. The Bessel functions of these solutions can be continued analytically into the nonphysical region $r < 0$. For positive r, the analytic continuation formula for integer-order Bessel function is

$$J_n(-\xi r) = (-1)^n J_n(\xi r). \tag{9.28}$$

By means of Eq. (9.28) and the preceding general solutions, it is straightforward to establish that

$$\tilde{\rho}_n(-r, x, t) = (-1)^n \rho_n(r, x, t)$$
$$p_n(-r, x, t) = (-1)^n p_n(r, x, t)$$
$$T_n(-r, x, t) = (-1)^n T_n(r, x, t)$$
$$u_n(-r, x, t) = (-1)^n u_n(r, x, t)$$
$$v_n(-r, x, t) = (-1)^{n+1} v_n(r, x, t)$$
$$w_n(-r, x, t) = (-1)^{n+1} w_n(r, x, t), \tag{9.29}$$

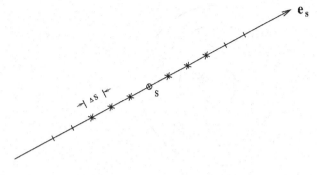

Figure 9.5. Approximating directional derivative in the $\mathbf{e}_s$ direction at s by a 7-point stencil finite difference quotient.

where ρ_n, p_n, etc. are the amplitude functions of the Fourier series expansions in ϕ.

Now, Eq. (9.29) may be used to extend the solution into the nonphysical negative r region to facilitate the computation of high-order large-stencil finite difference in r for points adjacent to the jet axis. That is, when a finite difference stencil extends into the negative r region, the values of the variables are found by Eq. (9.29). It is noted, however, that for $n = 1$,

$$\lim_{r \to 0} v_1\,(r, x, t) \neq 0, \qquad \lim_{r \to 0} w_1\,(r, x, t) \neq 0,$$

in general. For this reason, terms such as v_1/r and w_1/r cannot be computed at the jet axis $r = 0$. It is recommended that the values of v and w and all the other variables not be computed directly by the finite difference marching scheme at $r = 0$. Instead, the numerical solution at a new time level for all the other mesh points is first computed. The values of all the variables on mesh points in the region $r < 0$ is then calculated by analytical continuation. This leaves the values at $r = 0$ still to be determined. Here, it is suggested that they are to be found by high-order symmetric interpolation (see Chapter 11). Once the values of all physical variables at the axis $r = 0$ are found, the computation at the new time level is completed.

9.4.2 The General Case

In using cylindrical coordinates for two- or three-dimensional computation, there are two basic problems in computing the solution at and near $r = 0$. The first problem is how to approximate $\partial/\partial r$ by a finite difference quotient at mesh points close to $r = 0$. The second problem is how to calculate the solution at $r = 0$ where there is an apparent singularity. This section addresses these two issues in the general situation.

9.4.2.1 Directional Derivative
Let $\mathbf{e}_s$ be a unit vector in the s direction as shown in Figure 9.5. To approximate $\partial/\partial s$, one may use a 7-point finite difference quotient on a line in the $\mathbf{e}_s$ direction as shown; i.e.,

$$\left(\frac{\partial \Phi}{\partial s} \right)_0 = \frac{1}{\Delta s} \sum_{j=-3}^{3} a_j \Phi_j. \qquad (9.30)$$

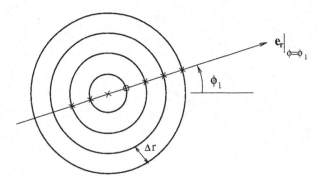

Figure 9.6. Approximating the r derivative at $\phi = \phi_1$, $r = \Delta r$ as directional derivative by a 7-point stencil finite difference quotient.

Here, $\partial \Phi / \partial s = \mathbf{e}_s \cdot \nabla \Phi$ is the directional derivative of Φ in the $\mathbf{e}_s$ direction. $j = 0$ is the point at which the derivative is to be computed.

9.4.2.2 r Derivative of a Scalar Variable Near $r = 0$

The r derivative of a variable Φ at $\phi = \phi_1$ and $r = \Delta r$ (see Figure 9.6) may be regarded as a directional derivative. That is,

$$
\left. \frac{\partial \Phi}{\partial r} \right|_{\phi=\phi_1, r=\Delta r} = \mathbf{e}_r \Big|_{\phi=\phi_1} \cdot \nabla \Phi \Big|_{r=\Delta r}. \tag{9.31}
$$

A finite difference approximation to the directional derivative of Eq. (9.31) may be formed in the same way as Eq. (9.30). Thus,

$$
\left. \frac{\partial \Phi}{\partial r} \right|_{\phi=\phi_1, r=\Delta r} = \frac{1}{\Delta r} \sum_{j=-3}^{3} a_j \Phi \left((j+1) \Delta r, \phi_1 \right), \tag{9.32}
$$

where the origin $r = 0$ is at $j = -1$. In this way, except for $r = 0$, there is no problem in computing the solution to the discretized Navier-Stokes equations on a cylindrical coordinate mesh.

9.4.2.3 Scalar Variables at $r = 0$

Because the Navier-Stokes equations in cylindrical coordinates have apparent singularities at $r = 0$, the values of scalar flow variables cannot be advanced in time by means of the discretized form of the equations. To find the values of the scalar variables at $r = 0$, a simple way is to compute first the solution at all other mesh points in the computational domain. After this is done, the values of the flow variables at $r = 0$ may be found by using high-order multidimensional optimized interpolation based on the values of all the mesh points on the first three rings of the cylindrical mesh. Just an average of all the values of the variable on the first ring is often a good approximation. For this purpose, the multidimensional optimized interpolation method of Chapter 13 would be very useful.

9.4.2.4 r Derivative of the Velocity Field

Let the (x, r, ϕ) components of the velocity field be denoted by (u, v, w). The r derivative of u at points near the cylindrical axis may be treated exactly the same as if u is a scalar. However, care must be exercised in forming the finite difference approximation for $\frac{\partial v}{\partial r}$ and $\frac{\partial w}{\partial r}$.

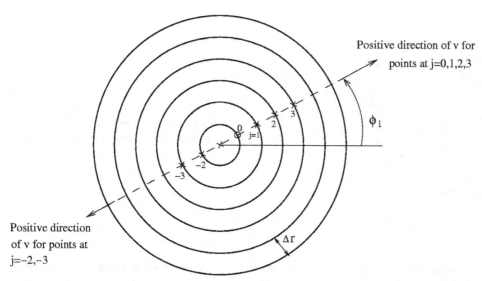

Figure 9.7. Forming the v and w r derivatives in the direction ϕ_1 at a point near $r = 0$ by finite difference approximation.

In a cylindrical coordinate system, v is positive in the positive direction of r. Now, to form a finite difference quotient to approximate the r derivative of v for a point at $r = \Delta r$ in the direction of ϕ_1, a 7-point symmetric stencil will extend to the other side of the origin at $r = 0$ as shown in Figure 9.7. For the points beyond $r = 0$, positive v is in the opposite direction. So, for the purpose of forming a directional derivative, we must use $-v$ for these points. Hence,

$$\frac{\partial v}{\partial r}\Big|_{\phi=\phi_1, r=\Delta r} \simeq \frac{1}{\Delta r}\left[\sum_{j=0}^{3} a_j v(j\Delta r, \phi_1) + a_{-1}v(r = 0, \phi_1) - \sum_{j=-3}^{-2} a_j v(j\Delta r, \phi_1)\right] \quad (9.33)$$

Similar reasoning leads to the formula for $\frac{\partial w}{\partial r}$ as follows:

$$\frac{\partial w}{\partial r}\Big|_{\phi=\phi_1, r=\Delta r} \simeq \frac{1}{\Delta r}\left[\sum_{j=0}^{3} a_j w(j\Delta r, \phi_1) + a_{-1}w(r = 0, \phi_1) - \sum_{j=-3}^{-2} a_j w(j\Delta r, \phi_1)\right].$$

$$(9.34)$$

9.4.2.5 The Values of v and w at $r = 0$

The cylindrical axis is a singular line of the cylindrical coordinates. All the other points have a v and w velocity component. However, at $r = 0$, there is no radial or any azimuthal velocity component. In the $r - \phi$ plane, there is only a velocity vector $\mathbf{v}_{r=0}$ at $r = 0$.

At $r = 0$, the cylindrical coordinate system is singular. It is not recommended to compute the velocity $\mathbf{v}_{r=0}$ by means of the governing equations of motion. A good approximation is to determine $\mathbf{v}_{r=0}$ by interpolation from values of v and w of the first few rings of points around the origin $r = 0$ (see Figure 9.8). For mesh

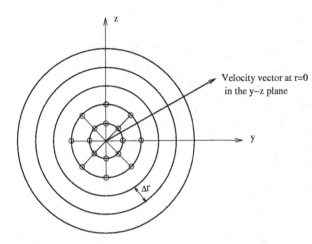

Figure 9.8. Determination of **v** at $r = 0$ by interpolation using values of (v, w) at mesh points in the first or first two rings.

points at angle $\phi = \phi_1$ and $r = n\Delta r$, the velocity components in the directions of the cylindrical coordiates and in the y and z direction are related by

$$v_y = v \cos \phi_1 - w \sin \phi_1, \quad v_z = v \sin \phi_1 + w \cos \phi_1.$$

Now, the value of v_y and v_z at $r = 0$ may be taken, as a first approximation, to be equal to the average of the values of v_y and v_z at all the mesh points of the first ring. A more accurate value of these two quantities may be obtained by applying the multidimensional interpolation procedure of Chapter 13 to the values at the mesh points on the first two rings of the grid.

9.5 Perfectly Matched Layer as an Absorbing Boundary Condition

Perfectly matched layer (PML) was invented by Berenger (1994, 1996) as an absorbing boundary condition for computational electromagnetics. The idea is to enclose a computational domain by a PML. In the PML, a modified set of governing equations are used. The modified equations, as designed, have the unusual characteristics that, when an outgoing disturbance impinges on the interface between the computational domain and the absorbing layer, no wave is reflected back. In other words, all outgoing disturbances are transmitted into the absorbing layer where they are damped out.

Hu (1996) was the first to apply PML successfully to aeroacoustic problems governed by the linearized Euler equations. In the beginning, the linearization was done over a uniform mean flow. Recently, Hu (2005) had extended his work to nonuniform mean flow. Tam *et al.* (1998) analyzed the original version of the PML equations. They pointed out that those equations supported spatially growing unstable solutions. This is because, in the presence of a mean flow, the PML equations give rise to dispersive waves. In the original version of PML, a small band of these waves has phase velocity in the opposite direction to the group velocity. Spatial damping in the PML is associated with the direction of the phase velocity. As a result, these waves grow in amplitude as they propagate across the PML instead of being damped. Hu (2001, 2004) has since resolved this instability problem. In this chapter, the more recent PML equations developed by Hu are presented and analyzed.

9.5.1 Derivation of the PML Equation

The PML equation for the linearized Euler equation (linearized over a uniform mean flow in the x direction) may be derived following three basic steps. To make the discussion of each of these steps as simple as possible, two-dimensional problems in Cartesian coordinates will first be considered. Let L be the length scale, a_0 (sound speed of the uniform mean flow) be the velocity scale, L/a_0 be the time scale, ρ_0 (gas density of mean flow) be the density scale, and $\rho_0 a_0^2$ be the pressure scale. The dimensionless linearized Euler equation may be written in the following form:

$$\frac{\partial \mathbf{u}}{\partial t} + \mathbf{A}\frac{\partial \mathbf{u}}{\partial x} + \mathbf{B}\frac{\partial \mathbf{u}}{\partial y} = 0, \tag{9.35}$$

where

$$\mathbf{u} = \begin{bmatrix} \rho \\ u \\ v \\ p \end{bmatrix}, \qquad \mathbf{A} = \begin{bmatrix} M & 1 & 0 & 0 \\ 0 & M & 0 & 1 \\ 0 & 0 & M & 0 \\ 0 & 1 & 0 & M \end{bmatrix}, \qquad \mathbf{B} = \begin{bmatrix} 0 & 0 & 1 & 0 \\ 0 & 0 & 0 & 0 \\ 0 & 0 & 0 & 1 \\ 0 & 0 & 1 & 0 \end{bmatrix}$$

M is the Mach number of the mean flow.

The first step is to make a change of variables from (x, y, t) to $(\bar{x}, \bar{y}, \bar{t})$. The variables are related by

$$\bar{x} = x, \qquad \bar{y} = (1 - M^2)^{\frac{1}{2}}y, \qquad \bar{t} = t + \frac{Mx}{1 - M^2}. \tag{9.36}$$

This transformation is necessary to make the PML equation stable. The linearized Euler equation now becomes

$$\left(\mathbf{I} + \frac{M}{1 - M^2}\mathbf{A}\right)\frac{\partial \mathbf{u}}{\partial \bar{t}} + \mathbf{A}\frac{\partial \mathbf{u}}{\partial \bar{x}} + (1 - M^2)^{\frac{1}{2}}\mathbf{B}\frac{\partial \mathbf{u}}{\partial \bar{y}} = 0. \tag{9.37}$$

In Eq. (9.37), $\mathbf{I}$ is the identity matrix. Assuming, for the time being, a time dependence of $e^{-i\bar{\omega}\bar{t}}$, Eq. (9.37) may be rewritten as

$$-i\bar{\omega}\left(\mathbf{I} + \frac{M}{1 - M^2}\mathbf{A}\right)\hat{\mathbf{u}} + \mathbf{A}\frac{\partial \hat{\mathbf{u}}}{\partial \bar{x}} + (1 - M^2)^{\frac{1}{2}}\mathbf{B}\frac{\partial \hat{\mathbf{u}}}{\partial \bar{y}} = 0, \tag{9.38}$$

where $\mathbf{u}(x, y, t) = \hat{\mathbf{u}}(x, y)e^{-i\bar{\omega}\bar{t}}$.

The second step is to introduce damping into the equation through a so-called *complex change of variables*. Let $\sigma_x(x) > 0$, $\sigma_y(y) > 0$ be the absorption coefficients. The complex change of variables is to let

$$\bar{x} \to \bar{x} + \frac{i}{\bar{\omega}}\int^{\bar{x}} \sigma_x(x)\,dx, \qquad \bar{y} \to \bar{y} + \frac{i}{\bar{\omega}}\int^{\bar{y}} \sigma_y(y)\,dy$$

so that

$$\frac{\partial}{\partial \bar{x}} \to \frac{1}{1 + \frac{i\sigma_x}{\bar{\omega}}}\frac{\partial}{\partial \bar{x}}, \qquad \frac{\partial}{\partial \bar{y}} \to \frac{1}{1 + \frac{i\sigma_y}{\bar{\omega}}}\frac{\partial}{\partial \bar{y}}.$$

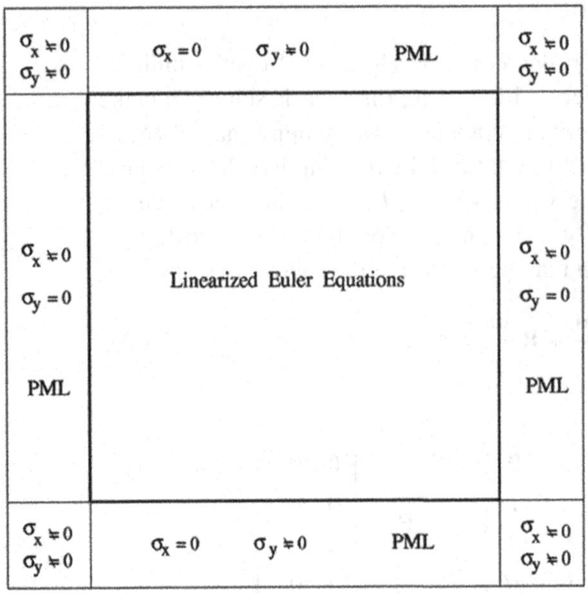

Figure 9.9. A Cartesian computational domain governed by the linearized Euler equations enclosed by PMLs.

On performing the complex change of variables on Eq. (9.38), it becomes

$$-i\overline{\omega}\left(\mathbf{I}+\frac{M}{1-M^2}\mathbf{A}\right)\hat{\mathbf{u}}+\frac{1}{1+\frac{i\sigma_x}{\overline{\omega}}}\mathbf{A}\frac{\partial\hat{\mathbf{u}}}{\partial\overline{x}}+\frac{(1-M^2)^{\frac{1}{2}}}{1+\frac{i\sigma_y}{\overline{\omega}}}\mathbf{B}\frac{\partial\hat{\mathbf{u}}}{\partial\overline{y}}=0. \qquad (9.39)$$

The final step is to recast Eq. (9.39) back into the original physical variables. This may be done by first multiplying the equation by $[1+(i\sigma_x/\overline{\omega})][1+(i\sigma_y/\overline{\omega})]$. In addition, it is convenient to introduce an auxiliary variable of $\hat{\mathbf{q}}$ defined by

$$\hat{\mathbf{q}}=\frac{i}{\overline{\omega}}\hat{\mathbf{u}}. \qquad (9.40)$$

This yields

$$\left(\mathbf{I}+\frac{M}{1-M^2}\mathbf{A}\right)(-i\overline{\omega}+\sigma_x)\hat{\mathbf{u}}+\mathbf{A}\frac{\partial\hat{\mathbf{u}}}{\partial\overline{x}}+\sigma_y\hat{\mathbf{u}}+\sigma_y\mathbf{A}\left(\frac{-i\overline{\omega}M}{1-M^2}\hat{\mathbf{q}}+\frac{\partial\hat{\mathbf{q}}}{\partial\overline{x}}\right)$$
$$+\sigma_x\sigma_y\left(\mathbf{I}+\frac{M}{1-M^2}\mathbf{A}\right)\hat{\mathbf{q}}+\sigma_x(1-M^2)^{\frac{1}{2}}\mathbf{B}\frac{\partial\hat{\mathbf{q}}}{\partial\overline{y}}+(1-M^2)^{\frac{1}{2}}\mathbf{B}\frac{\partial\hat{\mathbf{u}}}{\partial\overline{y}}=0. \qquad (9.41)$$

Upon restoring the time dependence and switching back to the original (x, y, t) coordinate system, Eq. (9.41) becomes

$$\frac{\partial\mathbf{u}}{\partial t}+\mathbf{A}\frac{\partial\mathbf{u}}{\partial x}+\mathbf{B}\frac{\partial\mathbf{u}}{\partial y}+(\sigma_x+\sigma_y)\mathbf{u}+\sigma_y\mathbf{A}\frac{\partial\mathbf{q}}{\partial x}+\sigma_x\mathbf{B}\frac{\partial\mathbf{q}}{\partial y}$$
$$+\sigma_x\sigma_y\mathbf{q}+\sigma_x\frac{M}{1-M^2}\mathbf{A}(\mathbf{u}+\sigma_y\mathbf{q})=0, \qquad (9.42)$$

where

$$\frac{\partial\mathbf{q}}{\partial t}=\mathbf{u}. \qquad (9.43)$$

Eqs. (9.42) and (9.43) are the PML equations for rectangular computational domain as shown in Figure 9.9.

It is worthwhile to point out that the auxiliary variable $\mathbf{q}$ is needed only in the PML domains. This is because the spatial derivative $\partial \mathbf{q}/\partial x$ disappears for the PML equation when $\sigma_y = 0$ and, similarly, $\partial \mathbf{q}/\partial y$ disappears when $\sigma_x = 0$. Thus, $\partial \mathbf{q}/\partial x$ is required only in a horizontal PML and $\partial \mathbf{q}/\partial y$ is required only a vertical PML. In short, there is no need to compute or store $\mathbf{q}$ in the Euler computation domain.

9.5.2 Perfectly Matching and Stability Consideration

Perfectly matching between the solutions of the PML equations and the linearized Euler equations is one of the most important properties of a PML. It is, however, not obvious from the derivation that the governing PML equation actually possesses this unusual property. Here, it is intended to confirm perfect matching. For this purpose, consider the interface between the Euler computational domain and the vertical PML on the right side of Figure 9.9. If will now be assumed that the computational domain is large compared with the size of the outgoing disturbances, so that the interface may be regarded as infinite in extent in the y direction as shown in Figure 9.10. For convenience, a Cartesian coordinate system centered at the interface will be used. In the Euler domain, the linearized Euler equation (9.35), when written out in full, becomes

$$\frac{\partial \rho}{\partial t} + M\frac{\partial \rho}{\partial x} + \frac{\partial u}{\partial x} + \frac{\partial v}{\partial y} = 0$$

$$\frac{\partial u}{\partial t} + M\frac{\partial u}{\partial x} + \frac{\partial p}{\partial x} = 0$$

$$\frac{\partial v}{\partial t} + M\frac{\partial v}{\partial x} + \frac{\partial p}{\partial y} = 0$$

$$\frac{\partial p}{\partial t} + M\frac{\partial p}{\partial x} + \frac{\partial u}{\partial x} + \frac{\partial v}{\partial y} = 0. \tag{9.44}$$

To allow the most general incident disturbances to impinge on the interface, one may choose to apply Fourier transform in y and Laplace transform in t to Eq. (9.44). This reduces Eq. (9.44) to a system of ordinary differential equations in x in the form ($\sim$ over a variable is used to indicate a Fourier-Laplace-transformed variable) as follows:

$$-i\omega\tilde{\rho} + M\frac{d\tilde{\rho}}{dx} + \frac{d\tilde{u}}{dx} + i\beta\tilde{v} = 0$$

$$-i\omega\tilde{u} + M\frac{d\tilde{u}}{dx} + \frac{d\tilde{p}}{dx} = 0$$

$$-i\omega\tilde{v} + M\frac{d\tilde{v}}{dx} + i\beta\tilde{p} = 0$$

$$-i\omega\tilde{p} + M\frac{d\tilde{p}}{dx} + \frac{d\tilde{u}}{dx} + i\beta\tilde{v} = 0. \tag{9.45}$$

In Eq. (9.45), β is the Fourier transform variable and ω is the Laplace transform variable. Eq. (9.45) is a fourth-order differential system in x. Four linearly

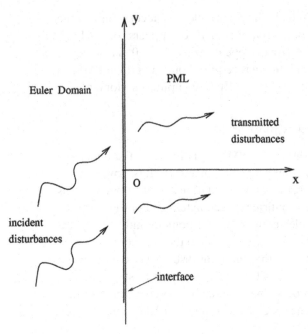

Figure 9.10. The interface region between the Euler computational domain and the PML on the right.

independent solutions can easily be found. They are given as follows:

(a) Entropy waves

$$\tilde{u} = \tilde{v} = \tilde{p} = 0, \qquad \tilde{\rho} = A(\beta, \omega)e^{i(\omega/M)x}. \tag{9.46}$$

The amplitude function $A(\beta, \omega)$ is arbitrary.

(b) Vorticity waves

$$\tilde{\rho} = \tilde{p} = 0, \qquad \tilde{u} = B(\beta, \omega)e^{i(\omega/M)x}, \qquad \tilde{v} = -\frac{\omega}{M\beta}B(\beta, \omega)e^{i(\omega/M)x}. \tag{9.47}$$

The amplitude function $B(\beta, \omega)$ is arbitrary.

(c) Acoustic waves

$$\begin{bmatrix} \tilde{\rho} \\ \tilde{u} \\ \tilde{v} \\ \tilde{p} \end{bmatrix} = C(\beta, \omega) \begin{bmatrix} (\omega - \lambda_+ M) \\ \lambda_+ \\ \beta \\ (\omega - \lambda_+ M) \end{bmatrix} e^{i\lambda_+ x}. \tag{9.48}$$

The amplitude function $C(\beta, \omega)$ is arbitrary and

$$\lambda_+ = \frac{-\omega M + [\omega^2 - \beta^2(1 - M^2)]^{\frac{1}{2}}}{1 - M^2}.$$

The branch cuts of the square root function in the ω plane are shown in Figure 9.11. Eq. (9.48) gives the acoustic waves that propagate or spread to the right. The left propagating acoustic wave solution is obtained by replacing the square root function by its negative value. Note that, for acoustic waves that propagate to the far field, λ_+ must be real. This requires $\omega^2 > \beta^2 (1 - M^2)$. The components for which $\omega^2 < \beta^2 (1 - M^2)$ are nonpropagating acoustic near field. They decay exponentially with increasing x.

The most general disturbances that exit the Euler domain on the right side to enter the PML are represented by a combination of entropy, vorticity, and acoustic

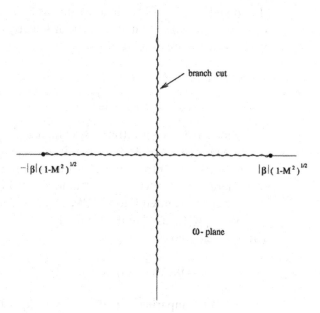

Figure 9.11. Branch cuts for the square root function $[\omega^2 - \beta^2(1 - M^2)]^{1/2}$ in the complex ω plane.

waves. It turns out that the PML equation supports a similar set of wave solutions. The PML equations, Eq. (9.42), when written out in full, are

$$\frac{\partial \rho}{\partial t} + M\frac{\partial \rho}{\partial x} + \frac{\partial u}{\partial x} + \frac{\partial}{\partial y}(v + \sigma_x q_3) + \sigma_x \rho + \frac{\sigma_x M}{1 - M^2}(M\rho + u) = 0$$

$$\frac{\partial u}{\partial t} + M\frac{\partial u}{\partial x} + \frac{\partial p}{\partial x} + \sigma_x u + \frac{\sigma_x M}{1 - M^2}(Mu + p) = 0$$

$$\frac{\partial v}{\partial t} + M\frac{\partial v}{\partial x} + \frac{\partial p}{\partial y} + \sigma_x \frac{\partial q_4}{\partial y} + \sigma_x v + \frac{\sigma_x M^2}{1 - M^2}v = 0$$

$$\frac{\partial p}{\partial t} + M\frac{\partial p}{\partial x} + \frac{\partial u}{\partial x} + \frac{\partial v}{\partial y} + \sigma_x \frac{\partial q_3}{\partial y} + \sigma_x p + \frac{\sigma_x M}{1 - M^2}(u + Mp) = 0$$

$$\frac{\partial q_3}{\partial t} = v$$

$$\frac{\partial q_4}{\partial t} = p. \quad (9.49)$$

The Fourier-Laplace transform of Eq. (9.49) in y and t is

$$-i\omega\tilde{\rho} + M\frac{d\tilde{\rho}}{dx} + \frac{d\tilde{u}}{dx} + i\beta\tilde{v}\left(\tilde{v} - \sigma_x\frac{\tilde{v}}{i\omega}\right) + \frac{\sigma_x}{1 - M^2}(\tilde{\rho} + M\tilde{u}) = 0$$

$$-i\omega\tilde{u} + M\frac{d\tilde{u}}{dx} + \frac{d\tilde{p}}{dx} + \frac{\sigma_x}{1 - M^2}(\tilde{u} + M\tilde{p}) = 0$$

$$-i\omega\tilde{v} + M\frac{d\tilde{v}}{dx} + i\beta\tilde{p} - \sigma_x\frac{\beta}{\omega}\tilde{p} + \frac{\sigma_x}{1 - M^2}\tilde{v} = 0$$

$$-i\omega\tilde{p} + M\frac{d\tilde{p}}{dx} + \frac{d\tilde{u}}{dx} + i\beta\tilde{v} - \sigma_x\frac{\beta}{\omega}\tilde{v} + \frac{\sigma_x}{1 - M^2}(\tilde{p} + M\tilde{u}) = 0. \quad (9.50)$$

It should be pointed out that at the interface $x = 0$, the y and t dependence of the transmitted waves must be the same as those of the incident waves. This requires that β and ω in Eq. (9.50) be the same as those in Eq. (9.45). Eq. (9.50) is a fourth-order differential system in x. The four linearly independent solutions, which can be verified easily, are as follows

(a) Entropy waves

$$\tilde{u} = \tilde{v} = \tilde{p} = 0, \quad \tilde{\rho} = \overline{A}(\beta, \omega)e^{\left(i\omega - \frac{\sigma_x}{1-M^2}\right)\frac{x}{M}}, \tag{9.51}$$

where $\overline{A}(\beta, \omega)$ is arbitrary. By comparing Eq. (9.46) and Eq. (9.51), it is clear by setting $\overline{A}(\beta, \omega) = A(\beta, \omega)$ that the entropy wave solution in the PML will match perfectly the entropy wave solution of the Euler domain at the interface $x = 0$. There is no reflected wave. In addition, Eq. (9.51) indicates that any entropy waves transmitted into the PML will be damped spatially with a damping rate of $\sigma_x/[M(1 - M^2)]$ in the x direction.

(b) Vorticity waves

$$\tilde{\rho} = \tilde{p} = 0, \quad \tilde{u} = \overline{B}(\beta, \omega)e^{\left(i\omega - \frac{\sigma_x}{1-M^2}\right)\frac{x}{M}}, \quad \tilde{v} = \frac{\omega}{M\beta}\overline{B}(\beta, \omega)e^{\left(i\omega - \frac{\sigma_x}{1-M^2}\right)\frac{x}{M}}. \tag{9.52}$$

By comparing Eq. (9.47) and Eq. (9.52), it is obvious that by setting $\overline{B}(\beta, \omega) = B(\beta, \omega)$, the vorticity waves in the Euler domain are completely transmitted into the PML without reflection at the interface $x = 0$. In the PML, the spatial damping rate for the vorticity waves is the same as that of the entropy waves.

(c) Acoustic waves

$$\begin{bmatrix} \tilde{\rho} \\ \tilde{u} \\ \tilde{v} \\ \tilde{p} \end{bmatrix} = \overline{C}(\beta, \omega) \begin{bmatrix} (\omega - \lambda_+ M) \\ \lambda_+ \\ \beta \\ (\omega - \lambda_+ M) \end{bmatrix} e^{i\lambda x}, \tag{9.53}$$

where $\overline{C}(\beta, \omega)$ is arbitrary and $\lambda = \lambda_+ + \frac{i\sigma_x}{\omega(1-M^2)}[\omega^2 - \beta^2(1 - M^2)]^{\frac{1}{2}}$.

By comparing Eq. (9.48) and Eq. (9.53), it is possible to conclude that there is perfect matching between the acoustic waves incident on the interface at $x = 0$ and the transmitted acoustic waves in the PML. Perfect matching is achieved by setting $\overline{C}(\beta, \omega) = C(\beta, \omega)$. This is true not only for propagating acoustic waves for which $\omega^2 > \beta^2(1 - M^2)$, but it is also true for nonpropagating near-field components with $\omega^2 < \beta^2(1 - M^2)$. The spatial damping rate in the PML is equal to $\frac{\sigma_x}{\omega(1-M)^2}[\omega^2 - \beta^2(1 - M^2)]^{\frac{1}{2}}$ in the x direction.

It was mentioned previously that the first version of PML equations were unstable. Now, it is easy to show that PML equation Eq. (9.49) does not support unstable solutions. To prove this assertion, one is required to show that there is no unstable acoustic wave solution in the PML. The simplest way to complete the proof is to take the Fourier transform of Eq. (9.50) in x. Eq. (9.50) is a homogeneous system. It is straightforward to find the condition for nontrivial solution is given by setting the determinant of the coefficient matrix to zero. This gives, after some algebraic simplifications,

$$F(\omega) = \frac{\beta^2}{1 - M^2}, \tag{9.54}$$

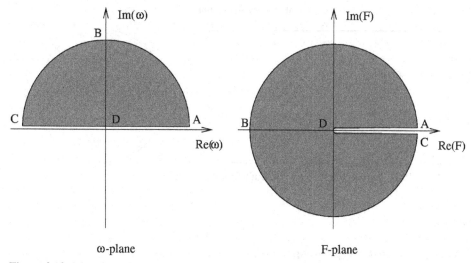

ω-plane   F-plane

Figure 9.12. Mapping of the upper half of the ω plane on to the F plane.

where

$$F(\omega) = \frac{\omega^2}{(1-M^2)^2} - \frac{\omega^2}{(\omega + i\sigma_x)^2}\left(\alpha + \frac{M\omega}{1-M^2}\right)^2. \tag{9.55}$$

In Eq. (9.55), α is the Fourier transform variable for x. Eq. (9.54) is the dispersion relation for the acoustic waves supported by the PML equations. To prove that there is no unstable acoustic wave solution, it is sufficient to show that dispersion relation (9.54) has no roots in the upper half ω plane for arbitrary α and β. This can be done by mapping the upper half ω plane into the complex F plane. Figure 9.12 gives the image in the F plane. Notice that there is no value in the upper half ω plane that makes F real and positive. Therefore, Eq. (9.54) cannot be satisfied because the right side of the equation is real and positive. Since both the entropy and vorticity waves are damped waves, PML Eq. (9.49) would only have damped wave solutions.

9.5.3 PML in Three Dimensions

In three dimensions, the linearized Euler equations may be written in a matrix form as

$$\frac{\partial \mathbf{u}}{\partial t} + \mathbf{A}\frac{\partial \mathbf{u}}{\partial x} + \mathbf{B}\frac{\partial \mathbf{u}}{\partial y} + \mathbf{C}\frac{\partial \mathbf{u}}{\partial z} = 0, \tag{9.56}$$

where

$$\mathbf{u} = \begin{bmatrix} \rho \\ u \\ v \\ w \\ p \end{bmatrix}, \quad \mathbf{A} = \begin{bmatrix} M & 1 & 0 & 0 & 0 \\ 0 & M & 0 & 0 & 1 \\ 0 & 0 & M & 0 & 0 \\ 0 & 0 & 0 & M & 0 \\ 0 & 1 & 0 & 0 & M \end{bmatrix}, \quad \mathbf{B} = \begin{bmatrix} 0 & 0 & 1 & 0 & 0 \\ 0 & 0 & 0 & 0 & 0 \\ 0 & 0 & 0 & 0 & 1 \\ 0 & 0 & 0 & 0 & 0 \\ 0 & 0 & 1 & 0 & 0 \end{bmatrix}, \quad \mathbf{C} = \begin{bmatrix} 0 & 0 & 0 & 1 & 0 \\ 0 & 0 & 0 & 0 & 0 \\ 0 & 0 & 0 & 0 & 0 \\ 0 & 0 & 0 & 0 & 1 \\ 0 & 0 & 0 & 1 & 0 \end{bmatrix}.$$

A straightforward application of the method of Section 9.5.1 yields the following PML equations:

$$\frac{\partial \mathbf{u}}{\partial t} + \mathbf{A}\frac{\partial \mathbf{u}_1}{\partial x} + \mathbf{B}\frac{\partial \mathbf{u}_2}{\partial y} + \mathbf{C}\frac{\partial \mathbf{u}_3}{\partial z} + \mathbf{u}_4 + \frac{\sigma_x M}{1-M^2}\mathbf{A}\mathbf{u}_1 = 0, \tag{9.57}$$

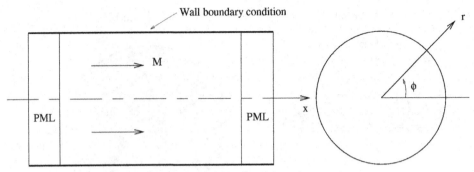

Figure 9.13. PML for circular ducted computational domain.

where

$$\mathbf{u_1} = \mathbf{u} + (\sigma_y + \sigma_z)\mathbf{q_1} + \sigma_y\sigma_z\mathbf{q_2}$$
$$\mathbf{u_2} = \mathbf{u} + (\sigma_z + \sigma_x)\mathbf{q_1} + \sigma_z\sigma_x\mathbf{q_2}$$
$$\mathbf{u_3} = \mathbf{u} + (\sigma_x + \sigma_y)\mathbf{q_1} + \sigma_x\sigma_y\mathbf{q_2}$$
$$\mathbf{u_4} = (\sigma_x + \sigma_y + \sigma_z)\mathbf{u} + (\sigma_y\sigma_z + \sigma_z\sigma_x + \sigma_x\sigma_y)\mathbf{q_1} + \sigma_x\sigma_y\sigma_z\mathbf{q_2}.$$

The auxiliary variables $\mathbf{q_1}$ and $\mathbf{q_2}$ are given by

$$\frac{\partial \mathbf{q_1}}{\partial t} = \mathbf{u}, \qquad \frac{\partial \mathbf{q_2}}{\partial t} = \mathbf{q_1}. \qquad (9.58)$$

Notice that the absorption coefficients σ_x, σ_y, and σ_z are arbitrary functions of x, y, and z, respectively. Again, as in the two-dimensional case, the auxiliary variables, $\mathbf{q_1}$ and $\mathbf{q_2}$ need to be computed only in the PML regions.

The PML absorbing boundary condition is especially useful in a ducted environment. It is very effective for simulating a long-duct termination. Figure 9.13 shows a computational domain inside a circular duct with rigid walls. In cylindrical coordinates, (x, r, ϕ), the linearized Euler equations may be written as

$$\frac{\partial \mathbf{u}}{\partial t} + \mathbf{A}\frac{\partial \mathbf{u}}{\partial x} + \mathbf{B}\frac{\partial \mathbf{u}}{\partial r} + \mathbf{C}\frac{1}{r}\frac{\partial \mathbf{u}}{\partial \phi} + \frac{1}{r}\mathbf{Du} = 0, \qquad (9.59)$$

where

$$\mathbf{u} = \begin{bmatrix} \rho \\ u \\ v \\ w \\ p \end{bmatrix}, \quad \mathbf{A} = \begin{bmatrix} M & 1 & 0 & 0 & 0 \\ 0 & M & 0 & 0 & 1 \\ 0 & 0 & M & 0 & 0 \\ 0 & 0 & 0 & M & 0 \\ 0 & 1 & 0 & 0 & M \end{bmatrix}, \quad \mathbf{B} = \begin{bmatrix} 0 & 0 & 1 & 0 & 0 \\ 0 & 0 & 0 & 0 & 0 \\ 0 & 0 & 0 & 0 & 1 \\ 0 & 0 & 0 & 0 & 0 \\ 0 & 0 & 1 & 0 & 0 \end{bmatrix},$$

$$\mathbf{C} = \begin{bmatrix} 0 & 0 & 0 & 1 & 0 \\ 0 & 0 & 0 & 0 & 0 \\ 0 & 0 & 0 & 0 & 0 \\ 0 & 0 & 0 & 0 & 1 \\ 0 & 0 & 0 & 1 & 0 \end{bmatrix}, \quad \mathbf{D} = \begin{bmatrix} 0 & 0 & 1 & 0 & 0 \\ 0 & 0 & 0 & 0 & 0 \\ 0 & 0 & 0 & 0 & 0 \\ 0 & 0 & 0 & 0 & 0 \\ 0 & 0 & 1 & 0 & 0 \end{bmatrix}.$$

The corresponding PML equation is

$$\frac{\partial \mathbf{u}}{\partial t} + \mathbf{A}\frac{\partial \mathbf{u}}{\partial x} + \mathbf{B}\frac{\partial \mathbf{u}_5}{\partial r} + \mathbf{C}\frac{1}{r}\frac{\partial \mathbf{u}_5}{\partial \phi} + \frac{1}{r}\mathbf{D}\mathbf{u}_5 + \mathbf{u}_6 = 0, \qquad (9.60)$$

where

$$\mathbf{u}_5 = \mathbf{u} + \sigma_x \mathbf{q}_1, \quad \mathbf{u}_6 = \sigma_x \mathbf{u}, \quad \text{and} \quad \frac{\partial \mathbf{q}_1}{\partial t} = \mathbf{u}.$$

In applying PML to a ducted domain, the wall boundary condition for the PML is the same as that for the Euler domain. For instance, for the rigid wall circular duct problem of Figure 9.13, the same rigid wall boundary condition, namely, $v = 0$, at $r = D/2$ is to be used in the PML regions.

In implementing PML, the damping coefficients, say, $\sigma_x(x)$, $\sigma_y(y)$, are often taken as smooth functions. A good practice is to set these functions equal to zero at the interface with the Euler domain. It is also a good practice to let $\sigma_x(x)$ increase smoothly to a constant level in the main part of the PML. At the termination of the PML, the computed variables are usually exponentially small so that reflection is not a concern. However, it is a good practice to impose a standard radiation boundary condition that is of minimal cost to the computation.

9.6 Boundaries with Discontinuities

In CAA, one often has to compute acoustic reflections or propagation over surfaces that are made up of different materials. At the junction of two materials, such as shown in Figure 9.14, the incident sound waves encounter a line discontinuity. Computationally, the grid size is, inevitably, finite. Because of the finite size mesh, experience indicates that spurious reflected waves are usually produced. To minimize spurious reflected waves, it is necessary to use a very fine computational mesh. This is sometimes a very costly remedy. It would be ideal to have a method that minimizes such spurious reflected waves without having to refine the computational mesh.

In jet engines, acoustic liners are, invariably, installed on the inside surface of the inlet and exhausted ducts, (see Motsinger and Kraft, 1991; Eversman, 1991). Acoustic liners are the most effective device for the suppression of engine fan noise at the present time. The junction between two acoustical liners with different impedance are examples of surface discontinuity. To facilitate the installation and maintenance of acoustic liners, hard wall splices are introduced. These splices are installed between two pieces of acoustic liners. Their presence creates additional surface discontinuities. To minimize the scattering of acoustic waves by hard wall splices, engine manufacturers resort to the use of very narrow splices. This poses a real dilemma for duct mode computation. In the absence of narrow splices, the use of a computational mesh with 7 to 8 mesh points per wave length in conjunction with a large-stencil CAA algorithm will have enough resolution for the computation. However, with narrow splices, the splice width is sometimes less than the size of a mesh spacing. This is as shown in Figure 9.15. In this case, the hard wall splice is invisible to the computation. Traditional wisdom will call for the use of much finer mesh. A factor of 10 or more mesh size reduction is often deemed necessary. This is to allow 8 to 10 mesh points to lie on the hard wall splice. This is, undoubtedly, very expensive. Not only the number of grid points has to increase substantially, but also the size of the time step used in the computation has to be significantly reduced. It

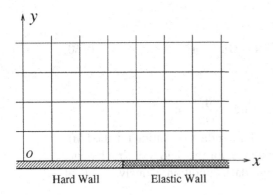

Figure 9.14. Discontinuity at the junction of two surface materials. Also shown is the computational mesh.

would, therefore, be extremely useful to have a method that can perform the narrow splice-scattering computation using the original coarse grid without mesh size reduction.

In this section, a method capable of computing acoustic wave propagation and reflection in the presence of surface discontinuities as shown in Figures (9.14) and (9.15) using a large-size computational mesh is discussed. The method is known as the "wave number truncation method" (see Tam and Ju, 2009).

Now consider the wall boundary condition shown in Figure 9.14. On the hard wall side of the junction, the rigid wall boundary condition is

$$v = 0, \tag{9.61}$$

where v is the velocity component normal to the wall. Let κ be the elastic constant of the elastic wall and ζ be the vertical displacement of the surface (positive in the direction of outward-pointing normal), then the elastic wall boundary condition is

$$\zeta = -\kappa (p - p_{\text{ref}}), \tag{9.62}$$

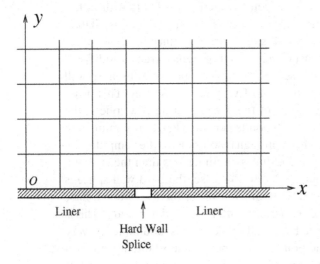

Figure 9.15. A narrow splice with width smaller than a grid spacing.

where p is the pressure and p_{ref} is the reference pressure for zero displacement. By differentiating Eq. (9.62) with respect to t, it is found that

$$\frac{\partial \zeta}{\partial t} = v = -\kappa \frac{\partial p}{\partial t}. \tag{9.63}$$

Boundary conditions (9.61) and (9.63) may be combined into a single boundary condition by means of the unit step function, $H(x)$. Let the wall junction be at $x = 0$, then the unified boundary condition at the wall is

$$v = -\kappa \frac{\partial p}{\partial t} H(x). \tag{9.64}$$

It is useful to consider boundary condition (9.64) in wave number space. The Fourier transform of boundary condition (9.64) is

$$\tilde{v}(\alpha) = -\frac{\kappa}{2\pi} \int_{-\infty}^{\infty} \tilde{H}(\alpha - k) \frac{\partial \tilde{p}}{\partial t}(k)\, dk. \tag{9.65}$$

The right side of Eq. (9.65) is a convolution integral. The Fourier transform of the unit step function is

$$\tilde{H}(\alpha) = -\frac{i}{2\pi \alpha}. \tag{9.66}$$

Thus, $\tilde{H}(\alpha)$ involves all wave numbers, including wave numbers in the short and ultrashort wave number range of the computational scheme (see Chapter 2, Section 2.2, and Figure 2.1). The convolution integral of Eq. (9.65) would, therefore, generate wave numbers outside the long wave range. In other words, because $\tilde{H}(\alpha)$ contains spurious waves with respect to the computational scheme, the surface discontinuity would scatter off spurious waves in an acoustic wave reflection computation.

A way to minimize the generation of spurious waves is to remove all the wave numbers in the short and ultrashort wave number range from $\tilde{H}(\alpha)$; i.e., wave number higher than the cutoff. Thus, a modified boundary condition is to replace $\tilde{H}(\alpha)$ by $\tilde{H}(\alpha)\,[H(\alpha + \alpha_c) - H(\alpha - \alpha_c)]$ in Eq. (9.65), where α_c is the cutoff wave number (see Figure 2.1). In physical space, this is tantamount to replacing $H(x)$ in Eq. (9.64) by $\hat{H}(x)$, where $\hat{H}(x)$ is the inverse Fourier transform of $\tilde{H}(\alpha)\,[H(\alpha + \alpha_c) - H(\alpha - \alpha_c)]$; i.e.,

$$\hat{H}(x) = \int_{-\infty}^{\infty} -\frac{i}{2\pi \alpha}[H(\alpha + \alpha_c) - H(\alpha - \alpha_c)]e^{i\alpha x}\, d\alpha$$

$$= \frac{1}{2} + \frac{1}{\pi} Si(\alpha_c x). \tag{9.67}$$

In Eq. (9.67), the $Si(z)$ function is defined as (see Abramowitz and Stegun, 1964 [Chapter 5, Section 5.2]),

$$Si(z) = \int_{0}^{z} \frac{\sin(y)}{y}\, dy. \tag{9.68}$$

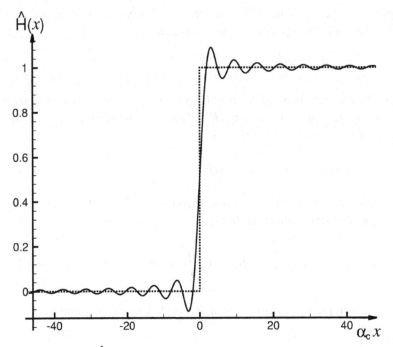

Figure 9.16. The $\hat{H}(x)$ function.

Figure 9.16 shows $\hat{H}(x)$ as a function of $\alpha_c x$. Now, it is proposed to use the following modified boundary condition instead of the exact boundary condition (9.64):

$$v = -\kappa \frac{\partial p}{\partial t} \hat{H}(x). \tag{9.69}$$

That is, $H(x)$ is replaced by $\hat{H}(x)$. The effectiveness and accuracy of the wave number truncation method and its extension (narrow scatterer) will be demonstrated by several examples in the last section of Chapter 10.

9.7 Internal Flow Driven by a Pressure Gradient

In many internal aeroacoustics problems, the mean flow is driven by a pressure gradient. For example, maintaining the nozzle flow in Figure 9.17 would require a pressure difference imposed at x_1 and x_2. Now, if it is required to impose at $x = x_1$ the boundary condition $\bar{p} = p_1$ and at $x = x_2$ the boundary condition $\bar{p} = p_2$ ($\bar{p}$ is the mean flow pressure), then it would be difficult to impose additional incoming wave or outgoing wave or absorbing boundary conditions at $x = x_1$ and $x = x_2$. To circumvent this pressure gradient boundary condition problem, it is recommended to perform a pressure gradient transformation to absorb the pressure gradient into the governing equations. In this way, it will remove the need for enforcing a static pressure boundary condition and thus allow the imposition of inflow/outflow or absorbing boundary conditions at the two ends of the computational domain.

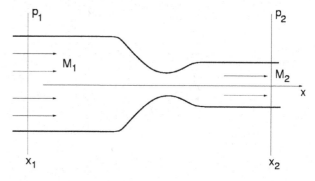

Figure 9.17. Nozzle flow problem requiring the imposition of a static pressure difference at x_1 and x_2.

Consider the nozzle flow problem of Figure 9.17. Mathematically, the pressure gradient field may be represented by

$$P(x) = p_1 + \frac{(p_2 - p_1)}{(x_2 - x_1)}(x - x_1) \equiv p_1 + \beta(x - x_1). \tag{9.70}$$

The pressure gradient is

$$\frac{dP}{dx} = \beta. \tag{9.71}$$

Now, the following transformation of variables will move the pressure gradient boundary condition into the differential equations. Let

$$\mathbf{v} = \mathbf{v}', \quad \rho = \rho', \quad p = P + p'. \tag{9.72}$$

Substitution of transformation (9.72) into the Navier-Stokes equations, the governing equations for $(\mathbf{v}', \rho', p')$ in Cartesian tensor notation are found to be

$$\frac{\partial \rho'}{\partial t} + \frac{\partial (\rho' v_j')}{\partial x_j} = 0$$

$$\rho' \left[\frac{\partial v_i'}{\partial t} + v_j' \frac{\partial v_i'}{\partial x_j} \right] = -\frac{\partial p'}{\partial x_i} - \beta \delta_{i1} + \frac{\partial \sigma_{ij}'}{\partial x_j}$$

$$\frac{\partial p'}{\partial t} + v_j' \frac{\partial p'}{\partial x_j} + v_j \delta_{j1} \beta + \gamma (P + p') \frac{\partial v_j'}{\partial x_j} = 0.$$

The velocity boundary conditions are unchanged by the transformation. The mean pressure boundary conditions for p' are

$$x = x_1, \quad p' = 0; \quad x = x_2, \quad p' = 0.$$

EXERCISES

9.1. Many aeroacoustic problems involve the scattering of vorticity or acoustic waves by an object. To simulate this class of problems in the time domain, the incoming waves are generated by the boundary conditions imposed on the artificial external boundaries of the computation domain.

Dimensionless variables with respect to length scale L (L will be chosen later), velocity scale a_∞ (ambient sound speed), time scale L/a_∞, density scale ρ_∞ (ambient density), and pressure scale $\rho_\infty a_\infty^2$ are used. Suppose there is a uniform mean flow

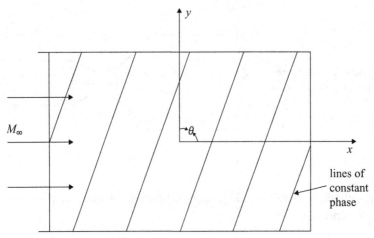

Figure 9.18. Lines of constant phase associated with plane waves in the computational plane.

of Mach number $M_\infty = 0.5$ in the x direction. For small-amplitude two-dimensional disturbances, the flow variables satisfy the linearized Euler equations as follows:

$$\frac{\partial u}{\partial t} + M_\infty \frac{\partial u}{\partial x} + \frac{\partial p}{\partial x} = 0$$

$$\frac{\partial v}{\partial t} + M_\infty \frac{\partial v}{\partial x} + \frac{\partial p}{\partial y} = 0$$

$$\frac{\partial p}{\partial t} + M_\infty \frac{\partial p}{\partial x} + \frac{\partial u}{\partial x} + \frac{\partial v}{\partial y} = 0. \tag{1}$$

Consider plane wave solutions of the form,

$$\begin{bmatrix} u \\ v \\ p \end{bmatrix} = \mathrm{Re} \left\{ \begin{bmatrix} \hat{u} \\ \hat{v} \\ \hat{p} \end{bmatrix} e^{i(\alpha x + \beta y - \omega t)} \right\}, \tag{2}$$

where Re{} is the real part of. Substitution of (2) into (1) gives

$$\begin{bmatrix} (\omega - M_\infty \alpha) & 0 & -\alpha \\ 0 & (\omega - M_\infty \alpha) & -\beta \\ -\alpha & -\beta & (\omega - M_\infty \alpha) \end{bmatrix} \begin{bmatrix} \hat{u} \\ \hat{v} \\ \hat{p} \end{bmatrix} = 0. \tag{3}$$

The determinant of matrix of Eq. (3) must be equal to zero. This yields the dispersion relation

$$(\omega - M_\infty \alpha)[(\omega - M_\infty \alpha)^2 - \alpha^2 - \beta^2] = 0. \tag{4}$$

The dispersion relation for vorticity waves is given by setting the first factor of (4) equal to zero, namely,

$$\omega - M_\infty \alpha = 0. \tag{5}$$

The lines of constant phase of the plane waves (see Figure 9.18) are given by

$$\alpha x + \beta y + \omega t = \text{constant}.$$

Let the slope of these lines be denoted by θ. Thus,

$$\frac{dy}{dx} = -\frac{\alpha}{\beta} = \tan\theta. \tag{6}$$

The solution of (5) and (6) yields

$$\alpha = \frac{\omega}{M_\infty}, \quad \beta = -\frac{\omega\cot\theta}{M_\infty} \tag{7}$$

The mode shape is given by the eigenfunction of Eq. (3). Upon replacing α and β by Eq. (7), the vorticity wave solution may be written as

$$\begin{bmatrix} u \\ v \\ p \end{bmatrix} = \varepsilon_v \begin{bmatrix} \cos\theta \\ \sin\theta \\ 0 \end{bmatrix} \cos\left[\frac{\omega}{M_\infty}(x - \cot\theta\, y - M_\infty t)\right]. \tag{8}$$

The total wave number of vorticity wave in Eq. (8) is equal to $\omega/M_\infty \sin\theta$. For convenience, let us set the total wave number to 2π or $\omega = 2\pi\, M_\infty \sin\theta$. This is the same as assigning the length scale L to be equal to the wavelength.

Use a 200×200 mesh as the computation domain. Choose a suitable size Δx and Δy. Propagate a plane vorticity wave across the computation domain with $\theta = 60°$ and $\varepsilon_v = 10^{-5}$. You may add artificial selective damping if needed. Compare your numerical result with the exact solution. Find the maximum error over the entire computation domain over a period of oscillation.

9.2. This is a continuation of Problem 9.1. Now suppose one is interested in acoustic wave propagation. The dispersion relation is

$$(\omega - M_\infty \alpha)^2 - \alpha^2 - \beta^2 = 0$$

or

$$\omega = M_\infty \alpha + (\alpha^2 + \beta^2)^{\frac{1}{2}} \tag{1}$$

(the negative sign of the square root gives a second acoustic mode).

Let V and ϕ be the velocity and direction of propagation. Then, from the group velocity, it is easy to find

$$\frac{\partial\omega}{\partial\alpha} = M_\infty + \frac{\alpha}{(\alpha^2 + \beta^2)^{1/2}} = V\cos\phi \tag{2}$$

$$\frac{\partial\omega}{\partial\beta} = \frac{\beta}{(\alpha^2 + \beta^2)^{1/2}} = V\sin\phi. \tag{3}$$

If the direction of propagation ϕ is specified (see Figure 9.19), then the solution of Eqs. (1) to (3) gives

$$V = M_\infty \cos\phi + \left(1 - M_\infty^2 \sin^2\phi\right)^{\frac{1}{2}} \tag{4}$$

$$\alpha = \frac{\omega(V\cos\phi - M_\infty)}{1 + M_\infty(V\cos\phi - M_\infty)} \tag{5}$$

$$\beta = \frac{\omega V\sin\phi}{1 + M_\infty(V\cos\phi - M_\infty)}. \tag{6}$$

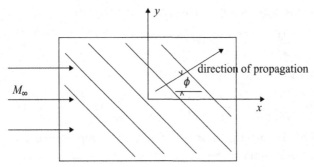

Figure 9.19. Direction of propagation of plane waves.

The eigenfunction for the acoustic wave is given by the solution of Eq. (3) of Problem (9.1). It may be written in the following form:

$$\begin{bmatrix} u \\ v \\ p \end{bmatrix} = \varepsilon_a \begin{bmatrix} V\cos\phi - M_\infty \\ V\sin\phi \\ 1 \end{bmatrix} \cos(\alpha x + \beta y - \omega t),\tag{7}$$

where V, α, and β are given by Eqs. (4), (5), and (6), and ε_a is the amplitude of the acoustic wave. It is easy to show that the total wave number is equal to $\omega/[1 + M_\infty(V\cos\phi - M_\infty)]$. Thus, by taking $\omega = 2\pi[1 + M_\infty(V\cos\phi - M_\infty)]$, the acoustic wavelength is chosen as the length scale.

Use a 200×200 mesh as the computation domain. Choose a suitable mesh size, propagate a plane acoustic wave train across the computation domain with $\varphi = 30°$ and $\varepsilon_a = 10^{-5}$. Compare your numerical results with the exact solution. Find the error distribution across the computation domain.

9.3. Many flow and acoustic problems are quasi-two-dimensional. This happens when the flow spreads out very slowly. A good example is the case of a jet. Locally, it makes sense to assume that the jet flow is parallel, that is, the velocity is parallel and independent of the axial coordinate x.

Consider the scattering of an incident plane acoustic wave by a jet. Let the mean axial velocity and the density of the jet be $\bar{u}(y, z)$ and $\bar{\rho}(y, z)$, locally parallel flow assumed. $\bar{p}$ is a constant. The linearized Euler equations are

$$\frac{\partial u}{\partial t} + \bar{u}\frac{\partial u}{\partial x} + v\frac{\partial \bar{u}}{\partial y} + w\frac{\partial \bar{u}}{\partial z} + \frac{1}{\bar{\rho}}\frac{\partial p}{\partial x} = 0$$

$$\frac{\partial v}{\partial t} + \bar{u}\frac{\partial v}{\partial x} + \frac{1}{\bar{\rho}}\frac{\partial p}{y} = 0$$

$$\frac{\partial w}{\partial t} + \bar{u}\frac{\partial w}{\partial x} + \frac{1}{\bar{\rho}}\frac{\partial p}{z} = 0$$

$$\frac{\partial p}{\partial t} + \bar{u}\frac{\partial p}{\partial x} + \lambda\bar{p}\left(\frac{\partial u}{\partial x} + \frac{\partial v}{\partial y} + \frac{\partial w}{\partial z}\right) = 0.\tag{1}$$

Let the incident plane acoustic wave be

$$p = e^{i\frac{\Omega}{a_0}(\cos\Theta x + \sin\Theta \sin\phi y + \sin\Theta \cos\phi z - a_0 t)},\tag{2}$$

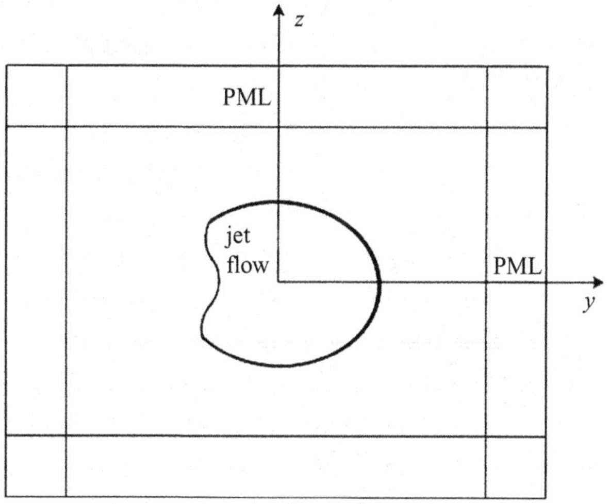

Figure 9.20. Computational domain enclosing the jet flow. Shown also are the PMLs.

where (Θ, ϕ) are the angular coordinates of the direction of propagation with respect to a spherical polar coordinate system centered at the jet exit, and a_0 is the ambient sound speed.

Since the coefficients of Eq. (1) are independent of x, the x dependence of the scattered field must be the same as the incident wave. By factoring out $\exp[i(\Omega/a_0)(\cos\Theta)x]$, the governing equations for the scattered sound field has spatial dependence on y and z only. Outside the jet (see Figure 9.20), where $\bar{u} = 0$, the equations reduced to

$$\frac{\partial \tilde{u}}{\partial t} + \frac{i\Omega \cos \Theta}{a_0 \rho_0} \tilde{p} = 0$$

$$\frac{\partial \tilde{v}}{\partial t} + \frac{1}{\rho_0} \frac{\partial \tilde{p}}{\partial y} = 0$$

$$\frac{\partial \tilde{w}}{\partial t} + \frac{1}{\rho_0} \frac{\partial \tilde{p}}{\partial z} = 0$$

$$\frac{\partial \tilde{p}}{\partial t} + \gamma p_e \left(\frac{i\Omega \cos \Theta}{a_0 \rho_0} \tilde{u} + \frac{\partial \tilde{v}}{\partial y} + \frac{\partial \tilde{w}}{\partial z} \right) = 0. \tag{3}$$

It is easy to show that (3) forms a dispersive wave system. The asymptotic solution of a dispersive wave system is not suitable for constructing radiation boundary condition.

Develop a set of PML absorbing boundary condition for Eq. (3) to be used in the boundary region of a computation domain as shown in the figure. The PML is to absorb the scattered waves. Show that your PML equations are stable.

9.4. This problem is intended to test the discrete formulation and computation of sound transmitted through a surface of discontinuity. Here, the surface of discontinuity is formed by the interface of two fluids with different densities and sound speeds as shown in Figure 9.21. An incident acoustic wave at an angle of incidence θ impinges on the interface. Part of the wave is transmitted and part of it is reflected. For computation purposes, we will use the following length, velocity, time, pressure,

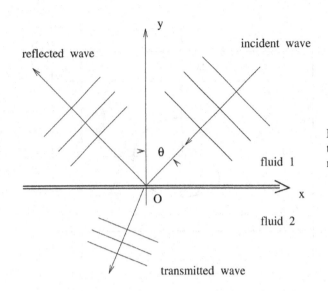

Figure 9.21. Schematic diagram showing the incident, the reflected, and the transmitted waves.

and density scales. Subscripts 1 and 2 indicate fluids 1 and 2.

$$\text{length scale} = L$$

$$\text{velocity scale} = a_1 \text{ (sound speed in region 1)}$$

$$\text{time scale} = \frac{L}{a_1}$$

$$\text{density scale} = \overline{\rho}_1 \text{ (density of 'uid in region 1)}$$

$$\text{pressure scale} = \overline{\rho}_1 a_1^2$$

$$\alpha = \frac{\rho_2}{\rho_1}, \qquad \lambda = \frac{\gamma_2}{\gamma_1}$$

The governing equations for small-amplitude disturbances in fluid 1 are

$$\frac{\partial \rho_1}{\partial t} + \left(\frac{\partial u_1}{\partial x} + \frac{\partial v_1}{\partial y} \right) = 0$$

$$\frac{\partial u_1}{\partial t} = -\frac{\partial p_1}{\partial x}$$

$$\frac{\partial v_1}{\partial t} = -\frac{\partial p_1}{\partial y}$$

$$\frac{\partial p_1}{\partial t} + \left(\frac{\partial u_1}{\partial x} + \frac{\partial v_1}{\partial y} \right) = 0.$$

The governing equations for small-amplitude disturbances in fluid 2 are

$$\frac{\partial \rho_2}{\partial t} + \alpha \left(\frac{\partial u_2}{\partial x} + \frac{\partial v_2}{\partial y} \right) = 0$$

$$\alpha \frac{\partial u_2}{\partial t} = -\frac{\partial p_2}{\partial x}$$

$$\alpha \frac{\partial v_2}{\partial t} = -\frac{\partial p_2}{\partial y}$$

$$\frac{\partial p_2}{\partial t} + \lambda \left(\frac{\partial u_2}{\partial x} + \frac{\partial v_2}{\partial y} \right) = 0.$$

The dynamic and kinematic boundary conditions at the fluid interface are

$$p_1 = p_2, \quad v_1 = v_2.$$

Now, consider a plane wave at an incident angle θ and frequency ω given below:

$$\begin{bmatrix} p_1 \\ u_1 \\ v_1 \\ \rho_1 \end{bmatrix}_{\text{incidence wave}} = \mathrm{Re} \begin{bmatrix} 1 \\ -\sin\theta \\ -\cos\theta \\ 1 \end{bmatrix} e^{-i\omega(\sin\theta x + \cos\theta y + t)}$$

where Re is the real part of. Determine the intensity and direction of the transmitted and reflected waves for the two cases with $\theta = 20°$ and $65°$. The frequency and other parameters are $\omega = 0.7$, $\alpha = 0.694$, and $\lambda = 1$. Plot contours of p at intervals of 0.25 at the beginning of a cycle.

The exact solution is given in Section 1.4. At $\theta = 65°$, there is a total internal reflection. The transmitted wave amplitude is complex indicating that there is no transmitted wave radiated to the far field.

(Hint: In this problem, although both p and v are continuous at the vortex sheet their y derivatives are, however, discontinuous. Because of the discontinuous derivatives, one may use the ghost point method to enforce the continuity of p and v.)

9.5. For the purpose of computing sound radiation from a conical surface with a half cone angle δ as shown in Figure 9.22, a set of natural coordinates to use is (ξ, η, ϕ), where ξ and η are related to the cylindrical coordinates (r, x, ϕ) by

$$\xi = x/\cos\delta, \quad \eta = r - x\tan\delta.$$

(ξ, η) are, in fact, the body-fitted oblique Cartesian coordinates in the $\phi = \text{constant}$ plane. By assuming the solution to have a $e^{im\phi}$ dependence, show that, when $e^{im\phi}$ is

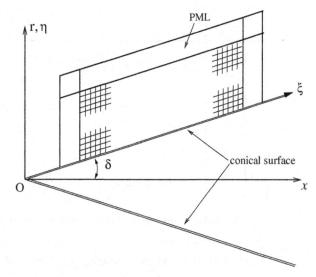

Figure 9.22. Oblique Cartesian coordinates (ξ, η) and a PML on a conical surface.

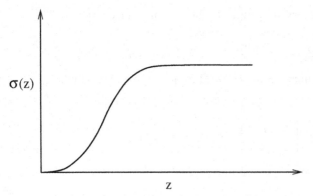

Figure 9.23. Profile of damping function $\sigma(z)$.

separated out, the governing equations are

$$\frac{\partial \mathbf{u}}{\partial t} + \mathbf{A}\frac{\partial \mathbf{u}}{\partial \xi} + \mathbf{B}\frac{\partial \mathbf{u}}{\partial \eta} + \frac{1}{\eta + \xi \sin \delta}\mathbf{C}\mathbf{u} = 0,$$

where

$$\mathbf{u} = \begin{pmatrix} \hat{u}_m \\ \hat{v}_m \\ \hat{w}_m \\ \hat{p}_m \end{pmatrix} \qquad \mathbf{A} = \begin{pmatrix} 0 & 0 & 0 & \frac{-1}{\cos \delta} \\ 0 & 0 & 0 & 0 \\ 0 & 0 & 0 & 0 \\ \frac{-1}{\cos \delta} & 0 & 0 & 0 \end{pmatrix}$$

$$\mathbf{B} = \begin{pmatrix} 0 & 0 & 0 & \tan \delta \\ 0 & 0 & 0 & -1 \\ 0 & 0 & 0 & 0 \\ \tan \delta & -1 & 0 & 0 \end{pmatrix} \qquad \mathbf{C} = \begin{pmatrix} 0 & 0 & 0 & 0 \\ 0 & 0 & 0 & 0 \\ 0 & 0 & 0 & im \\ 0 & -1 & im & 0 \end{pmatrix}.$$

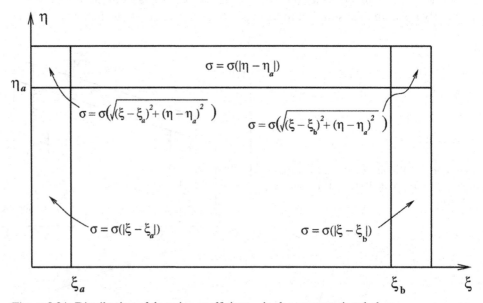

Figure 9.24. Distribution of damping coefficient σ in the computational plane.

A subscript m is used to denote the solution associated with $e^{im\phi}$. (u_m, v_m, w_m) are the velocity components in the cylindrical coordinates. For computing sound radiation, it is proposed to use PML as boundary condition. The PML encloses the computational domain as shown in Figure 9.22.

Show that an appropriate set of PML equation is

$$\frac{\partial \mathbf{u}}{\partial t} + \mathbf{A}\frac{\partial \mathbf{u}}{\partial \xi} + \mathbf{B}\frac{\partial \mathbf{u}}{\partial \eta} + \frac{1}{\eta + \xi \sin \delta}\mathbf{C}\mathbf{u} + \sigma\mathbf{u} = 0,$$

where $\sigma(z)$ is a damping function with a profile as shown in Figure 9.23. The distribution of this function in the computational plane is shown in Figure 9.24.

Implement the PML equations to test their effectiveness. You may use an oscillating point monopole located at the center line of the conical surface as the source of sound.

10 Time-Domain Impedance Boundary Condition

Nowadays, acoustic treatment is invariably used on the inside surface of commercial aircraft jet engines for fan noise reduction. Acoustic treatment panels or liners, when properly toned, are extremely effective noise suppressors. Because of structural integrity requirements, the acoustic liners are usually of the Helmholtz resonator type. Figure 10.1 shows a single-layer acoustic liner. The liner consists of a face sheet with holes at a regular pattern. Underneath the face sheet are cavities. The cavities control the frequency range within which damping is most effective. The dominant damping mechanism of this type of liner is attributed to the dissipation arising from vortex shedding at the mouths of the holes (see Melling, 1973; Tam and Kurbatskii, 2000; Tam et al., 2001). The vortex-shedding process converts acoustic energy into the rotational kinetic energy of the shed vortices. These vortices are subsequently dissipated by molecular viscosity. In the absence of vortex shedding, which is the case at a low sound pressure level, the principal damping mechanism is associated with viscous dissipation of the oscillatory flow at the mouths of the Helmholtz resonators. The flow and acoustic fields around the Helmholtz resonators of acoustic liners are exceedingly complicated, especially when there is a mean flow over the liner.

For engineering purposes, a gross macroscopic description of the effects of the acoustic liners on incident acoustic waves is definitely preferred over a more demanding microscopic description of the actual phenomenon. In the aeroacoustic community, it is an accepted practice to characterize the macroscopic properties of an acoustically treated surface by a single quantity Z called the impedance. Impedance is defined as the ratio of the acoustic pressure p to the acoustic velocity component normal to the treated surface v_n (positive when pointing into the surface). That is,

$$p = Zv_n. \tag{10.1}$$

Impedance is a complex quantity, $Z = R - iX$ ($e^{-i\omega t}$ time dependence is assumed). The use of a complex quantity is needed to account for the damping and phase shift imparted on the sound waves by the acoustically treated surface. The acoustic resistance R and the acoustic reactance X are generally frequency dependent. They also vary with the intensity of the incident sound waves and the adjacent mean flow velocity. These quantities are usually measured empirically, although some semiempirical formulas are available for their estimates. It has been found experimentally that R is positive and does not vary much with frequency. On the other hand, X may be

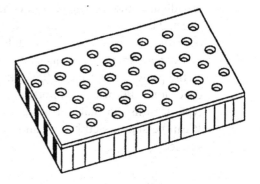

Figure 10.1. A single-layer Helmholtz resonator type acoustic liner.

both positive or negative depending on the frequency. For many acoustic liners, the dependence of X on angular frequency ω can be represented adequately by a simple analytical expression of the following form (Tam and Auriault, 1996):

$$\frac{X}{\rho a_0} = \frac{X_{-1}}{\omega} + X_1 \omega, \tag{10.2}$$

where X_{-1} and X_1 are two parameters, ρ is the gas density, and a_0 is the speed of sound. Figure 10.2 shows an accurate representation of the data of Motsinger and Kraft (1991) by the two-parameter formula. The values of X_{-1} and X_1 obtained by mean-least-square fit are -13.48 and 0.0739, respectively, where ω is measured in kiloradians per second.

The impedance boundary condition (10.1) is basically a boundary condition established for frequency-domain analysis. As it is, it cannot be used in time-domain computation. For time marching computation, a suitable equivalent of the impedance boundary condition in the time domain is needed. One significant advantage of time-domain methods over frequency-domain methods is that broadband noise problems can be handled relatively easily, almost without extra effort. For

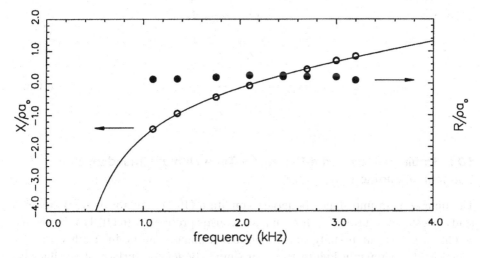

Figure 10.2. Measured dependence of the resistance and reactance of a 6.7% perforate treatment panel on frequency at low sound intensity from Motsinger and Kraft (1991). ○, reactance; ●, resistance.

broadband noise problems, frequency-domain methods are computationally inten-
sive and laborious. In addition, if the problem is nonlinear, frequency-domain methods would be exceedingly difficult to carry out.

Time-domain problems can be solved only if they are well posed and stable.
In the presence of a mean flow, standard formulation of the impedance boundary
condition is known to support spurious unstable solution of the Kelvin-Helmholtz
type. It turns out that, for acoustic liner problems, such instability is rather weak.
It can be suppressed by the imposition of artificial selective damping at the liner
surface.

10.1 A Three-Parameter Broadband Model

A simple model for the broadband impedance boundary condition is to take the
resistance R to be a constant and the reactance X to be given by Eq. (10.2). That is,

$$Z = R_0 - i\left[\frac{X_{-1}}{\omega} + X_1\omega\right], \tag{10.3}$$

where $R_0 > 0, X_{-1} < 0, X_1 > 0$.

For this model, the time-domain impedance boundary condition may be written
as follows:

$$\frac{\partial p}{\partial t} = R_0\frac{\partial v_n}{\partial t} - X_{-1}v_n + X_1\frac{\partial^2 v_n}{\partial t^2}. \tag{10.4}$$

It is straightforward to show, by assuming time dependence of the form $e^{-i\omega t}$
for all variables, that Eq. (10.4) is equivalent to the frequency-domain impedance
boundary condition with impedance given by Eq. (10.3).

For a single-frequency sound field of frequency Ω ($\Omega > 0$), this three-parameter
model may still be used. Let $Z_0 = R_0 - iX_0$ be the impedance of the acoustic liner at
frequency Ω, then the following choice of X_{-1} and X_1 will transform the broadband
model to a single-frequency model.

$$X_{-1} = \frac{-\left(1 - \frac{X_0}{|Z_0|}\right)^2\Omega}{\left[\frac{2}{|Z_0|} - \frac{X_0}{|Z_0|^2}\right]} \tag{10.5a}$$

$$X_1 = \frac{1}{\left[\frac{2}{|Z_0|} - \frac{X_0}{|Z_0|^2}\right]\Omega} \tag{10.5b}$$

10.2 Stability of the Three-Parameter Time-Domain Impedance Boundary Condition

The time-domain impedance boundary condition (10.4) may not be used unless it
leads to a stable solution. To show that an acoustic problem with (10.4) as boundary
condition is computationally stable, consider a plane acoustic liner adjacent to a
sound field as shown in Figure 10.3. For simplicity, let the surface of the liner be
the x–z plane. In terms of dimensionless variables with L (a typical length of the
problem) as the length scale, a_0 (the sound speed) as the velocity scale, L/a_0 as

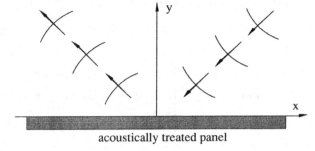

Figure 10.3. Sound field adjacent to an acoustically treated panel.

acoustically treated panel

the time scale, ρ_0 (the gas density) as the density scale, $\rho_0 a_0^2$ as the pressure scale, and $\rho_0 a_0$ as the scale for impedance, the acoustic field equations (the linearized momentum and energy equations) are as follows:

$$\frac{\partial \mathbf{v}}{\partial t} = -\nabla p \tag{10.6}$$

$$\frac{\partial p}{\partial t} = -\nabla \cdot \mathbf{v}. \tag{10.7}$$

Let $\bar{f}(\alpha, \beta, \omega)$ be the Fourier-Laplace transform of a function $f(x, z, t)$. The functions f and $\tilde{f}$ are related by

$$\tilde{f}(\alpha, \beta, \omega) = \frac{1}{(2\pi)^3} \int\limits_{-\infty}^{\infty} \int\limits_{-\infty}^{\infty} \int\limits_{0}^{\infty} f(x, z, t) e^{-i(\alpha x + \beta z - \omega t)} \, dt \, dx \, dz \tag{10.8a}$$

$$f(x, z, t) = \int\limits_{-\infty}^{\infty} \int\limits_{-\infty}^{\infty} \int\limits_{\Gamma} \tilde{f}(\alpha, \beta, \omega) e^{i(\alpha x + \beta z - \omega t)} \, d\omega \, d\alpha \, d\beta. \tag{10.8b}$$

By applying Fourier-Laplace transforms to Eqs. (10.6) and (10.7), it is easy to find that the solution that satisfies the radiation and boundedness condition as $y \to \infty$ is

$$\begin{bmatrix} \tilde{p} \\ \tilde{v} \end{bmatrix} = A \begin{bmatrix} 1 \\ \frac{(\omega^2 - k^2)^{\frac{1}{2}}}{\omega} \end{bmatrix} e^{i(\omega^2 - k^2)^{\frac{1}{2}} y}, \tag{10.9}$$

where $k = (\alpha^2 + \beta^2)^{1/2}$, and v is the velocity component in the y direction (note that $v = -v_n$). The branch cuts of the function $(\omega^2 - k^2)^{1/2}$ are taken to be $0 \le \arg(\omega^2 - k^2)^{1/2} \le \pi$, the left (right) equality is to be used if ω is real and positive (negative). The branch cut configuration in the ω plane is shown in Figure 10.4.

The Fourier-Laplace transform of Eq. (10.4) is

$$-\omega \tilde{p} = [-\omega R_0 + i(X_{-1} + \omega^2 X_1)] \tilde{v}_n. \tag{10.10}$$

Substitution of Eq. (10.9) into Eq. (10.10) results in the dispersion relation as follows:

$$\frac{\omega^2}{(\omega^2 - k^2)^{\frac{1}{2}}} + \omega R_0 - i X_1 \omega^2 = i X_{-1}. \tag{10.11}$$

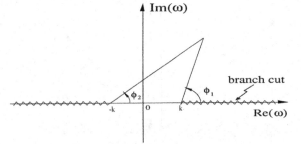

Figure 10.4. Branch cut configuration for $(\omega^2 - k^2)^{1/2}$. Argument of $(\omega^2 - k^2)^{1/2} = \frac{1}{2}(\phi_1 + \phi_2)$

Let

$$F(\omega) = \frac{\omega^2}{(\omega^2 - k^2)^{\frac{1}{2}}} + \omega R_0 - iX_1\omega^2. \tag{10.12}$$

That is, $F(\omega)$ is the left side of Eq. (10.11). By tracing over the contour $ABCDEFGH$ in the upper-half ω plane, it is easy to establish that the upper-half ω plane is mapped into the shaded region in the F plane as shown in Figure 10.5. The mapped region does not include the negative imaginary axis. Now X_{-1}, on the right side of Eq. (10.11), is negative. This means that no value of ω in the upper-half ω plane would satisfy dispersion relation (10.11). Thus, the solutions of the initial value problem

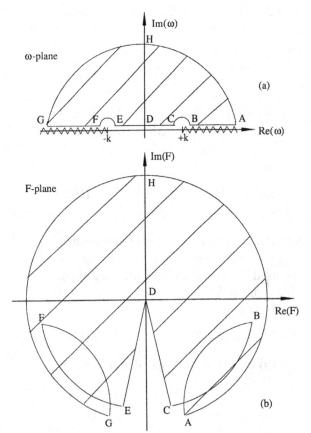

Figure 10.5. Map of the upper-half ω plane in the F plane. (a) ω plane, (b) F plane.

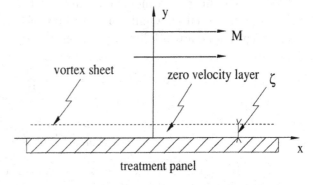

Figure 10.6. Schematic diagram showing a postulated zero-velocity fluid layer adjacent to an acoustically treated panel in the presence of a mean flow.

can only have ω with a negative imaginary part. These solutions are numerically stable.

10.3 Impedance Boundary Condition in the Presence of a Subsonic Mean Flow

In jet engines, there is invariably a mean flow adjacent to the acoustic liners. Traditionally, impedance boundary condition in the presence of a mean flow is formulated with the assumption of the existence of a very thin zero-velocity fluid layer at the surface of the acoustic liner (see Figure 10.6). At the interface of the zero-velocity fluid layer and the mean flow, the condition of continuity of particle displacement is used. In dimensionless variables the frequency-domain impedance boundary condition on liner surface $S_0(x, y, z) = 0$, derived by Myers (1980), is

$$i\omega v_n Z + (-i\omega + \mathbf{U} \cdot \nabla) p = p\mathbf{n} \cdot (\mathbf{n} \cdot \nabla)\mathbf{U},$$

where $\mathbf{U}$ is the mean flow velocity on the liner surface, $\mathbf{n}$ is the unit outward pointing normal of S_0; that is, $\mathbf{n} = \nabla S_0 / |\nabla S_0|$. v_n is the velocity normal to S_0 with positive pointing into the liner. If the three-parameter model for Z is used, the time-domain impedance boundary condition is

$$-R\frac{\partial v_n}{\partial t} + X_{-1}v_n - X_1\frac{\partial^2 v_n}{\partial t^2} + \left(\frac{\partial p}{\partial t} + \mathbf{U} \cdot \nabla p\right) = p\mathbf{n} \cdot (\mathbf{n} \cdot \nabla)\mathbf{U}.$$

For the special case for which S_0 is a plane as shown in Figure 10.6, the frequency-domain impedance boundary condition, after simplification, may be written as

$$-i\omega + M\frac{\partial p}{\partial x} = -i\omega Z v_n. \tag{10.13}$$

In an extensive numerical study of the normal modes of a duct with acoustic liners, Tester (1973) found that boundary condition (10.13) led to an unstable solution. The unstable solution is of the Kelvin-Helmholtz-type instability arising from the vortex sheet interface between the mean flow and the zero-velocity fluid layer. In standard duct acoustics analysis using the frequency-domain approach, this instability is either not mentioned or totally bypassed and ignored. Now, it is easy to show that the use of boundary condition (10.13) always gives rise to an unstable solution.

The linearized momentum and energy equations governing the sound field superimposed on a uniform mean flow of Mach number M in the x direction (see Figure 10.6) are as follows:

$$\frac{\partial \mathbf{v}}{\partial t} + M \frac{\partial \mathbf{v}}{\partial x} = -\nabla p \tag{10.14}$$

$$\frac{\partial p}{\partial t} + M \frac{\partial p}{\partial x} + \nabla \cdot \mathbf{v} = 0. \tag{10.15}$$

By applying Fourier-Laplace transforms to Eqs. (10.14) and (10.15), it is easy to find that the solution that satisfies outgoing wave and boundedness conditions at $y \to \infty$ is

$$\begin{bmatrix} \tilde{u} \\ \tilde{v} \\ \tilde{w} \\ \tilde{p} \end{bmatrix} = A \begin{bmatrix} \alpha/\tilde{\omega} \\ k(\hat{\omega}^2 - 1)^{1/2}/\hat{\omega} \\ \beta/\hat{\omega} \\ 1 \end{bmatrix} \exp\left[ik(\hat{\omega}^2 - 1)^{\frac{1}{2}} y \right], \tag{10.16}$$

where $\tilde{\omega} = \omega - M\alpha$, $k = (\alpha^2 + \beta^2)^{1/2}$, and $\hat{\omega} = \tilde{\omega}/k$. The branch cuts of the square root function $(\hat{\omega}^2 - 1)^{1/2}$ are the same as those stipulated in Figure 10.4 with $\hat{\omega}$ replacing ω and k set equal to 1.

Substitution of solution (10.16) into the Fourier-Laplace transforms of boundary condition (10.13) yields the following dispersion relation:

$$\frac{\hat{\omega}^2}{(\hat{\omega}^2 - 1)^{\frac{1}{2}}} + Z\hat{\omega} = -\frac{\alpha}{k} MZ. \tag{10.17}$$

Now, let the left side of Eq. (10.17) be denoted by $\overline{f}(\hat{\omega})$ as follows:

$$\overline{f}(\hat{\omega}) = \frac{\hat{\omega}^2}{(\hat{\omega}^2 - 1)^{\frac{1}{2}}} + Z\hat{\omega}. \tag{10.18}$$

Figure 10.7 shows the map of the upper-half $\hat{\omega}$ plane in the $\overline{f}$ plane (shaded region) for $X > 0$. Since α can be positive or negative, there will always be values of α for which the point representing the right side of Eq. (10.17) lies in the shaded region of the $\overline{f}$ plane. Therefore, there will always be an unstable solution. If X is negative, a similar mapping procedure will show that there is always an unstable solution.

The existence of a Kelvin-Helmholtz-type instability renders the boundary value problem ill-posed for the time-domain solution. The instability is, however, non-physical. Its origin is in the postulate of a vortex sheet discontinuity right next to the impedance boundary. In reality, no such vortex sheet exists in the flow. For time-domain problems, the instability is weak. It can be effectively suppressed by the addition of artificial selective damping near and at the surface of the acoustic liner. With the weak instability suppressed, boundary condition (10.13) has yielded solutions that are in agreement with frequency-domain calculations.

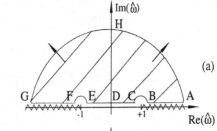

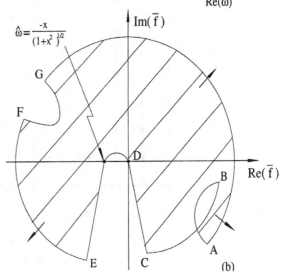

Figure 10.7. Map of the upper-half $\hat{\omega}$ plane in the $\bar{f}$ plane $(X > 0)$: (a) $\hat{\omega}$ plane, (b) $\bar{f}$ plane.

10.4 Numerical Implementation

On incorporating the effect of convection (see Eq. (10.13)) the time-domain impedance boundary condition (10.4) or the Myers (1980) boundary condition becomes

$$\frac{\partial p}{\partial t} + M\frac{\partial p}{\partial x} = R\frac{\partial v_n}{\partial t} - X_{-1}v_n + X_1\frac{\partial^2 v_n}{\partial t^2}. \tag{10.19}$$

The numerical implementation of this boundary condition will now be considered.

Let a large acoustically treated panel be lying on the x–z plane as shown in Figure 10.3. It will be assumed that there is a uniform flow adjacent to the panel. The governing equations of the acoustic field are the linearized Euler equations. In dimensionless form, these equations are

$$\frac{\partial \rho}{\partial t} + M\frac{\partial \rho}{\partial x} + \frac{\partial u}{\partial x} + \frac{\partial v}{\partial y} + \frac{\partial w}{\partial z} = 0 \tag{10.20}$$

$$\frac{\partial u}{\partial t} + M\frac{\partial u}{\partial x} = -\frac{\partial p}{\partial x} \tag{10.21}$$

$$\frac{\partial v}{\partial t} + M\frac{\partial v}{\partial x} = -\frac{\partial p}{\partial y} \tag{10.22}$$

$$\frac{\partial w}{\partial t} + M\frac{\partial w}{\partial x} = -\frac{\partial p}{\partial z} \tag{10.23}$$

$$\frac{\partial p}{\partial t} + M\frac{\partial p}{\partial x} + \frac{\partial u}{\partial x} + \frac{\partial v}{\partial y} + \frac{\partial w}{\partial z} = 0. \tag{10.24}$$

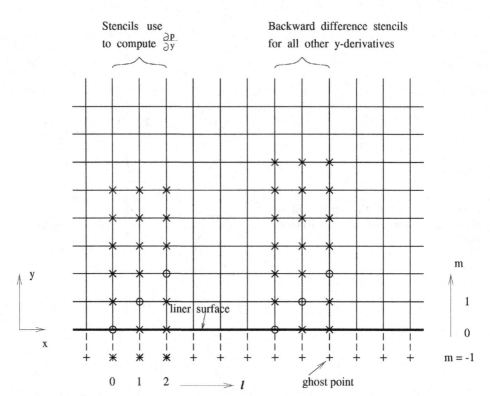

Figure 10.8. Backward difference stencils used in the boundary region adjacent to the liner surface.

Now consider a computational domain as shown in Figure 10.8. $y = 0$ is the surface of the acoustically treated panel. Eqs. (10.20) to (10.24) apply to every grid point on the mesh. This includes the boundary points at $y = 0$ or $m = 0$, where m is the mesh index in the y direction. However, boundary condition (10.19) also applies to the boundary mesh point at $y = 0$ or $m = 0$. This means that there are more equations than unknowns. To resolve this problem of an overdetermined system, a row of ghost values of pressure, $p_{\ell,-1,k}$ is introduced at $m = -1$, the ghost point. Here, subscripts ℓ and k are the mesh indices in the x and z directions, respectively. Including the ghost values, the number of equations and unknowns are exactly equal. Since the ghost values are not governed by a time-dependent equation, it is necessary to combine boundary condition and equations of motion to reduce one of the equations effectively to a form without a time derivative. This form of the boundary condition is then used to determine the ghost values.

To facilitate the implementation of the time-domain impedance boundary condition (10.19), an auxiliary variable $q(x, z, t)$, defined at $y = 0$ by

$$q = \frac{\partial v_n}{\partial t} = -\left.\frac{\partial v}{\partial t}\right|_{y=0} \tag{10.25}$$

is introduced. q is only defined on the acoustic liner surface; i.e., $y = 0$. Note: $v_n = -v$ for the flow configuration of Figure 10.3. Boundary condition (10.19) may be rewritten, after eliminating $\partial p/\partial t$ by Eq. (10.24), as follows:

$$\frac{\partial q}{\partial t} = -\frac{1}{X_1}\left[Rq + X_{-1}v + \frac{\partial u}{\partial x} + \frac{\partial v}{\partial y} + \frac{\partial w}{\partial z}\right]. \tag{10.26}$$

Now, at $y = 0$ or $m = 0$, Eq. (10.22) may be rewritten in the following form:

$$\frac{\partial p}{\partial y} = q - M\frac{\partial v}{\partial x}.\tag{10.27}$$

Eq. (10.27) does not contain time derivative. It is in a suitable form for use to determine the ghost value $p_{\ell,-1,k}$.

The discretized form of Eq. (10.27) using the backward difference stencil of Figure 10.8 is

$$\frac{1}{\Delta y}\sum_{j=-1}^{5} a_j^{51} p_{\ell,j,k}^{(n)} = q_{\ell,k}^{(n)} - \frac{M}{\Delta x}\sum_{j=-3}^{3} a_j v_{\ell+j,0,k}^{(n)},$$

where Δx and Δy are the mesh sizes. On solving for the ghost value, it is straightforward to find

$$p_{\ell,-1,k}^{(n)} = \frac{\Delta y}{a_{-1}^{51}}\left[q_{\ell,k}^{(n)} - \frac{M}{\Delta x}\sum_{j=-3}^{3} a_j v_{\ell+j,0,k}^{(n)} - \frac{1}{\Delta y}\sum_{j=0}^{5} a_j^{51} p_{\ell,j,k}^{(n)}\right].\tag{10.28}$$

On choosing the ghost value by Eq. (10.28), the governing Eq. (10.27) or its progenitor Eq. (10.22) is automatically satisfied. To compute the entire sound field, the values of ρ, u, w, and p at every mesh point (ℓ, m, k) are calculated by Eqs. (10.20), (10.21), (10.23), and (10.24). For velocity component v at all mesh points, except those on the acoustically treated panel surface, i.e., $m = 0$, the value is updated by means of Eq. (10.22). For mesh points at $m = 0$, $v_{\ell,0,k}$ are computed by Eq. (10.25). The auxiliary variable q is to be calculated by Eq. (10.26). In discretizing the spatial derivatives in y for mesh points in rows $m = 0, 1, 2$, the backward difference stencils shown in Figure 10.8 are to be used. Finally, artificial selective damping must be added to each time-dependent equation to suppress the weak instability of the impedance model.

10.5 A Numerical Example

To illustrate how well the time-domain impedance boundary condition works, consider an initial value problem associated with the normal-incidence impedance tube (see Figure 10.9). An impedance tube is designed to measure the impedance of an acoustic liner sample. Let the length of the impedance tube be 1.8 m. The speed of sound at room temperature is 340 m/s. To ensure that there are at least 7 mesh points per wavelength in the computation for sound frequency up to 4 kHz, the impedance tube is divided into 150 mesh spacings yielding $\Delta x = 0.012$ m. In this example Δx is used as the length scale; i.e., $L = \Delta x$. The impedance of the treatment panel is taken to be $R = 0.18$, $X_{-1} = -0.47567$, and $X_1 = 2.09236$ (ω is nondimensionalized by $a_0/\Delta x$). These are values corresponding to those of Figure 10.2. At time $t = 0$, a transient acoustic pulse is generated. This produces a broadband pressure field in the frequency domain. A part of the acoustic pulse propagates down the tube and impinges on the acoustic treatment panel on the right. This leads to a partially reflected pulse. It will be assumed that the disturbance is generated by the following initial conditions at $t = 0$:

$$u = 0,\quad p = e^{-0.00444(x+83.333)^2}\cos[0.444\,(x+83.333)].\tag{10.29}$$

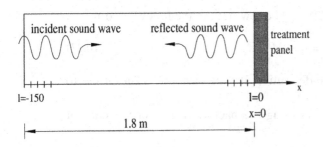

incident sound wave   reflected sound wave

treatment
panel

Figure 10.9. A normal incidence impedance tube.

$l=-150$ $l=0$
                                        $x=0$

1.8 m

This choice of initial conditions ensures that the center of the acoustic spectrum
of the incident wave at the surface of the acoustic treatment panel has a frequency
of 2 kHz and a spectrum half-width of 0.5 kHz The impedance tube problem has an
exact solution, which may be derived as follows.

The governing equations for the acoustic field are the one-dimensional version
of Eqs. (10.6) and (10.7). They are

$$\frac{\partial u}{\partial t} = -\frac{\partial p}{\partial x} \tag{10.30}$$

$$\frac{\partial p}{\partial t} = -\frac{\partial u}{\partial x}. \tag{10.31}$$

Upon eliminating u, the governing equation for p is

$$\frac{\partial^2 p}{\partial t^2} = \frac{\partial^2 p}{\partial x^2}. \tag{10.32}$$

The solution of Eq. (10.32) satisfying initial condition (10.29) is

$$p = \frac{1}{2} e^{-0.00444(x-t+83.333)^2} \cos\left[0.444\left(x - t + 83.333\right)\right]$$

$$+ \frac{1}{2} e^{-0.00444(x+t+83.333)^2} \cos\left[0.444\left(x + t + 83.333\right)\right]. \tag{10.33}$$

The first term of Eq. (10.33) is the part of the pulse that propagates in the positive
x direction. This is the sound pulse that impinges on the acoustically treated panel at
$x = 0$. For convenience, a subscript "inc" will be used to label the incident acoustic
field, a subscript "ref" will be used to label the reflected field. Thus, the incident
wave is

$$p_{\text{inc}} = u_{\text{inc}} = \frac{1}{2} e^{-0.00444(x-t+83.333)^2} \cos\left[0.444\left(x - t + 83.333\right)\right]. \tag{10.34}$$

Let

$$p_{\text{ref}} = -u_{\text{ref}} = \Phi(x + t). \tag{10.35}$$

Eq. (10.35) is a general solution of Eq. (10.32). At the impedance surface, $x = 0$,
the Laplace transforms of the incident and reflected acoustic field variables are

$$\tilde{p}_{\text{inc}} = \tilde{u}_{\text{inc}} = 2.11677 \left[e^{-\frac{(\omega+0.444)^2}{0.01776}} + e^{-\frac{(\omega-0.444)^2}{0.01776}} \right] e^{i83.333\omega} \tag{10.36}$$

$$\tilde{p}_{\text{ref}} = -\tilde{u}_{\text{ref}} = \tilde{\Phi}(\omega). \tag{10.37}$$

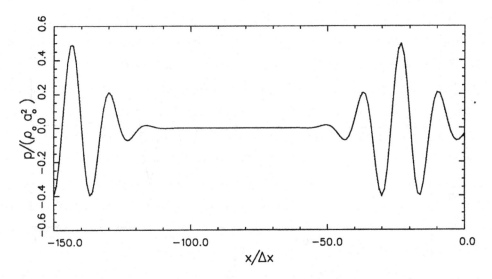

Figure 10.10. Pressure waveform of the incident acoustic pulse at $t = 60.1$.

On imposing the impedance boundary condition (10.1), one finds

$$(\tilde{p}_{\text{inc}} + \tilde{p}_{\text{ref}}) = Z(\tilde{u}_{\text{inc}} + \tilde{u}_{\text{ref}}). \tag{10.38}$$

On using Eqs. (10.36) and (10.37), it is found that

$$\tilde{\Phi} = \frac{Z-1}{Z+1}\tilde{p}_{\text{inc}}. \tag{10.39}$$

The exact solution is obtained by taking the inverse transform of (10.39) and adjust the argument according to Eq. (10.35). This gives

$$
p_{\text{ref}}(x, t) = \int_{-\infty}^{\infty} \frac{(Z-1)}{(Z+1)} 2.116 \left[e^{-56.265(\omega+0.444)^2} + e^{-56.265(\omega-0.444)^2} \right] e^{-i\omega(x+t-83.333)} d\omega
$$

$$
= 4.232 \, \text{Re} \left\{ \int_{0}^{\infty} \frac{(Z-1)}{(Z+1)} \left[e^{-56.265(\omega+0.444)^2} + e^{-56.265(\omega-0.444)^2} \right] e^{-i\omega(x+t-83.333)} d\omega \right\}.
$$

$$\tag{10.40}$$

Figure 10.10 shows the computed pressure distribution at time $t = 60.1$. At this time, the left half of the initial pulse is about to exit the computational domain, whereas the right half of the pulse is about to impinge on the surface of the treatment panel. The dotted curve is the exact solution. Figure 10.11 shows the reflected pulse propagating away from the impedance boundary of the tube at $t = 140.1$. The amplitude of the reflected pulse is considerably smaller than that of the incident pulse. Part of the acoustic energy is dissipated off the impedance surface during the reflection process. The exact solution is represented by the dotted curve (the dotted curve is difficult to see because it is almost identical to the computed result). There is excellent agreement between numerical results and the exact solution. This is true for both the wave amplitude and phase. This example provides confidence in the use of time-domain impedance boundary condition.

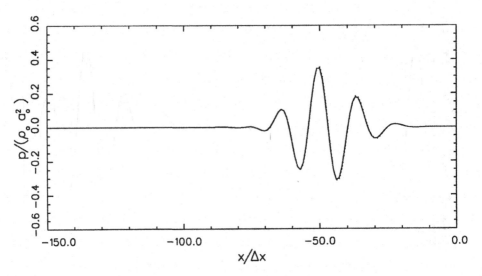

Figure 10.11. Pressure waveform of the reflected acoustic pulse at $t = 140.1$.

10.6 Acoustic Wave Propagation and Scattering in a Duct with Acoustic Liner Splices

Inside the inlet or the exhaust duct of a jet engine, the continuous reflections of acoustic waves by the walls lead to the formation of coherent propagating wave entities called duct modes or acoustic modes. These duct modes were first found by Tyler and Sofrin (1962) and have been the subject of numerous studies since. To suppress the duct modes, acoustic liners are commonly used. Because of installation and maintenance issues, a single large piece of liner is seldom used. A practical way is to install several smaller pieces of liners separated by hard wall splices. The presence of hard wall splices causes the scattering of the duct modes generated by the rotating fan blades of the engine. The scattered duct modes would, inevitably, include lower-order spinning azimuthal modes. The lower-order modes are not as heavily damped by the acoustic liner. This results in a higher level of radiated noise from the engine. This is extremely undesirable.

In this section, three numerical examples are presented to illustrate the effectiveness and accuracy of the time-domain impedance boundary condition and the wave number truncation method (see Section 9.6) for computing acoustic wave propagation and scattering in ducts with material discontinuities. These examples have practical relevance to jet engine duct acoustics. The inlet duct of a jet engine generally has a nearly circular geometry. The inside surface is covered by an acoustic liner. The liner occupies the space between two hard wall surfaces. One hard wall surface is the metallic surface surrounding the fan of the engine. The other hard wall surface is the lip of the engine inlet. Acoustic or duct modes are generated by the rotating fan of the engine. The duct modes propagate from the fan face upstream against the flow. They are radiated out from the inlet. A simplified model of the inlet duct is shown in Figure 10.12. This model has been used by numerous investigators for computing duct/acoustic mode propagation in jet engine ducts; e.g., Regan and Eaton (1999), McAlpine et al. (2003), Tester et al. (2006), and Tam et al. (2008). For a number of practical reasons, it is a standard practice to install hard wall splices

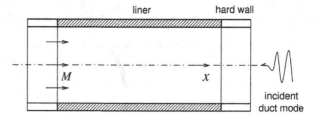

Figure 10.12. A computational model of
the inlet duct of a jet engine.

to separate a liner into smaller pieces. Two types of splices are used. The more
common type is the axial splice. An axial splice runs along the entire length of the
liner. Figure 10.13 shows an inlet duct with two axial splices. Instead of axial splices,
circumferential splices are often used. These splices form a ring around the inside
surface of the duct. A duct with a circumferential splice is shown in Figure 10.14.

In the following examples, dimensionless variables are used. The scales of the
various variables are: length scale D (diameter of the duct), velocity scale a_0 (speed
of sound), time scale D/a_0, density scale ρ_0 (mean flow density), pressure scale $\rho_0 a_0^2$,
and impedance scale $\rho_0 a_0$.

10.6.1 Scattering of Acoustic Duct Mode at the Entrance and Exit of an Inlet Duct

At the entrance or the exit of an inlet duct, there is a surface discontinuity at the
junction of the liner and hard wall (see Figure 10.12 or Figure 10.14). These junctions
invariably cause significant spurious scattering of a propagating duct mode. The
scattering is an artifact of discrete computation. It can be reduced by using a finer
size mesh or by implementing the wave number truncation method of Section 9.6.

The governing equations of motion of the gas inside an inlet duct are the linearized dimensionless Euler equations. In cylindrical coordinates, these equations
are

$$\frac{\partial \rho}{\partial t} + M\frac{\partial \rho}{\partial x} + \frac{\partial v}{\partial r} + \frac{v}{r} + \frac{1}{r}\frac{\partial w}{\partial \phi} + \frac{\partial u}{\partial x} = 0, \tag{10.41}$$

$$\frac{\partial v}{\partial t} + M\frac{\partial v}{\partial x} = -\frac{\partial p}{\partial r}, \tag{10.42}$$

$$\frac{\partial w}{\partial t} + M\frac{\partial w}{\partial x} = -\frac{1}{r}\frac{\partial p}{\partial \phi}, \tag{10.43}$$

$$\frac{\partial u}{\partial t} + M\frac{\partial u}{\partial x} = -\frac{\partial p}{\partial x}, \tag{10.44}$$

$$\frac{\partial p}{\partial t} + M\frac{\partial p}{\partial x} + \frac{\partial v}{\partial r} + \frac{v}{r} + \frac{1}{r}\frac{\partial w}{\partial \phi} + \frac{\partial u}{\partial x} = 0, \tag{10.45}$$

where $M = u/a_0$ is the flow Mach number.

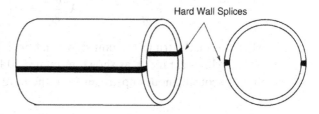

Figure 10.13. Computational model of
an inlet duct with two axial splices.

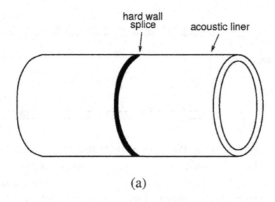

(a)

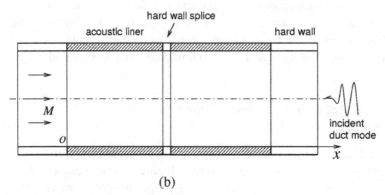

(b)

Figure 10.14. Computational model of an inlet duct with a circumferential splice.

On the duct surface, the hard wall boundary condition is

$$r = \frac{1}{2}, \qquad v = 0. \tag{10.46}$$

On the liner surface, the boundary condition is Eq. (10.19). At the left-hand junction of the computational model (Figure 10.12), the hard wall and liner boundary conditions can be combined to form a single boundary condition by means of the unit step function $H(x)$. The combined boundary condition is

$$R\frac{\partial v}{\partial t} - X_{-1}v + X_1\frac{\partial^2 v}{\partial t^2} = H(x)\left(\frac{\partial p}{\partial t} + M\frac{\partial p}{\partial x}\right). \tag{10.47}$$

Note: On the hard wall, $x < 0$, the $H(x)$ function of boundary condition (10.47) is zero. This reduces the right side of the equation to zero. With the initial condition $v = \partial v/\partial t = 0$, the only solution is $v = 0$. In implementing the wave number truncation method, boundary condition (10.47) is replaced by

$$R\frac{\partial v}{\partial t} - X_{-1}v + X_1\frac{\partial^2 v}{\partial t^2} = \hat{H}(x)\left(\frac{\partial p}{\partial t} + M\frac{\partial p}{\partial x}\right), \tag{10.48}$$

where $\hat{H}(x)$, the truncated unit step function, is given by Eq. (9.67).

A cylindrical mesh, as shown in Figure 10.15, is used to compute the propagation of acoustic modes upstream from the hard wall region on the right side of the

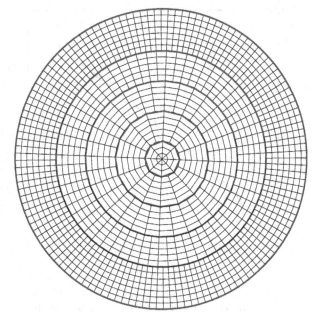

Figure 10.15. Cylindrical grid for inlet duct acoustic mode computation. Only half the mesh lines are shown.

inlet model of Figure 10.12 to the left side. A uniform size mesh in the axial direction is used. Now, consider an incident duct mode with azimuthal mode number $m = 26$ and radial mode number $n = 1$ and a dimensionless angular frequency $\Omega = 0.573$. For such an acoustic mode, an axial mesh with $\Delta x = 0.008$ will allow the computation to have more than 8 mesh points per axial wavelength. This is the mesh used in all the computations of this section unless explicitly stated otherwise. In the computation, the 7-point stencil dispersion-relation-preserving (DRP) scheme is used to approximate the derivatives. The multimesh-size, multitime-step DRP scheme (see Chapter 12) is used to march the solution in time. The duct wall boundary conditions (liner impedance, $Z = 2 + i$, time factor $\exp(-i\Omega t)$) are enforced by the ghost point method. Recently, this method has been used successfully by Tam et al. (2008) in their spliced liner study. On the left and right end of the computational domain, a perfectly matched layer (PML) is implemented. The PML absorbs all the outgoing waves. The incident acoustic mode, which enters the computational domain on the right side, is introduced into the computation by the split-variable method as discussed in Section 9.1.

Figure 10.16 shows the axial distribution of computed acoustic energy flux (PWL) associated with an incident duct mode with azimuthal mode number 26 and other parameters as stipulated above. PWL is the total energy flux of all the sound waves in the duct. It is defined as

$$\text{PWL} = \int_0^{2\pi} \int_0^{1/2} \langle (1 + M^2)pu + M(p^2 + u^2) \rangle \, r d\, r d\phi, \qquad (10.49)$$

where $\langle \, \rangle$ is the time average. $\text{PWL(dB)} = 10 \log (\text{PWL}/\text{PWL}_{\text{incident wave}})$. This form of energy flux was derived by Morfey (1971). Of particular interest here are the computed results near the left junction between acoustic liner and hard wall. In this region, the effect of scattering of spurious acoustic waves is most severe and

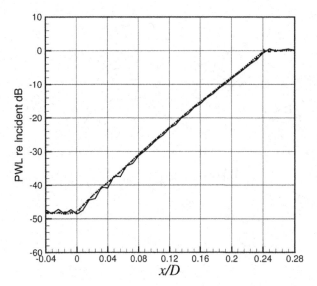

Figure 10.16. Axial distribution of PWL. Incident wave has azimuthal mode number $m = 26$, radial mode number $n = 1$, angular frequency $\Omega = 0.573$. Liner impedance $Z = 2 + i$ (time factor $e^{-i\Omega t}$). ———, Computed by boundary condition (10.47), $\Delta x = 0.008$; – – – –, computed by boundary condition (10.47), $\Delta x = 0.001$; .... computed by boundary condition (10.48), $\Delta x = 0.008$.

easily observable. The full line in this figure is the computed axial distribution of PWL using boundary condition (10.47) and axial mesh size 0.008. This curve exhibits strong spatial oscillations with a wavelength nearly equal to two axial mesh spacings. This is clear evidence of spurious scattering. The dotted line is the same computation, but with a modified boundary condition (10.48). The computed distribution of PWL is smooth and free of spatial oscillations. This indicates that the wave number truncation method is indeed capable of removing spurious numerical scattering. To check the accuracy of the computed results by this method, a series of computations using boundary condition (10.47) is carried out with smaller and smaller axial mesh size Δx. Δx equals 1/4, 1/8, and 1/16 of the original mesh size used. The results near the left side liner-hard wall junction are shown in Figure 10.17 in an enlarged scale. As can be seen, as Δx is reduced, the computed result approaches that using boundary condition (10.48) but with a much larger Δx. The fact that there is good

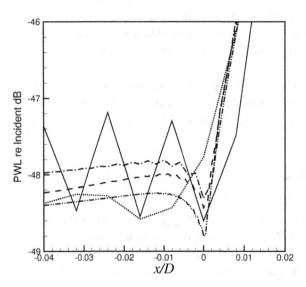

Figure 10.17. Enlarged bottom left-hand corner of Figure 10.16. Computed by boundary condition (10.47); ———, $\Delta x = 0.008$; — · —, $\Delta x = 0.002$; – – – –, $\Delta x = 0.001$; — · · —, $\Delta x = 0.0005$; Computed by boundary condition (10.48); .... $\Delta x = 0.008$.

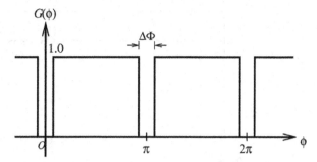

Figure 10.18. The periodic $G(\phi)$ function.

agreement between results using much smaller Δx and boundary condition (10.47), and that using boundary condition (10.48) but much larger Δx, is evidence that the spatial oscillations of the full curve is of numerical origin. Further, it shows that the method of wave number truncation is effective and accurate. It also offers large savings in computer memory and CPU time.

10.6.2 Scattering by Very Thin Axial Splices

This example involves acoustic mode scattering by two thin axial splices. The computational model is shown in Figure 10.13. The same computational grid (see Figure 10.15) as in the previous subsection is used. The azimuthal mesh size of the outermost ring of the cylindrical mesh is equal to 0.0123. Here, the acoustic scattering by two thin hard wall splices with a width of 0.006 is studied. It is noted that the splice width is approximately half of that of the mesh spacing (see Figure 9.15). In other words, the splices are invisible to the discretized computation.

Consider the periodic function, $G(\phi)$, of period 2π, defined by

$$G(\phi) = \begin{cases} 0, & -\frac{1}{2}\Delta\Phi < \phi < \frac{\Delta\Phi}{2}, \quad \pi - \frac{\Delta\Phi}{2} < \phi < \pi + \frac{\Delta\Phi}{2} \\ 1, & \text{otherwise} \end{cases} \qquad (10.50)$$

where $\Delta\Phi$ is the angle subtended by a splice. A graph of $G(\phi)$ is shown in Figure 10.18. When applied to the surface of the duct, $G(\phi) = 1$ where there is acoustic liner. It is equal to zero where there is hard wall splice. By means of the G-function, the wall boundary condition in the duct may be written as

$$R\frac{\partial v}{\partial t} - X_{-1}v + X_1\frac{\partial^2 v}{\partial t^2} = G(\phi)\left[\frac{\partial p}{\partial t} + M\frac{\partial p}{\partial x}\right]. \qquad (10.51)$$

Now, $G(\phi)$ is a period function; it may be expanded as a Fourier series. It is easy to find that

$$G(\phi) = \sum_{j=0}^{\infty} a_j \cos(j\phi), \qquad (10.52)$$

where

$$a_0 = \left(1 - \frac{\Delta\Phi}{\pi}\right), \quad a_j = -\frac{2}{\pi j}[1 + (-1)^j]\sin\left(\frac{j\Delta\Phi}{2}\right), \quad j = 1, 2, 3, \ldots. \qquad (10.53)$$

It is clear from Fourier expansion (10.52) that the $G(\phi)$ function contains components that are beyond the resolution of the 7-point stencil DRP scheme. To compute

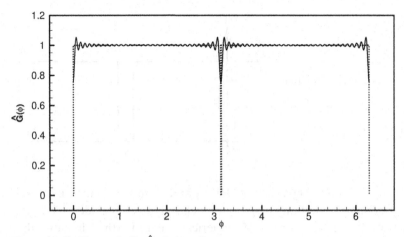

Figure 10.19. The periodic $\hat{G}(\phi)$ function.

acoustic wave propagation in the duct, these high wave number components essentially make the splice invisible to the computational mesh. A simple way to make the spliced liner compatible with the numerical scheme is to remove all Fourier components with azimuthal wavelengths shorter than the cutoff wavelength of the computational algorithm. Accordingly, all Fourier components with mode number m larger than m_c, where the wavelength of mode m_c is the same as or is closest to the smallest resolved wavelength of the computation scheme. For the cylindrical mesh used in the present computation, the azimuthal mesh size at the outermost ring is equal to 0.0123. For the 7-point stencil DRP scheme, $\alpha_c \Delta x = 1.2$, so that $\alpha_c = 1.2/0.0123 = 97.6$. Hence, by equating m_c to the integer value closest to $\alpha_c/2$, it is found that $m_c = 49$. Figure 10.19 shows the graph of the truncated Fourier series, $\hat{G}(\phi)$ (Fourier components with m greater than 49 are removed), i.e.,

$$\hat{G}(\phi) = \sum_{m=0}^{m_c} a_j \cos(j\phi). \qquad (10.54)$$

The modified spliced liner boundary condition is to replace $G(\phi)$ by $\hat{G}(\phi)$ in Eq. (10.51), i.e.,

$$R\frac{\partial v}{\partial t} - X_{-1}v + X_1\frac{\partial^2 v}{\partial t^2} = \hat{G}(\phi)\left[\frac{\partial p}{\partial t} + M\frac{\partial p}{\partial x}\right]. \qquad (10.55)$$

Figure 10.20 shows three computed PWL distributions along the length of the duct for the same incident duct mode as Section 10.6.1. In this computation, boundary condition (10.48) is used in the neighborhood of the liner hard wall junction at the two ends of the computational model. The dotted line in Figure 10.20 shows the numerical result when boundary condition (10.51) is used. There is no scattering in this case because the splices are invisible to the computational mesh. The full line is the computation using modified boundary condition (10.55). There is a large difference between these two curves. The difference is the energy of the scattered waves that are clearly quite substantial at the duct inlet. To check the accuracy of the wave number truncation method, a third computation using the quasi-two-dimension

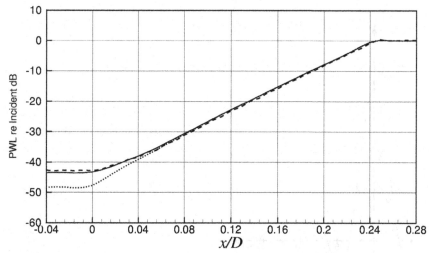

Figure 10.20. Axial distribution of PWL. Incident wave has azimuthal mode number $m = 26$, radial mode number $n = 1$, angular frequency $\Omega = 0.573$. Liner impedance $Z = 2 + i$ (time factor $e^{-i\Omega t}$). $\Delta x = 0.008$. ———, computed by boundary condition (10.55); ······· computed by boundary condition (10.51); – – – –, computed by Fourier expansion method (89 modes) of Tam et al. (2008).

azimuthal mode expansion method of Tam et al. (2008). This method expands the propagating acoustic waves in a large number of azimuthal modes. For the present computation, modes up to $(m_c + 40)$ are included. As a result of the large number of modes, the computation requires very large computer memory and long CPU time. The dashed curve of Figure 10.20 is the computed result using this method. It is evident that the result is very close to that of the wave number truncation method. This example offers a rigorous test of the effectiveness and accuracy of the wave number truncation method for computing the scattering of acoustic waves by narrow scatterers.

10.6.3 Scattering by Thin Circumferential Splices

In this third example, the scattering of acoustic modes by a circumferential splices is considered. The computational model is shown in Figure 10.14. For this test case, the width of the circumferential splices is taken to be $2\Delta x$. For this size splice, there are only 2 mesh points on the splice surface ($\Delta x = 0.008$).

Consider the function $F(x)$ defined by

$$F(x) = \left[H\left(x - x_c + \frac{W}{2} \right) - H\left(x - x_c - \frac{W}{2} \right) \right], \qquad (10.56)$$

where x_c is the location of the center of the splice and W is its width. By means of the $F(x)$ function, the wall boundary condition for a liner with a single circumferential splice may be written as

$$R\frac{\partial v}{\partial t} - X_{-1}v + X_1\frac{\partial^2 v}{\partial t^2} = [1 - F(x)]\left[\frac{\partial p}{\partial t} + M\frac{\partial p}{\partial x} \right]. \qquad (10.57)$$

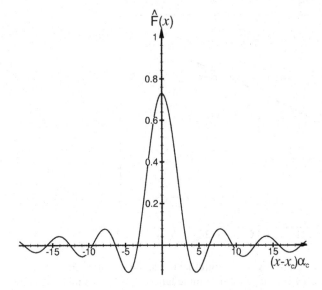

Figure 10.21. The $\hat{F}(x)$ function. $W\alpha_c = 2.5$.

Now, the Fourier transform of $F(x)$ is

$$\tilde{F}(\alpha) = \frac{e^{-i\alpha x_c}}{\pi\alpha}\sin\left(\frac{\alpha W}{2}\right). \tag{10.58}$$

Following the concept of the wave number truncated method, the high wave number components of Eq. (10.58) are to be removed at this stage. The truncated $\hat{F}(\alpha)$ function is

$$\hat{F}(\alpha) = \frac{e^{-i\alpha x_c}}{\pi\alpha}\sin\left(\frac{\alpha W}{2}\right)[H(\alpha + \alpha_c) - H(\alpha - \alpha_c)]. \tag{10.59}$$

The inverse of $\hat{F}(\alpha)$ is easily found to be

$$\hat{F}(x) = \frac{1}{\pi}[Si((x - x_c + 0.5W)\alpha_c) - Si((x - x_c - 0.5W)\alpha_c)]. \tag{10.60}$$

A graph of $\hat{F}(x)$ is given in Figure 10.21. The modified boundary condition to replace Eq. (10.57) is

$$R\frac{\partial v}{\partial t} - X_{-1}v + X_1\frac{\partial^2 v}{\partial t^2} = [1 - \hat{F}(x)]\left(\frac{\partial p}{\partial t} + M\frac{\partial p}{\partial x}\right). \tag{10.61}$$

Figure 10.22 shows the results of four computations for duct mode scattering by a circumferential splice of width 0.02 located between $x = 0.11$ and 0.13. The incident duct mode and the liner impedance are the same as in Section 10.6.1. The first three computations use boundary condition (10.57). The mesh size used in the first computation is $\Delta x = 0.002$. There is spurious acoustic scattering at the upstream end of the splice. The axial distribution of the acoustic PWL is the uppermost broken curve in Figure 10.22. The second and third computation use a mesh size of $\Delta x = 0.0005$ and 0.00025, respectively. It is clear from this figure that the spurious scattering diminishes with the use of a finer mesh. The fourth computation uses modified boundary condition (10.61) and a very coarse mesh of $\Delta x = 0.008$. This is 32 times the size of the finest mesh used. For this computation, there are only 2 mesh

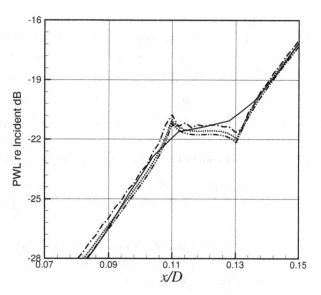

Figure 10.22. Duct mode scattering by a circumferential splice of width 0.02. Incident wave mode and liner impedance are the same as Figure 10.16. Comparison of transmitted PWL: computed by boundary condition (10.61) ——— $\Delta x = 0.008$; the following computed by boundary condition (10.57): — · — $\Delta x = 0.002$, · · · · · · · · $\Delta x = 0.0005$, — · · — $\Delta x = 0.00025$.

points on the splice. The computed axial distribution of PWL is shown as the full line in Figure 10.22. For the transmitted PWL, which is one of the important quantities of the duct mode transmission computation, it is clear that the fourth computation using a mesh size 32 times larger gives almost identical results similar to those of the finest mesh. This example further illustrates the advantage of the wave number truncation method.

EXERCISES

10.1. Consider a single-frequency sound field of frequency Ω adjacent to an acoustically treated panel or acoustic liner as shown in Figure 10.3. In other words, all the flow variables have the form $p(x, y, z, t) = \text{Re}\{\tilde{p}(x, y, z)e^{-i\Omega t}\}$ when the solution has reached a time periodic state. Let the impedance of the liner be $Z = R - iX$. Verify that the following time-domain impedance boundary condition

$$\frac{\partial p}{\partial t} = R\frac{\partial v_n}{\partial t} - X\Omega v_n \tag{A}$$

yields the correct frequency-domain impedance boundary condition $p = Zv_n$.

Show by examining the dispersion relation of the acoustic field that boundary condition (A) is stable only when $X < 0$.

10.2. An acoustic wave train of frequency $\Omega = 1.6$ kHz is sent down the normal incidence impedance tube shown in Figure 10.9. In dimensionless form, the incident wave is given by

$$p_{\text{incident}} = u_{\text{incident}} = \cos[\Omega(x - t)] = \text{Re}\left\{e^{i\Omega(x-t)}\right\}.$$

The dimensionless resistance and reactance of the treatment panel are $R = 0.18$, $X = -0.598$, respectively. Compute the reflected wave train using boundary condition (A) of Problem 10.1. For this purpose, set up an incoming wave boundary condition on the left side of the computational domain. On the right side, use the ghost point method to enforce the time-domain impedance boundary condition. Run your code long enough to attain a time periodic state before collecting numerical results.

Compare the pressure envelop inside the tube with that of the exact solution. The exact solution is

$$p_{\text{reflected}} = \text{Re}\left\{Ae^{i\Omega(x+t)}\right\}, \quad u_{\text{reflected}} = -\text{Re}\left\{Ae^{i\Omega(x+t)}\right\},$$

where $A = (Z-1)/(Z+1)$. Note: Inside the normal incidence impedance tube the incident and the reflected wave train form a standing wave pattern. The maximum pressure amplitude at every point inside the tube form the pressure envelop.

10.3 Repeat Exercise 10.1 but replace boundary condition (A) by the following

$$p = Rv_n + \frac{X}{\Omega}\frac{\partial v_n}{\partial t}. \tag{B}$$

Verify that time-domain impedance boundary condition (B) also leads to the correct frequency-domain impedance boundary condition $p = Z\,v_n$. Show that (B) is stable only when $X \geq 0$.

10.4 Repeat Exercise 10.2 but at a higher frequency of 3 kHz. At this frequency, the resistance and reactance of the treatment panel are $R = 0.18$, $X = 0.678$. Use time-domain impedance boundary condition (B) in your computation. Compare your computed pressure envelop with the exact solution.

11 Extrapolation and Interpolation

11.1 Extrapolation and Numerical Instability

Extrapolation and interpolation are often used in large-scale computation. Here, extrapolation and interpolation on a regular mesh will be considered first. Interpolation on an irregular mesh will be discussed in Chapter 13. Most textbooks treat interpolation in great depth, but few discuss extrapolation. Some books state "extrapolation should only be used with great caution." The warning is proper and real. It is a well-known fact in scientific computing that extrapolation often leads to numerical instability.

The most popular method for interpolation and extrapolation is the method of Lagrange polynomials. Suppose an N-point computation stencil with a mesh spacing Δx is used to extrapolate the values of a function $f(x)$ to the point $x_0 + \eta \Delta x$ ($\eta < 1$) (see Figure 11.1). Without the loss of generality, x_0 will be taken as the first stencil point. The Lagrange polynomials formed by the N mesh points are

$$\ell_k^{(N)}(x) = \prod_{\substack{j=0 \\ j \neq k}}^{N-1} \frac{(x - x_j)}{(x_k - x_j)}, \qquad k = 0, 1, 2, \ldots, (N-1) \tag{11.1}$$

It is easy to verify that

$$\ell_k^{(N)}(x_j) = 0 \begin{cases} j = 0, 1, 2, \ldots, (N-1) \\ j \neq k \end{cases}$$

$$\ell_k^{(N)}(x_k) = 1.$$

The extrapolated value of f at $\eta \Delta x$, according to the Lagrange polynomials method, is given by

$$f(x_0 + \eta \Delta x) \simeq \sum_{j=0}^{N-1} f(x_j) \ell_j^{(N)}(x_0 + \eta \Delta x), \tag{11.2}$$

where $f(x_j)$, $j = 0, 1, \ldots, (N-1)$, are the given values of f at the mesh points.

Despite the popularity of the Lagrange polynomials method, there is no known way to assess the errors quantitatively. Standard estimates given in textbooks are obtained by expanding the left and right sides of Eq. (11.2) by the Taylor series for

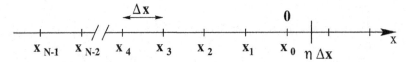

Figure 11.1. Schematic diagram showing a N-point extrapolation stencil at spacing Δx and the extrapolation point at $x_0 + \eta \Delta x$.

small Δx. This way yields an error of the order of $\partial^N f/\partial x^N|_{x_0}(\Delta x)^N$ if N mesh points are used for extrapolation. This error estimate is, however, not at all useful.

To provide an example of numerical instability arising from the use of the Lagrange polynomials extrapolation and to show that such an instability can be avoided by using an improved method, the problem of acoustic wave propagation and reflection in a long one-dimensional tube with a closed end will now be considered. The problem is shown schematically in Figure 11.2. Dimensionless variables with Δx as the length scale, a_0 (sound speed) as the velocity scale, $\Delta x/a_0$ as the time scale, ρ_0 (ambient gas density) as the density scale, $\rho_0 a_0^2$ as the pressure scale will be used. It will be assumed that the end wall of the tube is not at a mesh point, but at a distance η ($\eta < 1$) from the last mesh point at $\ell = 0$. The governing equations are the linearized momentum and energy equations.

$$\frac{\partial u}{\partial t} + \frac{\partial p}{\partial x} = 0 \tag{11.3}$$

$$\frac{\partial p}{\partial t} + \frac{\partial u}{\partial x} = 0. \tag{11.4}$$

The wall boundary condition is

$$u = 0 \quad \text{at} \quad x = \eta. \tag{11.5}$$

Suppose the problem is solved on a mesh as shown in Figure 11.2. To reduce dispersion error, the standard 7-point stencil central difference scheme (sixth order)

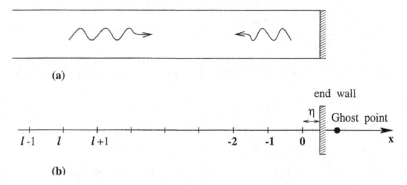

(a)

(b)

Figure 11.2. Schematic diagram showing (a) the propagation and reflection of one-dimensional acoustic waves in a long duct with a closed end, (b) the computational mesh and ghost point.

is selected to approximate the x derivatives. The semidiscretized forms of Eqs. (11.3) and (11.4) are

$$\frac{du_\ell}{dt} + \sum_{j=-3}^{3} a_j p_{\ell+j} = 0 \tag{11.6}$$

$$\frac{dp_\ell}{dt} + \sum_{j=-3}^{3} a_j u_{\ell+j} = 0. \tag{11.7}$$

Near the wall where a 7-point central difference scheme cannot fit, backward difference stencils are to be used. Since the wall is not at a mesh point, the value of u at the wall may be obtained by extrapolation using the values of u at the nearest 7 mesh points. Suppose Lagrange polynomials extrapolation is used. The wall boundary condition (11.5) becomes

$$\sum_{k=0}^{6} u_{-k} \ell_k^{(7)} = 0. \tag{11.8}$$

The ghost point method will be used to enforce this boundary condition. For this purpose, a ghost value of pressure is introduced at the ghost point $\ell = 1$. The ghost value is chosen so that Eq. (11.8) is satisfied at every time level of the computation.

Tam and Kurbatskii (2000) have solved this discretized problem analytically. Their solution reveals that for $\eta > 0.42$, this discrete system is numerically unstable. The system supports one boundary instability mode with time dependence $e^{-i\omega t}$ where $\omega = \omega_r + i\omega_i$ is the complex angular frequency. If ω_i is positive, the mode is an unstable mode. This instability can also be determined computationally by solving Eqs. (11.6) to (11.8) using a time marching scheme such as the fourth-order Runge-Kutta scheme. Figure 11.3 shows the computed value of p_ℓ at $\ell = -3$ as a function of time. By measuring the oscillation period and amplitude, the angular frequency ω_r and growth rate ω_i can be found. As shown in Figure 11.4, the numerically determined values of ω_r and ω_i are in good agreement with the analytical results.

The observed numerical instability of the acoustic wave reflection problem is due entirely to the use of Lagrange polynomials extrapolation formula. It will be shown later that, by using an improved extrapolation scheme, there is no numerical instability.

11.2 Wave Number Analysis of Extrapolation

It will be assumed that the function $f(x)$, to be extrapolated, has a Fourier transform $\tilde{f}(\alpha)$. $f(x)$ and $\tilde{f}(\alpha)$ are related by

$$f(x) = \int_{-\infty}^{\infty} \tilde{f}(\alpha) e^{i\alpha x} d\alpha. \tag{11.9}$$

For convenience, let the absolute value and argument of $\tilde{f}(\alpha)$ be A and ϕ, respectively, i.e.,

$$A(\alpha) = |\tilde{f}(\alpha)| \quad \text{and} \quad \phi(\alpha) = \arg[\tilde{f}(\alpha)]$$

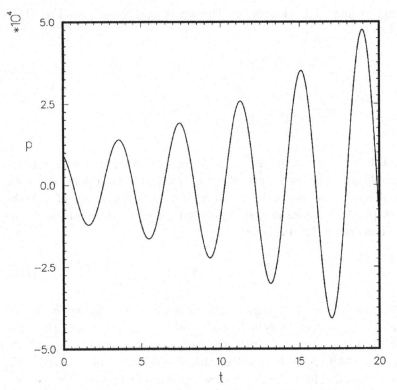

Figure 11.3. Computed pressure time-history at $\ell = -3$. The wall is located at $\eta = 0.45$.

so that Eq. (11.9) may be rewritten as

$$f(x) = \int\limits_{-\infty}^{\infty} A(\alpha)e^{i[\alpha x + \phi(\alpha)]}d\alpha. \tag{11.10}$$

A simple interpretation of Eq. (11.10) is that $f(x)$ is made up of a superposition of simple waves with wave number α and amplitude $A(\alpha)$. The primary goal here is to develop an extrapolation scheme that is highly accurate over the low wave number range, say, $-\kappa \leq \alpha \, \Delta x \leq \kappa$. For this purpose, it will be sufficient to consider waves with unit amplitude over the desired band of wave numbers. The simple wave with wave number α to be considered is

$$f_\alpha(x) = e^{i[\alpha x + \phi(\alpha)]}. \tag{11.11}$$

11.2.1 Extrapolation Error in Wave Number Space

Now, instead of Lagrange polynomials extrapolation formula (11.2), a more general formula,

$$f(x_0 + \eta\Delta x) = \sum_{j=0}^{(N-1)} S_j f(x_j), \qquad x_j = x_0 - j\Delta x, \tag{11.12}$$

will be used, where S_j $(j = 0, 1, 2, \ldots, (N-1))$ are the stencil coefficients. S_j is unknown at this stage and S_j is equal to $\ell_j^{(n)}(x_0 + \eta\Delta x)$ for the Lagrange polynomial

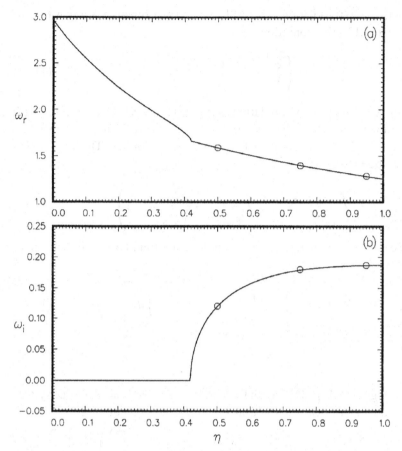

Figure 11.4. Boundary instability induced by numerical extrapolation using Lagrange poly-nomials. (a) Real part of unstable frequency ω versus the distance of extrapolation to the wall, (b) imaginary part of ω. ———, analytical; $\circ$, numerical simulation.

extrapolation method. Now, the local extrapolation error for wave number $\alpha \Delta x$, $E_{\text{local}}(\eta, \kappa, \alpha \Delta x, N)$, is defined as the square of the absolute value of the difference between the left and right side of Eq. (11.12) when a single Fourier component of Eq. (11.11) is substituted into the formula as follows:

$$E_{\text{local}}(\eta, \kappa, \alpha \Delta x, N) = \left| e^{i[\alpha(x_0+\eta\Delta x)+\phi]} - \sum_{j=0}^{N-1} S_j e^{i[\alpha(x_0-j\Delta x)+\phi]} \right|^2$$

$$= \left| e^{i\eta\alpha\Delta x} - \sum_{j=0}^{N-1} S_j e^{-ij\alpha\Delta x} \right|^2 \tag{11.13}$$

Note that $E_{\text{local}}(\eta, \kappa, -\alpha\Delta x, N) = E_{\text{local}}(\eta, \kappa, \alpha\Delta x, N)$.

The integrated error over the band of wave numbers from $\alpha\Delta x = 0$ to $\alpha\Delta x = \kappa$ is

$$E = \int_0^\kappa E_{\text{local}} \, dx = \int_0^\kappa \left| e^{i\eta\alpha\Delta x} - \sum_{j=0}^{N-1} S_j e^{-ij\alpha\Delta x} \right|^2 d(\alpha\Delta x). \tag{11.14}$$

One expects the extrapolation error to be zero if the function is a constant or $\alpha = 0$. From Eq. (11.13), this condition gives

$$\left(\sum_{j=0}^{N-1} S_j - 1 \right) = 0. \tag{11.15}$$

Now, S_j, the coefficients of extrapolation formula (11.12), is to be chosen so that E is a minimum subjected to constraint (11.15). This constrained optimization problem can easily be solved by the method of the Lagrange multiplier. The Lagrangian function may be defined as follows:

$$L = \int_0^{\kappa} \left| e^{i\eta y} - \sum_{j=0}^{N-1} S_j e^{-ijy} \right|^2 dy + \lambda \left(\sum_{j=0}^{N-1} S_j - 1 \right). \tag{11.16}$$

The coefficients S_j and the Lagrange parameter λ are found by solving the linear system as follows:

$$\frac{\partial L}{\partial S_j} = 0 \quad \text{or} \quad \mathrm{Re}\left[-\int_0^{\kappa} e^{ijy} \left(e^{i\eta y} - \sum_{k=0}^{N-1} S_k e^{-iky} \right) dy \right] + \frac{\lambda}{2} = 0 \tag{11.17}$$

$$\frac{\partial L}{\partial \lambda} = 0 \quad \text{or} \quad \sum_{j=0}^{N-1} S_j - 1 = 0. \tag{11.18}$$

Now, the integrals of Eq. (11.17) can be readily evaluated. The algebraic system of Eqs. (11.17) and (11.18) leads to the following linear matrix equation:

$$
\begin{bmatrix}
\kappa & \sin\kappa & \frac{\sin 2\kappa}{2} & \frac{\sin 3\kappa}{3} & \frac{\sin 4\kappa}{4} & \cdot & & \cdot & \frac{\sin[(N-1)\kappa]}{N-1} & \frac{1}{2} \\
\sin\kappa & \kappa & \sin\kappa & \frac{\sin 2\kappa}{2} & \frac{\sin 3\kappa}{3} & \cdot & & \cdot & \frac{\sin[(N-2)\kappa]}{N-2} & \frac{1}{2} \\
\frac{\sin 2\kappa}{2} & \sin\kappa & \kappa & \sin\kappa & \frac{\sin 2\kappa}{2} & \cdot & & \cdot & \frac{\sin[(N-3)\kappa]}{N-3} & \frac{1}{2} \\
\frac{\sin 3\kappa}{3} & \frac{\sin 2\kappa}{2} & \sin\kappa & \kappa & \sin\kappa & \cdot & & \cdot & \frac{\sin[(N-4)\kappa]}{N-4} & \frac{1}{2} \\
\cdot & \cdot & \cdot & \cdot & \cdot & \cdot & & \cdot & \cdot & \cdot \\
\cdot & \cdot & \cdot & \cdot & \cdot & \cdot & & \cdot & \cdot & \cdot \\
\frac{\sin[(N-1)\kappa]}{N-1} & \frac{\sin[(N-2)\kappa]}{N-2} & \cdot & \cdot & \cdot & \cdot & \frac{\sin 2\kappa}{2} & \sin\kappa & \kappa & \frac{1}{2} \\
1 & 1 & 1 & \cdot & & \cdot & 1 & 1 & 1 & 0
\end{bmatrix}
$$

$$
\cdot
\begin{bmatrix}
S_0 \\ S_1 \\ S_2 \\ S_3 \\ \cdot \\ \cdot \\ S_{N-1} \\ \lambda
\end{bmatrix}
=
\begin{bmatrix}
\frac{\sin(\eta\kappa)}{\eta} \\
\frac{\sin(\eta+1)\kappa}{\eta+1} \\
\frac{\sin(\eta+2)\kappa}{\eta+2} \\
\frac{\sin(\eta+3)\kappa}{\eta+3} \\
\cdot \\
\cdot \\
\frac{\sin(\eta+N-1)\kappa}{\eta+N-1} \\
1
\end{bmatrix}
\tag{11.19}
$$

In Eq. (11.19), κ is a free parameter. This parameter is to be selected so as to offer the best overall results. Once S_j's are found, the local error E_{local} may be

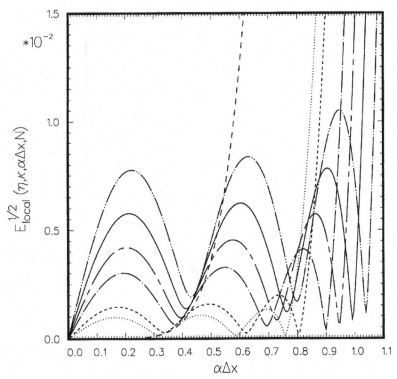

Figure 11.5. Dependence of local error on wave number over the long wave range for the case $\eta = 0.75$, $N = 7$. $\cdots\cdots \kappa = 0.8$, $----\kappa = 0.85$, $-\cdot-\kappa = 0.95$, $-\cdots-\kappa = 1.0$, ———$\kappa = 1.05$, $-\cdot-\cdots-\kappa = 1.1$, $----$ Lagrange polynomials extrapolation.

computed according to Eq. (11.13). Figure 11.5 shows the dependence of $E_{\text{local}}^{1/2}$ on wave number $\alpha \Delta x$ in the case $\eta = 0.75$ and $N = 7$ for several values of κ. This figure is typical of other values of η and N. Plotted in this figure is also the local error of the Lagrange polynomials extrapolation. It is readily seen that the Lagrange polynomials extrapolation is very accurate for low wave numbers, but the error becomes large for $\alpha \Delta x \geq 0.6$ or wavelengths shorter than $10.5\Delta x$. Many finite difference schemes used in large-scale computations and simulations can resolve waves longer than eight mesh spacings or $\alpha \Delta x \leq 0.85$. Therefore, it would be incompatible to use the Lagrange polynomials extrapolation with these schemes. On the other hand, the local error of the optimized extrapolation is larger at very low wave numbers. The error up to $\alpha \Delta x = 0.85$ is quite comparable to those of the high-order computational schemes. Figure 11.6 shows the distribution of local error over the full range of wave numbers, i.e., $0 \leq \alpha \Delta x \leq \pi$. Note that at $\alpha \Delta x = \pi$, the local error of the Lagrange polynomials method is exceedingly large.

To examine the variation of extrapolation error on η, let

$$E_{\max} (\eta, \kappa, N) = \max_{0 \leq \alpha \Delta x \leq 0.85} E_{\text{local}} (\eta, \kappa, \alpha \Delta x, N). \qquad (11.20)$$

This is the maximum error incurred in the extrapolation process if the function involved has a Fourier spectrum confined to the range $\alpha \Delta x \leq 0.85$. Figure 11.7 shows the dependence of $E_{\max}$ on η for the case $N = 7$. For $\kappa = 0.85$, $E_{\max}^{1/2}$ is less than

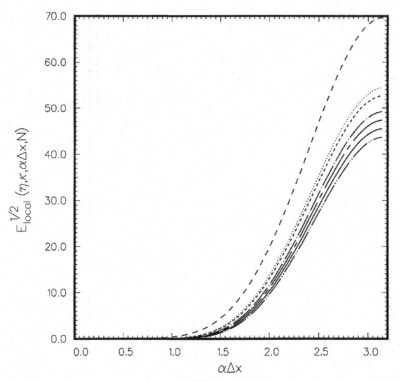

Figure 11.6. Dependence of local error on wave number (full range) for the case $\eta = 0.75$, $N = 7$. $\cdots\cdots$, $\kappa = 0.8$; $----$, $\kappa = 0.85$; $-\cdot-$, $\kappa = 0.95$; $-----$, $\kappa = 1.0$; $———$, $\kappa = 1.05$; $-\cdot\cdot\cdot\cdot-$, $\kappa = 1.1$; $----$, Lagrange polynomials extrapolation.

10^{-2} for η up to 1.0. A similar investigation for $N = 5, 6$, and 8 shows that E_{max} remains low if κ is taken to be 0.85. Based on this numerical study, it is recommended that the parameter κ be assigned the value 0.85 when used in connection with large-scale high-order finite difference computation.

11.2.2 Additional Constraint on Optimized Extrapolation

Figure 11.6 indicates that the extrapolation error for high wave numbers, particularly for $\alpha \Delta x$ near π, is very large. This means that, in a large-scale computation, the high wave number components can be greatly amplified by the extrapolation process. This observation suggests that it is this numerical error amplification mechanism that is responsible for the often encountered numerical instability associated with extrapolation. From this point of view, it would be desirable to keep the local error at $\alpha \Delta x = \pi$ and over the high wave number range smaller. This may be done by imposing an additional constraint on the optimization process. Suppose it is desired to fix the extrapolated error $E_{local}^{1/2}$ of the function at $\alpha \Delta x = \pi$ to be a prescribed value, say, $h(\eta)$. For this purpose, an additional condition,

$$e^{i(\alpha x_0 + \phi)} \left[\sum_{j=0}^{N-1} S_j e^{-ij\pi} - h(\eta) \right] = 0, \tag{11.21}$$

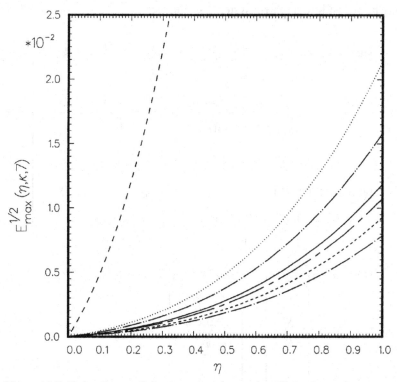

Figure 11.7. Dependence of maximum local error (maximized over $0 \leq \alpha \Delta x \leq 0.85$) on η for the case $N = 7. \cdots\cdots, \kappa = 0.8; ----, \kappa = 0.85; —\cdot—, \kappa = 0.95; -----, \kappa = 1.0; ———, \kappa = 1.05; -------, \kappa = 1.1; ----,$ Lagrange polynomials extrapolation.

is imposed. The first term of Eq. (11.21) is equal to the extrapolated value of the simple wave function $f_\alpha(x)$ of Eq. (11.11) at $\alpha \Delta x = \pi$. Thus, for all intents and purposes, it is equal to the square root of the local error. By specifying the function $h(\eta)$, one stipulates the maximum error allowed at $\alpha \Delta x = \pi$. Extensive numerical experiments have found that a good choice of $h(\eta)$ is

$$h(\eta) = 1.0 + 19\eta. \tag{11.22}$$

Upon including Eq. (11.21) as an additional constraint, the Lagrangian function to be minimized is

$$L = \int_0^\kappa \left| e^{iny} - \sum_{j=0}^{N-1} S_j e^{-ijy} \right|^2 dy + \lambda \left(\sum_{j=0}^{N-1} S_j - 1 \right) + \mu \left(\sum_{j=0}^{N-1} S_j e^{-ij\pi} - h(\eta) \right). \tag{11.23}$$

The conditions for minimization are as follows:

$$\frac{\partial L}{\partial S_j} = 0, \quad j = 0, \ 1, 2, \ldots, (N-1), \quad \frac{\partial L}{\partial \lambda} = 0, \quad \frac{\partial L}{\partial \mu} = 0. \tag{11.24}$$

Eq. (11.24) may be recast into a matrix system as follows:

$$
\begin{bmatrix}
\kappa & \sin\kappa & \frac{\sin 2\kappa}{2} & \frac{\sin 3\kappa}{3} & \frac{\sin 4\kappa}{4} & \cdot & \cdot & \frac{\sin[(N-1)\kappa]}{N-1} & \frac{1}{2} & \frac{1}{2} \\
\sin\kappa & \kappa & \sin\kappa & \frac{\sin 2\kappa}{2} & \frac{\sin 3\kappa}{3} & \cdot & \cdot & \frac{\sin[(N-2)\kappa]}{N-2} & \frac{1}{2} & -\frac{1}{2} \\
\frac{\sin 2\kappa}{2} & \sin\kappa & \kappa & \sin\kappa & \frac{\sin 2\kappa}{2} & \cdot & \cdot & \frac{\sin[(N-3)\kappa]}{N-3} & \frac{1}{2} & \frac{1}{2} \\
\frac{\sin 3\kappa}{3} & \frac{\sin 2\kappa}{2} & \sin\kappa & \kappa & \sin\kappa & \cdot & \cdot & \frac{\sin[(N-4)\kappa]}{N-4} & \frac{1}{2} & -\frac{1}{2} \\
\cdot & \cdot & \cdot & \cdot & \cdot & \cdot & \cdot & \cdot & \cdot & \cdot \\
\cdot & \cdot & \cdot & \cdot & \cdot & \cdot & \cdot & \cdot & \cdot & \cdot \\
\frac{\sin[(N-1)\kappa]}{N-1} & \frac{\sin[(N-2)\kappa]}{N-2} & \frac{\sin[(N-3)\kappa]}{N-3} & \cdot & \cdot & \cdot & \cdot & \kappa & \frac{1}{2} & (-1)^{N-1}\frac{1}{2} \\
1 & 1 & 1 & 1 & 1 & \cdot & \cdot & 1 & 1 & 0 \\
1 & -1 & 1 & -1 & 1 & \cdot & \cdot & (-1)^{N-1} & 0 & 0
\end{bmatrix}
$$

$$
\begin{bmatrix}
S_0 \\ S_1 \\ S_2 \\ S_3 \\ \cdot \\ \cdot \\ S_{N-1} \\ \lambda \\ \mu
\end{bmatrix}
=
\begin{bmatrix}
\frac{\sin(\eta\kappa)}{\eta} \\
\frac{\sin(\eta+1)\kappa}{\eta+1} \\
\frac{\sin(\eta+2)\kappa}{\eta+2} \\
\frac{\sin(\eta+3)\kappa}{\eta+3} \\
\cdot \\
\cdot \\
\frac{\sin(\eta+N-1)\kappa}{\eta+N-1} \\
1 \\
h(\eta)
\end{bmatrix}
\tag{11.25}
$$

Eq. (11.25) may be solved easily by a standard matrix equation solver. Once S_j's are found, the local error E_{local} can be calculated. Figure 11.8 shows the variation of $E_{\text{local}}^{1/2}$ with wave numbers $\alpha\Delta x$ for $\kappa = 0.85$, $N = 7$ at four values of η. For $\alpha\Delta x \leq 0.8$ or 8 mesh points or more per wavelength, the maximum extrapolation error is less than 1.5 percent. Shown in this figure also are the local errors of the Lagrange polynomials extrapolation method. It is evident that for $\alpha\Delta x > 0.6$, the error is huge. Figure 11.9 shows an identical plot but covers the entire range of wave numbers. Near $\alpha\Delta x = \pi$, there is significant difference in the errors between the two extrapolation procedures. For instance, for $\eta = 1.0$, the Lagrange polynomials method has an error about six times larger than that of the optimized scheme. One would, therefore, expect that the improved optimized scheme is less likely to induce numerical instability.

Figures 11.10 to 11.13 provide similar information as Figures 11.8 and 11.9 at stencil sizes of 5 and 8. They illustrate the effect of using a smaller or larger extrapolation stencil. Generally speaking, the use of a larger stencil reduces the error in the low wave number range but increases the error at high wave numbers. This is a piece of useful information in deciding the proper stencil size to use.

As an example that the added constraint in the optimized extrapolation method does reduce numerical instability, the one-dimensional acoustic wave propagation and reflection problem of Figure 11.2 is again considered. Now the computation is

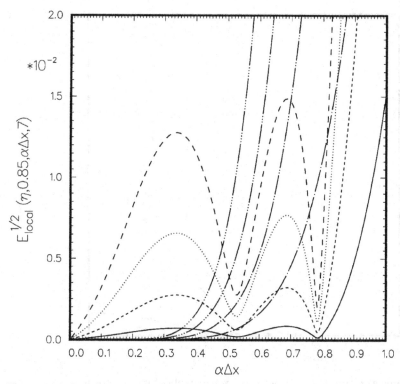

Figure 11.8. Local error as a function of wave number using the optimized extrapolation method with an added constraint. $\kappa = 0.85$, $N = 7$.

η	optimized extrapolation	Lagrange polynomials extrapolation
0.25	————	— · —
0.50	– – – –	— · · —
0.75	· · · · · · ·	— · · · —
1.00	— – – —	— · · · · —

repeated using the optimized extrapolation scheme to find the value of u at the wall, $x = \eta$,; i.e.,

$$\sum_{j=0}^{6} u_{-j} S_j(\eta) = 0. \tag{11.26}$$

It can be shown analytically and demonstrated computationally (see Tam and Kurbatskii, 2000) that there is no boundary instability. In other words, the optimized scheme with the additional constraint has effectively eliminated the boundary instability mode created by numerical extrapolation using the Lagrange polynomials method.

11.3 Optimized Interpolation Method

In this section, a wave number analysis of interpolation error is performed. At the same time, an optimized interpolation procedure is formulated. Interpolation is generally a numerically stable operation incurring relatively little error. For this

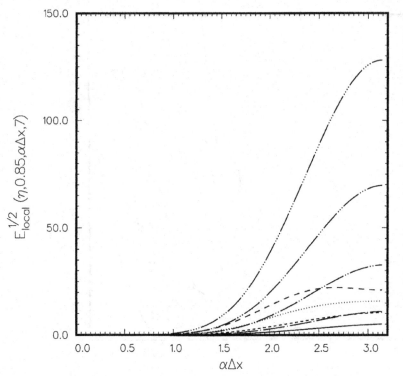

Figure 11.9. Same as Figure 11.8, except the full wave number range is shown.

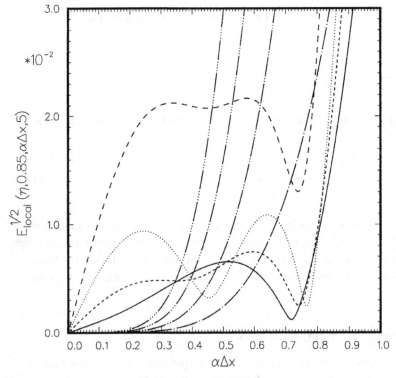

Figure 11.10. Same as Figure 11.8, except for $N = 5$.

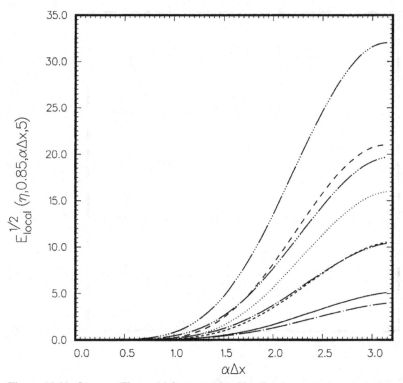

Figure 11.11. Same as Figure 11.9, except for $N = 5$.

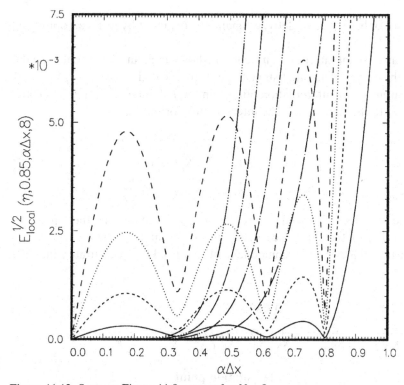

Figure 11.12. Same as Figure 11.8, except for $N = 8$.

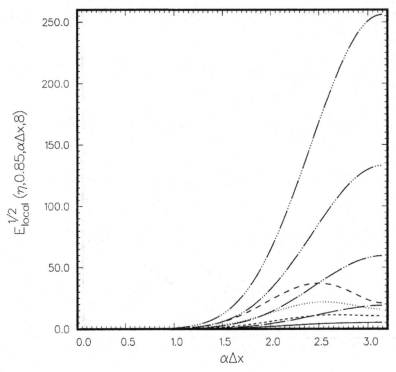

Figure 11.13. Same as Figure 11.9, except for $N = 8$.

reason, one can only expect a relatively small improvement by using an optimized scheme.

Consider an N-point interpolation stencil as shown in Figure 11.14. Let x_0 be the first point of the stencil. The other stencil points are located at $x_j = x_0 - j\Delta x$ ($j = 1$ to $(N - 1)$). Suppose the interpolation point is in the Kth interval of the stencil at a distance of $\eta\Delta x$ to the right of x_K. The interpolation formula is

$$f(x_0 - (K - \eta)\Delta x) = \sum_{j=0}^{N-1} S_j f(x_j). \tag{11.27}$$

As before, the local interpolation error, $\overline{E}_{\text{local}}$, is defined as the square of the absolute value of the difference between the left and the right sides of Eq. (11.27) when the single Fourier component of Eq. (11.11) is substituted into the

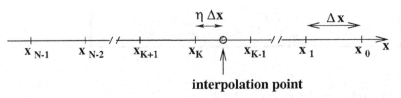

Figure 11.14. A N-point interpolation stencil showing the location of the interpolation point.

formula, i.e.,

$$\overline{E}_{\text{local}} = \left| e^{i[\alpha(x_0 - K\Delta x + \eta \Delta x) + \phi]} - \sum_{j=0}^{N-1} S_j e^{i[\alpha(x_0 - j\Delta x) + \phi]} \right|^2$$

$$= \left| e^{-i(K-\eta)\alpha \Delta x} - \sum_{j=0}^{N-1} S_j e^{-ij\alpha \Delta x} \right|^2 \tag{11.28}$$

Now, the local error $\overline{E}_{\text{local}}(\eta, \kappa, \alpha \Delta x, N, K)$ depends on the parameter K as well as $\eta, \kappa, \alpha \Delta x$, and N.

The integrated error over a band of wave numbers from $\alpha \Delta x = 0$ to $\alpha \Delta x = \kappa$ is

$$\overline{E} = \int_0^\kappa \left| e^{-i(K-\eta)\alpha \Delta x} - \sum_{j=0}^{N-1} S_j e^{-ij\alpha \Delta x} \right|^2 d(\alpha \Delta x). \tag{11.29}$$

Now, the interpolation coefficients S_j are chosen so that $\overline{E}$ is a minimum subjected to the condition that there is no error for zero wave number. The zero wave number condition may be written as

$$\overline{E}_{\text{local}}(\eta, \kappa, 0, K, N) = \left| 1 - \sum_{j=0}^{N-1} S_j \right|^2 = 0. \tag{11.30}$$

The Lagrangian function of this constrained minimization problem is

$$\overline{L} = \int_0^\kappa \left| e^{-i(K-\eta)y} - \sum_{k=0}^{N-1} S_k e^{-iky} \right|^2 dy + \lambda \left(\sum_{j=0}^{N-1} S_j - 1 \right). \tag{11.31}$$

The conditions for minimum are as follows:

$$\frac{\partial \overline{L}}{\partial S_j} = 0, \quad \frac{\partial \overline{L}}{\partial \lambda} = 0, \quad j = (0, 1, 2, \ldots, (N-1)). \tag{11.32}$$

Eq. (11.32) yields the following algebraic equations

$$\sum_{j=0}^{N-1} S_j \frac{\sin(\ell - j)\kappa}{(\ell - j)} + \frac{\lambda}{2} = \frac{\sin(\ell - K + \eta)\kappa}{(\ell - K + \eta)}, \quad \ell = 0, 1, 2, \ldots, (N-1) \tag{11.33a}$$

$$\sum_{j=0}^{N-1} S_j = 1. \tag{11.33b}$$

The linear system Eq. (11.33) may be rewritten in a matrix form similar to Eq. (11.19). The matrix equation can be solved easily to provide the interpolation coefficients, S_j $(j = 0, 1, 2, \ldots, (N-1))$.

The local interpolation error $\overline{E}_{\text{local}}$ as a function of wave number $\alpha \Delta x$ depends on a number of parameters. They are N (the stencil size), K (the interval number), η (the distance to the mesh point), and the free parameter κ. Numerical results are now provided to offer an idea on the magnitude of the error and how it is influenced by the various parameters.

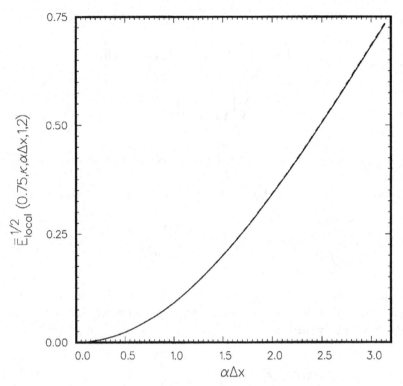

Figure 11.15. Dependence of local interpolation error on wave number. $N = 2$, $K = 1$, $\eta = 0.75$. $\cdots\cdots\cdots$, $\kappa = 0.85$; $----$, $\kappa = 0.9$; $-\cdot-$, $\kappa = 0.95$; $-----$, $\kappa = 1.0$; ———, $\kappa = 1.05$; $-------$, $\kappa = 1.1$; $----$, Lagrange polynomials interpolation.

Consider first the effect of N, the size of the stencil. Figure 11.15 shows the distribution of $\overline{E}_{\text{local}}^{1/2}$ in wave number space for $\eta = 0.75$ using only two interpolation points, $N = 2$. This is a very special case. The parameter κ appears to have a negligible effect on the local error. In addition, the optimized scheme yields almost identical results as the Lagrange polynomials interpolation. The error is relatively large in the high wave number range, but relatively small at low wave numbers. The interpolation point will now be kept in the first interval of the stencil, i.e., $K = 1$, but the stencil size is allowed to increase. Figures 11.16 and 11.17 show the local error distributions as functions of wave number at various values of κ with $N = 4$ or four interpolation points. On comparison with Figure 11.15, it is clear that, by increasing the size of the stencil, the error in the low wave number range drops rapidly. At the same time, the error in the high wave numbers increases. Figures 11.18 and 11.19 provide the results for an even larger stencil, $N = 7$. In this case, the low wave number error becomes negligible, but the error in the neighborhood of $\alpha\Delta x = \pi$ is much larger.

If the interpolation point lies in the interior of a large stencil, one intuitively expects much smaller error than when it is located at the first interval. Figures 11.20 and 11.21 show the numerical results for $N = 7$, $\eta = 0.75$, and $K = 2$. Figures 11.22 and 11.23 show the corresponding results for $K = 3$. These results completely confirm the intuitive expectation. Over the long wave range, $\alpha\Delta x \leq 0.9$, there is hardly any error when the interpolation point lies in the center interval for large N

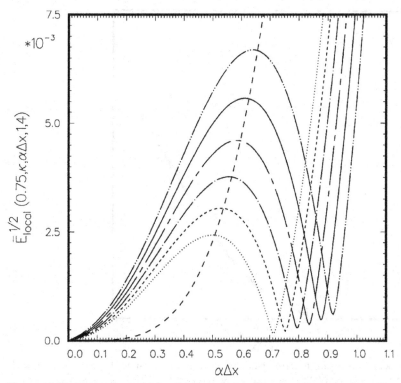

Figure 11.16. Dependence of local interpolation error on wave number. $N = 4$, $K = 1$, $\eta = 0.75$. $\cdots\cdots$, $\kappa = 0.85$; $----$, $\kappa = 0.9$; $-\cdot-$, $\kappa = 0.95$; $-----$, $\kappa = 1.0$; ———, $\kappa = 1.05$; $-------$, $\kappa = 1.1$; $----$, Lagrange polynomials interpolation.

The effect of the distance of the interpolation point from the stencil point, namely, the magnitude of η will now be considered. Figures 11.24 and 11.25 give the local errors for $\eta = 0.25$. The values of the other parameters are the same as those of Figures 11.18 and 11.19. As readily seen, the two sets of figures are about the same. This indicates that η has a relatively small effect on the local error.

Finally, the effect of free parameter κ is considered. After reviewing and comparing all the numerical results, it is found that a good overall choice for this parameter is 1.0. This value of κ provides a large range of low wave numbers over which the interpolation error is small. At the same time, the error at high wave numbers is still numerically small. In most large-scale computations, the high wave number components will remain unresolved. They are the spurious numerical waves, which are suppressed by the inclusion of artificial selective damping or filtering. As a result, the amplitude of the high wave number components is expected to be small. Any amplification by the interpolation process would still be small. They should not be a concern except in unusual circumstances.

11.4 A Numerical Example

As an example on the use of extrapolation and interpolation in large-scale computing, the problem of scattering of plane acoustic waves by a cylinder in two dimensions is considered. Let the diameter of the cylinder be D. The governing equations are

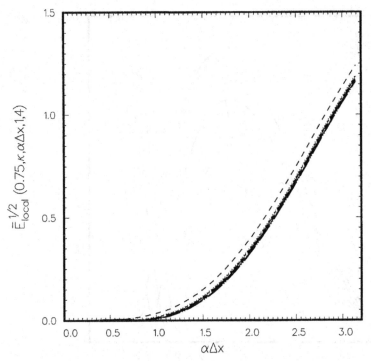

Figure 11.17. Local interpolation error over the wave number range $0 \le \alpha \Delta x \le \pi$. $N = 4$, $K = 1$, $\eta = 0.75$. $\cdots\cdots$, $\kappa = 0.85$; $----$, $\kappa = 0.9$; $-\cdot-$, $\kappa = 0.95$; $-\cdot\cdot\cdot-$, $\kappa = 1.0$; ———, $\kappa = 1.05$; $------$, $\kappa = 1.1$; $----$, Lagrange polynomials interpolation.

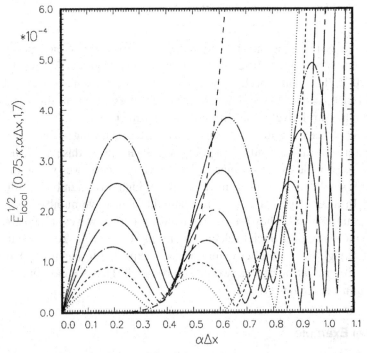

Figure 11.18. Dependence of local interpolation error on wave number. $N = 7$, $K = 1$, $\eta = 0.75$. $\cdots\cdots$, $\kappa = 0.85$; $----$, $\kappa = 0.9$; $-\cdot-$, $\kappa = 0.95$; $-----$, $\kappa = 1.0$; ———, $\kappa = 1.05$; $------$, $\kappa = 1.1$; $----$, Lagrange polynomials interpolation.

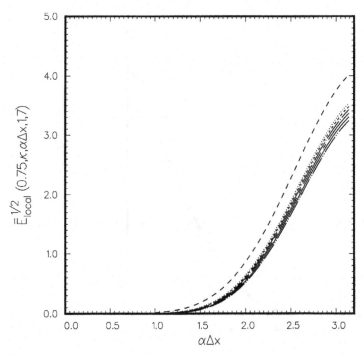

Figure 11.19. Local interpolation error over the wave number range $0 \le \alpha\Delta x \le \pi$. $N = 7$, $K = 1$, $\eta = 0.75$. $\cdots\cdots$, $\kappa = 0.85$; $----$, $\kappa = 0.9$; $-\cdot-$, $\kappa = 0.95$; $-----$, $\kappa = 1.0$; ———, $\kappa = 1.05$; $-\cdot-\cdot-\cdot-$, $\kappa = 1.1$; $----$, Lagrange polynomials interpolation.

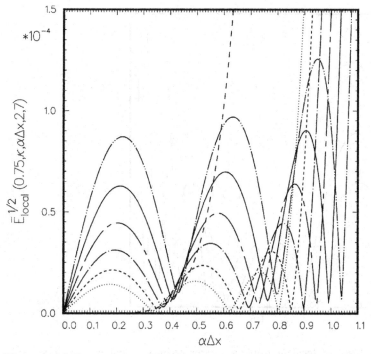

Figure 11.20. Dependence of local interpolation error on wave number. $N = 7$, $K = 2$, $\eta = 0.75$. $\cdots\cdots$, $\kappa = 0.85$; $----$, $\kappa = 0.9$; $-\cdot-$, $\kappa = 0.95$; $-----$, $\kappa = 1.0$; ———, $\kappa = 1.05$; $-\cdot-\cdot-\cdot-$, $\kappa = 1.1$; $----$, Lagrange polynomials interpolation.

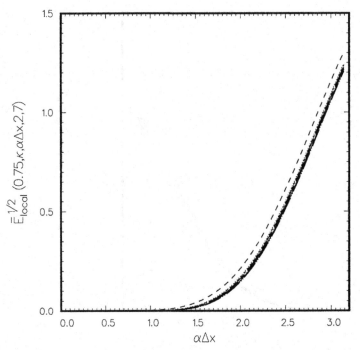

Figure 11.21. Local interpolation error over the wave number range $0 \le \alpha \Delta x \le \pi$. $N = 7$, $K = 2$, $\eta = 0.75$. $\cdots\cdots$, $\kappa = 0.85$; $----$, $\kappa = 0.9$; $-\cdot-$, $\kappa = 0.95$; $-----$, $\kappa = 1.0$; ———, $\kappa = 1.05$; $-------$, $\kappa = 1.1$; $----$, Lagrange polynomials interpolation.

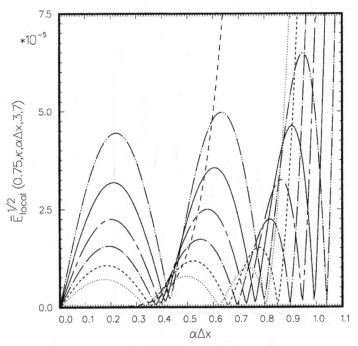

Figure 11.22. Dependence of local interpolation error on wave number. $N = 7$, $K = 3$, $\eta = 0.75$. $\cdots\cdots$, $\kappa = 0.85$; $----$, $\kappa = 0.9$; $-\cdot-$, $\kappa = 0.95$; $-----$, $\kappa = 1.0$; ———, $\kappa = 1.05$; $-------$, $\kappa = 1.1$; $----$, Lagrange polynomials interpolation.

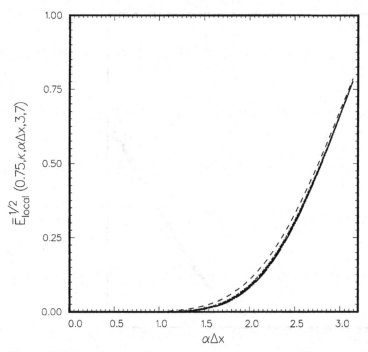

Figure 11.23. Local interpolation error over the wave number range $0 \le \alpha \Delta x \le \pi$. $N = 7$, $K = 3$, $\eta = 0.75$. $\cdots\cdots$, $\kappa = 0.85$; $----$, $\kappa = 0.9$; $-\cdot-$, $\kappa = 0.95$; $-\cdot--\cdot$, $\kappa = 1.0$; $\longrightarrow$, $\kappa = 1.05$; $------$, $\kappa = 1.1$; $----$, Lagrange polynomials interpolation.

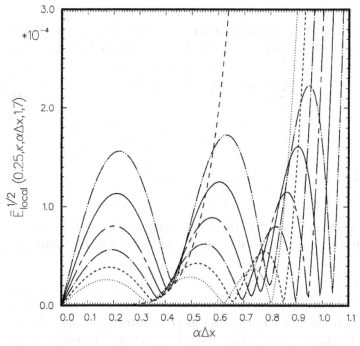

Figure 11.24. Dependence of local interpolation error on wave number. $N = 7$, $K = 1$, $\eta = 0.25$. $\cdots\cdots$, $\kappa = 0.85$; $----$, $\kappa = 0.9$; $-\cdot-$, $\kappa = 0.95$; $-\cdot--\cdot$, $\kappa = 1.0$; $\longrightarrow$, $\kappa = 1.05$; $------$, $\kappa = 1.1$; $----$, Lagrange polynomials interpolation.

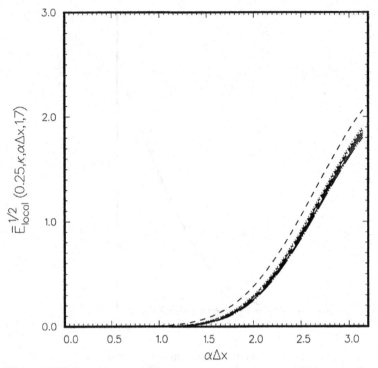

Figure 11.25. Local interpolation error over the wave number range $0 \leq \alpha \Delta x \leq \pi$. $N = 7$, $K = 1$, $\eta = 0.25$. $\cdots\cdots\cdots$, $\kappa = 0.85$; $----$, $\kappa = 0.9$; $-\cdot-\cdot$, $\kappa = 0.95$; $-----$, $\kappa = 1.0$; ———, $\kappa = 1.05$; $--------$, $\kappa = 1.1$; $----$, Lagrange polynomials interpolation.

the linearized Euler equations. In dimensionless form, they may be written as (with respect to length scale D, velocity scale a_0 (speed of sound), time scale D/a_0, density scale ρ_0, and pressure scale $\rho_0 a_0{}^2$).

$$\frac{\partial \mathbf{U}}{\partial t} + \frac{\partial \mathbf{E}}{\partial x} + \frac{\partial \mathbf{F}}{\partial y} = 0, \tag{11.34}$$

where

$$\mathbf{U} = \begin{bmatrix} \rho \\ u \\ v \\ p \end{bmatrix}, \qquad \mathbf{E} = \begin{bmatrix} u \\ p \\ 0 \\ u \end{bmatrix}, \qquad \mathbf{F} = \begin{bmatrix} v \\ 0 \\ p \\ v \end{bmatrix}.$$

On the surface of the cylinder, the boundary condition is $\mathbf{V} \cdot \mathbf{n} = 0$, where $\mathbf{n}$ is the unit normal to the surface. It can be shown easily that an equivalent form of the boundary condition is

$$\frac{\partial p}{\partial n} = 0. \tag{11.35}$$

Eq. (11.34) may be discretized on a Cartesian mesh using the dispersion-relation-preserving scheme. The scattered sound field may be found by time marching the numerical solution to a time periodic state. However, because the cylinder has a curved surface, a special boundary treatment is needed to enforce boundary condition (11.35). For this purpose, the ghost point method has been extended to treat

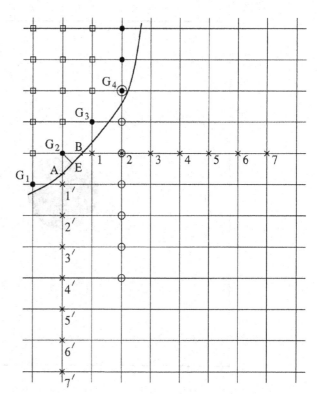

Figure 11.26. Cartesian boundary treatment of curved wall surfaces.

arbitrarily curved wall boundaries (see Kurbatskii and Tam, 1997). As an illustration of the boundary treatment, consider the curved boundary shown in Figure 11.26. First, the boundary curve is approximated by line segments joining the intersection points of the computation mesh and the wall boundary. For instance, the curved surface between points A and B in Figure 11.26 is replaced by a straight-line segment. G_2 is a ghost point. A ghost value of pressure is assigned to G_2 to enforce boundary condition (11.35). The enforcement point is at E; G_2E is perpendicular to AB. Since $\partial p/\partial x$ and $\partial p/\partial y$ are not known, except on the mesh points, their values at A and B are found by extrapolation from the points at $(1', 2', 3', 4', 5', 6', 7')$ and $(1, 2, 3, 4, 5, 6, 7)$, respectively. The corresponding values at E are then calculated by interpolation from those at A and B. Once $\partial p/\partial x$ and $\partial p/\partial y$ are found at E, Eq. (11.35) can be enforced readily.

To show the impact of the extrapolation scheme used on the computed results of the scattered waves, a computation domain of 320×320 with $\Delta x = \Delta y = D/32$ is used in the solution of the scattering problem. The wavelength of the incoming acoustic waves is equal to $8\Delta x$. The method of numerical treatment of incoming plane acoustic waves of Section 10.1 is followed in the computation. Because of the special boundary treatment needed to enforce the curved wall boundary conditions, the accuracy of the computed scattered wave is greatly influenced by the extrapolation formula used. Figure 11.27 shows the computed zero pressure contours of the scattered waves at the beginning of a cycle when the Lagrange polynomials extrapolation method is used in the computation. Shown in dotted lines are the zero pressure contours of the exact solution. It is obvious that there are significant differences between

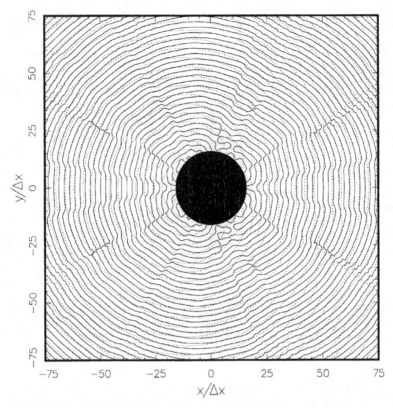

Figure 11.27. Zero pressure contours of the scattered sound field at the beginning of a cycle. Lagrange polynomials are used for extrapolation. ——, numerical results; · · · · · · ·, exact solution.

the two sets of contours. It is not difficult to understand why there are such large discrepancies. Wave number analysis indicates that, at 8 mesh points per wavelength, the Lagrange polynomials extrapolation method would give rise to large errors. Such errors contaminate the entire computed scattered wave field.

Figure 11.28 is an identical computation using the optimized extrapolation scheme with an added constraint. Clearly, there is excellent agreement between the numerical results and the exact solution. This is, of course, not surprising since the optimized extrapolation as well as the time marching algorithm are designed to yield fairly accurate results for waves with wavelengths of eight mesh spacings or longer.

EXERCISES

11.1. The optimized extrapolation method can be applied even when the stencil points are not spaced uniformly apart. Let the coordinates of the stencil points be $x_0, x_{-1}, x_{-2}, \ldots, x_{-7}$ (a 7-point stencil). Suppose the extrapolation point is at $x = x_0 + \eta$. The extrapolation formula is

$$f(x_0 + \eta) = \sum_{j=0}^{6} S_j f(x_{-j}).$$

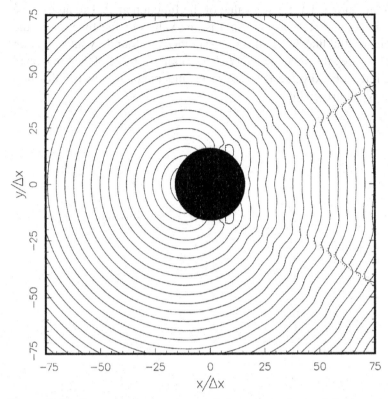

Figure 11.28. Zero pressure contours of the scattered sound field at the beginning of a cycle. Optimized extrapolation with an added constraint is used. ———, numerical results; ·······, exact solution.

Determine the matrix system by which optimized extrapolation coefficients, S_j, can be calculated. It is suggested that the averaged spacing between adjacent stencil points, Δ, be used as the length scale.

11.2. Consider two-dimensional interpolation using a rectangular stencil of N by M mesh points. To find the interpolated value at a point P with coordinates (x_0, y_0), one may first find the values of the function at the intersection of line AA' (see Figure 11.29) and the mesh lines by interpolation in the y direction. The next

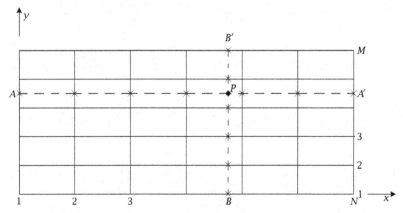

Figure 11.29. Interpolation in two dimensions.

step is to interpolate along AA' for the value of the function at P. The alternative is to interpolate in the x direction first. That is, the values of the function at the intersection points on the line BB' with the mesh lines are found by interpolation first. Then the values are interpolated along BB' to obtain the value of the function at P. Show that the interpolated value at P is the same independent of which direction the interpolation is performed first.

12 Multiscales Problems

Many aeroacoustics problems involve multiple length and time scales. This should not be difficult to understand. For, in addition to the intrinsic sizes and scales of the noise sources, the acoustic wavelength is an inherent length scale of the problem. In many instances, the length scale of the noise source differs greatly from the acoustic wavelength. This leads to a large disparity in length scales as in classical multiscales problems. For example, in supersonic jet noise, Mach wave radiation is generated by the instability waves of the jet flow. The instability waves are supported by the thin shear layer of the jet. In the region near the nozzle exit, the averaged shear layer thickness is about $0.1D$, where D is the jet diameter. The acoustic wavelength, on the other hand, is two or more jet diameters long. Thus, there is an order of magnitude difference between those characteristic lengths. In sound scattering problems, the length scale of the surface geometry of the scatterers may be much smaller than the acoustic wavelength. This occurs very often in edge scattering and diffraction problems. A concrete example is the radiation of fan noise from a jet engine inlet. The acoustic wavelength could be much longer than the radius of the lip of the engine inlet. To obtain an accurate numerical solution of the inlet diffraction problem, a fine mesh is needed around the lip region. Oftentimes, an aeroacoustics problem becomes a multiscales problem because of the change in the physics governing the different parts of the computational domain. An example is the shedding of vortices at the edge of a resonator or a sharp edge of a solid body induced by high-intensity incident sound waves. Away from the solid surface, the fluid is nearly inviscid, but close to the wall, the viscosity effect dominates. The oscillatory motion of the incident sound waves induces a very thin Stokes layer on the solid surface. The Stokes layer rolls up at the corner of a solid surface to form vortices that shed periodically. To simulate the vortex shedding process, therefore, it is necessary to use very fine mesh close to the solid surface and around the corner to resolve the Stokes layer. But away from the solid surface, a coarse mesh with 7 mesh points per acoustic wavelength is all that is needed to capture the sound waves accurately using the 7-point stencil dispersion-relation-preserving (DRP) scheme.

Computationally, a very effective way to treat a multiscales problem is to use a multisize mesh. Figure 12.1 shows a typical multisize mesh. In this example, the mesh size changes by a factor of two across the mesh-size-change interface. A factor-of-2 change in the mesh size is not an absolute necessity. It is recommended because a

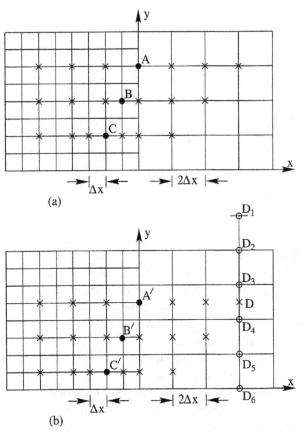

(a)

(b)

Figure 12.1. Special stencils for use in the mesh-size-change buffer regions. •, mesh points at which finite difference approximation of spatial derivative is to be found; ×, stencil points; ○, points used for interpolation.

larger change may result in a very strong artificial discontinuity inside the computational domain. On the other hand, a smaller change may require several steps of changes to achieve the same total change in mesh size. This will give rise to a large number of mesh-size-change interfaces, which is not desirable.

The time step used in a computation is dictated very often by the choice of spatial mesh size. Recall that numerical instability requirements link Δt, the time step, to the mesh size Δx. If a single time step marching scheme such as the Runge-Kutta method is used, then Δt is dictated by the smallest size mesh of the entire computational domain. This results in an inefficient computation. The optimum situation is to use a Δt consistent with the local numerical stability requirement. In such an arrangement, most of the computation is concentrated in regions with fine meshes, as it should be. In the coarse mesh region, the solution is updated only occasionally with a much larger time step. This type of algorithm is referred to as "multi-size-mesh multi-time-step" schemes. If the mesh-size change is a factor of 2 across the mesh block interface, the time step should also change by a factor of 2 to maintain numerical stability. For multilevel time marching schemes such as the DRP scheme, the use of multiple time steps with a factor-of-2 change between adjacent mesh blocks is easy to arrange. Figure 12.2 shows such an arrangement in which the computed solution advanced in steps of Δt in the fine mesh region and in steps of $2 \Delta t$ in the coarse mesh region.

In carrying out a multi-size-mesh multi-time-step computation, the computation scheme to be used in each of the uniform mesh blocks is the same as if there is a single

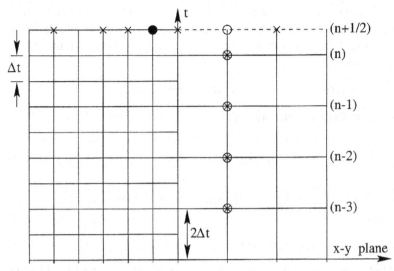

Figure 12.2. Special time marching stencil for use in the mesh-size-change buffer region. o, mesh points at which solution to be found at a half-time level; ⊗, stencil points.

size mesh throughout the entire computation domain. The exception is for a narrow buffer region at the block interface. In the buffer region, special stencils are to be used so that information on the solution can be transmitted through without significant degradation. Also, note that any surface of discontinuity, no matter whether it is a mesh-size-change interface or an internal or external boundary, is a likely source of short spurious numerical waves. To suppress the generation of spurious short waves, special artificial selective damping stencils should be incorporated in the computation scheme for mesh points adjacent to the block interface. The design of these special stencils is given in the next three sections.

12.1 Spatial Stencils for Use in the Mesh-Size-Change Buffer Region

If a factor-of-2 increase in the mesh size between adjacent blocks is used, then every other mesh line in the fine mesh block continues into the coarse mesh block as shown in Figure 12.1. The remaining set of mesh lines terminates at the mesh-size-change interface. To compute the x derivative for points on the coarse grid, including points on the interface such as point A, the 7-point central difference DRP scheme may be used even though part of the stencil is extended into the fine mesh region. For mesh points on the fine grid, again the 7-point central difference DRP scheme may be used except for the first three columns (or rows) of mesh points right next to the coarse grid. For points on the continuing mesh lines such as points B and C in Figure 12.1, special central difference stencils, as indicated, are to be used. These stencils may be written in the following form:

$$\left(\frac{\partial f}{\partial x}\right)_B \simeq \frac{1}{\Delta x} \sum_{j=-3}^{3} a_j^B f_j^B \tag{12.1}$$

$$\left(\frac{\partial f}{\partial x}\right)_C \simeq \frac{1}{\Delta x} \sum_{j=-3}^{3} a_j^C f_j^C. \tag{12.2}$$

To find the stencil coefficients a_j^B and a_j^C ($j = -3$ to 3), the optimization procedure of Chapter 2 may be followed. Now, first consider the stencil for point B. As in Section 11.2, it will be assumed that $f(x)$ has a Fourier transform $\tilde{f}(\alpha)$ with absolute value $A(\alpha)$ and argument $\phi(\alpha)$; i.e.,

$$A(\alpha) = |\tilde{f}(\alpha)|, \qquad \phi(\alpha) = \arg[\tilde{f}(\alpha)].$$

The Fourier transform may be written as

$$f(x) = \int_{-\infty}^{\infty} A(\alpha) e^{i[\alpha x + \phi(\alpha)]} d\alpha. \tag{12.3}$$

In other words, $f(x)$ is made up of a superposition of simple waves, $f_\alpha(x)$, of the following form:

$$f_\alpha(x) = e^{i[\alpha x + \phi(\alpha)]}, \tag{12.4}$$

weighted by $A(\alpha)$. Now the accuracy of finite difference approximation (12.1) of each simple wave component of $f(x)$ is examined. Upon substituting $f_\alpha(x)$ into Eq. (12.1), the finite difference approximation becomes

$$\alpha f_\alpha(x) \simeq \bar{\alpha} f_\alpha(x),$$

where

$$\bar{\alpha} = \frac{2}{\Delta x} \left[a_1^B \sin(\alpha \Delta x) + a_2^B \sin(3\alpha \Delta x) + a_3^B \sin(5\alpha \Delta x) \right] \tag{12.5}$$

is the wave number of the finite difference stencil. In deriving (12.5), the antisymmetric condition, $a_{-j}^B = -a_j^B$, has been invoked.

On following the steps of Chapter 2, the condition that Eq. (12.1) or Eq. (12.5) be accurate to order $(\Delta x)^4$ is imposed. By means of Taylor series expansion, this condition yields the following restrictions on the stencil coefficients:

$$2(a_1^B + 3a_2^B + 5a_3^B) = 1$$
$$a_1^B + 3^3 a_2^B + 5^3 a_3^B = 0 \tag{12.6}$$

The coefficients will now be chosen so that the error of using $\bar{\alpha} \Delta x$ to approximate $\alpha \Delta x$ over the band of wave number $|\alpha \Delta x| < \kappa$, E, is minimum subjected to conditions (12.6). For this purpose, the error is defined as

$$E = \int_0^\kappa [\bar{\alpha} \Delta x - \alpha \Delta x]^2 d(\alpha \Delta x) \tag{12.7}$$

The condition for a minimum is given by (after eliminating a_2^B and a_3^B by Eq. (12.6))

$$\frac{dE}{da_1^B} = 0. \tag{12.8}$$

An extensive numerical study of the effects of the choice of κ on $\bar{\alpha}(\alpha)$ and $d\bar{\alpha}/d\alpha$ has been carried out. Based on the numerical results, it is deemed that a good choice

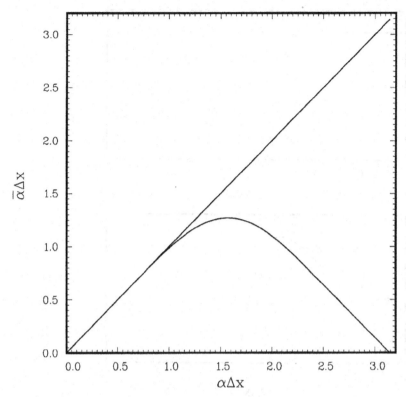

Figure 12.3. Dependence of $\bar{\alpha}\Delta x$ on $\alpha\Delta x$ for stencil B. $\kappa = 0.85$.

of the values of κ is 0.85. For this choice, the values of the stencil coefficients are

$$a_0^B = 0.0$$
$$a_1^B = a_{-1}^B = 0.595328177715$$
$$a_2^B = a_{-2}^B = -0.037247422191$$
$$a_3^B = a_{-3}^B = 0.003282817772.$$

Figures 12.3 and 12.4 show the corresponding $\bar{\alpha}(\alpha)$ and $d\bar{\alpha}/d\alpha$ curves. The resolution and dispersion characteristics of the stencil lie in between those of the central difference DRP scheme on the two sides of the interface.

On proceeding as for stencil B, the stencil coefficients for stencil C may be found. For stencil C, it is recommended that the value $\kappa = 1.0$ be used. The stencil coefficients are

$$a_0^C = 0.0$$
$$a_1^C = a_{-1}^C = 0.726325187522$$
$$a_2^C = a_{-2}^C = -0.120619908868$$
$$a_3^C = a_{-3}^C = 0.003728657553.$$

Figures 12.5 and 12.6 show the corresponding $\bar{\alpha}(\alpha)$ and $d\bar{\alpha}/d\alpha$ curves for this stencil. Again, the resolution and dispersion characteristics lie in between those of the central difference DRP scheme on the two sides of the interface.

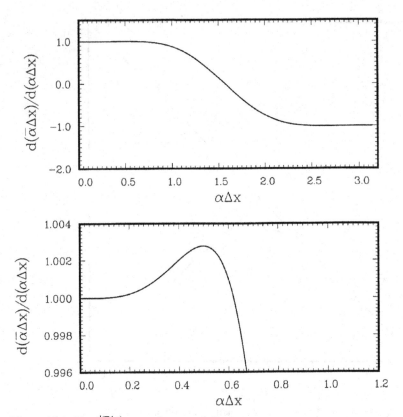

Figure 12.4. The $\frac{d(\bar{\alpha}\Delta x)}{d(\alpha\Delta x)}$ curve for stencil B.

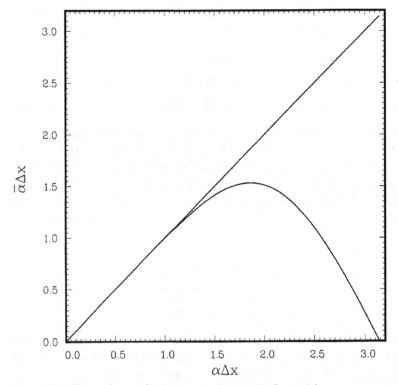

Figure 12.5. Dependence of $\bar{\alpha}\Delta x$ on $\alpha\Delta x$ for stencil C. $\kappa = 1.0$.

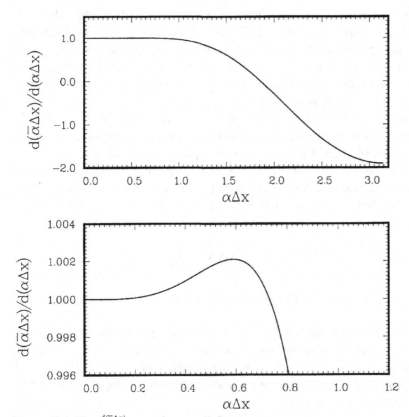

Figure 12.6. The $\frac{d(\bar{\alpha}\Delta x)}{d(\alpha\Delta x)}$ curve for stencil C.

For mesh points lying on the terminating mesh lines such as points A', B', and C' of Figure 12.1, the same stencils as for A, B, and C may be used, However, the stencils extend into points in the coarse mesh region where the solution is not computed. To obtain the values of the solution at these points, such as point D, it is recommended that interpolation be used. The interpolation stencil, a symmetric stencil is preferred, makes use of the values of the adjacent 6 mesh points, D_1 to D_6, as shown in Figure 12.1. In actual computation, the interpolation step is executed only after the solution on the coarse mesh region has been advanced by a time step. The interpolated values allow the solution on the fine mesh side to continue advancing in time.

12.2 Time Marching Stencil

The time marching step, Δt, is constrained by the numerical stability or accuracy requirements of a computation scheme. In general, the stability or accuracy requirements link Δt directly to the mesh size Δx. Thus, in a region with large mesh size, a larger Δt may be used. When more than one mesh size is used, the optimum strategy is to use the largest time step permissible in each region. It follows, if the mesh size changes by a factor of 2 in adjacent domains, the time step should also change by the same ratio. Figure 12.2 shows the time levels of the computation in the mesh-size-change buffer region. In the fine mesh region on the left, the time step is Δt. In the coarse mesh region on the right, the time step is $2\Delta t$. Effectively, the fine mesh region is computed twice as often as the coarse mesh region.

Suppose the solution is known at time level n (see Figure 12.2), the solution in the fine mesh region may be advanced by a time step Δt to time level $(n + 1/2)$ in the usual way. Once this is completed, the next step is to advance the solution to time level $(n + 1)$ in both the fine and coarse mesh regions. It is straightforward to carry out this step in the coarse mesh region by using the solution on time levels n, $(n - 1)$, $(n - 2)$, and $(n - 3)$. However, for points in the buffer region on the fine mesh side, the stencils extend into the coarse mesh region. There is no information at the $(n + 1/2)$ time level at these points on the coarse mesh side. To provide the needed information of the solution, it is necessary to compute the solution at the time level $(n + 1/2)$ for the first two rows or columns of the mesh point on the coarse mesh side based on the solution at time levels n, $(n - 1)$, $(n - 2)$, and $(n - 3)$. To advance the solution by a half time step as shown in Figure 12.2, the following four-level scheme may be used:

$$\mathbf{u}_{\ell,m}^{(n+\frac{1}{2})} = \mathbf{u}_{\ell,m}^{(n)} + (2\Delta t) \sum_{j=0}^{3} b_j^* \mathbf{K}_{\ell,m}^{(n-j)}, \tag{12.9}$$

where $\mathbf{K} = d\mathbf{u}/dt$ is given by the governing equation, and (ℓ, m) are the spatial indices.

To find the stencil coefficients b_j^* of Eq. (12.9), one may consider $\mathbf{u}$ to be made up of a Fourier spectrum of simple harmonic components of the form $e^{-i\omega t}$, where ω is the angular frequency. The effect of time marching algorithm (12.9) on each Fourier component of frequency ω may be analyzed by substituting $\mathbf{u}_{\ell,m}^{(n)} = \mathbf{u}_{\ell,m}(t) = Ae^{-i\omega n(2\Delta t)}$ into Eq. (12.9). It is easy to find

$$\frac{d\mathbf{u}_{\ell,m}}{dt} = \frac{(e^{-i\omega \Delta t} - 1)}{(2\Delta t) \sum_{j=0}^{3} b_j^* e^{2ij\omega \Delta t}} \mathbf{u}_{\ell,m}. \tag{12.10}$$

Since $d(\mathbf{u}_{\ell,m}/dt) = i\omega \mathbf{u}_{\ell,m}$, the factor, on the right side of Eq. (12.10), multiplying $\mathbf{u}_{\ell,m}$, must be equal to $-i\bar{\omega}$, where $\bar{\omega}$ is the angular frequency of the time marching finite difference scheme. Thus, one finds that

$$\bar{\omega}\Delta t = \frac{i(e^{-i\omega \Delta t} - 1)}{2 \sum_{j=0}^{3} b_j^* e^{2ij\omega \Delta t}}. \tag{12.11}$$

There are four coefficients, namely, b_0^*, b_1^*, b_2^*, and b_3^*. Here, the condition that $\bar{\omega} \simeq \omega$ be accurate to order $(\Delta t)^3$ is imposed. On expanding Eq. (12.11) by the Taylor series, it is straightforward to find that the four coefficients are related by the following three conditions:

$$\sum_{j=0}^{3} b_j^* = \frac{1}{2} \tag{12.12}$$

$$\sum_{j=0}^{3} j b_j^* = -\frac{1}{8} \tag{12.13}$$

$$\sum_{j=0}^{3} j^2 b_j^* = \frac{1}{24}. \tag{12.14}$$

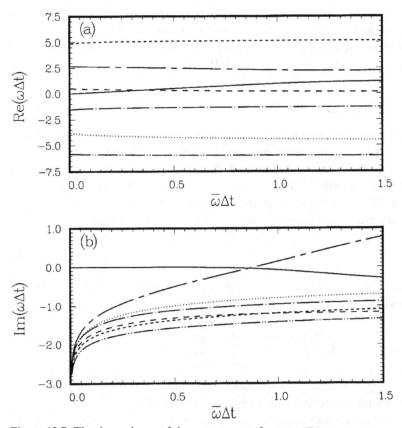

Figure 12.7. The dependence of the seven roots of $\omega\Delta t$ on $\overline{\omega}\Delta t$.

There is no loss of generality regarding b_0 as the remaining free parameter. The value of b_0 is chosen so that $\overline{\omega}$ is a good approximation of ω over the frequency band $-\lambda \leq \omega\Delta t \leq \lambda$. This is done by minimizing the integral as follows:

$$E(b_0) = \int_{-\lambda}^{\lambda} \left\{ \sigma[\text{Re}(\overline{\omega}\Delta t) - \omega\Delta t]^2 + (1-\sigma)[\text{Im}(\overline{\omega}\Delta t)]^2 \right\} d(\omega\Delta t), \quad (12.15)$$

where σ is the weight parameter. The condition for minimization is $dE/db_0 = 0$.

A numerical study of the real and imaginary parts of $(\overline{\omega}\Delta t - \omega\Delta t)$ for different choices of λ and σ has been carried out. Based on the results of this study, the values $\lambda = 0.5$ and $\sigma = 0.42$ are recommended for use. For this choice of parameters, the optimized coefficients are as follows:

$$b_0^* = 0.773100253426$$
$$b_1^* = -0.485967426944$$
$$b_2^* = 0.277634093611$$
$$b_3^* = -0.064766920092.$$

For a given $\overline{\omega}\Delta t$, Eq. (12.11) yields seven roots of $\omega\Delta t$. As in the original DRP scheme, only one of the roots yields the physical solution. All the other roots are spurious and should be damped if the scheme is to be stable. Figure 12.7 shows the dependence of the roots on $\bar{\omega}\Delta t$. The scheme is stable if $\bar{\omega}\Delta t < 0.88$. If $\bar{\omega}\Delta t$ is restricted to less than or equal to 0.19, then the damping rate, $\mathrm{Im}(\omega\Delta t)$, is less than 0.78×10^{-5}, which is quite small. The accuracy of this scheme when restricted to this range is comparable to that of the standard DRP scheme.

12.3 Damping Stencils

In numerical computation, surfaces of discontinuity, such as mesh-size-change interfaces, are potential sources of spurious numerical waves. For this reason, it is necessary to add artificial selective damping in the buffer region of the interface. For points A and A', the 7-point damping stencils discussed in Chapter 7 may be used. For points B or B', C or C' special damping stencils are needed.

Consider the x-momentum equation of the linearized Euler equations.

$$\frac{\partial u}{\partial t} + \cdots = 0. \tag{12.16}$$

The discretized form of Eq. (12.16) at point C, including the artificial selective damping terms, is

$$\left(\frac{\partial u}{\partial t}\right)_C + \cdots = \frac{v_a}{(\Delta x)^2} \sum_{j=-3}^{3} d_j^C u_{C,j}, \tag{12.17}$$

where the damping stencil coefficients satisfy the symmetric condition $d_{-j} = d_j$, v_a is the artificial kinematic viscosity, and Δx is the mesh size at C.

Consider a single Fourier component of u, i.e., $u = \hat{u}(t)e^{i\alpha\ell\Delta x}$. Substitution into Eq. (12.17) yields

$$\left(\frac{\partial u}{\partial t}\right)_C + \cdots = -\frac{v_a}{(\Delta x)^2} D^C (\alpha\Delta x) u, \tag{12.18}$$

where the damping function $D^C(\alpha\Delta x)$ is given by

$$D^C(\alpha\Delta x) = d_0^C + 2\left[d_1^C \cos(\alpha\Delta x) + d_2^C \cos(2\alpha\Delta x) + d_3^C \cos(4\alpha\Delta x)\right]. \tag{12.19}$$

As in Chapter 7, it is required that there is no damping if u is a constant (or $\alpha\Delta x \to 0$). This leads to

$$d_0^C + 2\sum_{j=1}^{3} d_j = 0. \tag{12.20}$$

The normalization condition is $D^C(\pi) = 1.0$ or

$$d_0^C + 2(-d_1^C + d_2^C + d_3^C) = 1. \tag{12.21}$$

In addition, it is intended to select the remaining parameters such that Eq. (12.19) is a good approximation to a Gaussian function of half-width σ centered at $\alpha\Delta x = \pi$. Specifically, the integral

$$\int_0^K \left[D^C(\alpha\Delta x) - e^{-(\ln 2)\left(\frac{\alpha\Delta x-\pi}{\sigma}\right)^2}\right] d(\alpha\Delta x)$$

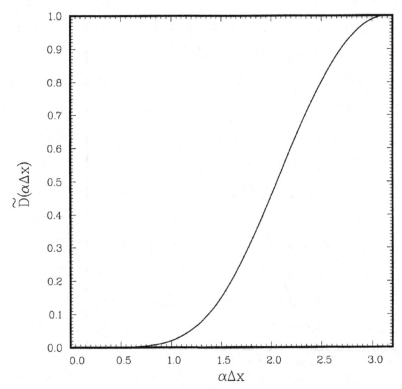

Figure 12.8. Damping function for stencil point C. $\kappa = 0.7$, $\sigma = 0.2\,\pi$.

is to be minimized. Numerical experiments suggest that a good choice of κ and σ are

$$\kappa = 0.7, \quad \sigma = 0.2\pi. \tag{12.22}$$

This yields the following damping stencil coefficients:

$$
\begin{aligned}
d_0^C &= 0.350576727483 \\
d_1^C &= d_{-1}^C = -0.25 \\
d_2^C &= d_{-2}^C = 0.0788677598279 \\
d_3^C &= d_{-3}^C = -0.004156123569.
\end{aligned}
\tag{12.23}
$$

The damping function corresponding to this stencil is shown in Figure 12.8.

On following these steps, the coefficients of the damping stencil for points B or B' are found. In this case, numerical experiments suggest the choice of $\kappa = 0.8$ and $\sigma = 0.2\pi$. The damping stencil coefficients are as follows:

$$
\begin{aligned}
d_0^B &= 0.5 \\
d_1^B &= d_{-1}^B = -0.294977493296 \\
d_2^B &= d_{-2}^B = 0.052389707989 \\
d_3^B &= d_{-3}^B = -0.007412214693.
\end{aligned}
\tag{12.24}
$$

The damping function is shown in Figure 12.9.

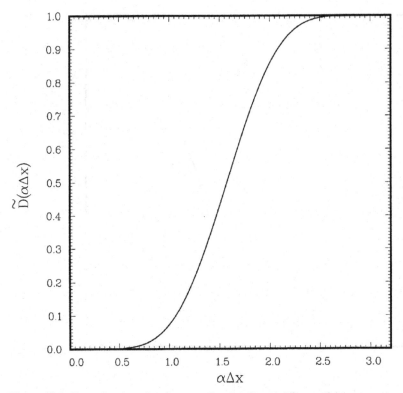

Figure 12.9. Damping function for stencil point B. $\kappa = 0.8$, $\sigma = 0.2\,\pi$.

One is often required to compute flows with a thin boundary layer adjacent to a wall. To calculate such rapidly changing flows, it is desirable to use higher spatial resolution in regions with rapid changes. Thus, the finest mesh is installed right next to the wall. In such a mesh design, one may have a cascade of mesh sizes through several layers from the outermost coarse layer to the finest mesh layer right at the wall. Experience has shown that it is possible for waves corresponding to grid-to-grid oscillations to be trapped in one of the nested mesh layers very close to the wall. This provides an environment for the growth of grid-to-grid oscillations as they bounce from one side of the layer to the other side, gaining amplitude at each reflection. It is, therefore, a good practice to impose stronger artificial selective damping in the computation for the fine meshes near a wall. For an estimate of the magnitude of mesh Reynolds number that should be imposed, it is noted that the time, T, needed by the short waves to propagate from one side of the mesh layer to the other side, is equal to the distance traveled divided by the speed of propagation. Here, the group velocity of the spurious grid-to-grid oscillations is taken to be approximately equal to twice the speed of sound a_0 (see Figure 2.4). Hence, one finds

$$T = \frac{N\Delta x}{2a_0},\tag{12.25}$$

where N is the number of mesh points across the width of the mesh layer, and Δx is the mesh size. The total damping in this time period is given by Eq. (7.15). Since $\tilde{D}(\pi) = 1.0$, the total damping factor is

$$e^{-\frac{\nu_a T}{(\Delta x)^2}} = e^{-\frac{N}{2R_\Delta}},\tag{12.26}$$

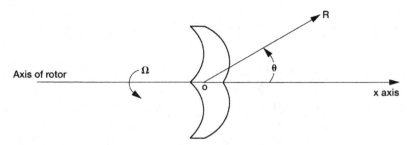

Figure 12.10. An open rotor.

where $R_\Delta = v_a/(\Delta x)a_0$ is the mesh Reynolds number. It is suggested that this factor should be equal to 10^{-2} or less to avoid accumulation of spurious grid-to-grid oscillations. On setting this factor equal to 10^{-2}, a criterion for the choice of the mesh Reynolds number is established, namely,

$$\frac{N}{R_\Delta} \geq 9.2. \tag{12.27}$$

For example, if the nested mesh layer has 20 mesh points, i.e., $N = 20$, then $R_\Delta^{-1} \geq 0.46$, say 0.5, should be used. This value is much larger than the value recommended for general background damping, which is $R_\Delta^{-1} \geq 0.05$. (Note: away from the solid boundaries a smaller and smaller inverse mesh Reynolds number should be used.) Formula (12.27) also suggests that any uniform mesh layer in a multisize-mesh computation should not have fewer than 20 mesh points. This is to avoid the necessity of using excessively large artificial damping.

12.4 Numerical Examples

To illustrate the implementation and the effectiveness of multi-size-mesh multi-time-step computing, two problems from the Third CAA Workshop on Benchmark Problems are considered here. The first is a two-dimensional rotor noise problem. The second is an automobile door cavity tone problem.

12.4.1 First Example: The Sound Field of an Open Rotor

For this problem, it is convenient to use nondimensional variables with respect to the following scales:

length scale $= b$(length of blade)
velocity scale $= a_\infty$(ambient sound speed)
time scale $= b/a_\infty$
density scale $= \rho_\infty$(ambient gas density)
pressure scale $= \rho_\infty a_\infty^2$
body force scale (per unit volume) $= \rho_\infty a_\infty^2/b$

A rotor (see Figure 12.10) exerts a rotating force on the fluid. As a model problem, the rotor is replaced by a distribution of rotating body force. The governing equations are the linearized Euler equations. In cylindrical coordinates (r, ϕ, x), they are as follows:

$$\frac{\partial v}{\partial t} = -\frac{\partial p}{\partial r} + F_r$$

$$\frac{\partial w}{\partial t} = -\frac{1}{r}\frac{\partial p}{\partial \phi} + F_\phi$$

$$\frac{\partial u}{\partial t} = -\frac{\partial p}{\partial x} + F_x$$

$$\frac{\partial p}{\partial t} + \frac{1}{r}\frac{\partial (vr)}{\partial r} + \frac{1}{r}\frac{\partial w}{\partial \phi} + \frac{\partial u}{\partial x} = 0, \qquad (12.28)$$

where (F_r, F_ϕ, F_x) are the components of the body force.

For simplicity, F_r is set equal to zero, while F_ϕ and F_x are assumed to consist of a single azimuthal and frequency component, i.e.,

$$\begin{bmatrix} F_\phi(r, \phi, x, t) \\ F_x(r, \phi, x, t) \end{bmatrix} = \text{Re}\left\{ \begin{bmatrix} \tilde{F}_\phi(r, x) \\ \tilde{F}_x(r, x) \end{bmatrix} e^{im(\phi - \Omega t)} \right\}, \qquad (12.29)$$

where Re{} is the real part of. For computation purposes, the following body force distributions in r and x are used:

$$\tilde{F}_\phi(r, x) = \begin{cases} F(x)rJ_m(\lambda_{mN}r), & r \le 1 \\ 0 & r > 1 \end{cases} \qquad (12.30)$$

$$\tilde{F}_x(r, x) = \begin{cases} F(x)J_m(\lambda_{mN}r), & r \le 1 \\ 0 & r > 1 \end{cases} \qquad (12.31)$$

$$F(x) = \exp[-(\ln 2)(10x)^2], \qquad (12.32)$$

where $J_m(\ )$ is the mth-order Bessel function, λ_{mN} is the Nth root of J'_m or $J'_m(\lambda_{mN}) = 0$. In this model, m is the number of blades, Ω is the angular velocity of the rotor. The choice of the Bessel functions in Eqs.(12.30) and (12.31) has no significance other than to make the analytical solution simple.

It is possible to reduce the three-dimensional problem of Eq. (12.28) to a two-dimensional problem by factoring out the azimuthal dependence. Let

$$\begin{bmatrix} u(r, \phi, x, t) \\ v(r, \phi, x, t) \\ w(r, \phi, x, t) \\ p(r, \phi, x, t) \end{bmatrix} = \text{Re}\left\{ \begin{bmatrix} \tilde{u}(r, x, t) \\ \tilde{v}(r, x, t) \\ \tilde{w}(r, x, t) \\ \tilde{p}(r, x, t) \end{bmatrix} e^{im\phi} \right\}. \qquad (12.33)$$

The governing equations for $(\tilde{u}, \tilde{v}, \tilde{w}, \tilde{p})$ are found by substituting Eqs. (12.29) to (12.33) into Eq. (12.28). Upon factoring out $e^{im\phi}$, these equations are

$$\frac{\partial \tilde{v}}{\partial t} = -\frac{\partial \tilde{p}}{\partial r}$$

$$\frac{\partial \tilde{w}}{\partial t} = -\frac{im}{r}\tilde{p} + F_\phi(r, x)e^{-im\Omega t}$$

$$\frac{\partial \tilde{u}}{\partial t} = -\frac{\partial \tilde{p}}{\partial x} + F_x(r, x)e^{-im\Omega t}$$

$$\frac{\partial \tilde{p}}{\partial t} + \frac{1}{r}\frac{\partial (\tilde{v}r)}{\partial r} + \frac{im\tilde{w}}{r} + \frac{\partial \tilde{u}}{\partial x} = 0. \qquad (12.34)$$

For an open rotor solution, it is only necessary to find the outgoing wave solution of Eq. (12.34) in the $r - x$ plane.

Now, the benchmark problem is to calculate the directivity, $D(\theta)$, of the radiated sound for a eight-blade rotor ($m = 8$) with $N = 1$ ($\lambda_{8,1} = 9.64742$). In a spherical polar coordinates system (R, θ, ϕ), with the x-axis as the polar axis, the directivity is defined by

$$D(\theta) = \lim_{R \to \infty} R^2 \overline{p^2 (R, \theta, \phi, t)},$$

where an overbar indicates a time average. The directivity $D(\theta)$ at one-degree interval at two rotational speeds

(a) $\Omega = 0.85$ (subsonic tip speed)
(b) $\Omega = 1.15$ (supersonic tip speed)

are to be computed.

To ensure that the computed solution is accurate, one must make provisions to take into account two important characteristics of this problem. First, the noise source is discontinuous at the blade tip. This could be a source of short spurious numerical waves. Second, the blades are slender, that is, the loading is concentrated over a narrow width. Computationally, this requires a finer spatial resolution in the source region than in the acoustic radiation field.

12.4.1.1 Grid Design
The half-width of the forcing function in Eqs. (12.30) and (12.31) is 0.2. To resolve this width, a minimum of 10 mesh points is necessary. In other words, the maximum mesh size in the source region is 0.02. This high resolution is not needed as one moves away into the acoustic field. The multi-size-mesh multi-time-step DRP algorithm is used for computation. This allows one to use coarser and coarser mesh starting from the source region radially outward. Figure 12.11 shows the computation domain. The domain is divided into three regions. The mesh size as well as the time step increases by a factor of 2 each time one crosses into an outer region.

12.4.1.2 Numerical Boundary Conditions
Two types of numerical boundary conditions are needed. Along the outer boundary of the computation domain, radiation boundary conditions are required. Along the axis of the cylindrical coordinates, i.e., $r = 0$, a special set of axis boundary conditions is needed. Here, the radiation boundary conditions are used as follows:

$$\left(\frac{\partial}{\partial t} + \frac{\partial}{\partial R} + \frac{1}{R} \right) \begin{bmatrix} \tilde{u} \\ \tilde{v} \\ \tilde{w} \\ \tilde{p} \end{bmatrix} = 0, \tag{12.35}$$

where $R = (r^2 + x^2)^{1/2}$.

As $r \to 0$, Eq. (12.34) has a numerical singularity and cannot be used as it is. The situation is similar to that discussed in Section 9.4. By adopting the numerical

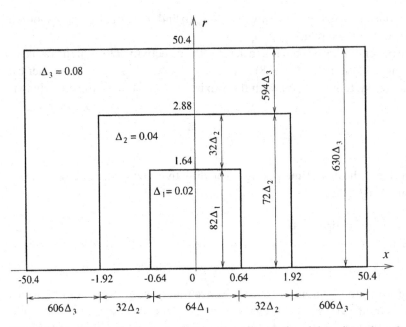

Figure 12.11. Schematic diagram showing the computational domain and mesh design.

treatment of Section 9.4, the solution is extended to the negative r part of the $x - r$ plane as follows:

$$\tilde{u}(-r, x) = (-1)^m \tilde{u}(r, x)$$
$$\tilde{v}(-r, x) = (-1)^{m-1} \tilde{v}(r, x)$$
$$\tilde{w}(-r, x) = (-1)^{m-1} \tilde{w}(r, x)$$
$$\tilde{p}(-r, x) = (-1)^m \tilde{p}(r, x) \tag{12.36}$$

Formula (12.36) allows one to extend the computed solution into the lower half of the $x - r$ plane as indicated in Figure 12.12. In this way, computation stencils for

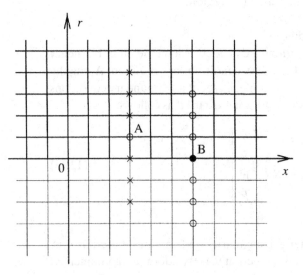

Figure 12.12. Extension of the computational domain, the upper part of the $x - r$ plane, to the nonphysical lower half plane.

points near the axis (but not on the axis such as point A) may be extended into the negative r half-plane as shown. For the points on the axis, such as point B, they will not be calculated by the time marching DRP scheme. They are to be found, after the values at all the other points have been updated to a new time level, by symmetric interpolation. A 7-point interpolation stencil for point B is shown in Figure 12.12.

12.4.1.3 Artificial Selective Damping
As discussed before, artificial selective damping is incorporated into the DRP computational algorithm for two purposes. First, it is used to provide background damping to eliminate short spurious waves to prevent them from propagating across the computation domain. Generally speaking, small-amplitude short spurious waves are just low-level pollutants of the numerical solution, but if these waves are allowed to impinge on an internal or external boundary of the computational domain, they could lead to the reflection of large-amplitude long waves. These spurious long waves are sometimes not distinguishable from the physical solution and are, therefore, extremely undesirable. The second reason to add artificial selective damping is to stabilize the numerical solution at a discontinuity. The damping prevents the buildup of spurious short waves, which are generated by the discontinuity, and this promotes stability.

In the present problem, the forcing functions are discontinuous at the blade tip. Thus, in addition to the general background damping with an inverse mesh Reynolds number 0.05, extra damping is added around the blade tip region. The mesh-size-change interface and the external boundary of the computation domain are also a form of discontinuity. Extra damping is added around these boundaries as well. To impose extra damping, a distribution of inverse mesh Reynolds number in the form of a Gaussian function with a half-width of four mesh spacings normal to the boundary is used. The maximum of the Gaussian is on the discontinuity with an assigned value of 0.05. At the tip of the blade, where the forcing function is discontinuous, more damping is required. A maximum value of 0.75 is used instead.

12.4.1.4 Asymptotic Solution
An asymptotic solution of the present problem may be found by a straightforward application of Fourier transform. Details of the analysis are given in the Proceedings of the Third CAA Workshop on Benchmark Problems (NASA CP-2000-209790, August 2000; Dahl, M.D. ed.). The directivity of sound radiation with respect to a spherical polar coordinate system (R, θ, ϕ) with the polar axis coinciding with the axis of the rotor is

$$D(\theta) = \lim_{R \to \infty} R^2 \overline{p^2}$$

$$= \left(\frac{\pi}{400 \ln 2}\right)^{1/2} \frac{m^2(1 + \Omega \cos \theta)\Omega \sin \theta}{\lambda_{mN}^2 - m^2\Omega^2 \sin^2 \theta} J_m(\lambda_{mN}) J_m'(mR \sin \theta) e^{-\frac{m^2\Omega^2 \cos^2 \theta}{400 \ln 2}}$$

12.4.1.5 Numerical Results
Eqs. (12.34) are discretized according to the multi-size-mesh mult-time-step DRP scheme and marched in time to a time periodic state. To start the computation, a zero initial condition is used. Figure 12.13 shows a comparison of the directivity at

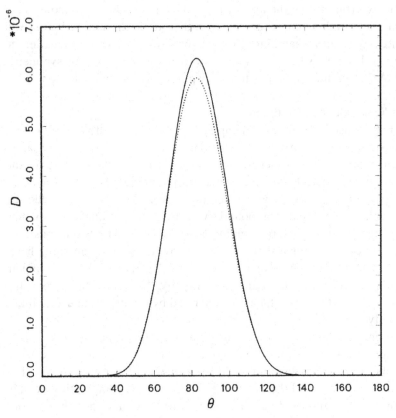

Figure 12.13. Directivity of sound field at $R = 50$; $\Omega = 0.85$; ——, numerical; ······, asymptotic.

$R = 50$ obtained computationally against the asymptotic solution for $\Omega = 0.85$, the subsonic tip speed case. As expected, most of the acoustic radiation is concentrated in the plane of rotation. There is good agreement between the numerical results and the asymptotic solution. Figure 12.14 shows the directivity at supersonic tip speed with $\Omega = 1.15$. At higher frequency, the acoustic wavelength is shorter. This case, therefore, offers a more stringent test of the accuracy of the entire computational algorithm. The agreement with asymptotic solution is reasonably good.

12.4.2 Second Example: Acoustic Resonances Induced by Flow over an Automobile Door Cavity

This problem is to compute the resonance tones induced by a turbulent boundary layer flow outside a scaled model automobile door cavity. To properly model and compute the turbulent boundary layer flow and its interaction with the cavity is a task that will require extensive time and effort. Here, a laminar boundary layer is considered instead. It is believed that the cavity tone frequencies would most likely be about the same whether the flow is turbulent or laminar, but the tone intensities are expected to be different.

A boundary layer flow will definitely be laminar if $R_{\delta*} < 600$. $R_{\delta*}$ is the Reynolds number based on the displacement thickness. This is the Reynolds number below

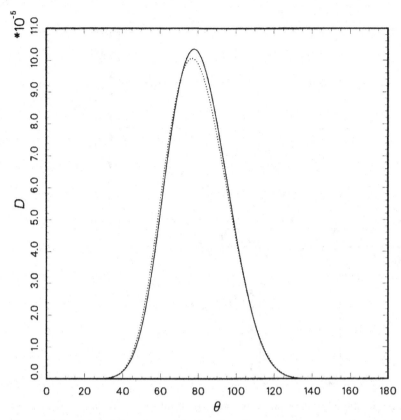

Figure 12.14. Directivity of sound field at $R = 50$; $\Omega = 1.15$; ———, numerical; ······, asymptotic.

the stability limit of the Tollmien-Schlichting waves. In modern facilities with low free-stream turbulence and sound, a boundary layer may remain laminar if $R_{\delta*}$ is larger than 600, but less than 3400. For a free-stream velocity of 50.9 m/s and 26.8 m/s (velocities prescribed by the benchmark problem), these correspond to a boundary layer thickness of 2.9 mm and 5.5 mm, respectively. In this example, consideration will be restricted to boundary layer thicknesses of 2 mm and 1 mm at flow velocities of 50.9m/s and 26.8m/s, respectively.

12.4.2.1 Computation Domain and Grid Design

The computation domain is shown in Figure 12.15. It is designed primarily for the case $U = 50.9$ m/s and a boundary layer thickness $\delta = 2$ mm. In the cavity opening region, viscous effects are important. To capture these effects, a fine mesh is needed. Away from the cavity, the disturbances are mainly acoustic waves. By using the DRP scheme in the computation, only a very coarse mesh would be necessary in the acoustic region. The mesh design is dictated by these considerations. The computation domain is divided into a number of subdomains as shown in Figure 12.15. The finest mesh with $\Delta x = \Delta y = 0.0825$ mm is used in the cavity opening region. The mesh size increases by a factor of 2 every time one crosses into the next subdomain. The mesh size in the outermost subdomain is thirty-two times larger than the finest mesh.

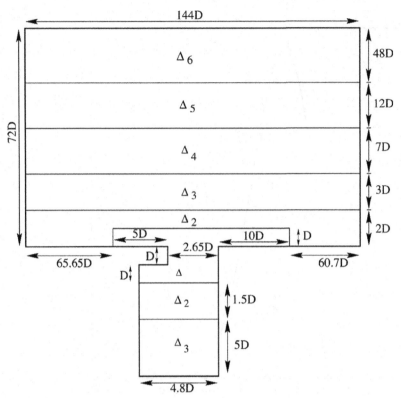

Figure 12.15. Computational domain showing the division into subdomains and their mesh size. $D = 3.3$ mm.

12.4.2.2 The Governing Equations and the Computational Algorithm

The governing equations are the compressible Navier-Stokes equations in two dimensions. In dimensionless form with $D(D = 3.3$ mm is the thickness of the cavity overhang) as length scale, U, the free-stream velocity (from left to right), as velocity scale, D/U as the time scale, ρ_0, the ambient gas density, as the density scale and $\rho_0 U^2$ as the pressure scale, they are

$$\frac{\partial \rho}{\partial t} + \rho \frac{\partial u_j}{\partial x_j} + u_j \frac{\partial \rho}{\partial x_j} = 0 \tag{12.37}$$

$$\frac{\partial u_i}{\partial t} + u_j \frac{\partial u_i}{\partial x_j} = -\frac{\partial p}{\partial x_i} + \frac{\partial \tau_{ij}}{\partial x_j} \tag{12.38}$$

$$\frac{\partial p}{\partial t} + u_j \frac{\partial p}{\partial x_j} + \gamma p \frac{\partial u_j}{\partial x_j} = 0 \tag{12.39}$$

$$\tau_{ij} = \frac{1}{R_D} \left(\frac{\partial u_i}{\partial x_j} + \frac{\partial u_j}{\partial x_i} \right), \tag{12.40}$$

where R_D is the Reynolds number based on D.

These equations are solved in time by the multi-size-mesh multi-time-step DRP algorithm. In each subdomain of Figure 12.15, the equations are discretized by the DRP scheme. At the mesh-size-change boundaries, special stencils as given in

Section 12.1 are used. The time steps of adjacent subdomains differ by a factor of 2 just as the mesh size does. With the use of the multi-size-mesh multi-time-step algorithm, most of the computation effort and time are spent in the opening region of the cavity where the resolution of the unsteady viscous layers is of paramount importance.

12.4.2.3 Numerical Boundary Conditions

Along the solid surfaces of the cavity and the outside wall, the no-slip boundary condition is enforced by the ghost point method. Along the vertical external boundary regions (3 mesh points adjacent to the boundary), the flow variables are split into a mean flow and a time-dependent component. The mean flow, with a given boundary layer thickness, is provided by the Blasius solution. The time-dependent part of the solution is the only portion of the solution that is computed by the time marching scheme. The boundary conditions used for the computation are as follows. Along the top and left external boundaries, the asymptotic radiation boundary conditions of Section 6.1 are imposed. Along the right boundary, the outflow boundary conditions of Section 9.3 are used.

12.4.2.4 Artificial Selective Damping

Artificial selective damping is added to the time marching DRP scheme to eliminate spurious short waves and to prevent the occurrence of numerical instability. The damping stencil with a damping curve of half-width 0.2π is used for background damping. Near the solid walls or the outer boundaries where a 7-point stencil does not fit, a 5- or 3-point stencil as provided in Section 7.4 is used instead. For general background damping, an inverse mesh Reynolds number of 0.05 is used everywhere. Along walls and mesh-change interfaces, additional damping is included. The added damping has an inverse mesh Reynolds number distribution in the form of a Gaussian function with the maximum value at the wall or interface and a half-width of 4 mesh points as in the first example. On the wall, the maximum value of R_Δ^{-1} is set equal to 0.15. The corresponding value at a mesh-size-change interface is 0.3. There are three external corners at the cavity opening. They are probably sites at which short spurious waves are generated. To prevent numerical instability from developing at these points, additional artificial selective damping is imposed. Again, a half-width of 4 mesh point Gaussian distribution of the inverse mesh Reynolds number centered at each of these points is used. The maximum value of R_Δ^{-1} at these points is set equal to 0.35. By implementing artificial selective damping distribution as described, no numerical instability nor excessive short spurious waves have been found in all the computations.

12.4.2.5 Numerical Results

To start the computation, the time-independent boundary layer solution without the cavity is used as the initial condition. The solution is time-marched until a time periodic state is reached.

The characteristic features of the flow in the vicinity of the cavity opening and the acoustic field are found by examining the instantaneous vorticity, steamlines, and pressure contours. Figure 12.16 shows a plot of the instantaneous vorticity contours for the case $U = 50.9$ m/s and $\delta = 2.0$ mm. As can easily be seen, vortices are

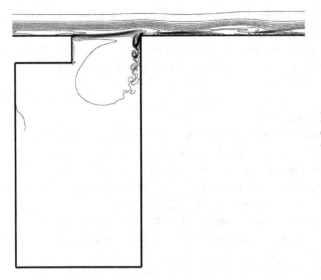

Figure 12.16. Instantaneous vorticity contours showing the shedding of small vortices at the trailing edge of the cavity. $U = 50.9$ m/s, $\delta = 2$ mm.

shed periodically at the trailing edge of the cavity. Some shed vortices move inside the cavity. Other vortices are shed into the flow outside. The shed vortices outside are convected downstream by the boundary layer flow. These convected vortices are clearly shown in the pressure contour plot of Figure 12.17. They form the low-pressure centers. These vortices persist over a rather long distance and are eventually dissipated by viscosity. Figure 12.18 shows the instantaneous streamline pattern. It is seen that the flow at the mouth of the cavity is completely dominated by a single large vortex. Below the large vortex, another vortex of opposite rotation often exists. The position of this vortex changes from time to time and does not always attach to the cavity wall. The near-field pressure contour pattern is shown in Figure 12.19. This pattern is the same as that of a monopole acoustic source in a low subsonic stream. That the noise source is a monopole, and not a dipole, is consistent with the model of Tam and Block (1978). The sound is generated by flow impinging periodically at the trailing edge of the cavity.

Experiments indicate that cavity resonance may consist of a single tone or multiple tones. The number of tones occured depend on the flow conditions and the cavity geometry. Figure 12.20 shows the noise spectrum measured at the center of the left wall of the cavity at $U = 50.9$ m/s, $\delta = 2$ mm. The spectrum consists of a single tone at 1.99 kHz.

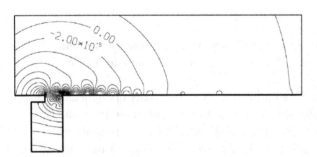

Figure 12.17. Near-field pressure contours showing the convection of shed vortices along the outside wall. $U = 50.9$ m/s, $\delta = 2$ mm.

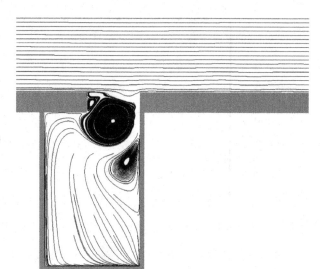

Figure 12.18. Instantaneous streamline pattern. $U = 50.9$ m/s, $\delta = 2$ mm.

Shown in Figure 12.20 also is the spectrum measured experimentally by Henderson (2000). There is good agreement between the tone frequency of the numerical simulation and that of the physical experiment. In the experiment, the boundary layer is turbulent; therefore, it is not expected that there would be good agreement in the tone intensity. Figure 12.21 shows the noise spectrum at the lower speed $U = 26.8$ m/s and $\delta = 1$ mm. In this case, there are two tones. One is at a frequency of 1.32 kHz and the other at 2.0 kHz. This is in fair agreement with the experimentally measured spectrum. Again, the tone frequencies are reasonably well reproduced in the numerical simulation, but the tone intensities are different. Figures 12.20 and 12.21 together suggest that as the flow velocity increases, one of the tones disappears. The strength of the remaining tone intensifies with flow speed.

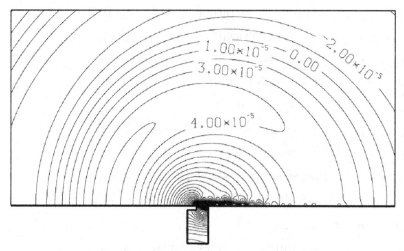

Figure 12.19. Far-field pressure contours showing a monopole acoustic field. $U = 50.9$ m/s, $\delta = 2$ mm.

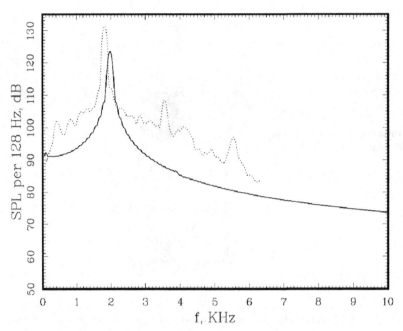

Figure 12.20. Noise spectrum at the center of the left wall of the cavity. $U = 50.9$ m/s; $\delta = 2$ mm; ———, numerical simulation; $\ldots\ldots\ldots$, experiment (Henderson, 2004).

12.5 Coefficients of Several Large Buffer Stencils

When high-quality, high-order computations are to be performed, one expects that large derivative stencils are to be used in the mesh-size-change buffer region. Optimized coefficients of several large buffer stencils are listed below. η is the range of optimization.

9-Point Stencils

1357 derivative stencil. $\eta = 0.885$, $a_0 = 0.0$

$$a_1 = -a_{-1} = 0.60902385609791059518785034631358 49$$
$$a_3 = -a_{-3} = -0.04690263734690552411752317530947456$$
$$a_5 = -a_{-5} = 7.459761574407952709049703450707 25\text{E-}3$$
$$a_7 = -a_{-7} = -8.0210741846196948293276251981393 66\text{E-}4$$

1246 derivative stencil. $\eta = 1.075$, $a_0 = 0.0$

$$a_1 = -a_{-1} = 0.75122469079529412627152275764 91948$$
$$a_2 = -a_{-2} = -0.13909917896817193620221722105 97338$$
$$a_4 = -a_{-4} = 7.6185409515963687179825703762642764\text{E-}3$$
$$a_6 = -a_{-6} = -5.834161108892881231697661724640704\text{E-}4$$

1235 derivative stencil. $\eta = 1.215$, $a_0 = 0.0$

$$a_1 = -a_{-1} = 0.80807017238010812681269898134 00916$$
$$a_2 = -a_{-2} = -0.20434323752480324834845092021 08968$$

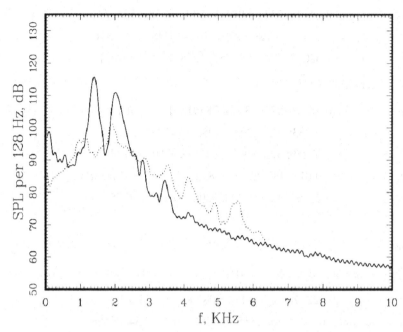

Figure 12.21. Noise spectrum at the center of the left wall of the cavity. $U = 26.8$ m/s; $\delta = 1$ mm; ———, numerical simulation; ...., experiment (Henderson, 2004).

$$a_3 = -a_{-3} = 0.035181913310815445565211731181155573$$
$$a_5 = -a_{-5} = -9.858874525895933622864668923529375\text{E-4}$$

11-Point Stencils

13579 derivative stencil. $\eta = 0.96$, $a_0 = 0.0$

$$a_1 = -a_{-1} = 0.6169610748807038603650623876266686$$
$$a_3 = -a_{-3} = -0.053037977159713435091000295888159493$$
$$a_5 = -a_{-5} = 0.0110576809675173923565132957079970069$$
$$a_7 = -a_{-7} = -2.2180071199244495954136156095099006\text{E-3}$$
$$a_9 = -a_{-9} = 2.6561128892451447702970341826597076\text{E-4}$$

12468 derivative stencil. $\eta = 1.15$, $a_0 = 0.0$

$$a_1 = -a_{-1} = 0.7631614437828716630397535144418575$$
$$a_2 = -a_{-2} = -0.14840307207850050785201259674573818$$
$$a_4 = -a_{-4} = 0.010272925705412492924195271253581683$$
$$a_6 = -a_{-6} = -1.4476431210717408970741183395007183\text{E-3}$$
$$a_8 = -a_{-8} = 1.5485703486372829374191300903706E-4$$

12357 derivative stencil. $\eta = 1.3$, $a_0 = 0.0$

$$a_1 = -a_{-1} = 0.82525430283448736663088909592187007$$
$$a_2 = -a_{-2} = -0.22300346076646621901283108687972073$$

$$a_3 = -a_{-3} = 0.043484732275884181589478378275243$$
$$a_5 = -a_{-5} = -2.16719116816279731554403481752948\text{1E-3}$$
$$a_7 = -a_{-7} = 1.6205395880093045772258815707159745\text{E-4}$$

12346 derivative stencil. $\eta = 1.4, a_0 = 0.0$

$$a_1 = -a_{-1} = 0.85596702705339407859130139757701\text{42}$$
$$a_2 = -a_{-2} = -0.26355038029346495866745731339388584$$
$$a_3 = -a_{-3} = 0.07259552278997621476486631628514457$$
$$a_4 = -a_{-4} = -0.012148236925938859899155655963543882$$
$$a_6 = -a_{-6} = 3.2335214456043900760615070158323756\text{E-4}$$

13-Point Stencils

1357911 derivative stencil. $\eta = 1.04, a_0 = 0.0$

$$a_1 = -a_{-1} = 0.62227448108845032080061617524023\text{52}$$
$$a_3 = -a_{-3} = -0.057457522835076529115000160548041\text{88}$$
$$a_5 = -a_{-5} = 0.014048807353574197207816006070727958$$
$$a_7 = -a_{-7} = -3.8223706624986957120774761605838786\text{E-3}$$
$$a_9 = -a_{-9} = 8.7732562820945351583743885082977\text{44E-4}$$
$$a_{11} = -a_{-11} = -1.1684412431690283206275822573907119\text{E-4}$$

1246810 derivative stencil. $\eta = 1.225, a_0 = 0.0$

$$a_1 = -a_{-1} = 0.77041596692040729875779075706601\text{12}$$
$$a_2 = -a_{-2} = -0.15423090409941798512572684114924175$$
$$a_4 = -a_{-4} = 0.012111428866866974752283356866705977$$
$$a_6 = -a_{-6} = -2.237909249412284471498739008161985\text{E-3}$$
$$a_8 = -a_{-8} = 4.551038697046676646061681355565937\text{7E-4}$$
$$a_{10} = -a_{-10} = -6.1324965020286200332741326983250\text{02E-5}$$

123579 derivative stencil. $\eta = 1.375, a_0 = 0.0$

$$a_1 = -a_{-1} = 0.83460075211983067803855772743553\text{69}$$
$$a_2 = -a_{-2} = -0.23351453982173775751336583621371804$$
$$a_3 = -a_{-3} = 0.048416204296670338536750729561625173$$
$$a_5 = -a_{-5} = -3.0732634681711003723284703258379413\text{E-3}$$
$$a_7 = -a_{-7} = 4.2764556235837229592649414282684104\text{E-4}$$
$$a_9 = -a_{-9} = -4.972077355769809243570567373051316\text{E-5}$$

123468 derivative stencil. $\eta = 1.5, a_0 = 0.0$

$$a_1 = -a_{-1} = 0.86950401471465643126157098245655\text{12}$$
$$a_2 = -a_{-2} = -0.28111202563583404799569450627628445$$

$$a_3 = -a_{-3} = 0.0845286783216550621158838448960038$$

$$a_4 = -a_{-4} = -0.0162793066924765656651937182199432$$

$$a_6 = -a_{-6} = 7.921732807092020750043354094040387E\text{-}4$$

$$a_8 = -a_{-8} = -6.272641528780892588558052108063413E\text{-}5$$

123457 derivative stencil. $\eta = 1.59$, $a_0 = 0.0$

$$a_1 = -a_{-1} = 0.8891744805205031441686220773749192$$

$$a_2 = -a_{-2} = -0.3096468670894452798278980518929917$$

$$a_3 = -a_{-3} = 0.10899054209331996393360155629344859$$

$$a_4 = -a_{-4} = -0.030279357225089779657034797121953563$$

$$a_5 = -a_{-5} = 5.0487756298281625592780951771430825E\text{-}3$$

$$a_7 = -a_{-7} = -1.3983169576488149741170426674042277E\text{-}4$$

15-Point Stencils

135791113 derivative stencil. $\eta = 1.11$, $a_0 = 0.0$

$$a_1 = -a_{-1} = 0.6258779354573266902120076920113189$$

$$a_3 = -a_{-3} = -0.060595621262294137510257150629882345$$

$$a_5 = -a_{-5} = 0.016419419315514504253832969339583166$$

$$a_7 = -a_{-7} = -5.332041836769511268442174297083009E\text{-}3$$

$$a_9 = -a_{-9} = 1.6673308740708772699299319145046008E\text{-}3$$

$$a_{11} = -a_{-11} = -4.2649198385963323058082868711331986E\text{-}4$$

$$a_{13} = -a_{-13} = 6.31968127067576950549124298229811E\text{-}5$$

124681012 derivative stencil. $\eta = 1.295$, $a_0 = 0.0$

$$a_1 = -a_{-1} = 0.7751653818251308369767154456316511$$

$$a_2 = -a_{-2} = -0.15810842159142594570323780970792248$$

$$a_4 = -a_{-4} = 0.01344368848930721405861880589370375$$

$$a_6 = -a_{-6} = -2.882509303736270724294138583564854E\text{-}3$$

$$a_8 = -a_{-8} = 7.737358357276028905705620340672378E\text{-}4$$

$$a_{10} = -a_{-10} = -2.0178401810068388499797391894808376E\text{-}4$$

$$a_{12} = -a_{-12} = 3.3309726507986522205418718975874384E\text{-}5$$

12357911 derivative stencil. $\eta = 1.45$, $a_0 = 0.0$

$$a_1 = -a_{-1} = 0.8405083211561195585119035254044648$$

$$a_2 = -a_{-2} = -0.24028234117501536764416518113371986$$

$$a_3 = -a_{-3} = 0.051710722871922495965649571941564053$$

$$a_5 = -a_{-5} = -3.7380739291593315864384143663748303E\text{-}3$$

$$a_7 = -a_{-7} = 6.732308091610869543962441130822362E\text{-}4$$

$$a_9 = -a_{-9} = -1.52128973798384168548041590956809\text{E-}4$$
$$a_{11} = -a_{-11} = 2.46461203634723316208053088355383\text{E-}5$$

12346810 derivative stencil. $\eta = 1.58$, $a_0 = 0.0$

$$a_1 = -a_{-1} = 0.876969751716291422272351256410025$$
$$a_2 = -a_{-2} = -0.291048550605084936500596344259554$$
$$a_3 = -a_{-3} = 0.0916072963679824385062388692443454$$
$$a_4 = -a_{-4} = -0.0188895863108165952426006222281686327$$
$$a_6 = -a_{-6} = 1.172537573103338497139038840768234\text{E-}3$$
$$a_8 = -a_{-8} = -1.76489721642551008468991037034160324\text{E-}4$$
$$a_{10} = -a_{-10} = 2.40497967717893265445008754533607\text{E-}5$$

1234579 derivative stencil. $\eta = 1.69$, $a_0 = 0.0$

$$a_1 = -a_{-1} = 0.89966669663763338312112615547268505$$
$$a_2 = -a_{-2} = -0.32482395031614055954216555781487774$$
$$a_3 = -a_{-3} = 0.12185450769709449174616083733794719$$
$$a_4 = -a_{-4} = -0.037322402940415078194007049492579432$$
$$a_5 = -a_{-5} = 7.19843055492260803999831858487961\text{E-}3$$
$$a_7 = -a_{-7} = -3.6954649492803779167797176495157803\text{E-}4$$
$$a_9 = -a_{-9} = 3.35217351341499724854951432937019537\text{E-}5$$

1234568 derivative stencil. $\eta = 1.765$, $a_0 = 0.0$

$$a_1 = -a_{-1} = 0.913206673001701850107184431788084$$
$$a_2 = -a_{-2} = -0.3458970880904337489997562609890365$$
$$a_3 = -a_{-3} = 0.142375240231816887547078180601490$$
$$a_4 = -a_{-4} = -0.0517593515672468110930752740016308566$$
$$a_5 = -a_{-5} = 0.0145561185019468885330897373167467$$
$$a_6 = -a_{-6} = -2.48173853257862295316555864085368666\text{E-}3$$
$$a_8 = -a_{-8} = 7.6128429804940584617413706678497267\text{E-}5$$

12.6 Large Buffer Selective Damping Stencils

Buffer regions, where a change in mesh size takes place, are favorite sites for the generation of spurious short numerical waves. These are also regions prone to numerical instability. It is, therefore, necessary to add artificial selective damping to these regions. Below are optimized coefficients of several large artificial selective damping stencils designed for use in buffer regions involving change in mesh size by a factor of 2.

9-Point Stencil. $\sigma = 0.1975\pi$

1235 stencil. $\beta = 1.215$

$$d_0 = 0.2708886738189505$$
$$d_1 = d_{-1} = -0.2206703054137707$$
$$d_2 = d_{-2} = 0.11455566309052476$$
$$d_3 = d_{-3} = -0.031235919680725313$$
$$d_5 = d_{-5} = 2.5094190624960092\text{E-}3$$

1246 stencil. $\beta = 1.075$

$$d_0 = 0.33027355532845215$$
$$d_1 = d_{-1} = -0.25$$
$$d_2 = d_{-2} = 0.09475341549663605$$
$$d_4 = d_{-4} = -0.011705918179843176$$
$$d_6 = d_{-6} = 1.815725018981057\text{E-}3$$

1357 stencil. $\beta = 0.885$

$$d_0 = 0.5$$
$$d_1 = d_{-1} = -0.3062891172075045$$
$$d_3 = d_{-3} = 0.07419885992060947$$
$$d_5 = d_{-5} = -0.022186172045155307$$
$$d_7 = d_{-7} = 4.276429332050351\text{E-}3$$

11-Point Stencil. $\sigma = 0.195\pi$

13579 stencil. $\beta = 0.96$

$$d_0 = 0.5$$
$$d_1 = d_{-1} = -0.3095004246385787$$
$$d_3 = d_{-3} = 0.08211616467055119$$
$$d_5 = d_{-5} = -0.03036277454777355$$
$$d_7 = d_{-7} = 9.650309477159707\text{E-}3$$
$$d_9 = d_{-9} = -1.9032749613586592\text{E-}3$$

12468 stencil. $\beta = 1.15$

$$d_0 = 0.32646005443343434$$
$$d_1 = d_{-1} = -0.25$$
$$d_2 = d_{-2} = 0.09831710703893635$$
$$d_4 = d_{-4} = -0.014318492283620247$$
$$d_6 = d_{-6} = 3.4468599861078082\text{E-}3$$
$$d_8 = d_{-8} = -6.755019581410823\text{E-}4$$

12357 stencil. $\beta = 1.3$

$$d_0 = 0.2635716529562409$$
$$d_1 = d_{-1} = -0.21772629224768608$$
$$d_2 = d_{-2} = 0.11821417352187955$$
$$d_3 = d_{-3} = -0.03499568161296116$$
$$d_5 = d_{-5} = 3.174478225315305\text{E-}3$$
$$d_7 = d_{-7} = -4.5250436466805904\text{E-}4$$

12346 stencil. $\beta = 1.4$

$$d_0 = 0.23453190899565207$$
$$d_1 = d_{-1} = -0.19992274952791061$$
$$d_2 = d_{-2} = 0.12196569374380435$$
$$d_3 = d_{-3} = -0.050077250472089385$$
$$d_4 = d_{-4} = 0.011317101840646436$$
$$d_6 = d_{-6} = -5.487500822768231\text{E-}4$$

13-Point Stencils. $\sigma = 0.1925\pi$

1357911 stencil. $\beta = 1.04$

$$d_0 = 0.5$$
$$d_1 = d_{-1} = -0.31182671961098623$$
$$d_3 = d_{-3} = 0.08793722059177961$$
$$d_5 = d_{-5} = -0.037310514270787353$$
$$d_7 = d_{-7} = 0.015220028715708166$$
$$d_9 = d_{-9} = -5.094690254969648\text{E-}3$$
$$d_{11} = d_{-11} = 1.0746748292554631\text{E-}3$$

1246810 stencil. $\beta = 1.225$

$$d_0 = 0.3240020968562207$$
$$d_1 = d_{-1} = -0.25$$
$$d_2 = d_{-2} = 0.10058161761678362$$
$$d_4 = d_{-4} = -0.016211394630950087$$
$$d_6 = d_{-6} = 4.751848956897231\text{E-}3$$
$$d_8 = d_{-8} = -1.4728623391755445\text{E-}3$$
$$d_{10} = d_{-10} = 3.497419683344133\text{E-}4$$

123579 stencil. $\beta = 1.375$

$$d_0 = 0.2595758998570965$$
$$d_1 = d_{-1} = -0.2161210442492141$$

$$d_2 = d_{-2} = 0.12021205007145175$$
$$d_3 = d_{-3} = -0.037173351241118195$$
$$d_5 = d_{-5} = 3.961825142400804E\text{-}3$$
$$d_7 = d_{-7} = -8.500469444013784E\text{-}4$$
$$d_9 = d_{-9} = 1.8261729233287355E\text{-}4$$

123468 stencil. $\beta = 1.5$

$$d_0 = 0.22856643029671658$$
$$d_1 = d_{-1} = -0.19658662660320927$$
$$d_2 = d_{-2} = 0.12336415951962278$$
$$d_3 = d_{-3} = -0.053413373396790725$$
$$d_4 = d_{-4} = 0.01316893300174682$$
$$d_6 = d_{-6} = -9.489310483476328E\text{-}4$$
$$d_8 = d_{-8} = 1.3262337861974504E\text{-}4$$

123457 stencil. $\beta = 1.59$

$$d_0 = 0.21457119275070027$$
$$d_1 = d_{-1} = -0.18683026262805987$$
$$d_2 = d_{-2} = 0.12266004527546252$$
$$d_3 = d_{-3} = -0.05945584021704768$$
$$d_4 = d_{-4} = 0.020054358349187353$$
$$d_5 = d_{-5} = -3.8748982424400496E\text{-}3$$
$$d_7 = d_{-7} = 1.610010875476098E\text{-}4$$

15-Point Stencils. $\sigma = 0.19\pi$

135791113 stencil. $\beta = 1.11$

$$d_0 = 0.5$$
$$d_1 = d_{-1} = -0.31339818418385334$$
$$d_3 = d_{-3} = 0.0921696857561028$$
$$d_5 = d_{-5} = -0.042740680643837763$$
$$d_7 = d_{-7} = 0.020359003550007293$$
$$d_9 = d_{-9} = -8.78888052362071E\text{-}3$$
$$d_{11} = d_{-11} = 3.1146836013276404E\text{-}3$$
$$d_{13} = d_{-13} = -7.156275561259213E\text{-}4$$

124681012 stencil. $\beta = 1.295$

$$d_0 = 0.32236714053321097$$
$$d_1 = d_{-1} = -0.25$$
$$d_2 = d_{-2} = 0.10216769227713358$$

$$d_4 = d_{-4} = -0.01756253154787246$$
$$d_6 = d_{-6} = 5.840424337495861\text{E-}3$$
$$d_8 = d_{-8} = -2.2168366108895102\text{E-}3$$
$$d_{10} = d_{-10} = 8.252438932473476\text{E-}4$$
$$d_{12} = d_{-12} = -2.3756261572031736\text{E-}4$$

12357911 stencil. $\beta = 1.45$

$$d_0 = 0.25689444748256046$$
$$d_1 = d_{-1} = -0.21501278581108324$$
$$d_2 = d_{-2} = 0.12155277625871977$$
$$d_3 = d_{-3} = -0.038648799797586175$$
$$d_5 = d_{-5} = 4.562916765604436\text{E-}3$$
$$d_7 = d_{-7} = -1.1632920312828478\text{E-}3$$
$$d_9 = d_{-9} = 3.7616806999180853\text{E-}4$$
$$d_{11} = d_{-11} = -1.1420719564397982\text{E-}4$$

12346810 stencil. $\beta = 1.58$

$$d_0 = 0.22513862785383467$$
$$d_1 = d_{-1} = -0.1946531938734711$$
$$d_2 = d_{-2} = 0.12416686947715667$$
$$d_3 = d_{-3} = -0.0553468061265289$$
$$d_4 = d_{-4} = 0.014280517691395201$$
$$d_6 = d_{-6} = -1.1996097976025642\text{E-}3$$
$$d_8 = d_{-8} = 2.4498737887899095\text{E-}4$$
$$d_{10} = d_{-10} = -6.207867674562712\text{E-}5$$

1234579 stencil. $\beta = 1.69$

$$d_0 = 0.21047015796352988$$
$$d_1 = d_{-1} = -0.18421029309503222$$
$$d_2 = d_{-2} = 0.12295164373519486$$
$$d_3 = d_{-3} = -0.06145360873722441$$
$$d_4 = d_{-4} = 0.02181327728304021$$
$$d_5 = d_{-5} = -4.569637725373214\text{E-}3$$
$$d_7 = d_{-7} = 2.7404296254122305\text{E-}4$$
$$d_9 = d_{-9} = -4.050340491136648\text{E-}5$$

1234568 stencil. $\beta = 1.765$

$$d_0 = 0.20494238199107911$$
$$d_1 = d_{-1} = -0.1800055838379142$$

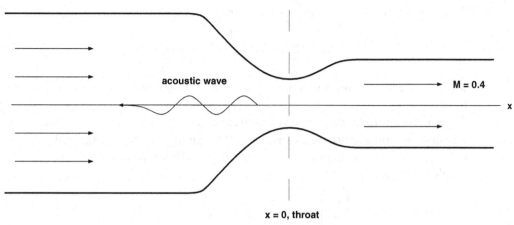

Figure 12.22. Nozzle configuration.

$$d_2 = d_{-2} = 0.12179720514762138$$
$$d_3 = d_{-3} = -0.06311640230450315$$
$$d_4 = d_{-4} = 0.024628464984657864$$
$$d_5 = d_{-5} = -6.878013857582646\text{E-}3$$
$$d_6 = d_{-6} = 1.14544595421895187\text{E-}3$$
$$d_8 = d_{-8} = -4.2307082037753\text{E-}5$$

EXERCISES

12.1. In a transonic cascade, the local Mach number of the flow in the narrow passages may be close to sonic. The computation of sound propagating through such regions presents a challenging problem. To reduce the complexity of the problem, but retaining the basic physics and difficulties, a one-dimensional acoustic wave transmission problem through a nearly choked nozzle is considered.

The following characteristic scales are used to form dimensionless flow variables.

length scale = diameter of nozzle in the uniform region downstream of the throat (see Figure 12.22), D
velocity scale = speed of sound in the same region, a_∞
time scale = D/a_∞
density scale = mean density of gas in the same region, ρ_∞
pressure scale = $\rho_\infty a_\infty^2$

Consider a one-dimensional nozzle with an area distribution as follows:

$$A(x) = \begin{cases} 0.536572 - 0.198086e^{-(\ln 2)(x/0.6)^2}, & x > 0 \\ 1.0 - 0.661514e^{-(\ln 2)(x/0.6)^2}, & x < 0 \end{cases}.$$

The governing equations in dimensionless form are

$$\frac{\partial \rho}{\partial t} + \frac{1}{A}\frac{\partial \rho u A}{\partial x} = 0$$

$$\rho \left(\frac{\partial u}{\partial t} + u \frac{\partial u}{\partial x} \right) + \frac{\partial p}{\partial x} = 0$$

$$A \frac{\partial p}{\partial t} + \frac{\partial (puA)}{\partial x} + (\gamma - 1) \, p \frac{\partial (uA)}{\partial x} = 0$$

where $\gamma = 1.4$. The Mach number in the uniform region downstream of the throat is 0.4.

Small-amplitude acoustic waves, with angular frequency $\omega = 0.6\pi$, are generated way downstream and propagate upstream through the narrow passage of the nozzle throat. Let the upstream propagating wave in the uniform region downstream of the nozzle throat be

$$\begin{bmatrix} \rho' \\ u' \\ p' \end{bmatrix} = \varepsilon \begin{bmatrix} 1 \\ -1 \\ 1 \end{bmatrix} \cos \left[\omega \left(\frac{x}{1 - M} + 1 \right) \right],$$

where $\varepsilon = 10^{-5}$. Use a computation domain of size 20, 10 upstream and 10 downstream of the nozzle throat, to calculate the distribution of maximum acoustic pressure inside the nozzle.

This problem can, of course, be calculated accurately if a very large number of mesh points is used. But this is not always practical. It is recommended that no more than 400 mesh points be used. In other words, use a multisize mesh to compute the solution. The boundary conditions developed in Problem 6.5 would be useful, as well.

The exact solution of this problem is available in the Proceedings of the Third Computational Aeroacoustics Workshop on Benchmark Problems, NASA CP-2000-209790, August 2000.

13 Complex Geometry

Computationally, there are two general ways to treat problems with complex geometry. One way is to use unstructured grids. The other is to use overset grids. Overset grids are formed by overlapping structured grids. In this chapter, the basic idea of overset grids methodology and its implementation are discussed.

13.1 Basic Concept of Overset Grids

To illustrate the basic idea of overset grids, consider the problem of computing the scattering of acoustic waves by a solid cylinder in two dimensions. In the space around the cylinder, the coordinates of choice for computing the solution is the cylindrical polar coordinates centered at the axis of the cylinder. This coordinate system provides a set of body-fitted coordinates and, hence, a body-fitted mesh when discretized around the cylinder. One significant advantage of using a body-fitted grid is the relative ease in enforcing the no-through-flow wall boundary condition using the ghost point method or other methods. Away from the cylinder, acoustic waves propagate with no preferred direction. The natural coordinate system to use is the Cartesian coordinates. Therefore, to take into account the advantages stated, one may use a polar mesh around the cylinder and a Cartesian mesh away from the cylinder with an overlapping mesh region. The overlapping mesh region is for data transfer from one set of grids to the other and vice versa.

Figure 13.1 shows a Cartesian mesh with a square hole around the cylinder. For the purpose of putting all variables in dimensionless form, the diameter, D, of the cylinder is taken as length scale, speed of sound a_0 as the velocity scale, D/a_0 as the time scale, ρ_0 (the density of the undisturbed gas) as the density scale, and $\rho_0 a_0^2$ as the pressure scale. The mesh size is $\Delta x = \Delta y = 1/32$. The size of the square hole is 1.6 by 1.6. Figure 13.2 shows a polar mesh with $\Delta r = 1/32$ and $\Delta\theta = \pi/150$. The polar mesh extends to a distance of $r = 1.5$. The mesh overlap region is between the square and the outside edge of the polar mesh as shown in Figure 13.2. Figure 13.3 is an enlarged view of the mesh overlap region. Both the cylindrical mesh and the Cartesian mesh are structured meshes. To compute the solution, it would be best to write the governing equations in the coordinate system used. For acoustic wave scattering problems under discussion, the governing equations are the full Euler equations. In dimensionless form, they are

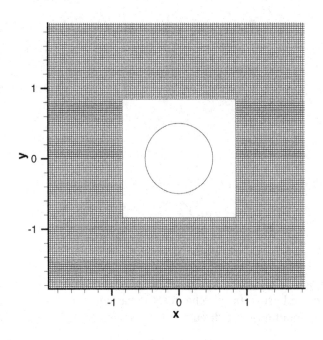

Figure 13.1. Cartesian grid used in acoustic wave scattering by a cylinder problem.

Cartesian coordinates

$$\frac{\partial \rho}{\partial t} + u\frac{\partial \rho}{\partial x} + v\frac{\partial \rho}{\partial y} + \rho\left(\frac{\partial u}{\partial x} + \frac{\partial v}{\partial y}\right) = 0$$

$$\frac{\partial u}{\partial t} + u\frac{\partial u}{\partial x} + v\frac{\partial u}{\partial y} + \frac{1}{\rho}\frac{\partial p}{\partial x} = 0$$

$$\frac{\partial v}{\partial t} + u\frac{\partial v}{\partial x} + v\frac{\partial v}{\partial y} + \frac{1}{\rho}\frac{\partial p}{\partial y} = 0$$

$$\frac{\partial p}{\partial t} + u\frac{\partial p}{\partial x} + v\frac{\partial p}{\partial y} + \gamma p\left(\frac{\partial u}{\partial x} + \frac{\partial v}{\partial y}\right) = 0 \tag{13.1}$$

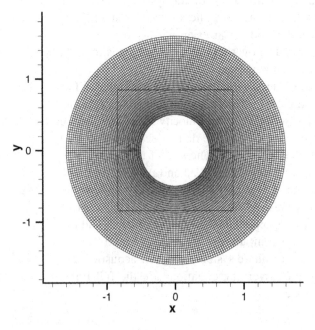

Figure 13.2. Polar grid used in acoustic wave scattering problem. Square indicates the inner boundary of the Cartesian grid.

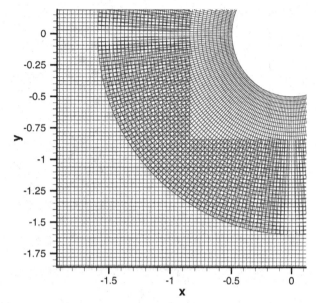

Figure 13.3. The overlapping region of the Cartesian and polar grids.

Polar coordinates

$$\frac{\partial \rho}{\partial t} + \frac{1}{r}\frac{\partial}{\partial r}(\rho v_r r) + \frac{1}{r}\frac{\partial(\rho v_\theta)}{\partial \theta} = 0$$

$$\frac{\partial v_r}{\partial t} + v_r\frac{\partial v_r}{\partial r} + \frac{v_\theta}{r}\frac{\partial v_r}{\partial \theta} - \frac{v_\theta^2}{r} + \frac{1}{\rho}\frac{\partial p}{\partial r} = 0$$

$$\frac{\partial v_\theta}{\partial t} + v_r\frac{\partial v_\theta}{\partial r} + \frac{v_\theta}{r}\frac{\partial v_\theta}{\partial \theta} + \frac{v_r v_\theta}{r} + \frac{1}{\rho}\frac{\partial p}{\partial \theta} = 0$$

$$\frac{\partial p}{\partial t} + v_r\frac{\partial p}{\partial r} + \frac{v_\theta}{r}\frac{\partial p}{\partial \theta} + \gamma p\left(\frac{1}{r}\frac{\partial v_r r}{\partial r} + \frac{1}{r}\frac{\partial v_\theta}{\partial \theta}\right) = 0. \tag{13.2}$$

The solution of (13.1) is computed on the Cartesian mesh. The solution of (13.2) is computed on the polar mesh. (u, v) and (v_r, v_θ) are related by

$$u = v_r \cos\theta - v_\theta \sin\theta$$
$$v = v_r \sin\theta + v_\theta \cos\theta$$
$$v_r = u \cos\theta + v \sin\theta.$$
$$v_\theta = -u \sin\theta + v \cos\theta$$

In computing the solution, the 7-point stencil dispersion-relation-preserving (DRP) scheme may be used. Since the 7-point stencil DRP scheme is a central difference scheme, the values of the computation variables of the three rows and columns of the Cartesian mesh around the square hole are not computed by the time marching scheme. They are found by interpolation from the values of the polar mesh using, say, a 16-point regular stencil. A typical stencil is shown in Figure 13.4. For the polar mesh, the values of the computation variables at the outer three rings are not computed. They are interpolated from the values of the Cartesian mesh using a 16-point regular stencil as shown in Figure 13.5.

Data transfer from one set of grids to the other is one of the crucial processes of the overset grids method. Since a high-resolution scheme is used for computation in each of the structured grids, it is important for the data transfer process to retain

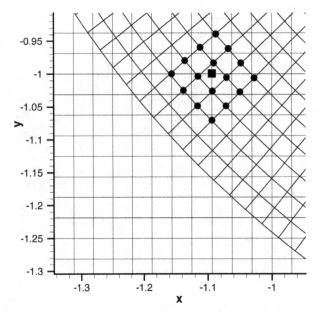

Figure 13.4. Interpolation from a polar grid to a Cartesian grid using a 16-point stencil. $\Delta x = \Delta y = \Delta r = 1/32$, $\Delta\theta = \pi/150$.

at least the same accuracy. In what follows, a highly accurate multidimensional interpolation scheme, suitable for use as a data transfer method, is discussed.

13.2 Optimized Multidimensional Interpolation

The idea and methodology of optimized multidimensional interpolation will be considered first. Numerical examples will be given later. For simplicity, only two-dimensional interpolation is discussed in detail. The generalization to three dimensions is fairly straightforward.

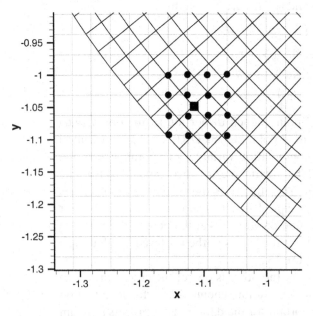

Figure 13.5. Interpolation from a Cartesian grid to a polar grid using a 16-point stencil. $\Delta x = \Delta y = \Delta r = 1/32$.

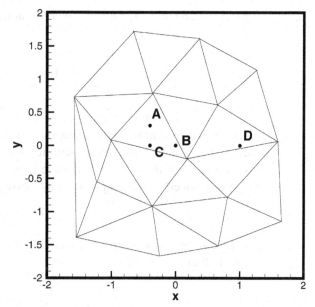

Figure 13.6. A 16-point irregular interpolation stencil.

The development of optimized multidimensional interpolation is, in many respects, similar to that in one dimension. Consider an interpolation stencil in a computational domain as shown in Figure 13.6. Without loss of generality, the interpolation is considered for implementation in the $x - y$ plane. However, x and y may be a set of curvilinear coordinates in the physical domain. For interpolation purposes, the values of a function $f(x, y)$ are given at the stencil points (x_j, y_j), $j = 1, 2, \ldots, N$ for an N-point stencil. The objective is to estimate the value $f(x_o, y_o)$ at a general point P at (x_o, y_o). By definition of interpolation (not extrapolation), the point P lies within the boundary of the stencil as defined by Figure 13.6. Let d_j be the distance between neighboring points of the stencil. A useful length scale is

$$\Delta = \frac{1}{K} \sum_{j=1}^{K} d_j, \tag{13.3}$$

where K is the number of links between stencil points.

The most general interpolation formula in two dimensions has the following form:

$$f(x_o, y_o) = \sum_{j=1}^{N} S_j f(x_j, y_j). \tag{13.4}$$

Different interpolation methods determine the coefficient S_j in different ways.

It will be assumed that the function $f(x, y)$ to be interpolated has a Fourier inverse transform, such that

$$f(x, y) = \int_{-\infty}^{\infty} \int_{-\infty}^{\infty} \tilde{f}(\alpha, \beta) e^{i(\alpha x + \beta y)} \, d\alpha \, d\beta, \tag{13.5}$$

where (α, β) are the wave numbers in the x and y directions, respectively.

Let $A(\alpha, \beta) = |\tilde{f}(\alpha, \beta)|$ and $\phi(\alpha, \beta) = \arg[\tilde{f}(\alpha, \beta)]$, Eq. (13.5) may be rewritten as

$$f(x, y) = \int\limits_{-\infty}^{\infty} \int\limits_{-\infty}^{\infty} A(\alpha, \beta) e^{i[\alpha x + \beta y + \varphi(\alpha, \beta)]} d\alpha \, d\beta. \tag{13.6}$$

This equation may be interpreted in the sense that $f(x, y)$ is made up of a superposition of simple waves $\exp[i(ax + by + \phi(a, b))]$ with amplitude $A(\alpha, \beta)$. It is useful to define the local interpolation error, E_{local}, for wave number (α, β) as the absolute value of the difference between the left and right side of Eq. (13.4) when $f(x, y)$ is replaced by the simple wave as follows:

$$f_{\alpha\beta} = e^{i[\alpha x + \beta y + \varphi(\alpha, \beta)]}. \tag{13.7}$$

Here all simple waves are assumed to have unit amplitude; i.e., $A(\alpha, \beta) = 1.0$. It is easy to find as follows:

$$E_{\text{local}}^2 = \left| e^{i(\alpha x_o + \beta y_o + \phi)} - \sum_{j=1}^{N} S_j e^{i(\alpha x_j + \beta y_j + \phi)} \right|^2$$

$$= \left| 1 - \sum_{j=1}^{N} S_j e^{i\left[\alpha\Delta\left(\frac{x_j - x_o}{\Delta}\right) + \beta\Delta\left(\frac{y_j - y_o}{\Delta}\right)\right]} \right|^2. \tag{13.8}$$

The total error, E, over the region $-\kappa \leq \alpha\Delta, \beta\Delta \leq \kappa$, in wave number space is obtained by integrating E_{local}^2 of Eq. (13.8) over this region, i.e.,

$$E^2 = \int\limits_{-\kappa}^{\kappa} \int\limits_{-\kappa}^{\kappa} \left| 1 - \sum_{j=1}^{N} S_j e^{i\left[\alpha\Delta\left(\frac{x_j - x_o}{\Delta}\right) + \beta\Delta\left(\frac{y_j - y_o}{\Delta}\right)\right]} \right|^2 d(\alpha\Delta) d(\beta\Delta). \tag{13.9}$$

When the function to be interpolated is a constant, which corresponds to setting $\alpha = \beta = 0$ in Eq. (13.7), it is reasonable to require the local interpolation error to be zero. By Eq. (13.8), this requirement leads to

$$\sum_{j=1}^{N} S_j - 1 = 0. \tag{13.10}$$

It is now proposed that the interpolation coefficients, S_j ($j = 1, 2, \ldots, N$) be chosen such that E^2 is a minimum subjected to constraint (13.10). This can readily be done by using the method of the Lagrange multiplier. Let

$$L = \int\limits_{-\kappa}^{\kappa} \int\limits_{-\kappa}^{\kappa} \left| 1 - \sum_{j=1}^{N} S_j e^{i\left[\xi\left(\frac{x_j - x_o}{\Delta}\right) + \eta\left(\frac{y_j - y_o}{\Delta}\right)\right]} \right|^2 d\xi \, d\eta + \lambda \left(\sum_{j=1}^{N} S_j - 1 \right), \tag{13.11}$$

where λ is the Lagrange multiplier. The conditions for L to be a minimum are

$$\frac{\partial L}{\partial S_j} = 0, \quad \frac{\partial L}{\partial \lambda} = 0, \quad j = 1, 2, \ldots, N. \tag{13.12}$$

It is easy to find that the condition $\partial L/\partial S_j = 0$ yields

$$2\mathrm{Re}\left\{\int_{-\kappa}^{\kappa}\int_{-\kappa}^{\kappa} e^{-i\left[\xi\frac{(x_j-x_0)}{\Delta}+\eta\frac{(y_j-y_0)}{\Delta}\right]}\left[1-\sum_{k=1}^{N}S_k e^{i\left[\xi\frac{(x_k-x_0)}{\Delta}+\eta\frac{(y_k-y_0)}{\Delta}\right]}\right]d\xi\,d\eta\right\}+\lambda = 0, \quad (13.13)$$

where Re{} is the real part of {}. Similarly it is easy to find that $\partial L/\partial\lambda$ leads to

$$\sum_{j=1}^{N} S_j - 1 = 0, \tag{13.14}$$

which is constraint (13.10).

Eqs. (13.13) and (13.14) form a linear matrix system of $(N + 1)$ unknowns for S_j $(j = 1, 2, \dots, N)$ and λ. It turns out that all the integrals of Eq. (13.13) can be evaluated in closed form. This allows the matrix elements to be written out explicitly. The values of S_j may be found by solving the matrix system as follows:

$$\mathbf{AS} = \mathbf{b}, \tag{13.15}$$

where vector $\mathbf{S}^{\mathrm{T}} = (S_1\ S_2\ S_3\ \dots\ S_N\ \lambda)$. Note: Superscript $\mathbf{T}$ denotes the transpose. The elements of the coefficient matrix $\mathbf{A}$ and vector $\mathbf{b}$ are given by

$$A_{jk} = \begin{cases} 4\kappa^2, & j = k \\ \frac{4\Delta^2}{(x_k-x_j)(y_k-y_j)}\sin\left[\kappa\frac{(x_k-x_j)}{\Delta}\right]\sin\left[\kappa\frac{(y_k-y_j)}{\Delta}\right], & j \neq k \end{cases}$$
$$(j, k = 1, 2, \dots, N) \tag{13.16}$$

$$A_{j(N+1)} = \frac{1}{2}, \qquad (j = 1, 2, \dots, N) \tag{13.17}$$

$$A_{(N+1)k} = 1, \qquad (k = 1, 2, \dots, N) \tag{13.18}$$

$$A_{(N+1)(N+1)} = 0 \tag{13.19}$$

$$b_j = \frac{4\Delta^2}{(x_j - x_0)(y_j - y_0)}\sin\left[\kappa\frac{(x_j - x_0)}{\Delta}\right]\sin\left[\kappa\frac{(y_j - y_0)}{\Delta}\right] \tag{13.20}$$

$$b_{(N+1)} = 1. \tag{13.21}$$

Note: In the case $x_j = x_k$ and/or $y_j = y_k$ or $x_j = x_0$ or $y_j = y_0$ the limit forms of these formulas are to be used. Computationally, the limit forms are used wherever $|x_j - x_k| < \varepsilon$ or $|y_j - y_k| < \varepsilon$. This is discussed in Appendix 13A at the end of of this chapter.

It is useful to point out that A_{jk} depends on Δ and κ as well as the coordinates of the stencil points only. The coordinates of the point to be interpolated to, (x_0, y_0), appears only in b_j. Thus, if many values of f are to be interpolated using the same stencil, $\mathbf{A}$ and $\mathbf{A}^{-1}$ need only be computed once.

For error estimate purposes, the following symmetry properties of the local error, given by Eq. (13.8), can easily be derived.

$$E_{\mathrm{local}}(\alpha, \beta) = E_{\mathrm{local}}(-\alpha, -\beta) \tag{13.22}$$

$$E_{\mathrm{local}}(-\alpha, \beta) = E_{\mathrm{local}}(\alpha, -\beta). \tag{13.23}$$

Thus, once S_j's are found, E_{local} can be calculated by formula (13.8). Upon invoking Eqs. (13.22) and (13.23), it is sufficient to examine $E_{local}(\alpha, \beta)$ over the upper half of the $\alpha\Delta - \beta\Delta$ plane for $-\pi \le \alpha\Delta, \beta\Delta \le \pi$.

13.2.1 Order Constraints

For small Δ, it is often desirable to require interpolation formula (13.4) to be accurate to a specified order of Taylor series expansion in Δ. This requirement imposes a set of constraints on the interpolation coefficients S_j. To find these constraints in their simplest form, first expand $f(x_j, y_j)$ about (x_o, y_o) to obtain

$$f(x_j, y_j) = f(x_o, y_o) + \sum_{p,q} \frac{1}{p!q!} \left(\frac{\partial^{p+q} f}{\partial x^p \partial y^q} \right)_{x_o, y_o} (x_j - x_o)^p (y_j - y_o)^q. \quad (13.24)$$

On substituting (13.24) into (13.4), it is found that

$$f(x_o, y_0) = f(x_o, y_o) \left(\sum_{j=1}^{N} S_j \right)$$

$$+ \sum_{p,q} \frac{1}{p!q!} \left(\frac{\partial^{p+q} f}{\partial x^p \partial y^q} \right)_{x_o, y_o} \left[\sum_{j=1}^{N} (x_j - x_o)^p (y_j - y_o)^q S_j \right]. \quad (13.25)$$

Thus, if Eq. (13.25) is truncated to order Δ^M, S_j must satisfy the following conditions:

$$\sum_{j=1}^{N} S_j = 1, \qquad M = 0 \quad (13.26)$$

$$\sum_{j=1}^{N} (x_j - x_o)^p (y_j - y_o)^q S_j = 0, \qquad 1 \le p + q \le M. \quad (13.27)$$

From Eq. (13.25) it is easy to see that, if the order of truncation is increased from $(M - 1)$ to M, $(M + 1)$ more constraints are added. Therefore, the total number of constraints, T, to be satisfied, if Eq. (13.24) is truncated at the Mth order, is

$$T = 1 + 2 + 3 + \cdots + (M + 1) = \tfrac{1}{2}(M + 1)(M + 2). \quad (13.28)$$

It is possible if $(x_j, y_j), j = 1, 2, \ldots, N$ are points on a highly regular grid, the total number of independent constraints given by Eqs. (13.26) and (13.27) is less than that calculated by Eq. (13.28). Degeneracy would occur when a number of x_j's or y_j's are the same. In this case, only the linearly independent constraints need to be imposed.

Now for a given interpolation stencil with a given set of points $(x_j, y_j), j = 1, 2, \ldots,$ N, the interpolation coefficients S_j for the point (x_o, y_o) may be found by minimizing the total error E^2 given by Eq. (13.9), subjected to order constraints (13.26) and

(13.27). This constrained minimization problem can again be easily solved by the method of Lagrange multipliers. Let

$$L = \int_{-\kappa}^{\kappa} \int_{-\kappa}^{\kappa} \left| 1 - \sum_{j=1}^{N} S_j e^{\left[i \xi \frac{(x_j - x_0)}{\Delta} + \eta \frac{(y_j - y_0)}{\Delta} \right]} \right|^2 d\xi \, d\eta + \lambda \left(\sum_{j=1}^{N} S_j - 1 \right)$$

$$+ \sum_{n,m} \mu_{mn} \left[\sum_{j=1}^{N} S_j (x_j - x_0)^n (y_j - y_0)^m \right], \tag{13.29}$$

where $n, m = 0, 1, 2, \ldots, M$; $n + m \leq M$; n and m are not both equal to zero. M is the order of truncation and μ_{mn} are Lagrangian multipliers. Since the total number of constraints (T as given by Eq. (13.28)) cannot exceed the number of stencil points, it is understood that $T < N$. The conditions for a minimum are as follows:

$$\frac{\partial L}{\partial S_j} = 0 \tag{13.30}$$

$$\frac{\partial L}{\partial \lambda} = 0 \tag{13.31}$$

$$\frac{\partial L}{\partial \mu_{mn}} = 0. \tag{13.32}$$

Again, Eqs. (13.30) to (13.32) lead to a linear matrix equation for the unknowns S_j, λ, and μ_{mn}. Let the transpose of the vector $\mathbf{X}$ and $\mathbf{d}$ be defined by

$$\mathbf{X}^T = (S_1 \, S_2 \, \cdots \, S_N \, \lambda \, \mu_{10} \, \mu_{01} \, \mu_{20} \, \mu_{11} \, \mu_{02} \, \cdots \, \mu_{mn} \, \cdots \, \mu_{0M}) \tag{13.33}$$

$$\mathbf{d}^T = (b_1 \, b_2 \, \cdots \, b_N \, 1 \, 0 \, 0 \, 0 \, \cdots \cdots \, 0 \, 0), \tag{13.34}$$

where b_j's are given by Eqs. (13.20) and (13.21). It is easy to establish by means of Eqs. (13.30) to (13.32) that $\mathbf{X}$ is the solution of the matrix equation

$$\mathbf{BX} = \mathbf{d}. \tag{13.35}$$

The coefficient matrix $\mathbf{B}$ may be partitioned into four submatrices as

$$\mathbf{B} = \begin{bmatrix} \mathbf{A} & \mathbf{C} \\ \mathbf{C}^T & \mathbf{0} \end{bmatrix}. \tag{13.36}$$

$\mathbf{A}$ is a $(N+1)$ by $(N+1)$ square matrix. It is the same as that of Eq. (13.15). $\mathbf{C}$ is a $(N+1)$ by $(T-1)$ matrix. $\mathbf{C}^T$ is the transpose of $\mathbf{C}$. The zero matrix $\mathbf{0}$ is a square matrix of size $(T-1)$. On writing out in full, the elements of matrix $\mathbf{C}$ are as follows:

$$\mathbf{C} = \begin{bmatrix}
(x_1 - x_0) & (y_1 - y_0) & (x_1 - x_0)^2 & (x_1 - x_0)(y_1 - y_0) & (y_1 - y_0)^2 & (x_1 - x_0)^3 & (x_1 - x_0)^2(y_1 - y_0) & \cdots & (x_1 - x_0)(y_1 - y_0)^{M-1} & (y_1 - y_0)^M \\
(x_2 - x_0) & (y_2 - y_0) & (x_2 - x_0)^2 & (x_2 - x_0)(y_2 - y_0) & (y_2 - y_0)^2 & (x_2 - x_0)^3 & (x_2 - x_0)^2(y_2 - y_0) & \cdots & (x_2 - x_0)(y_2 - y_0)^{M-1} & (y_2 - y_0)^M \\
\vdots & \vdots & \vdots & \vdots & \vdots & \vdots & \vdots & & \vdots & \vdots \\
(x_j - x_0) & (y_j - y_0) & (x_j - x_0)^2 & (x_j - x_0)(y_j - y_0) & (y_j - y_0)^2 & (x_j - x_0)^3 & (x_j - x_0)^2(y_j - y_0) & \cdots & (x_j - x_0)(y_j - y_0)^{M-1} & (y_j - y_0)^M \\
\vdots & \vdots & \vdots & \vdots & \vdots & \vdots & \vdots & & \vdots & \vdots \\
(x_N - x_0) & (y_N - y_0) & (x_N - x_0)^2 & (x_N - x_0)(y_N - y_0) & (y_N - y_0)^2 & (x_N - x_0)^3 & (x_N - x_0)^2(y_N - y_0) & \cdots & (x_N - x_0)(y_N - y_0)^{M-1} & (y_N - y_0)^M \\
0 & 0 & 0 & 0 & 0 & 0 & 0 & & 0 & 0
\end{bmatrix}. \tag{13.37}$$

Again, once Eq. (13.35) is solved, the interpolation coefficients S_j's are found. By substituting the values of S_j's into Eq. (13.8), the local interpolation error, E_{local}, at wave number $(\alpha\Delta, \beta\Delta)$ may be computed.

13.2.2 Interpolation Errors in Wave Number Space

An interpolation stencil may be regular or irregular depending on the configuration of the stencil relative to the coordinate system used for the computation. Generally speaking, interpolation error is small compared with extrapolation error. Figure 13.6 shows a 16-point irregular interpolation stencil. For reference purposes, the coordinates of the interpolation points of this stencil are as follows:

$(-0.70, 1.72)$
$(0.34, 1.58)$
$(1.24, 1.14)$
$(-1.62, 0.72)$
$(-0.38, 0.79)$
$(0.64, 0.60)$
$(-1.02, 0.08)$
$(0.18, -0.20)$
$(1.58, 0.06)$
$(-1.24, -0.56)$
$(-0.38, -0.92)$
$(0.80.-0.78)$
$(-1.56, -1.40)$
$(-0.26, -1.68)$
$(0.66, -1.52)$
$(1.64, -1.16)$

The points to be interpolated are as follows:

$A\ (-0.40, 0.30)$
$B\ (0.00, 0.00)$
$C\ (-0.40, 0.00)$
$D\ (1.00, 0.00)$

Figure 13.7 shows a 16-point regular stencil in the form of a square. Interpolation error analysis for the 4 points A, B, C, and D shown in both Figures 13.6 and 13.7 have been carried out. The finding is that the errors for points A, B, and C are very similar. For this reason, only the interpolation errors of A and D will be discussed. It is useful to consider point A as typical of all the points located around the center of the stencil and point D as typical of all the points located close to the boundary of the stencil.

Figure 13.8 shows the local error, E_{local}, contours in the $\alpha\Delta - \beta\Delta$ plane. In calculating E_{local}, κ has been set equal to 1.2. It has been found through numerical experimentation that $\kappa = 1.2$ is a reasonably good value to use. As readily seen in Figure 13.8, unlike extrapolation, there is no large error even for high wave numbers.

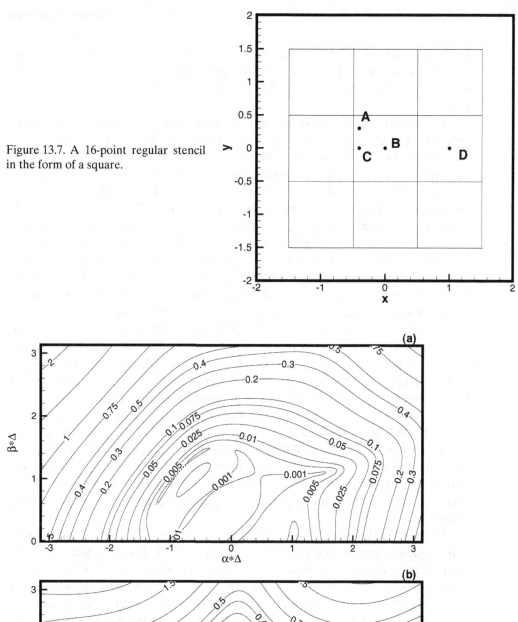

Figure 13.7. A 16-point regular stencil in the form of a square.

Figure 13.8. Contours of local interpolation error, E_{local}, in the $\alpha\Delta - \beta\Delta$ plane for the irregular stencil of Figure 13.6 using the optimized interpolation scheme. (a) at A, (b) at D.

In most aeroacoustics applications, the range of wave numbers that is of interest is $-1.0 \leq \alpha\Delta, \beta\Delta \leq 1.0$. In order to show error maps with more clarity, only the low wave number portion of the $\alpha\Delta - \beta\Delta$ plane is shown in the discussions to be followed.

The imposition of order constraints has a significant impact on the local error distribution in the wave number space. Generally speaking, imposing high-order constraints is beneficial in reducing the error in the low wave number region. On the other hand, it tends to cause an increase in error in the high wave number region.

Figures 13.9a–13.9c show the error contour distributions in the $\alpha\Delta - \beta\Delta$ plane for interpolation to point A using the irregular stencil of Figure 13.6. Figure 13.9a is obtained with the imposition of a second-order constraint. Figures 13.9b and 13.9c are the results with the imposition of a third- and fourth-order constraint, respectively. It is clear from these figures that the best result or least error in the region $-0.5 < \alpha\Delta, \beta\Delta < 0.5$ is realized by enforcing the highest-order constraint. On the other hand, if one is interested in keeping the error small over a larger region in wave number space, then the low-order constraint becomes attractive. For point A only the second-order constraint interpolation has an error less than 1 percent over the wave number region of $-1.0 \leq \alpha\Delta, \beta\Delta < \leq 1.0$. Figures 13.10a–13.10c are for interpolation to point D of the irregular stencil. Again, the error contours of these figures point to the benefit of minimizing interpolation error over the low wave number range by using the maximum order constraints. The error is less than 0.25 percent for $|\alpha\Delta|$, $|\beta\Delta| < 0.55$ (wave length longer than 12.5 mesh points). Again, for the larger wave number range, say $-1.0 \leq \alpha\Delta, \beta\Delta \leq 1.0$, the second-order constraint has an interpolation error less than 1 percent. Thus, the choice of order constraints depends on the range of wave numbers one wishes to have the error below a specified level. For general usage, the imposition of a medium-order constraint (say second or third order for a 16-point irregular stencil) may be a good compromise.

Figure 13.11 shows contours of local interpolation error, E_{local}, at A using the 16-point regular stencil of Figure 13.7. Figures 13.11a–13.11c are subjected to second-, third-, and fourth-order constraints, respectively. Because of regularity of the stencil points, $x_j = x_k$ and $y_j = y_k$ occur repeatedly, there are only four independent values of x_j and y_j ($j = 1, 2, \ldots, 16$). This immediately leads to degeneracy in the fourth-order constraints. To show this is, indeed, the case, recall that the interpolation coefficients are to be found by solving Eq. (13.35) $\mathbf{BX} = \mathbf{d}$. Here, N is 16. Including the last row of submatrix $\mathbf{A}$, the last 15 rows of matrix $\mathbf{B}$ are as follows:

$$
\begin{bmatrix}
1 & 1 & 1 & \cdots & 1 & 0 & \cdots & \cdots & \cdots & 0 \\
x_1 - x_0 & x_2 - x_0 & x_3 - x_0 & \cdots & x_N - x_0 & 0 & \cdots & \cdots & \cdots & 0 \\
y_1 - y_0 & y_2 - y_0 & y_3 - y_0 & \cdots & y_N - y_0 & 0 & \cdots & \cdots & \cdots & 0 \\
(x_1 - x_0)^2 & (x_2 - x_0)^2 & (x_3 - x_0)^2 & \cdots & (x_N - x_0)^2 & 0 & \cdots & \cdots & \cdots & 0 \\
\vdots & \vdots & \vdots & \vdots & \vdots & \vdots & \vdots & \vdots & \vdots & \vdots \\
(y_1 - y_0)^4 & (y_2 - y_0)^4 & (y_3 - y_0)^4 & \cdots & (y_N - y_0)^4 & 0 & \cdots & \cdots & \cdots & 0
\end{bmatrix}.
$$

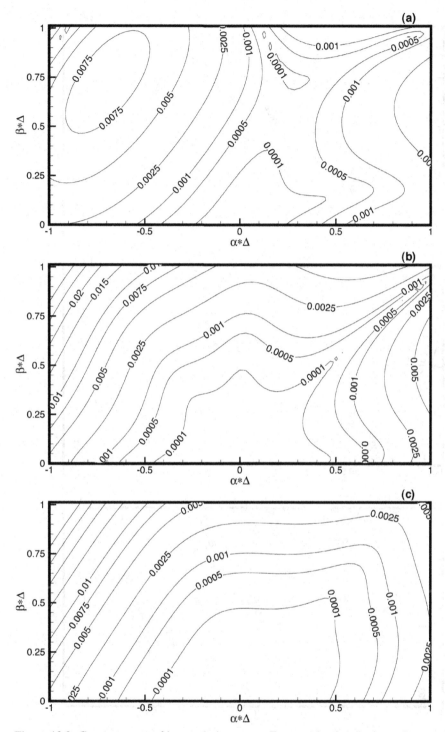

Figure 13.9. Contour maps of interpolation error, E_{local}, at A using the irregular stencil with order constraints. (a) Second order. (b) Third order. (c) Fourth order.

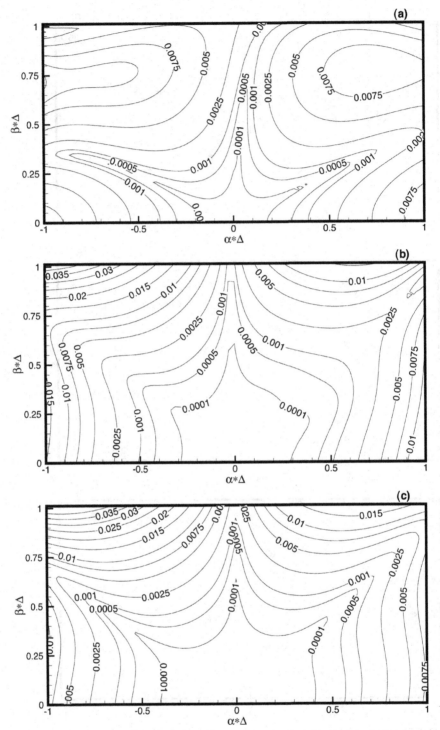

Figure 13.10. Contour maps of interpolation error, E_{local}, at D using the irregular stencil with order constraints. (a) Second order. (b) Third order. (c) Fourth order.

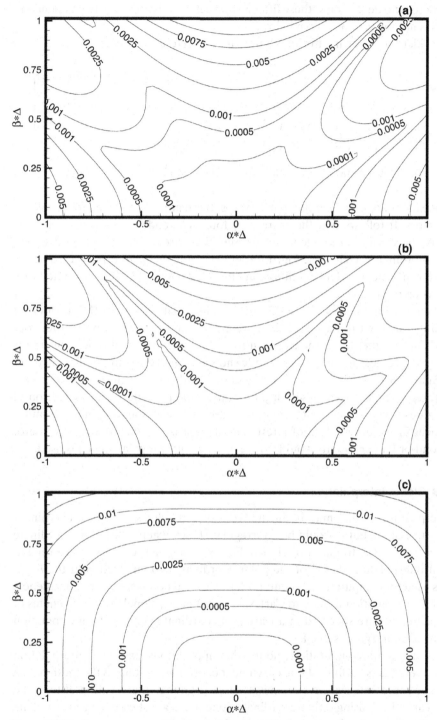

Figure 13.11. Contour maps of interpolation error, E_{local}, at A using the regular stencil with order constraints. (a) Second order. (b) Third order. (c) Fourth order.

If there is no degeneracy, these fifteen row vectors must be linearly independent. But this is not the case as shown below.

Consider the following five of the fifteen row vectors:

$$
\begin{bmatrix}
1 & 1 & 1 & \cdots & 1 & 0 & \cdots & \cdots & \cdots & 0 \\
x_1 - x_0 & x_2 - x_0 & x_3 - x_0 & \cdots & x_N - x_0 & 0 & \cdots & \cdots & \cdots & 0 \\
(x_1 - x_0)^2 & (x_2 - x_0)^2 & (x_3 - x_0)^2 & \cdots & (x_N - x_0)^2 & 0 & \cdots & \cdots & \cdots & 0 \\
(x_1 - x_0)^3 & (x_2 - x_0)^3 & (x_3 - x_0)^3 & \cdots & (x_N - x_0)^3 & 0 & \cdots & \cdots & \cdots & 0 \\
(x_1 - x_0)^4 & (x_2 - x_0)^4 & (x_3 - x_0)^4 & \cdots & (x_N - x_0)^4 & 0 & \cdots & \cdots & \cdots & 0
\end{bmatrix}.
$$

Since there are only four independent x_j's, there are only four independent column vectors. It follows that there are only four independent row vectors as well. Similarly, there is also one degeneracy because there are only four independent values of y coordinates. To impose fourth-order constraint on the interpolation scheme, only 13 linearly independent conditions are imposed. Figure 13.11c is computed with only 13 constraints.

On comparing Figures 13.11a–13.11c, it is clear that the case with fourth-order constraints turns out to have slightly more error than those with lower-order constraints. On the other hand, the cases with second-order and third-order constraints are about the same. Figures 13.12a–13.12c show similar error contours for the point D of Figure 13.7. Again because of degeneracy with the fourth-order constraints only thirteen conditions are used in the optimization procedure. An inspection of these figures suggests that there are only minor differences among them. If minor differences in the distribution of interpolation error is acceptable, then the use of the second-order constraints would suffice when a highly regular stencil is used.

13.2.3 Global Interpolation

An interpolation stencil may be in the form of a regular stencil in a curvilinear coordinate system, but it would be an irregular stencil when a different coordinate system is used as the frame of reference. For example, Figure 13.4 shows a regular 16-point interpolation stencil of the polar coordinate system. That is, if the $r - \theta$ plane is used for computation, the interpolation stencil points form a regular stencil. But, if the computation is done in Cartesian coordinates, namely the coordinates of the stencil points are specified in a Cartesian coordinate system, the interpolation points form an irregular stencil.

In overset grids computation and in other applications, extensive interpolation from one set of mesh points of one curvilinear coordinate system to the mesh points of another coordinate system becomes necessary. Global interpolation of this type may be carried out using either coordinate system. As a test case, consider a plane wave with wave vector $\mathbf{k}$ inclined at 30° to the x-axis in the $x - y$ plane. The wave function ϕ may be written as

$$
\phi(x, y) = \mathrm{Re}\left\{e^{ik[x\cos(\pi/6) + y\sin(\pi/6)]}\right\}, \tag{13.39}
$$

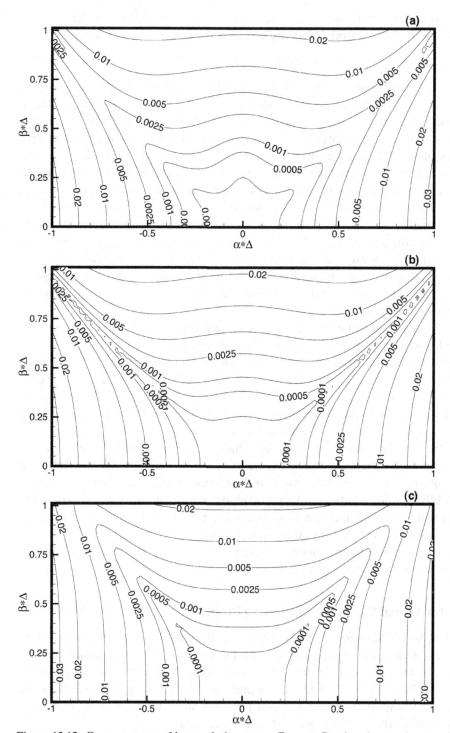

Figure 13.12. Contour maps of interpolation error, E_{local}, at D using the regular stencil with order constraints. (a) Second order. (b) Third order. (c) Fourth order.

where Re{ } is the real part of { }. The wave function can also be written in polar coordinates (r, θ) by a straightforward coordinate transformation, thus,

$$\phi(r, \theta) = \mathrm{Re}\{e^{ikr\cos(\theta - \pi/6)}\}. \tag{13.40}$$

Suppose a mesh of $\Delta x = \Delta y = 1/32$ in the Cartesian coordinates and $\Delta r = 1/32$, $\Delta \theta = \pi/150$ in the polar coordinate as shown in Figure 13.4 are used for interpolation purpose. Let the value of ϕ be known on the polar mesh points. The objective is to find ϕ at the mesh points of the Cartesian coordinates by interpolation using a 16-point stencil. That is, the aim is to interpolate the values of ϕ from the polar grid to the Cartesian grid. One obvious way is to refer the coordinates of all points to the polar coordinates system and perform the interpolation in the (r, θ) plane. In the (r, θ) plane the 16 interpolation points form a regular stencil. If the optimized interpolation scheme is used, it is easy to show that the matrix $\mathbf{A}$ (for interpolation in polar coordinate, see Eq. (13.15)) is the same for all interpolation stencils. For this reason, the inverse coefficient matrix $\mathbf{A}^{-1}$ needs to be computed only once. On the other hand, if the interpolations are carried out in the Cartesian coordinate system, the stencils are irregular. It follows that the matrix $\mathbf{A}$ for each point to be interpolated to is different.

One piece of information one would like to know about this global interpolation is which way of interpolation, using regular or irregular stencils, would yield the least global error. For the example under consideration, the exact wave function on the Cartesian mesh is given by Eq. (13.39). Figure 13.13a shows the interpolation error for the plane wave along the line $y = 1.0$ using the polar coordinates as the reference coordinates (regular stencils). Figure 13.13b shows the corresponding interpolation error when the interpolation is performed in the Cartesian coordinates (irregular stencil). It is evident from these figures that the error in Figure 13.13b is more than twice as large as that in Figure 13.13a. This strongly suggests that when global interpolation is needed, it would be best to use regular stencils. Figures 13.14a and 13.14b provide a similar comparison of global interpolation error arising from the use of regular and irregular stencils along the line $x = 1.0$. Again, the use of irregular stencils results in considerably larger interpolation error. Although, intuitively, the use of a regular stencil may appear to be the preferred choice, it is reassuring to see its confirmation by a concrete example.

13.3 Numerical Examples: Scattering Problems

As examples to illustrate the use of overset grid method, two problems involving acoustic scattering by a rigid cylinder in two dimensions discussed in Section 13.1 will now be considered. The governing equations in Cartesian and cylindrical coordinates are Eqs. (13.1) and (13.2). They are discretized, and the numerical solutions are advanced in time by the 7-point stencil DRP scheme. The ghost point method is used to enforce the no-through-flow boundary condition.

13.3.1 Transient Scattering Problem

First, consider the scattering of an initial pressure pulse by the cylinder. For this problem, the Euler equations are solved with the initial conditions as follows:

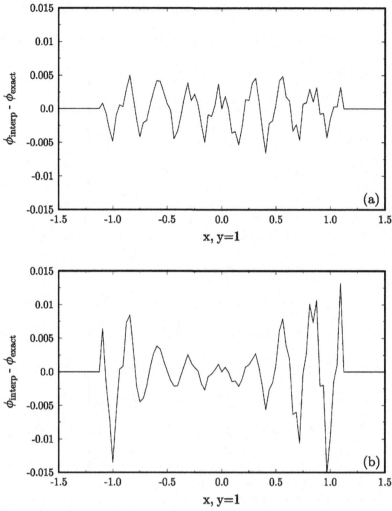

Figure 13.13. Distribution of interpolation error along the line $y = 1$. (a) Using a 16-point regular stencil in the $r - \theta$ plane. (b) Using a 16-point irregular stencil in the $x - y$ plane.

$$t = 0,$$

$$u = v = 0, \quad \rho = \left(1 - \frac{1}{\gamma}\right) + p$$

$$p = \frac{1}{\gamma} + \varepsilon \exp\left[-(\ln 2)\left(\frac{(x-4)^2 + y^2}{0.2^2}\right)\right]. \qquad (13.41)$$

With a very small initial pulse amplitude ($\varepsilon = 10^{-4}$), the computed solution is identical to the solution of the linearized Euler equations. This cylinder scattering problem governed by the linearized Euler equation is a benchmark problem of the Second Computational Aeroacoustics (CAA) Workshop on Benchmark Problems. The exact solution given in the conference proceedings unfortunately has an error. The correct solution is provided in Appendix 13B at the end of this chapter.

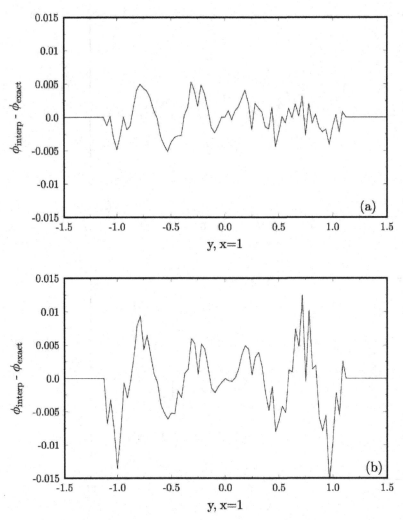

Figure 13.14. Distribution of interpolation error along the line $x = 1$. (a) Using a 16-point regular stencil in the $r - \theta$ plane. (b) Using a 16-point irregular stencil in the $x - y$ plane.

Figure 13.15 shows the computed pressure contours at $t = 1.6$, 4.0, and 6.4. The computed contours and the contours of the exact solution are indistinguishable (the difference is less than the thickness of the contour lines). The direct acoustic pulse and the scattered pulse can clearly be seen in Figure 13.15. For a more comprehensive comparison between the numerical and exact solution, the pressure time history at three locations with Cartesian coordinates $(-5, 0)$, $(-4, 4)$, and $(0, 5)$ are plotted in Figure 13.16. Plotted also in these figures is the exact solution. The differences between the exact (dashed line) and computed solution are again very small.

13.3.2 Plane Wave Scattering Problem

In a paper by Kurbatskii and Tam (1997), the problem of plane wave scattering by a cylinder was solved using a Cartesian boundary treatment method. Here, the same problem is solved using the overset grids method. One advantage of the overset grids method in this case is that the simple ghost point method may be used to enforce the

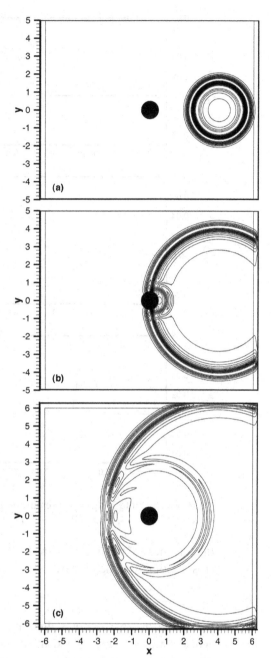

Figure 13.15. Pressure contours showing the scattering of an acoustic wave pulse by a circular cylinder. (a) $t = 1.6$. (b) $t = 4.0$. (c) $t = 6.4$. Dashed lines are the exact solution.

wall boundary condition. Again, the 7-point stencil DRP scheme is used to solve the Euler equations. In order to be able to validate the numerical solution, the amplitude of the incoming waves with wave front perpendicular to the x-axis and wavelength equal to $8 \, \Delta x$ is set to be very small. The problem, therefore, is essentially linear and the accuracy of the computed solution can be ascertained by comparison with the exact linear solution.

Figure 13.17 shows the computed zero pressure contours at the beginning of a cycle. Plotted in this figure also are the contours of the exact solution in dashed line. The difference between the computed and the exact solution is very small; less than

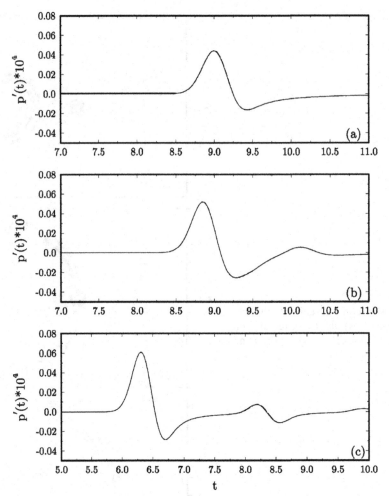

Figure 13.16. Time history of pressure perturbation $p' = p - (1/\gamma)$ at (a) $x = -5$, $y = 0$, (b) $x = -4$, $y = -4$, (c) $x = 0$, $y = -5$. Dashed line is the exact solution.

the thickness of the lines. Figure 13.18 shows the computed and the exact scattering cross section $\sigma(\theta)$ defined by

$$\sigma(\theta) = \lim_{r \to \infty} \left(r\overline{p_s^2} \right)^{\frac{1}{2}}, \tag{13.42}$$

where the overbar denotes time average and p_s is the pressure of the scattered acoustic waves. There is good agreement between the two. A major part of the small discrepancies arises because the computed result is measured at a finite distance from the center of the cylinder, whereas the exact solution is determined by taking the asymptotic limit $r \to \infty$.

13.4 Sliding Interface Problems

13.4.1 Sliding Interface Problem in Two Dimensions

Aircraft engine noise is a significant environmental and certification issue. Noise generated by the fan rotor wake impinging on the stator is an important noise component at both takeoff and landing. In the Second CAA Workshop on Benchmark

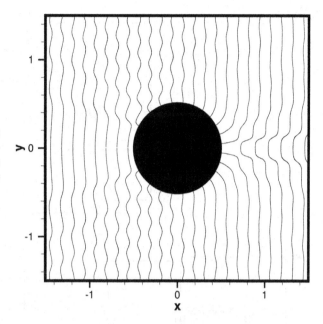

Figure 13.17. Zero pressure contours at the beginning of a cycle. Pressure pattern formed by the scattering of a plane wave by a cylinder. Wavelength equal to quarter diameter of cylinder.

Problems, a simplified model of this noise mechanism was proposed as a benchmark problem. One important feature of the problem is a sliding interface imitating the relative motion between a fan blade fixed computation domain and a stator blade fixed computation domain. Here, a less demanding sliding interface problem is considered. It will be shown that the use of an optimized interpolation scheme and overset grids method yields accurate numerical results.

Figure 13.19 shows two computation planes with six columns of overlapping meshes. The left computation plane is stationary. The right computation plane moves upward at a velocity of v_g. The sliding interface is the line at the center of the overlapping meshes. (x_1, y_1) will be used to denote the coordinates of a point in the stationary computation plane on the left, and (x_2, y_2) will be used to denote the coordinates of a point in the moving computation plane on the right. The relationships between the two coordinate systems are $x_2 = x_1$, $y_2 = y_1 - v_g t$.

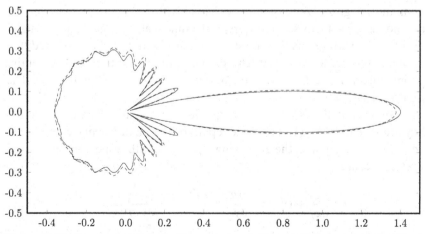

Figure 13.18. Comparison between computed and exact scattering cross section. Dashed curve is the exact solution.

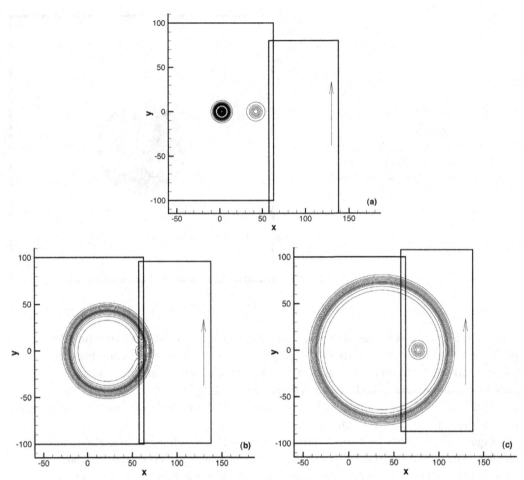

Figure 13.19. Density contours showing the transmission of an acoustic and an entropy pulse across a sliding interface. Dashed lines are the exact solution. (a) $t = 5.6$. (b) $t = 44.9$. (c) $t = 73.0$.

For computation purposes, square meshes are used in each computation plane, i.e., $\Delta x_1 = \Delta y_1$, $\Delta x_2 = \Delta y_2 = 0.78\Delta x_1$. However, the mesh sizes are not the same in the two computation grids. Dimensionless variables with Δx_1 as length scale, a_0 (the speed of sound) as the velocity scale, $\Delta x_1/a_0$ as the time scale, ρ_0 (mean flow density) as the density scale, and $\rho_0 a_0^2$ as the pressure scale are used. It will be assumed that there is a uniform mean flow at 0.5 Mach number in the x direction over the entire computation domain. The x direction is the horizontal direction in Figure 13.19. The sliding interface is at $x = 60$.

   The governing equations are the Euler equations. In the stationary computation plane, they are given by Eq. (13.1) with x replaced by x_1 and y replaced by y_1. In the moving computation plane, the governing equations, with respect to a grid fixed frame of reference, are:

$$\frac{\partial \rho}{\partial t} + u\frac{\partial \rho}{\partial x_2} + (v - v_g)\frac{\partial \rho}{\partial y_2} + \rho\left(\frac{\partial u}{\partial x_2} + \frac{\partial v}{\partial y_2}\right) = 0$$

$$\frac{\partial u}{\partial t} + u\frac{\partial u}{\partial x_2} + (v - v_g)\frac{\partial u}{\partial y_2} + \frac{1}{\rho}\frac{\partial p}{\partial x_2} = 0$$

$$\frac{\partial v}{\partial t} + u \frac{\partial v}{\partial x_2} + (v - v_g) \frac{\partial v}{\partial y_2} + \frac{1}{\rho} \frac{\partial p}{\partial y_2} = 0$$

$$\frac{\partial p}{\partial t} + u \frac{\partial p}{\partial x_2} + (v - v_g) \frac{\partial p}{\partial y_2} + \gamma p \left(\frac{\partial u}{\partial x_2} + \frac{\partial v}{\partial y_2} \right) = 0 \qquad (13.43)$$

At time equal to zero, a pressure pulse centered at $x_1 = y_1 = 0$ is released. At the same time, a vorticity pulse and an entropy pulse centered at $x_1 = 40$, $y_1 = 0$ are released. Mathematically, the initial conditions at $t = 0$ are as follows:

$$p = \frac{1}{\gamma} + \varepsilon \exp\left[-(\ln 2) \left(\frac{x_1^2 + y_1^2}{9} \right) \right]$$

$$\rho = 1 + \varepsilon \exp\left[-(\ln 2) \left(\frac{x_1^2 + y_1^2}{9} \right) \right]$$

$$+ 0.1\varepsilon \exp\left[-(\ln 2) \left(\frac{(x_1 - 40)^2 + y_1^2}{25} \right) \right]$$

$$u = 0.5 + 0.04\varepsilon y_1 \exp\left[-(\ln 2) \left(\frac{(x_1 - 40)^2 + y_1^2}{25} \right) \right]$$

$$v = -0.04\varepsilon (x_1 - 40) \exp\left[-(\ln 2) \left(\frac{(x_1 - 40)^2 + y_1^2}{25} \right) \right], \qquad (13.44)$$

where $\varepsilon = 0.005$ and $\gamma = 1.4$. In the course of time, the acoustic, the vorticity, and the entropy pulse will all be propagating or convected downstream. They will all reach the sliding interface at nearly the same time. They will then move across the sliding interface into the moving grid farther downstream. The exact solution of this problem was given in Section 6.4.

In the computation, the 7-point stencil DRP scheme is again used. For points on the stationary grid, the unknowns on every grid point including those on the sliding interface are determined at every time step by solving Eq. (13.1). Similarly, for points on the moving grid, the unknowns on every grid point including those on the sliding interface are updated according to Eq. (13.43). The unknowns of the three extra columns of grid points extending beyond the sliding interface are not calculated by the 7-point stencil DRP scheme. They are found by optimized interpolation using a 16-point stencil from the values of the variables on the other grid. In this way, information of the solution is passed between the two computation grids.

Figures 13.19a–13.19c show the computed density contours associated with the acoustic and entropy pulses at time $t = 5.6$, 44.9, and 73.0. In this computation v_g is set equal to 0.4. Also plotted in these figures are contours of the exact solution in dashed lines. Because the difference between the computed and the exact solution is small, the dashed lines may not be easily detected. In Figures 13.19a, the two pulses are in the stationary grid. Figures 13.19b shows that the acoustic pulse, moving downstream at three times the velocity of the entropy pulse, catches up with the entropy pulse right at the sliding interface. Figures 13.19c shows that the entropy pulse has now been convected past the sliding interface onto the moving grid. At this time, the acoustic pulse has already propagated further downstream. Figure 13.20 shows the computed and the exact density waveform along the x-axis at four different times. The sliding interface is located at $x = 60$. It is clear from these figures that

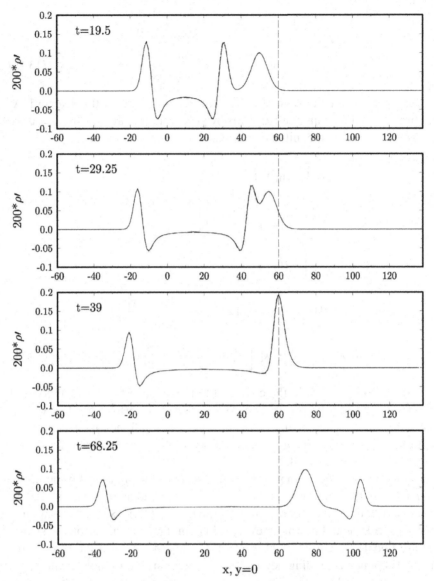

Figure 13.20. Spatial distribution of density perturbation $\rho' = \rho - 1$ associated with an acoustic and an entropy pulse propagating across a sliding interface located at $x = 60$. Dashed line is the exact solution.

the treatment of a sliding interface using optimized interpolation is effective and accurate.

Figures 13.21a and 13.21b show the computed and the exact fluid velocity excluding the mean flow, $(u'^2 + v'^2)^{1/2}$, contours of the acoustic and vorticity pulses as they propagate through the sliding interface. Again, the difference between the computed and exact contours, being less than the thickness of the contour lines, is difficult to detect. Figure 13.22 shows the waveform along the x-axis at four instances of time. At $t = 29.25$ to $t = 39$, the two pulses merge and propagate across the sliding interface together. At time $t = 68.25$, the acoustic pulse has passed through the entropy

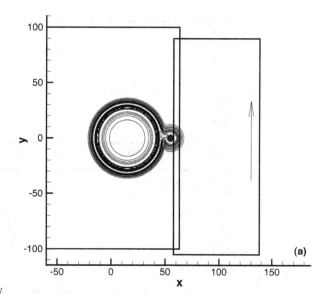

Figure 13.21. Contours of fluid velocity $(u'^2 + v'^2)^{1/2}$ associated with the transmission of an acoustic and a vorticity pulse across a sliding interface. Dashed lines are the exact solution. (a) $t = 28.1$. (b) $t = 67.3$.

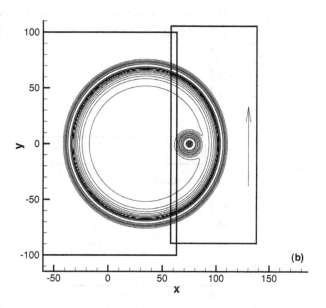

pulse and the two pulses are now separated. The computed waveforms are in good agreement with the exact solution.

In this computation, the moving grid moves at a subsonic velocity. In a similar computation the same results are recovered when the moving grid slides past the stationary grid at a supersonic velocity.

13.4.2 Sliding Interface Problem in Three Dimensions

As an example of the use of overset grids in three dimensions, consider the propagation of a small-amplitude three-dimensional spherically symmetric acoustic pulse across a sliding interface. This problem is chosen because an exact solution

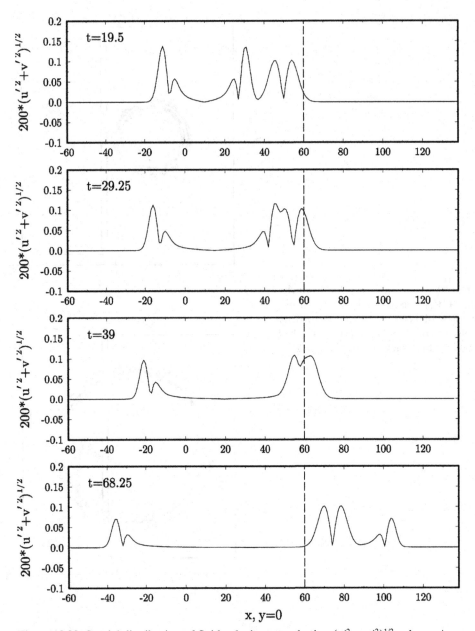

Figure 13.22. Spatial distribution of fluid velocity perturbation $(u'^2 + v'^2)^{1/2}$, where $u' = u - 0.5$, $v' = v$, associated with the transmission of an acoustic and a vorticity pulse across a sliding interface at $x = 60$. Dashed line is the exact solution.

(linearized) is available for comparison. On taking into account spherical symmetry, the linearized dimensionless momentum and energy equations are

$$\frac{\partial v}{\partial t} = -\frac{\partial p}{\partial R} \tag{13.45}$$

$$\frac{\partial p}{\partial t} + \frac{1}{R^2}\frac{\partial}{\partial R}(R^2 v) = 0, \tag{13.46}$$

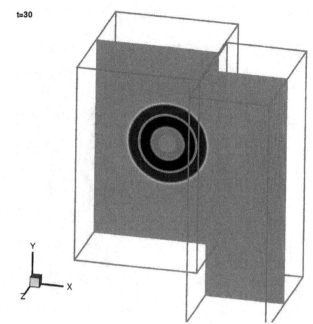

Figure 13.23. Computed pressure contours in the $x - y$ plane at $t = 30$.

where R is the radial coordinate and v is the radial velocity. The initial conditions at $t = 0$ corresponding to a radially symmetric Gaussian pressure pulse with a half-width b are

$$v = 0 \tag{13.47}$$

$$p = \varepsilon \exp\left[-(\ln 2)\left(\frac{R}{b}\right)^2\right] \tag{13.48}$$

This problem may be recast into a problem involving p alone, that is,

$$\frac{\partial^2 p}{\partial t^2} - \frac{1}{R}\frac{\partial^2}{\partial R^2}(Rp) = 0 \tag{13.49}$$

$$t = 0, \quad p = \varepsilon \exp\left[-(\ln 2)\left(\frac{R}{b}\right)^2\right] \tag{13.50}$$

$$\frac{\partial p}{\partial t} = 0. \tag{13.51}$$

The exact solution of Eqs. (13.49), (13.50), and (13.51), which is bounded at $R = 0$, is

$$p = \left[\frac{R-t}{2R}e^{-(\ln 2)\left(\frac{R-t}{b}\right)^2} + \frac{R+t}{2R}e^{-(\ln 2)\left(\frac{R+t}{b}\right)^2}\right]. \tag{13.52}$$

Numerical computation of this problem in three-dimensional Cartesian coordinates using the 7-point stencil DRP scheme and the optimized multidimensional interpolation method for data transfer across the sliding interface has been carried out. This may be regarded as a three-dimensional extension of the acoustic pulse problem of Section 13.4.1. The major difference is that there are now 64 points, $4 \times 4 \times 4$, in the interpolation stencil. Again, as in the two-dimensional problem, a

t=60

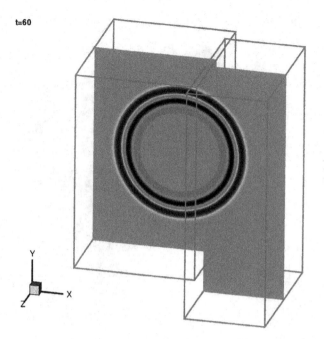

Figure 13.24. Computed pressure contours in the $x - y$ plane at $t = 60$.

uniform mean flow at Mach 0.5 in the x direction is included (the exact solution is obtained by applying a moving coordinate transformation to solution (13.52)). The sliding grid is taken to be moving in the z direction at a Mach number of 0.4. The full Euler equations are computed. The initial conditions are as given in Eqs. (13.47) and (13.48). The initial pressure pulse has a half-width of six mesh spacings and an intensity with $\varepsilon = 0.005$. The pulse is centered at the origin of the coordinate system. The sliding interface is located at $x = 60$.

Figures 13.23 and 13.24 show the computed pressure contours in the $x - y$ plane at time $t = 30$ and 60. The acoustic pulse expands spherically at a dimensionless speed

t=60

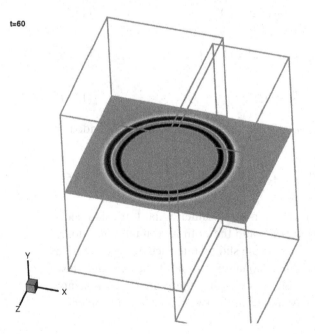

Figure 13.25. Computed pressure contours in the $x - z$ plane at $t = 60$.

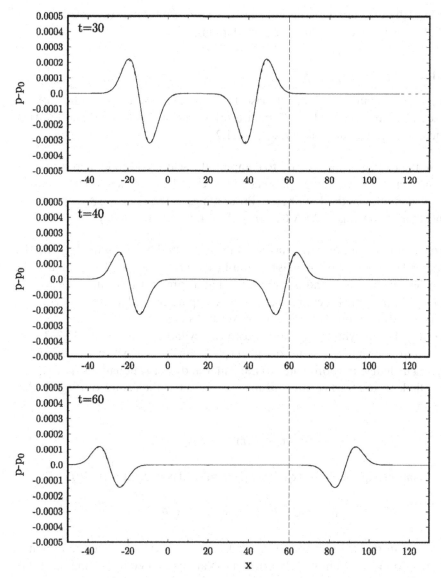

Figure 13.26. Comparisons of computed waveforms with exact solution at $t = 30, 40,$ and 60.

of unity and is simultaneously convected in the downstream direction by the mean flow at a speed of 0.5. At time $t = 30$, the pulse has not reached the sliding interface. At $t = 60$, part of the pulse has crossed the sliding interface into the downstream moving grid. The computed pressure contours remain circular. Figure 13.25 shows the pressure contours at $t = 60$ in the $x - z$ plane. Again, the contours are circles centered at $x = 30$, $z = 0$. Figure 13.26 shows the computed wave form in the $x - y$ plane at $t = 30, 40,$ and 60. Plotted in this figure also are the exact wave forms, but they cannot be detected because they differ from the computed solution by less than the thickness of the lines.

Finally, it is worthwhile to point out that overset grids method may also be used for computing the solutions of moving boundary problems. In such applications, a stationary grid and a moving grid attached to a moving object are used. For this type of problem, the overlapping grids change constantly with time. As a result, a good deal of effort is needed in locating the overlapping grids.

EXERCISES

13.1. To find the values of P in Exercise 11.2, one may use the multidimensional optimized interpolation method. Compare the result of this method and the two-step line interpolation method of Exercise 11.2.

13.2. Use the overset grids method to compute the scattered sound field generated by a plane incident wave train impinging on two cylinders. An exact solution can be found following the method of Sherer (Proceedings of the Fourth CAA Workshop on Benchmark Problems, NASA CP 2004-212954, September 2004).

13.3. (A moving boundary problem) Sound is generated by an oscillating cylinder of diameter D immersed in an inviscid, non-heat-conducting gas. The center of the cylinder lies on the x-axis and is moving so that at time t it is located at $x = F(t)$, where $F(t)$ is a periodic function. One way to compute the radiated sound is to use the overset grids method. This will involve several steps.

Step (1). Use a cylindrical grid fixed to the cylinder as in Figure 13.2. In the cylinder-fixed coordinate system, the governing equations have to be derived by performing a change in coordinates to the Euler and energy equations. Let (x_c, y_c) be the cylinder-fixed Cartesian coordinates centered at the center of the cylinder. Then,

$$x = x_c + F(t), \quad y = y_c.$$

The relationships between (x_c, y_c) and cylindrical coordinates (r_c, θ_c) are

$$x_c = r_c \cos(\theta_c), \quad y_c = r_c \sin(\theta_c).$$

Now, transform the Euler and energy equations to the cylinder-fixed coordinates. The value of (u, v, p, ρ) on the cylindrical mesh are computed by solving this set of equations. Note that u is still the velocity component in the x direction with respect to the background-fixed coordinates. If u_c is the component relative to the cylinder, then, $u_c = u - \frac{dF}{dt}$.

Step (2). Use a Cartesian coordinates system to compute the values of (u, v, p, ρ) in the background coordinates. The background mesh has a square hole similar to that of Figure 13.1. This hole moves with the cylinder in the course of time.

Step (3). The values of (u, v, p, ρ) on the outer most three rings of the cylindrical mesh are not computed. They are obtained by interpolation from the values of those on the background mesh. Similarly, those values at the innermost three rows and three columns of the square hole are not computed. They are to be found by interpolation from the values on the cylindrical mesh.

Develop a computational strategy for performing the acoustic radiation computation.

Appendix 13A. Computation of the Values of the Elements of Matrix A as $x_j \to x_k$

As alluded to previously, when the stencil points are aligned on a coordinate line, the limit form of the matrix elements as given by Eqs. (13.16) and (13.20) are to be used; e.g.,

$$\lim_{x_j \to x_k} \frac{\Delta \sin\left(\frac{\kappa(x_j - x_k)}{\Delta}\right)}{(x_j - x_k)} = \kappa.$$

However, for points very nearly aligned, direct numerical evaluation of the matrix or vector elements is not recommended because of the division by a small number. Instead, we advise that the limit values be used whenever $|x_j - x_k| < \varepsilon$. To find a suitable value of ε, let us consider the error incurred when $(\Delta/(x_j - x_k))\sin(\kappa(x_j - x_k)/\Delta)$ is approximated by its limit value κ, where $|x_j - x_k| < \varepsilon$. Let $\zeta = (x_j - x_k)/\Delta$, we have by Taylor series expansion,

$$\sin(\kappa\zeta) = \kappa\zeta - \frac{(\kappa\zeta)^3}{3!} + \frac{(\kappa\zeta)^5}{5!} - \cdots.$$

It is easy to establish

$$|\sin(\kappa\zeta) - \kappa\zeta| \le \frac{(\kappa\zeta)^3}{3!}$$

or

$$\left| \frac{\frac{\sin(\kappa\zeta)}{\zeta} - \kappa}{\kappa} \right| \le \frac{(\kappa\zeta)^2}{3!}.$$

Thus, the relative error of using the limit value to approximate the relevant parts of a matrix element is less than $(\kappa^2/3!)(\varepsilon/\Delta)^2$. We recommend to take $\varepsilon/\Delta = 10^{-5}$ for $\kappa = 1.2$. In this case, the relative error is less than 2.4×10^{-11}. This is an extremely small error.

Appendix 13B. Derivation of an Exact Solution for the Scattering of an Acoustic Pulse by a Circular Cylinder

In this appendix, an exact analytical solution for the scattering of an acoustic pulse by a circular cylinder is derived. The solution provided in the Proceedings of the Second CAA Workshop on Benchmark Problems has an error.

Let the cylinder be centered at $(x, y) = (0, 0)$ with a radius r_0 and the acoustic pulse be initialized as

$$p(x, y, 0) = e^{-b[(x-x_s)^2 + (y-y_s)^2]}, \qquad u = v = 0 \quad \text{at } t = 0, \qquad (13.53)$$

where p is the pressure, (u, v) is the velocity, and $b = (\ln 2)/d^2$. Here, d is the half-width of the pulse, and (x_s, y_s) is the center of the initial pulse. The solution will be found in terms of a velocity potential ϕ defined as

$$p = -\frac{\partial \phi}{\partial t}, \qquad u = \frac{\partial \phi}{\partial x}, \qquad v = \frac{\partial \phi}{\partial y}. \qquad (13.54)$$

Let

$$\phi(x, y, t) = \phi_i(x, y, t) + \phi_r(x, y, t),$$

where $\phi_i(x, y, t)$ represents the wave generated by the pulse in the free space and $\phi_r(x, y, t)$ represents the wave reflected by the cylinder. The solution for $\phi_i(x, y, t)$ is

$$\phi_i(x, y, t) = \text{Im}\left\{\int_0^\infty A_i(x, y, \omega) e^{-i\omega t} d\omega\right\}, \tag{13.55}$$

where

$$A_i(x, y, \omega) = \frac{1}{2b} e^{-\frac{\omega^2}{(4b)}} J_0(\omega r_s), \tag{13.56}$$

and $r_s = \sqrt{(x - x_s)^2 + (y - y_s)^2}$. Here, Im$\{\ \}$ and Re$\{\ \}$ denote the imaginary and real part of $\{\ \}$, respectively, and J_k denotes the Bessel function of order k.

To find ϕ_r, on following Eq. (13.55), the solution may be taken to have the following form:

$$\phi_r(x, y, t) = \text{Im}\left\{\int_0^\infty A_r(x, y, \omega) e^{-i\omega t} d\omega\right\}, \tag{13.57}$$

in which $A_r(x, y, \omega)$ satisfies the Helmholtz equation

$$\left(\frac{\partial^2 A_r}{\partial x^2} + \frac{\partial^2 A_r}{\partial y^2}\right) + \omega^2 A_r = 0$$

and the following solid wall boundary condition on the cylinder,

$$\frac{\partial A_r}{\partial r} = -\frac{\partial A_i}{\partial r} \quad \text{at} \quad r = r_0. \tag{13.58}$$

A_r can be solved by separation of variables in the polar coordinates (r, θ), which gives the following expansion: $(y_s = 0)$

$$A_r(x, y, \omega) = \sum_{k=0}^\infty C_k(\omega) H_k^{(1)}(r\omega) \cos(k\theta). \tag{13.59}$$

In Eq. (13.59), $H_k^{(1)}$ is the Hankel function of the first kind of order k. The expansion in Eq. (13.59) ensures that ϕ_r will satisfy the far-field radiation condition. Then, the boundary condition Eq. (13.58) leads to

$$\sum_{k=0}^\infty C_k(\omega)\omega H_k^{(1)'}(r_0\omega) \cos(k\theta) = \frac{1}{2b} e^{-\frac{\omega^2}{(4b)}} \omega J_1(\omega r_{s0}) \frac{r_0 - x_s \cos\theta - y_s \sin\theta}{r_{s0}}, \tag{13.60}$$

where

$$r_{s0} = r_s|_{r=r_0} = \sqrt{r_0^2 - 2r_0 x_s \cos\theta - 2r_0 y_s \sin\theta + x_s^2 + y_s^2}.$$

On considering the left side of Eq. (13.60) as a Fourier cosine series in θ, it is straightforward to find that

$$C_k(\omega) = \left(\frac{\omega}{2b} e^{-\frac{\omega^2}{(4b)}}\right) \frac{\varepsilon_k}{\pi \omega H_k^{(1)'}(r_0\omega)} \int_0^\pi J_1(\omega r_{s0}) \frac{r_0 - x_s \cos\theta - y_s \sin\theta}{r_{s0}} \cos(k\theta)d\theta,$$

where $\varepsilon_0 = 1$ and $\varepsilon_k = 2$ for $k \neq 0$.

Finally, the total pressure field can be found as follows:

$$p(x, y, t) = -\frac{\partial\phi}{\partial t} = -\frac{\partial}{\partial t}(\phi_i + \phi_r)$$

$$= \mathrm{Re}\left\{\int_0^\infty (A_i(x, y, \omega) + A_r(x, y, \omega))\omega e^{-i\omega t}d\omega\right\},$$

where $A_i(x, y, \omega)$ is given by Eq. (13.56) and $A_r(x, y, \omega)$ is given by Eq. (13.59).

14 Continuation of a Near-Field Acoustic Solution to the Far Field

By now, it is abundantly clear that all computational aeroacoustics (CAA) simulations can only be carried out in a finite, and in most cases, as small as possible, computational domain. A critical problem is how to continue the numerical solution to the far field. In some cases, the acoustic field consists of discrete frequency sound or tones. But in most cases, such as jet noise, the sound field is broadband and random. In aeroacoustics, most broadband noise is the result of turbulence. Turbulence is random and definitely nondeterministic, and the same is true of the radiated noise. The nondeterministic nature of the noise of turbulence makes its continuation to the far field a much more challenging task.

Mathematically, the problem of extending a near-field acoustic solution to the far field is akin to the mathematical procedure of "analytic continuation" in complex variables. Analytic continuation extends an analytic function defined in a limited domain to a large domain. Because the ideas behind the two problems are similar, the word "continuation" is used here to indicate the objective and intent.

At high Reynolds numbers, a flow is inevitably turbulent. As noted earlier, a turbulent flow is chaotic and random. For a turbulent flow such as that of a high Reynolds number jet, the solution is nonunique. Nonuniqueness is a characteristic of turbulence. Only the statistical averaged quantities are stationary with respect to time. This being the case, it brings forth the question of what physical variables should be continued to the far field. There is also the question of cost in performing the continuation. One may wish to continue the entire turbulence information to the far field, but the computational cost is likely to be prohibitive. Thus, this may not be a good idea. From a practical standpoint, the crucial information one needs in the far field is the noise directivity and spectra. If, indeed, the noise spectra and directivity are the information required in the far field, then a not so expensive computational procedure may be developed to continue this information from the near to the far field. In this chapter only the continuation of the spectra and directivity of turbulence noise are considered.

Turbulent flows are nonlinear, but outside the turbulence region, the velocity and pressure fluctuations are essentially linear. According to the investigation of Phillips (1955) and Stewart (1956), the mean squared fluctuating velocity associated with a turbulent jet or a similar flow decays with an inverse fourth power of the distance from the turbulence region. This prediction has been confirmed by Kibens (1968),

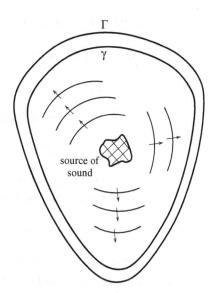

Figure 14.1. Sound generated by a localized source.

Bradbury (1965), and Maestrello (1973) experimentally. Outside the turbulence region, where the linearized Euler equations are valid, the mean squared velocity fluctuations decay with a second power with distance. This is the beginning of the acoustic near field. It will be assumed that the turbulent flow computation extends beyond the turbulence-irrotational boundary to the acoustic near field in the physical domain. The task of interest and the principal objective of this chapter are to develop a robust computational method to continue the numerical solution from the acoustic near field to the far field.

14.1 The Continuation Problem

Consider the acoustic field generated by some localized noise sources as shown in Figure 14.1. The sound field is governed by the compressible Navier-Stokes equation. Suppose the solution is known. To identify that it is the solution of the Navier-Stokes equations, a subscript NS will be added to the pressure and velocity field, i.e., p_{NS} and $\mathbf{v}_{NS}$.

Now let γ be a convex surface enclosing the sources and the near sound field. It will be assumed that γ is far enough away from the sources that the disturbances at γ are, for all intents and purposes, linear and inviscid. Under these conditions, the equations of motion outside γ are the linearized Euler equations as follows:

$$\rho_0 \frac{\partial \mathbf{v}}{\partial t} + \nabla p = 0 \tag{14.1}$$

$$\frac{\partial p}{\partial t} + \gamma p_0 \nabla \cdot \mathbf{v} = 0. \tag{14.2}$$

Thus, in the region outside γ, p_{NS}, $\mathbf{v}_{NS}$ and the solution of the linearized Euler equations, denoted by p_{Euler}, $\mathbf{v}_{Euler}$ are practically the same, namely $p_{NS} \to p_{Euler}$, $\mathbf{v}_{NS} \to \mathbf{v}_{Euler}$.

Suppose Γ is a closed convex surface enclosing γ as shown in Figure 14.1. On Γ, $p = p_{Euler}$ is known. Now, as a first step toward establishing a way to continue the solution from surface Γ to the far field, consider the following initial boundary value problem in the space outside Γ. The governing equations are the linearized Euler equations (14.1) and (14.2). The boundary and initial conditions are as follows:

$$\text{on } \Gamma, \quad p = p_{\text{Euler}} \tag{14.3}$$

$$\text{At } |\mathbf{x}| \to \infty, \quad p, v \text{ behave like outgoing waves} \tag{14.4}$$

$$\text{At } t = 0, \quad p = p_{\text{Euler}}(\mathbf{x}, 0), \quad \mathbf{v} = \mathbf{v}_{\text{Euler}}(\mathbf{x}, 0) \tag{14.5}$$

By eliminating $\mathbf{v}$ from Eqs. (14.1) and (14.2), the equation for p is the simple wave equation as follows:

$$\frac{\partial^2 p}{\partial t^2} - a_0^2 \nabla^2 p = 0. \tag{14.6}$$

Now, it is known that the solution of the simple wave equation (14.6)) satisfying boundary conditions (14.3) and (14.4) and initial conditions (14.5) is unique, but the original solution p_{NS}, $\mathbf{v}_{\text{NS}}$ which become p_{Euler}, $\mathbf{v}_{\text{Euler}}$ outside Γ is also a solution of Eq. 14.6 and boundary and initial conditions (14.3) to (14.5). Therefore, the two solutions must be equal. In other words, the continuation solution from surface Γ to the far field is given by the solution of Eqs. (14.1) to (14.5). Hence, a way to continue a near-field solution to the far field is to solve the initial boundary value problem stated above. How to construct such a solution is the subject of the next few sections of this chapter.

Instead of specifying $p = p_{\text{Euler}}$ on Γ as boundary condition, the solution of the simple wave equation with $\frac{\partial p}{\partial n} = \frac{\partial p_{\text{Euler}}}{\partial n}$ (the normal derivative) specified as boundary condition on Γ is also unique. By Eq. (14.1), the normal derivative of p is related to the velocity component in the normal direction as follows:

$$\nabla p \cdot \hat{n} = -\rho_0 \frac{\partial \mathbf{v}}{\partial t} \cdot \hat{n} = -\rho_0 \frac{\partial v_n}{\partial t}. \tag{14.7}$$

However, if $v_n(\mathbf{x}, t)$ is known on Γ, then $\frac{\partial v_n}{\partial t}$ is known. Therefore, it is possible to extend or continue a solution beyond surface Γ by solving Eqs. (14.1) and (14.2) with appropriate initial conditions and matching v_n to $v_{n,\text{Euler}}$ on boundary Γ. In other words, there are two ways to continue a solution from surface Γ to the far field. The first way is to match p on Γ. The second way is to match v_n on Γ.

It is worthwhile to note that the solution to the initial boundary value problem defined by Eqs. (14.1) to (14.5) consists of two parts. One part is associated with the initial conditions and zero boundary values on Γ. The other part is the zero initial conditions and nonzero boundary value on Γ. In most problems, interest is on the second part of the solution at large time. At large time, the transient solution (the first part) would propagate away. So for a long time solution, it is possible to ignore the initial condition altogether. In the rest of this chapter, attention is focused only on continuing the solution from surface Γ without reference to initial conditions.

In the literature, there are currently two favorite methods for continuing a solution to the far field. They are the Kirchhoff method (see, e.g., Lyrintzis (2003)) and the Ffowcs-Williams and Hawkings (1969) integral method. Mathematically, the Kirchhoff method yields a far-field acoustic pressure in terms of an integral over a closed surface. The integral involves the values of the pressure, its normal derivative, and the time derivative on the closed surface (three quantities altogether). This is well known to physicists and mathematicians. The Ffowcs-Williams and Hawkings method is similarly a surface integral representation. It solves the Lighthill equation instead of the simple wave equation. If the surface under consideration is outside all volume sources (quadrupoles), then the Kirchhoff method and the Ffowcs-Williams and Hawkings method are essentially similar. It is to be emphasized that both are integral representations, not solutions of the governing equations. To evaluate the integrals, the values of the pressure, its normal derivative, and time derivative on the surface must be known. It will, however, be shown later that it is sufficient to continue the pressure field from surface Γ to the far field by using only the pressure fluctuations on the surface. Alternatively, it is sufficient to determine the far-field pressure by using the fluctuating pressure gradient or the velocity component normal to the surface. It is not necessary to have three or more sets of information. This, nevertheless, does not mean that the Kirchhoff and the Ffowcs-Williams and Hawkings methods are in error, only that when these methods are used, the prescribed pressure, its normal derivative, and time derivative on surface Γ must be accurate. Otherwise, because of the overspecification of the boundary data (Note: in the theory of partial differential equations, when two sets of boundary data are prescribed, whereas one set is sufficient for determining a unique solution, it is referred to as overspecification of boundary data), the computed far-field pressure is likely to incur significant error.

14.2 Surface Green's Function: Pressure as the Matching Variable

Eqs. (14.1) and (14.2) are linear. The solution of these equations satisfying boundary condition $p = p_{\text{Euler}}$ on surface Γ can be found by means of a Green's function. For clarity, a superscript "g" is used to denote the variables of the Green's function. Let (ξ, η, ς) be a set of body-fitted curvilinear coordinates. The surface Γ corresponds to $\varsigma = \varsigma_0$. The Green's function $(\mathbf{v}^{(g)}, p^{(g)})$ satisfies the following boundary value problem:

$$\rho_0 \frac{\partial \mathbf{v}^{(g)}}{\partial t} + \nabla p^{(g)} = 0 \tag{14.8}$$

$$\frac{\partial p^{(g)}}{\partial t} + \gamma p_0 \nabla \cdot \mathbf{v}^{(g)} = 0 \tag{14.9}$$

On Γ, i.e., $\varsigma = \varsigma_0$,

$$p^{(g)}(\mathbf{x}, t; \xi_0, \eta_0, \varsigma_0, t_0) = \delta(\xi - \xi_0)\delta(\eta - \eta_0)\delta(t - t_0) \tag{14.10}$$

(Note: The notation that the first set of spatial and time argument of the Green's function represents the location and time of the observer while the second set represents that of the source is adopted here.)

When the Green's function is found, the far-field pressure is given by

$$p(\mathbf{x}, t) = \int_{-\infty}^{\infty} \int \int p^{(g)}(\mathbf{x}, t; \xi_0, \eta_0, \varsigma_0, t_0) p_{Euler}(\xi_0, \eta_0, \varsigma_0, t_0) d\xi_0 \, d\eta_0 \, dt_0. \tag{14.11}$$

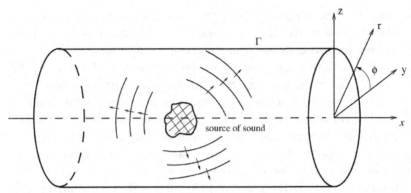

Figure 14.2. Source of sound enclosed by an infinite cylindrical surface of diameter D.

As an example of the use of surface Green's function, the case of Γ in the form of an infinite circular cylindrical surface of diameter D (see Figure 14.2) is considered. The cylindrical coordinates (r, ϕ, x) are the natural body-fitted coordinates of this problem. By eliminating $\mathbf{v}^{(g)}$ from Eqs. (14.8) and (14.9) in favor of $p^{(g)}$ and upon applying Fourier transform to t, the time variable, the governing equation for $\tilde{p}^g$ (the Fourier transform of $p^{(g)}$), is found to be

$$\frac{\partial^2 \tilde{p}^{(g)}}{\partial r^2} + \frac{1}{r}\frac{\partial \tilde{p}^{(g)}}{\partial r} + \frac{1}{r^2}\frac{\partial^2 \tilde{p}^{(g)}}{\partial \phi^2} + \frac{\partial^2 \tilde{p}^{(g)}}{\partial x^2} + \frac{\omega^2}{a_0^2}\tilde{p}^{(g)} = 0 \tag{14.12}$$

At $r = D/2$,

$$\tilde{p}^{(g)} = \frac{\delta(x - x_0)\delta(\phi - \phi_0)e^{i\omega t_0}}{\pi D}. \tag{14.13}$$

Now, by applying Fourier transform to x (denoting the transform variable by k) in Eqs. (14.12) and (14.13) and then expanding the solution in a Fourier series in ϕ, i.e.,

$$\tilde{p}^{(g)} = \sum_{n=-\infty}^{\infty} \int_{-\infty}^{\infty} \hat{p}_n(r, k, \omega)e^{ikx + in(\phi - \phi_0)} dk, \tag{14.14}$$

and upon making use of the expansion,

$$\delta(\phi - \phi_0) = \frac{1}{2\pi} \sum_{n=-\infty}^{\infty} e^{in(\phi - \phi_0)},$$

the equation for $\hat{p}_n(r, k, \omega)$ is found to be

$$\frac{d^2 \hat{p}_n}{dr^2} + \frac{1}{r}\frac{d\hat{p}_n}{dr} + \left(\frac{\omega^2}{a_0^2} - k^2\right)\hat{p}_n - \frac{n^2}{r^2}\hat{p}_n = 0. \tag{14.15}$$

The boundary condition at surface Γ, i.e., $r = D/2$ is

$$\hat{p}_n = \frac{1}{4\pi^3 D}e^{-ikx_0 + i\omega t_0}. \tag{14.16}$$

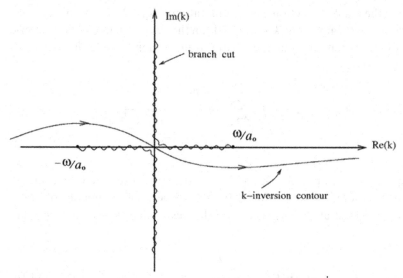

Figure 14.3. Branch cuts for the square root function $(k^2 - \frac{\omega^2}{a_0^2})^{\frac{1}{2}}$ in the k plane.

The solution of Eq. (14.15) satisfying boundary condition (14.16) can easily be found in terms of Hankel function $H_n^{(1)}[]$. On inverting the Fourier transform in x and the sum over n, it is found that

$$\tilde{p}^{(g)} = \sum_{n=-\infty}^{\infty} \int_{-\infty}^{\infty} \frac{H_n^{(1)}\left[i\left(k^2 - \frac{\omega^2}{a_0^2}\right)^{\frac{1}{2}} r\right]}{4\pi^3 D H_n^{(1)}\left[i\left(k^2 - \frac{\omega^2}{a_0^2}\right)^{\frac{1}{2}} \frac{D}{2}\right]} e^{ik(x-x_0)+in(\phi-\phi_0)+i\omega t_0} \, dk. \qquad (14.17)$$

The branch cuts for the square root function $(k^2 - \frac{\omega^2}{a_0^2})^{\frac{1}{2}}$ in the k plane is shown in Figure (14.3).

For radiation to the far field, it is advantageous to switch to polar coordinates (R, θ, ϕ) (see Figure 14.4) with

$$x = R\cos\theta, \qquad r = R\sin\theta$$

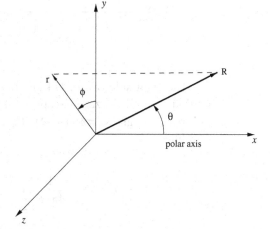

Figure 14.4. Spherical polar coordinates (R, θ, ϕ), cylindrical coordinates (r, ϕ, x), and Cartesian coordinates (x, y, z).

For large R, the Hankel function in the numerator of Eq. (14.17) may be replaced by its large argument formula. The k − integral can then be evaluated by the method of stationary phase. Upon inverting the Fourier transform in t, the far field Green's function is found to be

$$p^{(g)}_{R\to\infty} (R, \theta, \phi, t; x_0, \phi_0, t_0) = \frac{1}{2\pi^3 DR} \int_{-\infty}^{\infty} \sum_{n=-\infty}^{\infty} \frac{e^{i\frac{\omega}{a_0}(R-x_0\cos\theta)-i(n+1)\frac{\pi}{2}+in(\phi-\phi_0)-i\omega(t-t_0)}}{H_n^{(1)}\left(\frac{\omega D}{2a_0}\sin\theta\right)} d\omega.$$

$$(14.18)$$

As a simple demonstration that Eq. (14.18) is the correct surface Green's function, the case of a sound field produced by a time periodic monopole of angular frequency Ω located at the origin of coordinates is considered. The pressure field is

$$p(R, t) = \frac{e^{i\left(\frac{R}{a_0}-t\right)\Omega}}{4\pi R},$$

$$(14.19)$$

where R is the radial distance. On the cylindrical surface of diameter D, the polar distance is $R = (\frac{D^2}{4} + x_0^2)^{1/2}$ for a point at $x = x_0$. Thus, the fluctuating surface pressure is given by

$$p(x_0, \phi_0, t_0) = \frac{e^{i\left[\frac{\left(\frac{D^2}{4}+x_0^2\right)^{1/2}}{a_0}-t_0\right]\Omega}}{4\pi \left(\frac{D^2}{4}+x_0^2\right)^{1/2}}.$$

$$(14.20)$$

By inserting $p^{(g)}$ from Eq. (14.18) and $p(x_0, \phi_0, t_0)$ from Eq. (14.20) into Eq. (14.11), the far-field pressure due to surface pressure on the cylindrical surface is

$$p(R, \theta, \phi, t) = \int_{-\infty}^{\infty} \int_{-\infty}^{\infty} \int_{-\infty}^{\infty} \int_{0}^{2\pi} \frac{1}{8\pi^4 DR}$$

$$\times \sum_{n=-\infty}^{\infty} \frac{e^{i\frac{\omega}{a_0}(R-x_0\cos\theta)-i(n+1)\frac{\pi}{2}+in(\phi-\phi_0)-i\omega(t-t_0)+i\left(\frac{D^2}{4}+x_0^2\right)^{1/2}\frac{\Omega}{a_0}-it_0\Omega}}{\left(\frac{D^2}{4}+x_0^2\right)^{1/2} H_n^{(1)}\left[\frac{\omega D}{2a_0}\sin\theta\right]} \frac{D}{2} d\phi_0\, dt_0\, d\omega\, dx_0.$$

$$(14.21)$$

The right side of Eq. (14.21) may be integrated step by step as follows. Integration over $d\phi_0$ is zero except for $n = 0$. In this case, the integral is 2π. Integration over dt_0 gives $2\pi\delta(\omega - \Omega)$. Therefore, on integrating over $d\omega$, the right side simplifies to

$$p(R, \theta, \phi, t) = \frac{1}{4\pi^2 R} \int_{-\infty}^{\infty} \frac{e^{i\frac{\Omega}{a_0}(R-x_0\cos\theta)+i\left(\frac{D^2}{4}+x_0^2\right)^{1/2}\frac{\Omega}{a_0}-i\frac{\pi}{2}-i\Omega t}}{\left(\frac{D^2}{4}+x_0^2\right)^{1/2} H_0^{(1)}\left(\frac{\Omega D}{2a_0}\sin\theta\right)} dx_0.$$

$$(14.22)$$

Now, from the extensive compilation of Fourier integrals of Erdelyi *et al* (1954), the following closed-form integral is found:

$$\int_{-\infty}^{\infty} \frac{e^{ib(x^2+a^2)^{1/2}-iyx}}{(x^2+a^2)^{1/2}}dx = i\pi H_0^{(1)}\left[a(b^2-y^2)^{1/2}\right]. \tag{14.23}$$

By means of Eq. (14.23), it is straightforward to simplify Eq. (14.22) to

$$p(R,\theta,\phi,t) = \frac{e^{i\Omega\left(\frac{R}{a_0}-t\right)}}{4\pi R}. \tag{14.24}$$

Thus, the monopole sound field at large R is recovered.

14.3 Surface Green's Function: Normal Velocity as the Matching Variable

Instead of using surface pressure as the variable for continuation to the far field, an equally good variable to use is the velocity component normal to the surface or the normal pressure gradient. If this choice is made, the appropriate surface Green's function is given by solution of the following problem (a superscript "G" will be used to denote this Green's function):

$$\rho_0 \frac{\partial \mathbf{v}^{(G)}}{\partial t} + \nabla p^{(G)} = 0 \tag{14.25}$$

$$\frac{\partial p^{(G)}}{\partial t} + \gamma p_0 \nabla \cdot \mathbf{v}^{(G)} = 0, \tag{14.26}$$

On surface Γ, i.e., $\varsigma = \varsigma_0$, the boundary condition is

$$v_n^{(G)}(\mathbf{x},t;\xi_0,\eta_0,\varsigma_0,t_0) = \delta(\xi-\xi_0)\delta(\eta-\eta_0)\delta(t-t_0). \tag{14.27}$$

To find $v_n^{(G)}(\mathbf{x},t;\xi_0,\eta_0,\varsigma_0,t)$, it is advantageous to introduce a velocity potential $\Phi^{(G)}$ defined by

$$\mathbf{v}^{(G)} = \nabla\Phi^{(G)}, \quad p^{(G)} = -\rho_0\frac{\partial\Phi^{(G)}}{\partial t}. \tag{14.28}$$

Eq. (14.28) satisfies Eq. (14.25) identically. Substitution of Eq. (14.28) into Eq. (14.26), the governing equation for $\Phi^{(G)}$ is found to be

$$\frac{1}{a_0^2}\frac{\partial^2\Phi^{(G)}}{\partial t^2} - \nabla^2\Phi^{(G)} = 0. \tag{14.29}$$

The surface boundary condition on $\varsigma = \varsigma_0$ is found by inserting expression (14.28) into Eq. (14.27). This yields

$$\frac{\partial\Phi^{(G)}}{\partial n} = \delta(\xi-\xi_0)\delta(\eta-\eta_0)\delta(t-t_0), \tag{14.30}$$

where $\hat{\mathbf{n}}$ is the unit outward pointing normal of surface Γ.

Suppose $\Phi^{(G)}(\mathbf{x}, t; \xi_0, \eta_0, \varsigma_0, t_0)$ is found. Let $v_n(\xi_0, \eta_0, t_0)$ be the prescribed normal velocity on Γ, then the velocity potential ($\mathbf{v} = \nabla\Phi$ and $p = -\frac{1}{\rho_0}\frac{\partial\Phi}{\partial t}$) is given by

$$\Phi(\mathbf{x}, t) = \int\limits_{-\infty}^{\infty}\int\limits_{-\infty}^{\infty}\int\limits_{-\infty}^{\infty} \Phi^{(G)}(\mathbf{x}, t; \xi_0, \eta_0, \varsigma_0, t_0)v_n(\xi_0, \eta_0, t_0)\, d\xi_0\, d\eta_0\, dt_0. \quad (14.31)$$

The pressure field may be computed by differentiating Eq. (14.31) with respect to t, i.e.,

$$p(\mathbf{x}, t) = -\int\int\int \frac{1}{\rho_0}\frac{\partial\Phi^{(G)}(\mathbf{x}, t; \xi_0, \eta_0, \varsigma_0, t_0)}{\partial t}v_n(\xi_0, \eta_0, t_0)\, d\xi_0\, d\eta_0\, dt_0. \quad (14.32)$$

By means of Eq. (14.28), this expression may be rewritten as

$$p(\mathbf{x}, t) = \int\int\int p^{(G)}(\mathbf{x}, t; \xi_0, \eta_0, t_0)v_n(\xi_0, \eta_0, t_0)\, d\xi_0\, d\eta_0\, dt_0. \quad (14.33)$$

To illustrate the construction and use of surface Green's function $p^{(G)}$, consider again the case of Γ in the form of an infinite circular cylindrical surface as shown in Figure (14.2). The governing equation for $\Phi^{(G)}$ is Eq. (14.29). In cylindrical coordinates, the equation and boundary condition for $\Phi^{(G)}$ are

$$\frac{1}{a_0^2}\frac{\partial^2\Phi^{(G)}}{\partial t^2} - \left[\frac{\partial^2\Phi^{(G)}}{\partial r^2} + \frac{1}{r}\frac{\partial\Phi^{(G)}}{\partial r} + \frac{1}{r^2}\frac{\partial^2\Phi^{(G)}}{\partial\phi^2} + \frac{\partial^2\Phi^{(G)}}{\partial x^2}\right] = 0 \quad (14.34)$$

At $r = \frac{D}{2}$,

$$\frac{\partial\Phi^{(G)}}{\partial r} = \frac{\delta(\phi - \phi_0)\delta(x - x_0)\delta(t - t_0)}{D/2}. \quad (14.35)$$

Just as in Section 14.2, the solution of this problem may be constructed by first applying Fourier transforms to x and t and then expanding the solution as a Fourier series in $(\phi - \phi_0)$. On denoting the Fourier transform by a $\sim$ over the variable and the amplitude function of the mth term in the Fourier series by a subscript m, the solution takes on the following form:

$$\tilde{\Phi}^{(G)} = \sum_{-\infty}^{\infty} \tilde{\Phi}_m e^{im(\phi - \phi_0)}. \quad (14.36)$$

It is straightforward to find that

$$\tilde{\Phi}_m = \frac{H_m^{(1)}\left[i\left(k^2 - \frac{\omega^2}{a_0^2}\right)^{1/2}r\right]e^{-ikx_0 + i\omega t_0}}{4\pi^3 iD\left(k^2 - \frac{\omega^2}{a_0^2}\right)^{1/2}H'^{(1)}_m\left[i\left(k^2 - \frac{\omega^2}{a_0^2}\right)^{1/2}\frac{D}{2}\right]},$$

where $H_m^{(1)}$ and $H'^{(1)}_m$ are the mth order Hankel function and its derivative. The branch cuts of the square root function $(k^2 - \frac{\omega^2}{a_0^2})^{1/2}$ are as shown in Figure 14.3. On inverting the Fourier transforms, it is readily found that

$$
\Phi^{(G)}(r, \phi, x; \phi_0, x_0, t_0)
$$

$$
= \sum_{m=-\infty}^{\infty} \int_{-\infty}^{\infty} \int_{-\infty}^{\infty} \frac{H_m^{(1)}\left[i\left(k^2 - \frac{\omega^2}{a_0^2}\right)^{1/2} r\right] e^{i[k(x-x_0)-\omega(t-t_0)+m(\phi-\phi_0)]}}{4\pi^3 i D (k^2 - \frac{\omega^2}{a_0^2})^{1/2} H_m'^{(1)}\left[i\left(k^2 - \frac{\omega^2}{a_0^2}\right)^{1/2}\frac{D}{2}\right]} dk\, d\omega. \quad (14.37)
$$

For radiation to the far field, the k integral in Eq. (14.37) may be evaluated asymptotically (method of stationary phase) by first switching to spherical polar coordinates as in Section 14.2. In so doing, and upon using Eq. (14.28), it is found that

$$
\underset{R\to\infty}{p}^{(G)}(R, \theta, \phi, t; \phi_0, x_0, t_0)
$$

$$
= \frac{i\rho_0 a_0}{2\pi^3 RD} \sum_{m=-\infty}^{\infty} \int_{-\infty}^{\infty} \frac{e^{i\left[\frac{\omega}{a_0}(R-x_0\cos\theta)-\omega(t-t_0)+m(\phi-\phi_0)-(m+1)\frac{\pi}{2}\right]}}{H_m'^{(1)}\left(\frac{\omega D}{2a_0}\sin\theta\right)\cdot\sin\theta} d\omega. \quad (14.38)
$$

To demonstrate that Eq. (14.38) is the correct surface Green's function with v_n as the matching variable, the time periodic monopole source problem of Section 14.2 is again considered. For a monopole source located at the origin with angular frequency Ω, the pressure field is given by Eq. (14.19). By means of the radial (R direction) momentum equation, it is easy to find that the radial velocity v_R associated with the acoustic field is

$$
v_R = \frac{1}{4\pi \rho_0 \Omega R}\left(\frac{\Omega}{a_0} + \frac{i}{R}\right) e^{i\Omega\left(\frac{R}{a_0}-t\right)}. \quad (14.39)
$$

The component of v_R in the direction normal to the cylindrical surface and on the cylindrical surface, $r = D/2$ is

$$
v_n = \frac{\sin\hat\theta}{4\pi \rho_0 \Omega R_0}\left(\frac{\Omega}{a_0} + \frac{i}{R_0}\right) e^{i\Omega\left(\frac{R_0}{a_0}-t_0\right)}, \quad (14.40)
$$

where $R_0 = (\frac{D^2}{4} + x_0^2)^{1/2}$ and $\sin\hat\theta = \frac{D}{2R_0}$. x_0 and t_0 are the source coordinate and time.

Now, on inserting $p^{(G)}$ from Eq. (14.38) and v_n from Eq. (14.40) into Eq. (14.33), the radiated pressure field at $R\to\infty$ is given by

$$
\underset{R\to\infty}{p}(R, \theta, \phi, t)) =
$$

$$
\frac{1}{16\pi^4 R\Omega} \int_{-\infty}^{\infty}\int_{-\infty}^{\infty}\int_0^{2\pi}\int_{-\infty}^{\infty} \sum_{m=-\infty}^{\infty} \frac{1}{H_m'^{(1)}\left(\frac{\omega D}{2a_0}\sin\theta\right)\cdot\sin\theta}\left[\frac{i\Omega}{\left(\frac{D^2}{4}+x_0^2\right)} - \frac{a_0}{\left(\frac{D^2}{4}+x_0^2\right)^{3/2}}\right]\cdot
$$

$$
e^{i\left[\frac{\omega}{a_0}(R-x_0\cos\theta)-\omega(t-t_0)+m(\phi-\phi_0)-(m+1)\frac{\pi}{2}+\frac{\Omega}{a_0}\left(\frac{D^2}{4}+x_0^2\right)^{1/2}-\Omega t_0\right]}\frac{D}{2}\, dt_0\, d\phi_0\, d\omega\, dx_0. \quad (14.41)
$$

The integrals over dt_0 and $d\phi_0$ can easily be evaluated to yield

$$\int_{-\infty}^{\infty} e^{i(\omega-\Omega)t_0}\, dt_0 = 2\pi\delta(\omega-\Omega), \qquad \int_{0}^{2\pi} e^{im(\phi-\phi_0)}\, d\phi_0 = \begin{cases} 2\pi, & m=0 \\ 0, & m\neq 0 \end{cases}.$$

Thus, upon integration over $dt_0, d\phi_0,$ and $d\omega$ and on summing over m, Eq. (14.41) becomes

$$p(R,\theta,\phi,t) = \frac{e^{i\Omega\left(\frac{R}{a_0}-t\right)}}{8\pi^2 R\Omega\sin\theta} \frac{Da_0}{H_0^{\prime(1)}\left(\frac{\Omega D}{2a_0}\sin\theta\right)} \int_{-\infty}^{\infty}\left[\frac{\Omega}{\left(\frac{D^2}{4}+x_0^2\right)a_0} + \frac{i}{\left(\frac{D^2}{4}+x_0^2\right)^{3/2}}\right].$$
$$e^{-i\frac{\Omega}{a_0}\left[x_0\cos\theta-\left(\frac{D^2}{4}+x_0^2\right)^{1/2}\right]}\, dx_0. \tag{14.42}$$

The remaining integral of Eq. (14.42) can be evaluated in closed form by differentiating formula (14.23) with respect to parameter a. By means of this formula, it is easy to find that the pressure field becomes

$$p(R,\theta,\phi,t) = \frac{e^{i\Omega\left(\frac{R}{a_0}-t\right)}}{4\pi R}.$$

This is just the acoustic field of a time periodic monopole.

14.4 The Adjoint Green's Function

Oftentimes, the surface geometry one encounters does not fit a separable coordinate system. For these cases, the surface Green's function usually cannot be found easily. For such a surface geometry, one may use the adjoint Green's function and the reciprocity relation.

In many branches of mechanics, the reciprocity principle applies. However, the existence of a reciprocity principle has not been fully exploited in the fields of acoustics and fluid dynamics. Some earlier works that utilized reciprocity are Cho (1980), Howe (1975, 1981), Dowling (1983), and Tam and Auriault (1998) in acoustics; Roberts (1960), Eckhaus (1965), and Chardrasekhar (1989) in hydrodynamic stability; and Hill (1995) in receptivity problems. To fix ideas on reciprocity, consider a time periodic point source of sound located at $\mathbf{x}_s$ as shown in Figure 14.5(a). Let $G(\mathbf{x}_0, \mathbf{x}_s, \omega)$ be the pressure associated with the sound field measured by an observer at $\mathbf{x}_0$. The time factor $e^{-i\omega t}$ has been omitted. Mathematically, $G(\mathbf{x}_0, \mathbf{x}_s, \omega)$ is the Green's function of the Helmholtz equation and ω is the angular frequency of oscillations. (Note: the notation that the first argument of the Green's function is the location of the observer and the second argument is the location of the source is retained here.) Now, if the location of the sound source and that of the observer is interchanged as shown in Figure 14.5(b). Clearly, by symmetry, the pressure measured by the observer now at $\mathbf{x}_s$ while the source is at $\mathbf{x}_0$ is the same as before. That is,

$$G(\mathbf{x}_0, \mathbf{x}_s, \omega) = G(\mathbf{x}_s, \mathbf{x}_0, \omega). \tag{14.43}$$

Eq. (14.43) is simply a statement that the Green's function $G(\mathbf{x}_0, \mathbf{x}_s, \omega)$ is self-adjoint. It is the reciprocity relation.

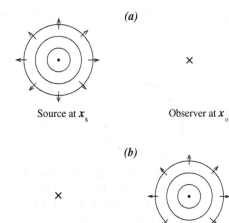

Figure 14.5. An acoustic source and an observer form a
self-adjoint system.

For a surface source and a far-field observer as shown in Figure 14.6(a) a reciprocity relation exists. The problem is, however, not self-adjoint. The adjoint Green's function is not governed by Eqs. (14.8) and (14.9). The boundary condition is not Eq. (14.10). These equations, referred to as the adjoint equations, may easily be derived in the frequency domain.

There is a significant advantage in using adjoint Green's function instead of the direct Green's function when an analytical formula for the Green's function cannot be found. Suppose the far-field sound in the direction of θ produced by surface sources as shown in Figure 14.7a is to be found. To determine the total far-field radiation, the radiation from each surface sources such as A, B, and C in Figure 14.7 has to be calculated and then summed. That is, the surface Green's functions at A, B, and C and other surface points have to be separately computed. In the absence of an analytical formula, this would be a tedious and laborious effort. On the other hand, if

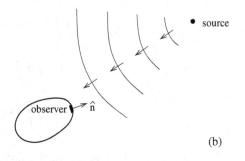

Figure 14.6. Surface sources and an observer form an
non-self-adjoint system.

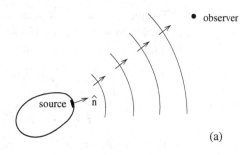

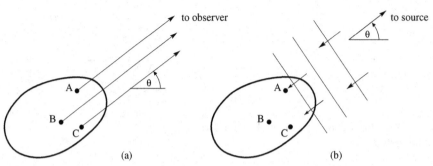

Figure 14.7. (a) Direct surface Green's function has to be found for every surface point on surface Γ for radiation in the direction θ. (b) The adjoint Green's function for all surface points is the solution of a single scattering problem with the incident wave coming from a distant source in direction θ.

the adjoint formulation is used the situation is different. Since the source point is now in the far field, the sound waves from the far-field source near surface Γ are plane waves. The adjoint Green's function is, therefore, the solution of a simple scattering problem by surface Γ. The scattering problem needs to be computed only once. By the reciprocity relation, the values of the direct Green's function for radiation from every point on surface Γ to the far-field point is found simultaneously in one single calculation of the scattered wave solution.

The Fourier transform of Eqs. (14.8), (14.9), and boundary condition (14.10) in time t are

$$-i\omega\rho_0\tilde{\mathbf{v}}^{(g)} + \nabla\tilde{p}^{(g)} = 0 \tag{14.44}$$

$$-i\omega\tilde{p}^{(g)} + \gamma p_0\nabla\cdot\tilde{\mathbf{v}}^{(g)} = 0, \tag{14.45}$$

On Γ or $\varsigma = \varsigma_0$,

$$\tilde{p}^{(g)}(\mathbf{x}, \omega; \xi_0, \eta_0, \varsigma_0, \tau) = \tfrac{1}{2\pi}\delta(\xi - \xi_0)\delta(\eta - \eta_0)e^{i\omega\tau} \tag{14.46}$$

To find the adjoint system of equations to Eq. (14.44) and (14.45), multiply Eq. (14.44) by $\mathbf{v}^{(a)}\cdot$ and Eq. (14.45) by $p^{(a)}$ (the superscript "a" denotes the adjoint) and integrate over all space outside Γ. This yields, after rearranging the terms,

$$\iiint\limits_{\substack{space \\ outside\ \Gamma}} \left[-i\omega\rho_0\mathbf{v}^{(a)}\cdot\tilde{\mathbf{v}}^{(g)} + \nabla\cdot(\tilde{p}^{(g)}\mathbf{v}^{(a)}) - \tilde{p}^{(g)}\nabla\cdot\mathbf{v}^{(a)} - i\omega\tilde{p}^{(g)}p^{(a)}\right)$$

$$+ \gamma p_0\nabla\cdot(p^{(a)}\tilde{\mathbf{v}}^{(g)}) - \gamma p_0\tilde{\mathbf{v}}^{(g)}\cdot\nabla p^{(a)}\right]dx\,dy\,dz = 0. \tag{14.47}$$

The two divergence terms in Eq. (14.47) may be integrated by means of the Divergence Theorem to become surface integrals over Γ. This leads to

$$\iiint\limits_{outside\ \Gamma} [-i\omega\rho_0\mathbf{v}^{(a)} - \gamma p_0\nabla p^{(a)}]\cdot\tilde{\mathbf{v}}^{(g)}dx\,dy\,dz$$

$$+ \iiint\limits_{outside\ \Gamma} [-i\omega p^{(a)} - \nabla\cdot\mathbf{v}^{(a)}]\tilde{p}^{(g)}dx\,dy\,dz$$

$$- \iint\limits_{surface\ \Gamma} [\tilde{p}^{(g)}\mathbf{v}^{(a)} + \gamma p_0 p^{(a)}\tilde{\mathbf{v}}^{(g)}]\cdot\hat{\mathbf{n}}\,dS = 0. \tag{14.48}$$

where $\hat{\mathbf{n}}$ is the unit outward pointing normal of surface Γ.

Now, the adjoint system is chosen to satisfy the following equations and boundary conditions:

$$-i\omega\rho_0 \mathbf{v}^{(a)} - \gamma p_0 \nabla p^{(a)} = 0 \tag{14.49}$$

$$-i\omega p^{(a)} - \nabla \cdot \mathbf{v}^{(a)} = \tfrac{1}{2\pi}\delta(\mathbf{x}-\mathbf{x}_1)e^{i\omega\tau} \tag{14.50}$$

$$\text{On } \Gamma \text{ or } \varsigma = \varsigma_0, \quad p^{(a)} = 0. \tag{14.51}$$

By means of this choice of the adjoint system, the integrals of Eq. (14.48) may be easily evaluated. This gives the reciprocity relation as follows:

$$\tilde{p}^{(g)}(\mathbf{x}_1;\xi_0, \eta_0, \varsigma_0;\omega, \tau) = v_n^{(a)}(\xi_0, \eta_0, \varsigma_0;\mathbf{x}_1;\omega, \tau), \tag{14.52}$$

where $v_n^{(a)} = \mathbf{v}^{(a)} \cdot \hat{\mathbf{n}}$ is the component of the adjoint velocity in the direction of outward pointing unit normal $\hat{\mathbf{n}}$. Thus, once the adjoint problem is solved, the direct surface Green's function on the entire surface is found.

14.4.1 Computation of the Adjoint Green's Function: Free Field Solution

By eliminating $\mathbf{v}^{(a)}$ from Eqs. (14.49) and (14.50), the governing equation for $p^{(a)}$ is found to be

$$\nabla^2 p^{(a)} + \frac{\omega^2}{a_0^2}p^{(a)} = \frac{i\omega}{2\pi a_0^2}\delta(\mathbf{x}-\mathbf{x}_1)e^{i\omega\tau}. \tag{14.53}$$

The solution of this equation in free space is

$$p^{(a)}(\mathbf{x};\mathbf{x}_1;\omega,\tau) = \frac{-i\omega}{8\pi^2 a_0^2}\frac{e^{i\frac{\omega}{a_0}|\mathbf{x}-\mathbf{x}_1|+i\omega\tau}}{|\mathbf{x}-\mathbf{x}_1|}. \tag{14.54}$$

Let (R, θ, ϕ) be the coordinates of a spherical polar coordinates system with the x-axis as the polar axis (see Figure 14.4). Also, let (r, ϕ, x) be the coordinates of a cylindrical coordinate system with the x-axis as its axis. For convenience, the source vector $\mathbf{x}_1$ is taken to lie on the plane $\phi = \phi_1$, then

$$\mathbf{x}_1 = R_1\cos\theta_1\hat{\mathbf{e}}_x + R_1\sin\theta_1\cos\phi_1\hat{\mathbf{e}}_y + R_1\sin\theta_1\sin\phi_1\hat{\mathbf{e}}_z.$$

Also, the position vector $\mathbf{x}$ is

$$\mathbf{x} = x\hat{\mathbf{e}}_x + r\cos\phi\,\hat{\mathbf{e}}_y + r\sin\phi\,\hat{\mathbf{e}}_z.$$

For large $R_1 (R_1 \to \infty)$, the distance between $\mathbf{x}_1$ and $\mathbf{x}$ is

$$|\mathbf{x}-\mathbf{x}_1| = R_1 - \frac{|\mathbf{x}\cdot\mathbf{x}_1|}{|\mathbf{x}_1|} = R_1 - x\cos\theta_1 - r\cos(\phi-\phi_1)\sin\theta_1.$$

Thus, Eq. (14.54) may be rewritten as

$$\begin{aligned}
p_{R_1\to\infty}^{(a)}(\mathbf{x};\mathbf{x}_1;\omega,\tau) &= \frac{-i\omega}{8\pi^2 a_0^2 R_1}e^{i\frac{\omega}{a_0}[R_1-x\cos\theta_1-r\sin\theta_1\cos(\phi-\phi_1)+a_0\tau]} \\
&= \frac{-i\omega}{8\pi^2 a_0^2 R_1}e^{i\frac{\omega}{a_0}(R_1-x\cos\theta_1)+i\omega\tau}\sum_{m=0}^{\infty}(-i)^m\varepsilon_m J_m\left(\frac{\omega}{a_o}\sin\theta_1 r\right) \\
&\quad \times \cos[m(\phi-\phi_1)],
\end{aligned} \tag{14.55}$$

where $\varepsilon_0 = 1$, $\varepsilon_m = 2$ for $m \geq 1$. The first line of Eq. (14.55) is just the plane wave solution for a source at a great distance. This is illustrated in Figure 14.7. The second line of Eq. (14.55) is the cylindrical wave expansion (see chapter IX of Magnus and Oberhettinger, 1949). Eq. (14.55) is only a part of the adjoint Green's function. The second part is the scattered wave. It will be shown that the scattered wave part of the adjoint Green's function may be computed as a two-dimensional problem.

14.4.2 Adjoint Green's Function for a Cylindrical Surface

Eq. (14.55) may be regarded as a particular solution of Eq. (14.53). There is also a homogeneous solution. When combining the particular and the homogeneous solution, the full solution satisfies boundary condition (14.51) on the cylindrical surface at $r = D/2$. In cylindrical coordinates, the appropriate homogeneous solution (denoted by a subscript "h") is

$$p_h^{(a)} = \frac{-i\omega}{8\pi^2 a_0^2 R_1} \left[\sum_{m=0}^{\infty} (-i)^m \varepsilon_m C_m H_m^{(1)} \left(\frac{\omega}{a_0} \sin\theta_1 r \right) \cos[m(\phi - \phi_1)] \right] e^{i\frac{\omega}{a_0}(R_1 - x\cos\theta_1) + i\omega\tau}$$

(14.56)

where C_m is a set of unknown constants. $H_m^{(1)}()$ is the mth order Hankel function of the first kind. By adding Eqs. (14.55) and (14.56), the adjoint Green's function is

$$p^{(a)} = \frac{-i\omega e^{i\frac{\omega}{a_0}(R_1 - x\cos\theta_1) + i\omega\tau}}{8\pi^2 a_0^2 R_1}$$

$$\times \sum_{m=0}^{\infty} (-i)^m \varepsilon_m \left[J_m \left(\frac{\omega}{a_0} \sin\theta_1 r \right) + C_m H_m^{(1)} \left(\frac{\omega}{a_0} \sin\theta_1 r \right) \right] \cos[m(\phi - \phi_1)].$$

Upon imposing boundary condition (14.51), it is found that

$$C_m = -\frac{J_m \left(\frac{\omega}{a_0} \sin\theta_1 \frac{D}{2} \right)}{H_m^{(1)} \left(\frac{\omega}{a_0} \sin\theta_1 \frac{D}{2} \right)}.$$

Thus, the adjoint Green's function $p^{(a)}$ and the radial velocity $v^{(a)}$ are

$$p^{(a)}(r, x, \phi; R_1, \theta_1, \phi_1; \omega, \tau) = \frac{-i\omega}{8\pi^2 a_0^2 R_1} e^{i\frac{\omega}{a_0}(R_1 - x\cos\theta_1) + i\omega\tau} \sum_{m=0}^{\infty} \frac{(-i)^m \varepsilon_m}{H_m^{(1)} \left(\frac{\omega}{a_0} \sin\theta_1 \frac{D}{2} \right)} \cdot$$

$$\left[J_m \left(\frac{\omega}{a_0} \sin\theta_1 r \right) H_m^{(1)} \left(\frac{\omega}{a_0} \sin\theta_1 \frac{D}{2} \right) - J_m \left(\frac{\omega}{a_0} \sin\theta_1 \frac{D}{2} \right) H_m^{(1)} \left(\frac{\omega}{a_0} \sin\theta_1 r \right) \right]$$

$$\times \cos[m(\phi - \phi_1)].$$

(14.57)

$$v^{(a)}(r, x, \phi; R_1, \theta_1, \phi_1; \omega, \tau) = \frac{\omega \sin\theta_1}{8\pi^2 a_0 R_1} e^{i\frac{\omega}{a_0}(R_1 - x\cos\theta_1) + i\omega\tau} \sum_{m=0}^{\infty} \frac{(-i)^m \varepsilon_m}{H_m^{(1)} \left(\frac{\omega}{a_0} \sin\theta_1 \frac{D}{2} \right)} \cdot$$

$$\left[J'_m \left(\frac{\omega}{a_0} \sin\theta_1 r \right) H_m^{(1)} \left(\frac{\omega}{a_0} \sin\theta_1 \frac{D}{2} \right) - J_m \left(\frac{\omega}{a_0} \sin\theta_1 \frac{D}{2} \right) H'^{(1)}_m \left(\frac{\omega}{a_0} \sin\theta_1 r \right) \right]$$

$$\times \cos[m(\phi - \phi_1)].$$

(14.58)

On the cylindrical surface, the terms inside the square bracket of Eq. (14.58) may be simplified by using the Wronskian relationship of Bessel functions. This gives

$$v^{(a)}\left(\tfrac{D}{2}, x, \phi; R_1, \theta_1, \phi_1; \omega, \tau\right)$$

$$= \frac{-i}{2\pi^3 R_1 D} \sum_{m=0}^{\infty} (-i)^m \varepsilon_m \frac{e^{i\frac{\omega}{a_0}(R_1 - x\cos\theta_1) + i\omega\tau}}{H_m^{(1)}\left(\frac{\omega}{a_0}\sin\theta_1\frac{D}{2}\right)} \cos[m(\phi - \phi_1)].$$

By means of the reciprocity relation (14.52), the Fourier transform of the pressure of the surface Green's function is given by the right-hand side of this equation. This expression may be further simplified by noting that

$$\sum_{m=0}^{\infty} (-i)^m \varepsilon_m \frac{\cos[m(\phi - \phi_1)]}{H_m^{(1)}\left(\frac{\omega}{a_0}\sin\theta_1\frac{D}{2}\right)} = \sum_{m=-\infty}^{\infty} \frac{e^{-i\frac{\pi}{2}m + im(\phi - \phi_1)}}{H_m^{(1)}\left(\frac{\omega}{a_0}\sin\theta_1\frac{D}{2}\right)}.$$

Therefore, by inverting the Fourier transform, the surface Green's function becomes

$$p^{(g)}\left(R, \theta, \phi, t; \tfrac{D}{2}, x_0, \phi_0, \tau\right)$$

$$= \frac{1}{2\pi^3 DR} \sum_{m=-\infty}^{\infty} \int_{-\infty}^{\infty} \frac{e^{i\frac{\omega}{a_0}(R - x_0\cos\theta) - i(m+1)\frac{\pi}{2} + im(\phi - \phi_0) - i\omega(t - \tau)}}{H_m^{(1)}\left(\frac{\omega}{a_0}\sin\theta\frac{D}{2}\right)} d\omega. \quad (14.59)$$

Eq. (14.59) is the same as the direct Green's function of Eq. (14.18).

14.5 Adjoint Green's Function for a Conical Surface

The noise of high-speed jets is an important aeroacoustic problem. Because the mixing of the jet and ambient gases, the jet expands laterally in the downstream direction. The use of a cylindrical surface Γ for this type of flows would not be appropriate. A conical surface with a well-chosen half-angle δ is a good choice. Figure 14.8 shows a conical surface Γ enclosing a high-speed spreading jet. The natural coordinates to use outside the conical surface is the spherical polar coordinates (R, θ, ϕ).

Let $p^{(g)}(R, \theta, \phi, t; R_0, \phi_0, t_0)$ be the direct surface Green's function with the source point located at (R_0, ϕ_0) and source time t_0. The Fourier transform of time t is

$$\tilde{p}^{(g)}(R, \theta, \phi; R_0, \phi_0; \omega; t_0) = \frac{1}{2\pi} \int_{-\infty}^{\infty} p^{(g)}(R, \theta, \phi, t; R_0, \phi_0, t_0) e^{i\omega t} dt. \quad (14.60)$$

The far-field sound pressure due to a surface pressure distribution of $\Phi(R_0, \phi_0, t_0)$ on the conical surface is then given by

$$p(R, \theta, \phi, t) =$$

$$\int_{-\infty}^{\infty}\int_{0}^{2\pi}\int_{0}^{\infty}\int_{-\infty}^{\infty} \tilde{p}^{(g)}(R, \theta, \phi; R_0, \phi_0; \omega; t_0)\Phi(R_0, \phi_0, t_0) e^{-i\omega t} R_0 \sin\delta\, d\omega\, dR_0\, d\phi_0\, dt_0.$$

$$(14.61)$$

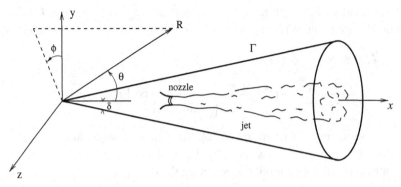

Figure 14.8. A conical surface enclosing a spreading high-speed jet.

The surface element of a conical surface is $dS = R\sin\delta\,dR\,d\phi$. By the reciprocity relation of Eq. (14.52), $\tilde{p}^{(g)}$ is related to the adjoint Green's function $v_n^{(a)}$ by

$$\tilde{p}^{(g)}(R_1, \theta_1, \phi_1; R_0, \phi_0; \omega, t_0) = v_n^{(a)}(R_0, \phi_0; R_1, \theta_1, \phi_1; \omega, t_0), \qquad (14.62)$$

where $v_n^{(a)}$ is the adjoint velocity component normal to the conical surface Γ. Positive is in the outward pointing direction.

As for the case of a cylindrical surface, a particular solution of the adjoint equations is given by Eq. (14.55). The relationship between cylindrical coordinates and spherical coordinates is

$$r = R\sin\theta, \quad x = R\cos\theta.$$

On replacing r and x by this relation and upon using Eq. (14.55), the particular solution (denoted by a subscript "p") may be written as

$$p_p^{(a)} = \frac{-i\omega}{8\pi^2 a_0^2 R_1} e^{i\frac{\omega}{a_0}(R_1 - R\cos\theta\cos\theta_1) + i\omega t_0} \sum_{m=-\infty}^{\infty} J_m\left(\frac{\omega}{a_0} R\sin\theta\sin\theta_1\right) e^{-im\frac{\pi}{2} + im(\phi_1 - \phi)}.$$

$$(14.63)$$

The adjoint velocity is related to the pressure by Eq. (14.49). Thus, the component normal to and on the conical surface is

$$v_{n,p}^{(a)} = \frac{\omega}{8\pi^2 a_0 R_1} e^{i\frac{\omega}{a_0}(R_1 - R\cos\delta\cos\theta_1) + i\omega t_0} \sum_{m=-\infty}^{\infty} \left[\sin\theta_1 \cos\delta J'_m\left(\frac{\omega}{a_0} R\sin\delta\sin\theta_1\right)\right.$$

$$\left. + i\cos\theta_1 \sin\delta J_m\left(\frac{\omega}{a_0} R\sin\delta\sin\theta_1\right)\right] e^{-im\frac{\pi}{2} + im(\phi_1 - \phi)}. \qquad (14.64)$$

Now, let the homogeneous solution (denoted by a subscript "h") of Eq. (14.49) and (14.50) in cylindrical coordinates to have the following form:

$$\begin{bmatrix} u_h^{(a)} \\ v_h^{(a)} \\ w_h^{(a)} \\ p_h^{(a)} \end{bmatrix} = \frac{-i\omega}{8\pi^2 a_0 R_1} e^{i\frac{\omega}{a_0} R_1 + i\omega t_0} \sum_{m=-\infty}^{\infty} \begin{bmatrix} \hat{u}_m^{(a)} \\ \hat{v}_m^{(a)} \\ \hat{w}_m^{(a)} \\ \hat{p}_m^{(a)} \end{bmatrix} e^{-im\frac{\pi}{2} + im(\phi_1 - \phi)}. \qquad (14.65)$$

This is a Fourier series expansion in angular variable $(\phi - \phi_1)$. The governing equations for the amplitude functions $(\hat{u}_m^{(a)}, \hat{v}_m^{(a)}, \hat{w}_m^{(a)}, \hat{p}_m^{(a)})$ can readily be found by substituting Eq. (14.65) into Eqs. (14.49) and (14.50) with the nonhomogeneous term omitted. These equations are as follows:

$$- i\omega \hat{u}_m^{(a)} - a_0^2 \frac{\partial \hat{p}_m^{(a)}}{\partial x} = 0 \tag{14.66}$$

$$- i\omega \hat{v}_m^{(a)} - a_0^2 \frac{\partial \hat{p}_m^{(a)}}{\partial r} = 0 \tag{14.67}$$

$$- i\omega \hat{w}_m^{(a)} + i a_0^2 \frac{m}{r} \hat{p}_m^{(a)} = 0 \tag{14.68}$$

$$- i\omega \hat{p}_m^{(a)} - \frac{\partial \hat{v}_m^{(a)}}{\partial r} - \frac{\hat{v}_m^{(a)}}{r} + i\frac{m}{r}\hat{w}_m^{(a)} - \frac{\partial \hat{u}_m^{(a)}}{\partial x} = 0. \tag{14.69}$$

The boundary condition from Eq. (14.51) is

$$\theta = \delta, \quad \hat{p}_m^{(a)} = -J_m\left(\frac{\omega}{a_0}R\sin\delta\sin\theta_1\right)e^{-i\frac{\omega}{a_0}R\cos\delta\cos\theta_1}. \tag{14.70}$$

For future reference, it is easy to show that by taking the complex conjugate of Eq. (14.66) to Eq. (14.70) that

$$\hat{u}_{-m}^{(a)}(r, x; \theta_1, -\omega) = \hat{u}_m^{(a)*}(r, x; \theta_1, \omega) \tag{14.71}$$

$$\hat{v}_{-m}^{(a)}(r, x; \theta_1, -\omega) = \hat{v}_m^{(a)*}(r, x; \theta_1, \omega), \tag{14.72}$$

where * denotes the complex conjugate.

Once $\hat{u}_m^{(a)}$ and $\hat{v}_m^{(a)}$ are found, it is straightforward to find, by combining with particular solution Eq. (14.64), that the normal velocity of the adjoint Green's function on conical surface $\theta = \delta$ with (R, ϕ) replacing by (R_0, θ_0) is

$$v_n^{(a)}(R_0, \phi_0; R_1, \theta_1, \phi_1; t_0, \omega)$$

$$= \frac{\omega}{8\pi^2 a_0 R_1} e^{i\frac{\omega}{a_0}R_1 + i\omega t_0} \sum_{m=-\infty}^{\infty} \left\{ \left[\sin\theta_1\cos\delta J_m'\left(\frac{\omega}{a_0}R_0\sin\theta_1\sin\delta\right)\right.\right.$$

$$\left.+ i\cos\theta_1\sin\delta J_m\left(\frac{\omega}{a_0}R_0\sin\theta_1\sin\delta\right)\right] e^{-i\frac{\omega}{a_0}R_0\cos\theta_1\cos\delta}$$

$$- i\left[\hat{v}_m^{(a)}(R_0, \delta; \omega)\cos\delta - \hat{u}_m^{(a)}(R_0, \delta; \omega)\sin\delta\right]\right\} \cdot e^{-im\frac{\pi}{2}+im(\phi_1-\phi_0)}. \tag{14.73}$$

By means of the reciprocity relation (14.62), the pressure associated with the direct Green's function is now known, i.e.,

$$\tilde{p}^{(g)}(R, \theta, \phi; R_0, \phi_0; t_0, \omega) = v_n^{(a)}(R_0, \phi_0; R, \theta, \phi; t_0, \omega). \tag{14.74}$$

For later application, it is useful to expand $\tilde{p}^{(g)}$ as a Fourier series in $(\phi - \phi_0)$ in the following form:

$$\tilde{p}^{(g)}(R, \theta, \phi; R_0, \phi_0; t_0, \omega) = \sum_{m=-\infty}^{\infty} \hat{p}_m(R, \theta; R_0, \omega)e^{i\omega t_0+im(\phi-\phi_0)}. \tag{14.75}$$

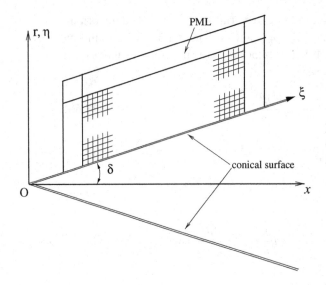

Figure 14.9. An oblique Cartesian coordinate system for computing the homogeneous adjoint Green's function.

By means of Eqs. (14.74) and (14.73), it is easy to find that

$$\hat{p}_m(R, \theta; R_0, \omega) = \frac{(-i)^m \omega}{8\pi^2 a_0 R} e^{i\frac{\omega}{a_0}R} \left\{ \left[\sin\theta \cos\delta J'_m\left(\frac{\omega}{a_0}\sin\theta \sin\delta R_0\right) \right. \right.$$

$$\left. + i\cos\theta \sin\delta J_m\left(\frac{\omega}{a_0}\sin\theta \sin\delta R_0\right) \right] e^{-i\frac{\omega}{a_0}R_0 \cos\theta \cos\delta}$$

$$\left. - i\left[\hat{v}_m^{(a)}(R_0, \delta, \omega)\cos\delta - \hat{u}_m^{(a)}(R_0, \delta, \omega)\sin\delta \right] \right\}. \quad (14.76)$$

Note: The dependence of $\hat{p}_m$ on R is in the factor $\frac{e^{i\frac{\omega}{a_0}R}}{R}$. An important property of $\hat{p}_m(R, \theta; R_0, \omega)$ is that it becomes its complex conjugate if $m \to -m$ and $\omega \to -\omega$, i.e.,

$$\hat{p}_{-m}(R, \theta; R_0, -\omega) = \hat{p}_m^*(R, \theta; R_0, \omega). \quad (14.77)$$

14.5.1 Computation of the Adjoint Green's Function for a Conical Surface

A relatively simple way to compute the homogenous solution is to convert Eqs. (14.66) to (14.69) into time-dependent equations by replacing $-i\omega$ by $\frac{\partial}{\partial t}$ and add a factor of $e^{-i\omega t}$ to the right side of boundary condition (14.70). The resulting problem may then be time marched to a periodic state, which is the nonhomogeneous adjoint solution. For this purpose, it is advantageous to switch to an oblique Cartesian coordinate system (ξ, η) as shown in Figure 14.9. The oblique Cartesian coordinates and the cylindrical coordinates are related by

$$\eta = r - x\tan\delta, \quad \xi = \frac{x}{\cos\delta} \quad \text{or} \quad r = \eta + \xi\sin\delta, \quad x = \xi\cos\delta. \quad (14.78)$$

With respect to the oblique Cartesian coordinates, the adjoint equations (14.66) to (14.69) in the time domain may be rewritten in the following form:

$$\frac{\partial \mathbf{U}_m}{\partial t} + \mathbf{A}_m \frac{\partial \mathbf{U}_m}{\partial \xi} + \mathbf{B}_m \frac{\partial \mathbf{U}_m}{\partial \eta} + \frac{1}{\eta + \xi\sin\delta}\mathbf{C}_m \mathbf{U}_m = 0, \quad (14.79)$$

where

$$
\mathbf{U}_m = \begin{bmatrix} \hat{u}_m^{(a)} \\ \hat{v}_m^{(a)} \\ \hat{w}_m^{(a)} \\ \hat{p}_m^{(a)} \end{bmatrix} e^{-i\omega t}, \quad
\mathbf{A_m} = \begin{pmatrix} 0 & 0 & 0 & -a_0^2 \sec\delta \\ 0 & 0 & 0 & 0 \\ 0 & 0 & 0 & 0 \\ -\sec\delta & 0 & 0 & 0 \end{pmatrix},
$$

$$
\mathbf{B}_m = \begin{pmatrix} 0 & 0 & 0 & a_0^2 \tan\delta \\ 0 & 0 & 0 & -a_0^2 \\ 0 & 0 & 0 & 0 \\ \tan\delta & -1 & 0 & 0 \end{pmatrix}, \quad
\mathbf{C_m} = \begin{pmatrix} 0 & 0 & 0 & 0 \\ 0 & 0 & 0 & 0 \\ 0 & 0 & 0 & ia_0^2 m \\ 0 & -1 & im & 0 \end{pmatrix}
$$

The boundary condition on the conical surface is

$$
\eta = 0, \quad \hat{p}_m^{(a)} = -J_m\left(\frac{\omega}{a_0}\xi \sin\theta_1 \sin\delta\right) e^{-i\omega(\xi \cos\delta \cos\theta_1 + t)}. \tag{14.80}
$$

To avoid the reflection of outgoing disturbances back into the computational domain, it is recommended to install a perfectly matched layer (PML) around the computational domain as shown in Figure 14.9. It is straightforward to develop a set of PML equations. A split-variable version of these equations is

$$
\mathbf{U}_m = \mathbf{U}_{m1} + \mathbf{U}_{m2} \tag{14.81a}
$$

$$
\frac{\partial \mathbf{U}_{m1}}{\partial t} + \mathbf{A}_m \frac{\partial \mathbf{U}_m}{\partial \xi} + \sigma \mathbf{U}_{m1} = 0 \tag{14.81b}
$$

$$
\frac{\partial \mathbf{U}_{m2}}{\partial t} + \mathbf{B}_m \frac{\partial \mathbf{U}_m}{\partial \eta} + \frac{1}{\eta + \xi \sin\delta} \mathbf{C}_m \mathbf{U}_m + \sigma \mathbf{U}_{m2} = 0. \tag{14.81c}
$$

In Eq. (14.81), σ is the damping function. Figure 14.10 shows a distribution of $\sigma(z)$ in the $\xi - \eta$ plane. For best results, it is recommended that a profile of $\sigma(z)$ resembling that in Figure 14.11 be used.

The solution of Eq. (14.79) together with PML boundary condition (14.81) and nonhomogeneous boundary condition (14.80) may be computed by using the 7-point stencil dispersion-relation-preserving (DRP) scheme. Boundary condition (14.80) may be enforced by the ghost point method. This boundary condition is responsible for introducing the correct disturbances into the computational domain. Since a time periodic solution is sought, a zero initial condition is the simplest way to get the solution started. The solution is to be marched in time until a time periodic solution is reached. Note: the adjoint Green's function is spatially oscillatory because boundary condition (14.80) is an oscillatory function. The period of oscillation of boundary condition (14.80) may be used to estimate the spatial resolution required for computing the adjoint Green's function. Experience suggests that it is prudent to turn the nonhomogeneous boundary term on gradually. This may be done by multiplying the right side of boundary condition (14.80) by a factor $(1 - e^{-t/T})$ and taking T to be a number of oscillatory periods long. The reason for this is that, after discretization, the response of the finite difference system to a sudden imposition of boundary condition is sometimes not the same as that for a partial differential system.

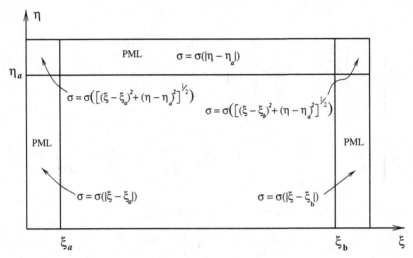

Figure 14.10. The computational domain in the $\xi - \eta$ plane surrounding by PML. Also shown is the distribution of damping functions.

14.5.2 Time Periodic Sources

For time periodic sources with time dependence $e^{-i\Omega t}$, the source function on the conical surface may be written in the following general form:

$$\theta = \delta, \quad p_{source}(R_0, \phi_0, t_0) = \Psi(R_0, \phi_0)e^{-i\Omega t_0}. \tag{14.82}$$

By Eq. (14.61), the far-field pressure field $(R \to \infty)$ associated with the source is given by

$$p(R, \theta, \phi, t)$$
$$= \int_0^{2\pi} \int_0^{\infty} \int_{-\infty}^{\infty} \int_{-\infty}^{\infty} \tilde{p}^{(g)}(R, \theta, \phi; R_0, \phi_0; \omega, t_0)\Psi(R_0, \phi_0)e^{-i\omega t - i\Omega t_0}R_0 \sin \delta d\omega \, dt_0 dR_0 \, d\phi_0.$$

$$\tag{14.83}$$

On using $\tilde{p}^{(g)}$ in a Fourier expansion in $(\phi - \phi_0)$, i.e., Eq. (14.75), Eq. (14.83) becomes after integrating over dt_0 (giving rise to $2\pi\delta(\omega - \Omega)$) and then upon integrating over $d\omega$,

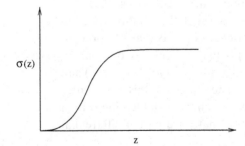

$\sigma(z)$

Figure 14.11. A typical profile of damping function $\sigma(z)$.

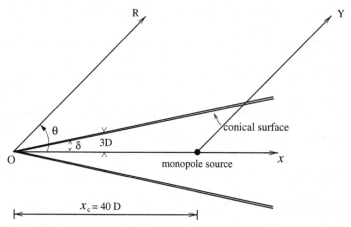

Figure 14.12. A monopole acoustic source surrounded by a conical surface.

$$p(R, \theta, \phi, t)$$

$$= \int_0^{2\pi} \int_0^\infty \sum_{m=-\infty}^{\infty} 2\pi \hat{p}_m(R, \theta; R_0, \Omega) \Psi(R_0, \phi_0) e^{im(\phi-\phi_0)-i\Omega t} R_0 \sin \delta dR_0 \, d\phi_0,$$

$$(14.84)$$

where $\hat{p}_m(R, \theta; R_0, \Omega)$ is given by Eq. (14.76). In this form, only two integrations need be evaluated.

As a concrete example, consider a time periodic monopole source located at a distance x_c from the apex of a conical surface as shown in Figure 14.12. Let Y be the distance of a point from the monopole source. For a point on the conical surface with a spherical polar coordinates (R_0, δ, ϕ_0), Y for this point is given by

$$Y_0 = \left(R_0^2 + x_c^2 - 2R_0 x_c \cos \delta\right)^{\frac{1}{2}}. \qquad (14.85)$$

The pressure field on the conical surface associated with the monopole acoustic source is

$$p_{source}(R_0, \phi_0, t_0) = \frac{e^{i\Omega\left(\frac{Y_0}{a_0} - t_0\right)}}{4\pi Y_0}. \qquad (14.86)$$

Thus, in the notation of Eq. (14.82), the Ψ function is

$$\Psi(R_0, \phi_0) = \frac{e^{i\Omega Y_0/a_0}}{4\pi Y_0}. \qquad (14.87)$$

Substitution into Eq. (14.84), the far pressure field of the monopole source, according to the continuation method, is as follows:

$$p(R, \theta, \phi, t) \underset{R \to \infty}{=} \int_0^\infty \pi \hat{p}_0(R, \theta; R_0, \Omega) \frac{e^{i\Omega\left(\frac{Y_0}{a_0} - t\right)}}{Y_0} R_0 \sin \delta dR_0. \qquad (14.88)$$

In deriving Eq. (14.88), the integral over ϕ_0 has been performed. This integral is zero except for $m = 0$. Note that $\hat{p}_0$ is given by Eq. (14.76). The remaining integral

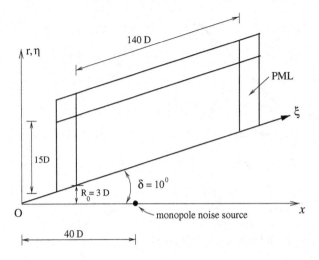

Figure 14.13. Computational   domain for the monopole acoustic source.

over R_0 has to be evaluated numerically. The first step is to compute the adjoint Green's function $\hat{v}_0^{(a)}$ and $\hat{u}_0^{(a)}$ in the $\xi - \eta$ plane as discussed before. These two quantities are required to compute $\hat{p}_0$. For the present example, a computational domain with a size as shown in Figure 14.13 is used. The half-apex angle δ is 10°. Note that a conical surface has no intrinsic length scale. Here, D in Figure 14.13 is taken as the length scale. In the far field, $R \rightarrow \infty$, the directivity factor is defined by

$$\tilde{D}\left(\theta, \frac{\Omega D}{a_0}\right) = \operatorname*{Lim}_{R \rightarrow \infty} \left[R\, p\, e^{-i\Omega\left(\frac{R}{a_0} - t\right)}\right], \tag{14.89}$$

where p is given by Eq. (14.88). For $\frac{\Omega D}{a_0} = 1.0$, the computed results are shown in Figures 14.14 and 14.15 for a source located at $x_c/D = 40.0$. The exact solution

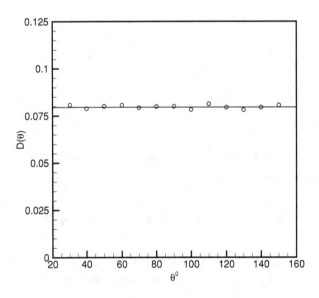

Figure 14.14. The computed directivity of a periodic monopole. The straight line is the exact solution.

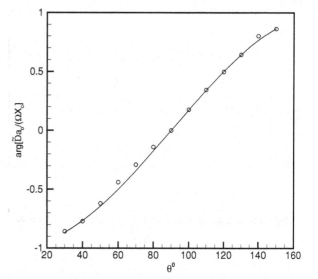

Figure 14.15. The computed phase of an off-center periodic monopole. Full line is the exact solution.

is $D(\theta) = |\tilde{D}(\theta, \frac{\Omega D}{a_0})| = \frac{1}{4\pi}$ and $\arg[\frac{a_0 \tilde{D}(\theta, \frac{\Omega D}{a_0})}{\Omega x_c}] = -\cos\theta$. The phase factor arises because the source is not located at the origin of the spherical polar coordinate system. As can be seen, the agreements between the magnitude and phase of the computed results and the exact solution are good. This provides confidence in the numerical method.

14.6 Generation of a Random Broadband Acoustic Field

Many aeroacoustics problems involve noise generated by turbulence. For this class of problems, both the source and the sound field are random and consist of a broad spectrum of frequencies. This is the case of high-speed jet noise. At a high Reynolds number, the jet flow is turbulent with a wide range of length scales. The noise emitted is broadband typically spread over three decades of frequencies. At the present time, there are not enough computational resources to perform direct numerical simulations of the jet flow. Many investigators choose to use large eddy simulation (LES) as an alternative, but the computational domain is finite and often is minimized to reduce computational cost. For this reason, without exception, only the acoustic near field is computed. However, for community noise purposes, interest is in the far field. Therefore, a method capable of continuing broadband near acoustic field to the far field is very much needed.

In Appendix F, a method to produce real-time pressure waves from a random broadband noise source is described. For a broadband monopole source, the acoustic field generated is spherically symmetric. Consider such a localized source as shown in Figure 14.12. Suppose the noise spectrum $S(\frac{\omega D}{a_0})$ is given where D is the length scale and a_0 is the speed of sound. The sound field emitted by the monopole source following Appendix F is as follows:

$$p(Y, t) = \frac{1}{Y} \int_{-\infty}^{\infty} A(\omega) \cos\left[\omega\left(\frac{Y}{a_0} - t\right) + \chi(\omega)\right] d\omega.$$

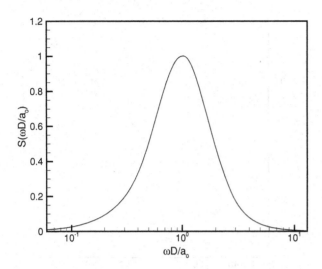

Figure 14.16. The similarity spectrum of the noise of large turbulence structures of high-speed jets. This spectrum is used as the noise spectrum of a broadband monopole source measured at $Y/D = 1.0$.

On following the energy-conserving discretization procedure, the discretized form of this equation is (see Eq. (F9)) as follows:

$$p(Y,t) = \frac{1}{Y} \sum_{j=-N}^{N} \left[2S\left(\frac{\omega D}{a_0}\right) \frac{(\Delta \omega) D}{a_0} \right]^{\frac{1}{2}} \cos\left[\omega_j \left(\frac{Y}{a_0} - t\right) + \chi_j\right]. \quad (14.90)$$

As a numerical example, suppose the noise spectrum $S(\frac{\omega D}{a_0})$ has the same shape as the similarity spectrum of the noise from the large turbulence structures of high-speed jets (see Tam, Golebiowski, and Seiner (1996); Tam, Viswanathan, Ahuja, and Panda (2008)). The similarity spectrum is shown in Figure 14.16. Note: the spectrum is assumed to be symmetric, i.e., $S(-\frac{\omega D}{a_0}) = S(\frac{\omega D}{a_0})$. Upon dividing the spectrum into 250 bands, the spatial distribution of pressure in the wave field at a selected instant of time, according to Eq. (14.90), is shown in Figure 14.17. Figure 14.18 shows the time history of pressure fluctuations measured at a distance of $Y = 5D$ from the monopole source. The random fluctuations are time stationary, that is, the stochastic properties of the broadband sound field is independent of time. An arbitrary starting time is used for the display in Figure 14.18.

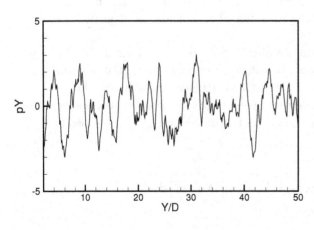

Figure 14.17. Spatial distribution of the pressure field of a broadband monopole source with a spectrum given by the similarity spectrum of Figure 14.16.

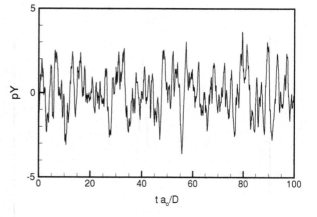

Figure 14.18. Time history of pressure fluctuations at $Y = 5D$ from the broadband monopole noise source of Figure 14.17.

The two-point space-time correlation function of the broadband sound field of a monopole source, following Eq. (F5), is given by

$$\overline{Y_1 Y_2 p(Y_1, t') p(Y_2, t' + \tau)} = \sum_{j=-N}^{N} S\left(\frac{\omega_j D}{a_0}\right) \frac{(\Delta \omega) D}{a_0} \cos\left[\frac{\omega_j}{a_0}(Y_2 - Y_1 - a_0 \tau)\right]$$

$$= F\left(\frac{Y_2 - Y_1 - a_0 \tau}{D}\right), \tag{14.91}$$

where the overbar means the time average. The function $F(\frac{Y_2 - Y_1 - a_0 \tau}{D})$ can also be computed directly from the acoustic field of the monopole source, that is,

$$F\left(\frac{Y_2 - Y_1 - a_0 \tau}{D}\right) = \lim_{T \to \infty} \frac{1}{T} \int_0^T \sum_{j=-N}^{N} \sum_{k=-N}^{N} 2 \left[S\left(\frac{\omega_j D}{a_0}\right) S\left(\frac{\omega_k D}{a_0}\right) \frac{(\Delta \omega_j) D}{a_0} \frac{(\Delta \omega_k) D}{a_0} \right]^{\frac{1}{2}}$$

$$\times \cos\left[\omega_j\left(\frac{Y_1}{a_0} - t\right) + \chi_j\right] \cdot \cos\left[\omega_k\left(\frac{Y_2}{a_0} - t - \tau\right) + \chi_k\right] dt. \tag{14.92}$$

Figure 14.19 shows the two-point space-time correlation function $F(\frac{Y_2 - Y_1 - a_0 \tau}{D})$ as computed directly according to Eq. (14.92). As a check on the accuracy of the energy-conserving discretization method and this formulation, the noise spectrum $S(\frac{\omega D}{a_0})$ is calculated by taking the Fourier transform of $F(\frac{a_0 \tau}{D})$, obtained by setting $Y_1 = Y_2$. This should yield the original prescribed similarity spectrum. A comparison of the two spectra is given in Figure 14.20. This figure indicates that there is good agreement between the two spectra. This confirms that the energy-conserving discretization method can, indeed, generate a broadband sound field with a given spectrum.

14.7 Continuation of Broadband Near Acoustic Field on a Conical Surface to the Far Field

An important CAA problem is to compute the noise of a high-speed turbulent jet. Since a computational domain is finite, it extends only to the near acoustic field. To determine the far-field sound, the continuation method developed in the Section 14.6 may now be used. The noise of a turbulent jet is broadband and random. It contains an enormous amount of information. However, for practical purposes, the

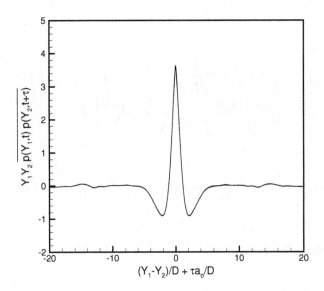

Figure 14.19. The two-point space-time correlation function computed according to Eq. (14.92).

quantities of interest in the far field are the noise spectra and directivity. For this reason, only the continuation of the noise spectra from the near field to the far field at different angular direction is considered.

In Section 14.5, it was suggested that a good choice of a matching surface Γ is a conical surface as shown in Figure 14.8. A method to compute the surface Green's function $p^{(g)}(R, \theta, \phi, t; R_0, \phi_0, t_0)$ through the use of the adjoint Green's function has also been developed. Let $p_s(R_0, \phi_0, t_0)$ be the pressure of a turbulent jet measured on the conical surface. The conical surface has a half-apex angle of δ. The far-field pressure is then given by

$$p(R, \theta, \phi, t)$$
$$= \int_{-\infty}^{\infty} \int_{0}^{2\pi} \int_{0}^{\infty} p^{(g)}(R, \theta, \phi, t; R_0, \phi_0, t_0) p_s(R_0, \phi_0, t_0) R_0 \sin \delta \, dR_0 \, d\phi_0 \, dt_0. \quad (14.93)$$

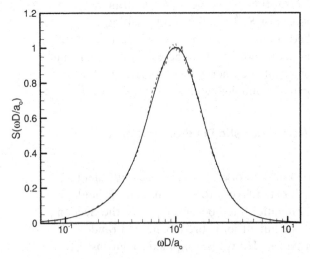

Figure 14.20. Comparison between the original (full line) and the computed spectrum (dotted line). The computed spectrum is the Fourier transform of the correlation function $F(\frac{a_0 \tau}{D})$ given in Figure 14.19.

By Eq. (14.75), the Fourier transform of the surface Green's function may be expanded in a Fourier series in $(\phi - \phi_0)$, that is,

$$p^{(g)}(R, \theta, \phi, t; R_0, \phi_0, t_0) = \sum_{m=-\infty}^{\infty} \int_{-\infty}^{\infty} \hat{p}_m(R, \theta; R_0, \omega) e^{im(\phi-\phi_0)-i\omega(t-t_0)} d\omega. \quad (14.94)$$

Now, the far-field noise spectrum, $S(R, \theta, \phi, \omega)$, is the Fourier transform of the autocorrelation function, i.e.,

$$S(R, \theta, \phi, \omega) = \frac{1}{2\pi} \int_{-\infty}^{\infty} \langle p(R, \theta, \phi, t) p(R, \theta, \phi, t+\tau) \rangle e^{i\omega\tau} d\tau. \quad (14.95)$$

In Eq. (14.95), the angular brackets, $\langle \rangle$, are the ensemble average of the random variables inside the brackets. Since the sound field is stationary random, the ensemble average and the time average are equal. By means of Eqs. (14.93) to (14.95), the following relation between the spectrum function and the source function can easily be established.

$$S(R, \theta, \phi, \omega) = \frac{1}{2\pi} \int \int \int \cdots \int \sum_{m=-\infty}^{\infty} \sum_{n=-\infty}^{\infty} \langle p_s(R', \phi', t') p_s(R'', \phi'', t'') \rangle$$
$$\times \hat{p}_m(R, \theta; R', \omega') \hat{p}_n(R, \theta; R'', \omega'')$$
$$\times e^{-i\omega'(t-t')-i\omega''(t-t'')+im(\phi-\phi')+in(\phi-\phi'')-i\omega''\tau+i\omega\tau}$$
$$\times R'R'' \sin^2 \delta \, d\tau \, d\omega' \, d\omega'' \, dt' \, dt'' \, dR' \, dR'' \, d\phi' \, d\phi''. \quad (14.96)$$

For jet noise, the surface pressure p_s is a stochastic variable that is stationary random in time and homogeneous in the azimuthal variable ϕ. These properties allow the two-point space-time correlation function to be written as

$$\langle p_s(R', \phi', t') p_s(R'', \phi'', t'') \rangle = \Phi(R', R'', \phi' - \phi'', t' - t''). \quad (14.97)$$

On inserting Eq. (14.97) into Eq. (14.96), the far-field spectrum function becomes

$$S(R, \theta, \phi, \omega) = \frac{1}{2\pi} \int \int \int \cdots \int \sum_{m=-\infty}^{\infty} \sum_{n=-\infty}^{\infty} \Phi(R', R'', \phi' - \phi'', t' - t'')$$
$$\times \hat{p}_m(R, \theta; R', \omega') \hat{p}_n(R, \theta; R'', \omega'')$$
$$\times e^{-i\omega'(t-t')-i\omega''(t-t'')+im(\phi-\phi')+in(\phi-\phi'')-i\omega''\tau+i\omega\tau}$$
$$\times R'R'' \sin^2 \delta \, d\tau \, d\omega' \, d\omega'' \, dt' \, dt'' \, dR' \, dR'' \, d\phi' \, d\phi''. \quad (14.98)$$

In the following, it will be shown that, regardless of what the two-point space-time correlation function $\Phi(R', R'', \phi' - \phi'', t' - t'')$ is, five of the integral and one summation in Eq. (14.98) can be evaluated.

The $d\tau$ integral may be evaluated in a straightforward manner as follows:

$$\int_{-\infty}^{\infty} e^{i(\omega-\omega'')\tau} d\tau = 2\pi \delta(\omega - \omega''). \quad (14.99)$$

The dt' and dt'' integrals may be evaluated by a change of variables to $\lambda = t' - t'', \xi = t''$. This gives

$$
\int_{-\infty}^{\infty} \int_{-\infty}^{\infty} \Phi(R', R'', \phi' - \phi'', t' - t'') e^{-i\omega' t' - i\omega'' t''} \, dt' \, dt''
$$

$$
= \int_{-\infty}^{\infty} \int_{-\infty}^{\infty} \Phi(R', R'', \phi' - \phi'', \lambda) e^{-i\omega' \lambda - i(\omega' + \omega'')\xi} \, d\xi \, d\lambda
$$

$$
= 2\pi \delta(\omega' + \omega'') \int_{-\infty}^{\infty} \Phi(R', R'', \phi' - \phi'', \lambda) e^{-i\omega' \lambda} \, d\lambda. \tag{14.100}
$$

The $d\phi'$ and $d\phi''$ integrals may also be evaluated by a change of variables to $\Theta = \phi' - \phi''$ and $\Psi = \phi''$, thus,

$$
\int_{0}^{2\pi} \int_{0}^{2\pi} \Phi(R', R'', \Theta, \lambda) e^{-im\phi' - in\phi''} \, d\phi' \, d\phi''
$$

$$
= \int_{0}^{2\pi} \int_{0}^{2\pi} \Phi(R', R'', \Theta, \lambda) e^{-im\Theta + i(m+n)\Psi} \, d\Psi \, d\Theta
$$

$$
= 2\pi \delta_{m,-n} \int_{0}^{2\pi} \Phi(R', R'', \Theta, \lambda) e^{-im\Theta} \, d\Theta. \tag{14.101}
$$

Substitution of Eqs. (14.99) to (14.101) into Eq. (14.98), and upon integrating over $d\omega'$ and $d\omega''$ and summing over m, the mathematical formula for the noise spectrum is

$$
S(R, \theta, \phi, \omega)
$$

$$
= 4\pi^2 \int_{0}^{\infty} \int_{0}^{\infty} \int_{0}^{2\pi} \int_{-\infty}^{\infty} \sum_{n=-\infty}^{\infty} \Phi(R', R'', \Theta, \lambda) \hat{p}_{-n}(R, \theta; R', -\omega) \hat{p}_{n}(R, \theta; R'', \omega)
$$

$$
\times e^{i\omega\lambda + in\Theta} R' R'' \sin^2 \delta \, d\lambda \, d\Theta \, dR' \, dR''. \tag{14.102}
$$

Eq. (14.102) is the principal result of this section. Physically, it signifies that the two-point space-time correlation function is the equivalent noise source on the conical surface. If this function is known, either by experimental measurement or by numerical simulation, then the far-field noise spectrum can be computed. Note: the $d\lambda$ and $d\Theta$ integration apply only to the equivalent noise source function. These two operations effectively decompose the source into frequency and azimuthal components.

Eq. (14.102) has been used in the experimental work of Tam, Viswanathan and Pastouchenko (2010) (see also Tam, Pastouchenko and Viswanathan (2010)). They measured the near-field two-point space-time pressure correlation function $\Phi(R', R'', \Theta, \lambda)$ of a Mach 1.66 hot jet on a conical surface of 10^0 half-angle. They used Eq. (14.102) to determine the far-field noise spectra. They showed that the noise spectra continued from the near-field were in good agreement with the measured far-field spectra.

14.7.1 A Test Case: A Broadband Monopole Source

To provide a simple validation test for formula (14.102), reconsider the case of the monopole noise source problem discussed in section 14.5.2 (see Figure 14.12). Instead of a time periodic source, it is replaced by the broadband monopole source of Section 14.6. The half-apex angle of the conical surface δ is again taken to be 10°. In this case, if Y' is a point on the conical surface, then

$$Y' = (R'^2 + x_c^2 - 2R'x_c \cos \delta)^{\frac{1}{2}}.$$

The two-point space-time correlation function has the following form (see Eq. (14.91)):

$$\Phi(R', R'', \Theta, \lambda) = F\left(\frac{Y'' - Y' - a_0\lambda}{D}\right) \frac{1}{Y'Y''}.$$

The F-function as computed directly is shown in Figure 14.19. In this special case, Φ is independent of Θ. For this reason, the integral over $d\Theta$ in Eq. (14.102) is zero except for $n = 0$. Thus the far-field noise spectrum obtained by continuation of the pressure field on the conical surface is as follows:

$$S(R, \theta, \phi, \omega)$$
$$= 8\pi^3 \int_0^\infty \int_0^\infty \int_{-\infty}^\infty \frac{1}{Y'Y''} F\left(\frac{Y'' - Y' - a_0\lambda}{D}\right) \hat{p}_0(R, \theta; R', -\omega) \hat{p}_0(R, \theta; R'', \omega)$$
$$\times e^{i\omega\lambda} R'R'' \sin^2 \delta \, d\lambda dR' \, dR''. \tag{14.103}$$

A change of integration variable from λ to $\eta = (Y'' - Y' - a_0\lambda)/D$ transforms the triple integrals in Eq. (14.103) into three separate integrals. This yields

$$S(R, \theta, \phi, \omega) \frac{a_0}{D}$$
$$= 8\pi^3 \left[\int_{-\infty}^\infty F(\eta)e^{-i\frac{\omega D}{a_0}\eta} \, d\eta\right] \left|\sin \delta \int_0^\infty \frac{\hat{p}_0(R, \theta; R'', \omega)}{Y''} e^{i\frac{\omega}{a_0}Y''} R''dR''\right|^2 \tag{14.104}$$

Figure 14.21 shows a comparison of the far-field noise spectra at $\theta = 150°, 90°$, and 30° (exhaust angle) computed numerically according to Eq. (14.104) and the original noise spectrum of the broadband monopole source (the similarity spectrum of the noise of large turbulence structures of high-speed jets). The agreement is excellent. The excellent agreement provides strong support for the validity of the present continuation method.

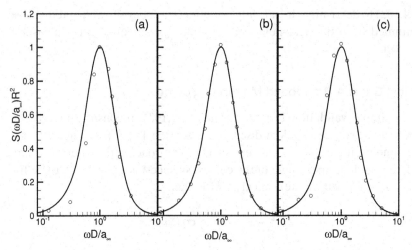

Figure 14.21. Comparisons between computed spectra (circles) and original spectra (full lines) at (a) $\theta = 150°$, (b) 90°, and (c) 30°.

EXERCISE

14.1. For a spherical matching surface, it is convenient and simple to use spherical harmonics to construct a surface Green's function.

    a. Develop a mathematical representation of the surface Green's function and the adjoint Green's function for a spherical surface of radius R.

    b. If $\Phi(\theta', \theta'', \phi', \phi'', t', t'')$ is the two-point space-time correlation function of a broadband random sound field on the spherical surface of radius R, find the relationship between $\Phi(\theta', \theta'', \phi', \phi'', t', t'')$ and the spectrum of the far-field radiated sound. (r, θ, ϕ) are the spherical polar coordinates.

15 Design of Computational Aeroacoustic Codes

The objective of this chapter is to discuss how to design a computation code to simulate an aeroacoustic phenomenon. In previous chapters, many methods and elements of numerical computation were discussed. In this chapter, they are to be synthesized to form a computer simulation code. The basic elements/ingredients of a good simulation algorithm in computational aeroacoustics would consist of the following:

1. A computational model containing all essential physics of the aeroacoustic phenomenon
2. A properly chosen computational domain
3. A well-designed computational grid
4. A least dispersive and dissipative high-resolution time marching algorithm
5. A time step that ensures numerical stability and good resolution
6. A set of high-quality boundary conditions for both exterior and interior boundaries
7. A properly chosen distribution of artificial selective damping to suppress the generation and propagation of spurious short waves
8. A set of properly prescribed initial conditions

15.1 Basic Elements of a CAA Code

Each of the eight elements in this list affects the simulation in some way. Some exert an influence on numerical stability. Some affect the accuracy and quality of the computed solution. Others control and influence the computation time. A more detailed consideration of some of these elements in this list is provided next.

15.1.1 Computational Model

The first step in performing a computational aeroacoustics (CAA) simulation is to formulate the problem. This means the development of a computational model. Usually this will include the selection of appropriate governing equations and physical boundary conditions.

In most practical CAA problems, the flow involved is turbulent (a discussion of turbulence modeling will be given in Section 15.5). Since direct numerical simulation of turbulent flow is prohibitively expensive in terms of computer resources required, some simplified method to treat the turbulence of the phenomenon is generally adopted. One may use a Reynolds Averaged Navier-Stokes (RANS) approach. The main advantage of using a RANS model is that it is easy to implement and not very expensive. The disadvantage is that it is not very accurate for time-dependent problems. Recently, the use of large eddy simulation (LES) and very large eddy simulation (VLES) has become quite popular. However, there is still a debate on how to close the LES model. One way is to use a subgrid scale model. The intent of a subgrid scale model is to model the effects of the spectrum of unresolved wave number on the resolved wave number spectrum of the computation, but it has been found that a subgrid scale model often performs only as a dissipation mechanism for the small-scale turbulence in the computation. For this reason, many investigators choose an alternative way. The alternative way is to use artificial numerical damping to remove the very small scale turbulent motion from the computation to avoid accumulation of energy in this wave number range (note: energy is cascaded up the wave number space to the highest wave number range supported by the computation). Excess energy accumulation in the smallest scales could cause a numerical simulation to blow up.

In many CAA problems, viscous effect is important only in a part of the computation domain. This may be because of a change in the dominant physics of the phenomenon in different parts of the computation domain. For example, near a wall, the viscous effect is important, while away from the wall, the compressibility effect may be most important. In such cases, the computation model will consist of the use of the Navier-Stokes equation in a part of the computation domain and the Euler equations in the remaining region.

Because a computational domain is finite, the influence of the flow and acoustic fields or sources outside may not be negligible. In such cases, the boundary conditions are forced to reproduce the outside influence as precisely as possible. In such a situation, the boundary condition is a crucial part of the computational model.

15.1.2 Computational Domain

The choice of a computational domain should not be taken lightly. A computational domain is, inevitably, finite in size. A natural tendency is to make the domain as small as possible. However, the domain should not be too small so as to leave some important noise sources outside. In some cases, the simulation involves flows with energetic disturbances. For this type of problem, it is advisable to choose a larger computation domain to allow the disturbances to decay to a less energetic state before leaving the computational domain.

CAA problems often involve the generation of tones by flow resonances. For example, the flow over a cavity or a resonator mounted flush to a flat wall often leads to the generation of strong tones. Since the tones are radiated to infinite space outside, the problem to simulate is an external radiation problem. The question often asked is, "How large should the external (external to the cavity) computational domain be?" Experience indicates that, if the domain is too small, it would lead to a higher computed frequency. This could be due to partial reflection because the external computational boundary is too close to the sources. To prevent this from happening, a rule of thumb is to use an external computational domain no smaller than 1.5 times the acoustic wavelength. Apparently no external radiation boundary condition is perfect. The reflections from the external boundary, even though small, could affect the noise generation processes resulting in a shift in tone frequency.

15.1.3 Computational Grid

It is not possible to overemphasize the importance of using a well-designed grid if a highly accurate numerical simulation is required. Most CAA problems include solid surfaces and bodies. In these cases, a body-fitted grid is most desirable. If a single set of body-fitted grids cannot be found for the entire domain, one may use local body-fitted grids and then transfer numerical data from one local grid to another through the overset grids method.

As mentioned previously, many aeroacoustic problems involve multiple scales. For problems of this kind, the use of multisize mesh is most appropriate. In order to design the mesh properly so as to offer adequate spatial resolution, one must have some idea of the governing physics in different parts of the computation domain. The mesh size is dictated by the dominant physics of the flow and acoustics. Once the spatial resolution required in different parts of the computational domain is known approximately, the domain may then be divided into subdomains. The mesh size of adjacent subdomains is allowed to change by a factor of 2. Overall, the mesh sizes of the entire computation are determined by the subdomain requiring the highest resolution.

In certain types of fluid flows, such as boundary layers or strongly sheared mixing layers, the flow field is characterized by two disparate length scales. The spatial rate of change of the flow field in the flow direction is mild compared with that in the perpendicular direction. Because of this disparity, it is possible to use computation grids with a fairly large aspect ratio for mean flow calculation. However, for aeroacoustic problems, sound waves usually propagate without a preferred direction. Therefore, it is recommended that the meshes, in physical domain, should have an aspect ratio close to unity. This will avoid any mesh-related anisotropy being introduced into the computation.

15.1.4 Computational Algorithm

CAA phenomena are, by definition, time-dependent. Thus, a time marching algorithm will naturally be required. In previous chapters, the 7-point dispersion-relation-preserving (DRP) scheme has been shown to have good dispersion and dissipative properties. It provides good resolution when using 7 or more mesh points per wavelength. In some situations, however, in order to meet the 7 mesh points per wavelength requirement, it will result in the use of an extremely small mesh. This is not desirable. Under this circumstance, one may use a larger stencil DRP scheme in just the part of the computational domain where the extra resolution is required. The 15-point stencil DRP scheme has a good resolution when using as few as 3.5 mesh points per wavelength. So it is a good strategy to use the 15-point stencil DRP scheme in the subdomain with very stringent resolution requirement and then transition to a 7-point stencil DRP scheme outside this subdomain.

15.1.5 Boundary Conditions

What boundary conditions to use in a CAA simulation should be given a good deal of thought. In previous chapters, a variety of numerical boundary conditions was discussed. So, for a given aeroacoustic phenomenon, more than one type of boundary condition may be used. For example, for a radiation boundary, one may use radiation boundary conditions based on asymptotic solution. An equally good, and maybe even better boundary condition, is to use the perfectly matched layer (PML) absorbing boundary condition.

Sometimes, the choice of which numerical boundary condition to use is severely limited, unless some new way to enforce the radiation boundary condition is found. For instance, to compute the noise and flow of an imperfectly expanded supersonic jet, an external boundary condition must be imposed that allows the ambient pressure, p_a, to be specified. At the nozzle exit, the static pressure of the jet, p_{exit}, is specified by the nozzle flow. Now, the enforcement of the nozzle exit boundary condition $p = p_{exit}$ is relatively straightforward. At the external boundary of the computational domain, the boundary condition must perform two different roles. It must allow the outgoing acoustic waves to exit with little reflection. At the same time, it must require the mean static pressure of the numerical solution to take on the value p_a. Most known absorbing boundary conditions do not have this capability. Among all the boundary conditions discussed in this book, it seems that only the asymptotic radiation boundary condition, discussed in Chapter 6 and Chapter 9, is capable of performing the dual functions.

Although it has been mentioned before, it is worthwhile to reemphasize that the external boundary conditions of a CAA simulation must exert the same influence on the computation as in the physical problem. This includes waves generated outside the computational domain, but propagated into the domain as incident waves. The same is true with mean flow and entrainment flow. It is not possible to anticipate what boundary condition requirement one might encounter. Thus, it is necessary to be creative when computing unusual aeroacoustic phenomena.

15.1.6 Distribution of Artificial Selective Damping

Inclusion of artificial selective damping in a computation code is absolutely necessary if a central difference scheme is used. Central difference schemes have no intrinsic damping. Without artificial selective damping, such a computation would fail either because the solution is heavily polluted by spurious waves or through a blowup. In a typical numerical simulation, artificial selective damping has two roles to play. First and foremost, it eliminates spurious waves, especially grid-to-grid oscillations as they propagate across the computation domain. For this purpose, a standard practice is to add a general background damping throughout the entire computation domain. The mesh Reynolds number should be chosen so that any grid-to-grid oscillations are damped by several orders of magnitude when propagating from one side to the other side of a subdomain. The second purpose of imposing artificial selective damping is to suppress the generation of spurious waves. Spurious waves are produced at surfaces of discontinuities. Thus, at solid surfaces, fluid interfaces, or mesh-size-change interfaces, extra artificial selective damping should be imposed. When a simulation contains solid surfaces with sharp edges and corners, one must recognize that these are sites for the generation of strong grid-to-grid oscillations. Grid-to-grid oscillations are one of the main causes of numerical instability. To suppress the generation of grid-to-grid oscillations, an effective method is to add additional artificial selective damping at and near sharp edges and corners.

15.2 Spatial Resolution Requirements

In designing a computational grid, the first question one faces is "what size mesh to use." This question cannot be answered until one has some idea of the flow and acoustic fields as well as the size of acoustic sources involved. Even if they are known, it is still not straightforward to decide what size mesh to use. For instance, if the flow contains a free shear layer of vorticity thickness, δ_ω, what mesh size to use is not completely obvious unless one has previous experience. The purpose of this section is to develop quantitative criteria to assist in selecting a proper mesh size for aeroacoustics computation.

15.2.1 Free Shear Flows

As an illustration of how one may formulate a mesh size selection procedure, consider the problem of computing a low Mach number two-dimensional wake flow of half-width b. According to the book by White (1991), the velocity deficit, $\tilde{u}$, of such a wake flow, whether it is laminar or turbulent, has the profile of a Gaussian function, i.e.,

$$\tilde{u}\left(\frac{y}{b}\right) = \frac{u - u_{min}}{u_{max} - u_{min}} = e^{-(\ln 2)\left(\frac{y}{b}\right)^2}. \tag{15.1}$$

The velocity profile is shown in Figure 15.1.

For computational purposes, the mesh size is to be assigned so as to be capable of resolving the Gaussian function part of the flow. Now, note that the Fourier transform of a Gaussian function is a Gaussian function, i.e.,

$$S(\alpha) = \frac{1}{2\pi} \int_{-\infty}^{\infty} e^{-(\ln 2)\left(\frac{y}{b}\right)^2} e^{i\alpha y} dy = \frac{b}{2(\pi \ln 2)^{\frac{1}{2}}} e^{-\frac{\alpha^2 b^2}{4\ln 2}}, \tag{15.2}$$

where α is the wave number. It is easy to show that the area under the wave number spectrum of Eq. (15.2) is unity, i.e.,

$$\int_{-\infty}^{\infty} S(\alpha) d\alpha = 1. \tag{15.3}$$

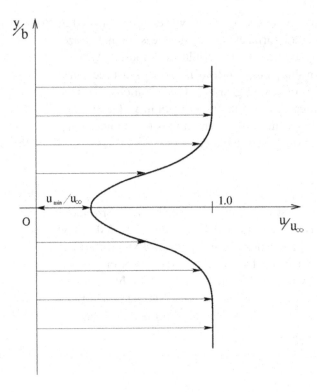

Figure 15.1. Velocity profile of a two-dimensional turbulent wake.

The fraction of the wave number spectrum up to a wave number k, denoted by $F(kb)$, is

$$F(kb) = 2 \int_0^k S(\alpha)d\alpha = erf\left[\frac{kb}{2(\ln 2)^{\frac{1}{2}}}\right],$$

where $erf[\,]$ is the error function. If the mesh size is chosen to resolve the wake flow up to a wake number k, then the fractional error, $E(kb)$, is equal to

$$E(kb) = 1 - F(kb) = 1 - erf\left[\frac{kb}{2(\ln 2)^{\frac{1}{2}}}\right]. \tag{15.4}$$

A plot of $E(kb)$ is shown in Figure 15.2.

Suppose the 7-point stencil DRP scheme is chosen to do the computation. This scheme is acceptable if the mesh size Δy is chosen so that $k\Delta y \le 0.95$ (see Chapter 2). Now consider that a fractional error β ($\beta << 1.0$) is acceptable for the computation; then, by using the 7-point stencil DRP scheme, the mesh size would be given by equating $E(kb)$ to β with $k = (0.95/\Delta y)$. This gives

$$\beta = 1 - erf\left[\frac{0.95}{2(\ln 2)^{\frac{1}{2}}}\frac{b}{\Delta y}\right]. \tag{15.5}$$

Figure 15.3 gives the numerical solution of Eq. (15.5), namely, β as a function

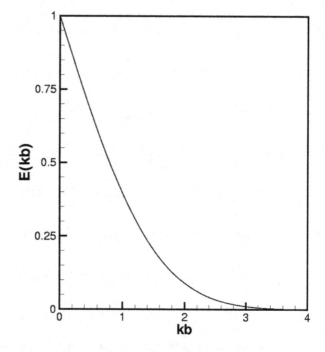

Figure 15.2. The error function E (kb).

of $2b/\Delta y$. Now, for an assigned value of β, there is a corresponding value of $2b/\Delta y$, say σ. Thus, the maximum mesh size allowed is

$$(\Delta y)_{\text{maximum}} = \frac{2b}{\sigma}. \tag{15.6}$$

By following this procedure, the maximum mesh size for a given acceptable error for other shear flows can be readily established. For the cases of a two-dimensional

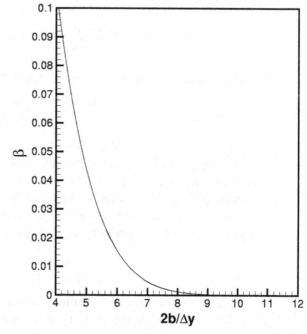

Figure 15.3. Spatial resolution curve for two-dimensional wake flow.

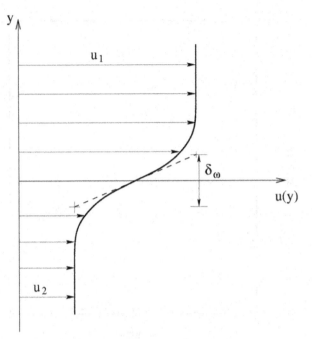

Figure 15.4. Velocity profile of a two-dimensional turbulent mixing layer.

mixing layer, a two-dimensional turbulent jet, a laminar boundary layer, and a turbulent boundary layer, the spatial resolution curves are provided in the following sections.

15.2.1.1 Two-Dimensional Turbulent Mixing Layer

The velocity profile of a low Mach number two-dimensional turbulent mixing layer between two fluid layers of velocities u_1 and u_2 is shown in Figure 15.4. In nondimensional form, the mean velocity profile may be represented analytically (see the book by White) by an error function.

$$\tilde{u} = \frac{u - u_2}{u_1 - u_2} = \frac{1}{2}\left[1 + erf\left(\frac{\pi^{\frac{1}{2}} y}{\delta_\omega}\right)\right], \tag{15.7}$$

where δ_ω is the maximum vorticity thickness defined by

$$\frac{u_1 - u_2}{\delta_\omega} = \left(\frac{du}{dy}\right)_{maximum}.$$

The velocity profile of a mixing layer tends to a constant as $|y| \to \infty$. For this reason, the profile has no Fourier transform in the usual sense. However, the velocity gradient du/dy is a well-behaved function. It is a well-known fact that the integral of a function is smoother and better behaved than its derivative. Therefore, if the mesh size is chosen to be able to resolve du/dy, then the mesh should be able to provide adequate resolution for computing $u(y)$. Thus, the function to be considered is

$$\frac{\delta_\omega}{u_1 - u_2}\frac{du}{dy} = e^{-\pi\left(\frac{y}{\delta_\omega}\right)^2}.$$

Figure 15.5 shows the resolution curve β as a function of $\delta_\omega/\Delta y$ assuming the 7-point stencil DRP scheme is to be used for computation.

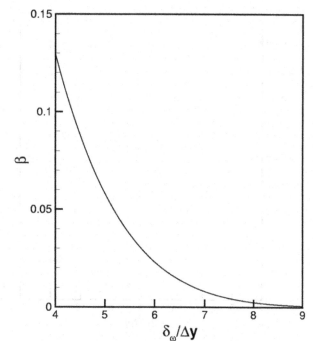

Figure 15.5. Spatial resolution curve for two-dimensional mixing layers.

15.2.1.2 Two-Dimensional Turbulent Jet

Figure 15.6 shows the mean velocity profile of a low-speed two-dimensional turbulent jet. According to White (1991), a good analytical representation of the velocity profile is

$$\frac{u}{u_{max}} = \text{sech}^2\left[\frac{0.88y}{b}\right], \tag{15.8}$$

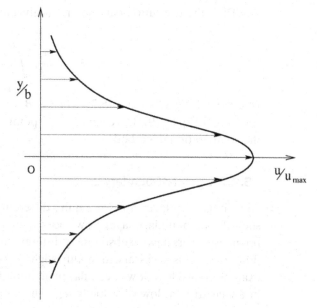

Figure 15.6. Velocity profile of a two-dimensional turbulent jet.

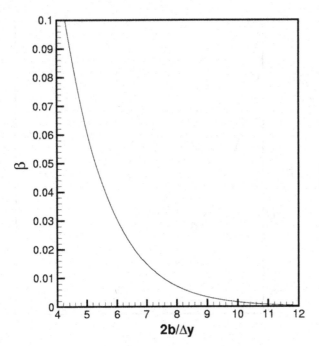

Figure 15.7. Spatial resolution curve for two-dimensional turbulent jets.

where b is the half-width of the jet. It is easy to show that the Fourier transform of the function on the right side of Eq. (15.3) is

$$S(\alpha) = u_{\max} \frac{\alpha}{2} \left(\frac{b}{0.88} \right)^2 \csc h \left(\frac{\pi}{2} \frac{\alpha b}{0.88} \right)$$

and that

$$2 \int_0^\infty S(\alpha) d\alpha = u_{\max}.$$

Thus, the resolution parameter β is given by

$$\beta = 1 - \frac{4}{\pi^2} \int_0^{\frac{\pi}{2} \frac{\alpha b}{0.88}} \frac{z dz}{\sinh(z)}, \tag{15.9}$$

where $\alpha = 0.95/\Delta y$. Figure 15.7 is a plot of β as a function of $2b/\Delta y$. It is the numerical solution of Eq. (15.9). Here again, the 7-point stencil DRP scheme is assumed to be the computational method.

15.2.2 Laminar Boundary Layer

Now, boundary layer flows are quite different from free shear layers. Unlike free shear flows, boundary layer flows occupy only a half-infinite space. Furthermore, boundary layers have a substantially different mean velocity profile. Because of these differences, it is necessary to modify slightly the mesh size selection procedure. For example, consider a low-speed flat-plate boundary layer of thickness δ (99 percent) in a free stream flow of velocity u_∞. The mean flow has a well-known similarity

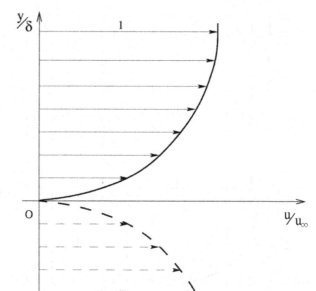

Figure 15.8. Velocity profile of a laminar boundary layer and its reflection.

profile. It is given by the Blasius solution (see Figure 15.8). In standard notation (White, 1991), it is as follows:

$$\frac{u}{u_\infty} = f'(\eta), \qquad \eta = \frac{3.5y}{\delta} (f'(3.5) = 0.99).$$

To facilitate the computation of the wave number spectrum, it is recommended to regard the problem as one to compute the velocity deficit as follows:

$$\tilde{u}\left(\frac{3.5y}{\delta}\right) = [1 - f'(|\eta|)] \tag{15.10}$$

instead of the mean velocity. Note: Eq. (15.10) extends the velocity deficit to the full range $-\infty < y < \infty$ by a simple reflection.

Now, the Fourier transform of $\tilde{u}$, denoted by S, is

$$S\left(\frac{\alpha\delta}{3.5}\right) = \frac{1}{2\pi} \int_{-\infty}^{\infty} \tilde{u}\left(\frac{3.5y}{\delta}\right) e^{-i\alpha y}\,dy = \frac{\delta}{7\pi} \int_{-\infty}^{\infty} [1 - f'(|\eta|)] e^{i\frac{\omega\delta}{3.5}\eta}\,d\eta. \tag{15.11}$$

It is easy to show that the area under the wave number spectrum is unity. Therefore, the normalized wave number spectrum is given by Eq. (15.11). Thus, the fractional error by using the 7-point stencil DRP scheme is

$$\beta = 1 - 2 \int_{0}^{\frac{0.95}{\Delta y}} S\left(\frac{\alpha\delta}{3.5}\right) d\alpha. \tag{15.12}$$

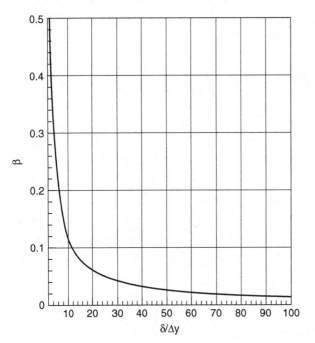

Figure 15.9. Spatial resolution curve for two-dimensional laminar boundary layer.

Figure 15.9 is the numerical solution of Eq. (15.12). On comparing Figure 15.9 and the resolution curves for free shear flows, it becomes clear that a much finer mesh is required to compute a boundary layer for the same resolution. This point is well worth remembering.

15.2.3 Turbulent Boundary Layer

The velocity profile of a turbulent boundary layer over a flat plate consists of three distinct layers. They are the viscous sublayer, the log layer, and the outer wake layer. Figure 15.10 shows the velocity profile of a turbulent boundary layer at low Mach number. The velocity profile is constructed by using the law of the wake profile for the outer layer and Spalding's formula for the inner layer. The viscous sublayer is very thin and has a very large velocity gradient. It accounts for less than 1 percent of the boundary layer thickness. The log layer lies on top of the viscous layer. It makes up about 20 percent of the boundary layer thickness. The log layer may be regarded as a transition layer. The top layer is the wake flow. It constitutes the bulk ($\sim$80 percent) of the boundary layer. In this layer, the velocity gradient is the mildest. Because of the large variation of mean flow velocity gradient across the boundary layer, it is not feasible to use a single size mesh for its computation. The mesh size needed to resolve each of the three layers will be considered separately below.

Let the velocity profile of a low-speed flat-plate turbulent boundary layer be $u(y/\delta)$, where δ is the boundary layer thickness. As in the case of laminar boundary layer, the velocity deficit $\tilde{u}$ (see Figure 15.11a) may be written in the following form:

$$\frac{\tilde{u}}{u_\infty} = \left[1 - \frac{u\left(\frac{|y|}{\delta}\right)}{u_\infty}\right]. \tag{15.13}$$

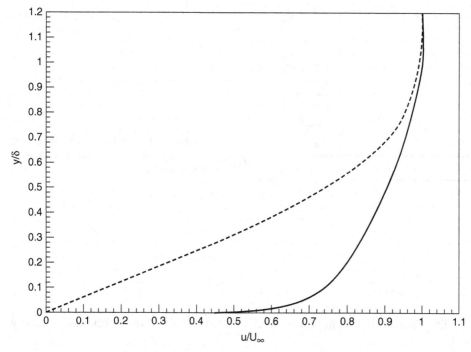

Figure 15.10. Velocity profile of a turbulent boundary over a flat plate (dark curve) and that of a laminar boundary layer (dotted curve).

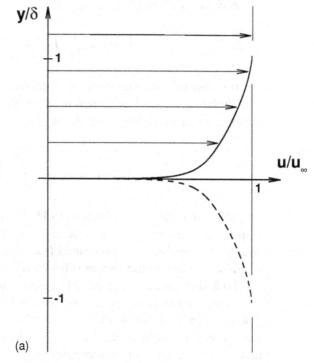

Figure 15.11a. Profile of velocity deficit $\tilde{u}/u_\infty$ of a low-speed turbulent boundary layer.

(a)

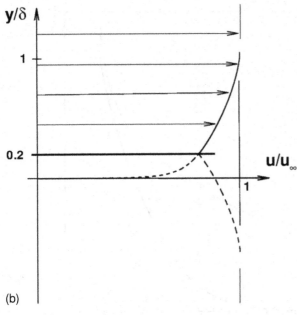

Figure 15.11b. Profile of velocity deficit $u_{-20\%}/u_\infty$ of a low-speed turbulent boundary layer.

(b)

The Fourier transform of the velocity deficit, extended to the full range $-\infty < y < \infty$, is

$$\tilde{S}(\alpha) = \frac{1}{2\pi} \int_{-\infty}^{\infty} \left[1 - \frac{u\left(\frac{|y|}{\delta}\right)}{u_\infty} \right] e^{i\alpha y} dy.$$

A change of variable leads to

$$\tilde{S}(\alpha\delta) = \frac{\delta}{2\pi} \int_{-\infty}^{\infty} \left[1 - \frac{u\left(|\xi|\right)}{u_\infty} \right] e^{i\alpha\delta\xi} d\xi. \tag{15.14}$$

It is easy to show that the area of the wave number spectrum is unity. Suppose the 7-point stencil DRP scheme is used for computation, then the fractional error β for using a computational mesh size Δy is

$$\beta = 1 - \frac{2}{\delta} \int_{0}^{0.95\left(\frac{\delta}{y}\right)} \tilde{S}(\zeta)d\zeta. \tag{15.15}$$

The numerical solution of Eq. (15.15) is given in Figure 15.12. By means of this figure, it is a simple matter to make a quantitative estimate of the maximum mesh size Δy permissible for a prescribed fractional error β. In this way, the mesh size required to compute the viscous sublayer is found.

To find the mesh size needed to resolve the log layer, one may use essentially the same procedure as before. Consider a turbulent boundary layer with the bottom 2 percent removed, that is, the viscous sublayer is taken out from consideration. Let the new mean velocity profile be $u_{-2\%}\left(\frac{y}{\delta}\right)$. Now, apply this analysis to $u_{-2\%}\left(\frac{y}{\delta}\right)$, the new velocity deficit with a symmetric extension to the negative y plane is

$$\tilde{u}_{-2\%} = \left[1 - u_{-2\%}\left(\frac{y}{\delta}\right) \right]. \tag{15.16}$$

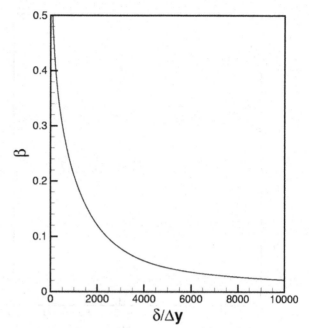

Figure 15.12. Spatial resolution curve for the viscous sublayer of a turbulent boundary layer.

The mesh size requirement for computing velocity deficit profile Eq. (15.16) by the 7-point stencil DRP scheme may be determined in the same way as velocity deficit profile Eq. (15.13). Figure 15.13 shows the mesh size resolution curve, β versus $\delta/\Delta y$, for computing the log layer of a turbulent boundary layer. This procedure may be applied to the wake layer as well. This is done by deleting the lower 20 percent of the boundary layer velocity profile as shown in Figure 15.11b. On proceeding as

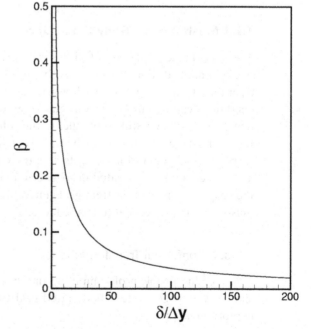

Figure 15.13. Spatial resolution curve for the log layer of a turbulent boundary layer.

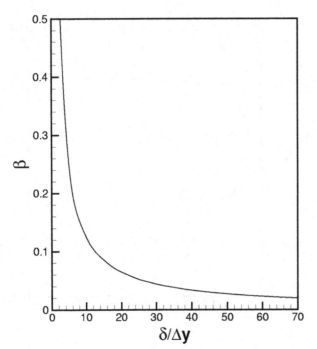

Figure 15.14. Spatial resolution curve for the wake layer of a turbulent boundary layer.

above, it is straightforward to derive the mesh resolution curve for computing the outer wake layer. Figure 15.14 is the β versus $\delta/\Delta y$ curve.

On comparing Figures 15.12, 15.13, and 15.14, it is evident that there are orders of magnitude of difference in the mesh size requirement in computing the three layers of a turbulent boundary layer. This is probably not too surprising. In a realistic computation, a multisize mesh is absolutely necessary. Such a computation can be handled by the multi-size-mesh multi-time-step method discussed in Chapter 12.

15.3 Mesh Design: Body-Fitted Grid

The importance of having a well-designed grid in a numerical simulation cannot be overemphasized. For problems involving solid surfaces, a body-fitted grid is most desirable. Computation on such a grid has the advantage that the wall boundary conditions can be enforced with relative ease and accuracy. With body-fitted grid, one set of grid lines will be parallel to the solid surface. This is ideal for computing the boundary layer adjacent to the wall. Developing a body-fitted grid is not straightforward. Grid generation is, by itself, a topic that has been extensively researched. The subject matter is treated in specialized books and advanced texts. It is beyond the scope of this book to treat grid generation other than providing an example to illustrate some aspects of the basic ideas.

15.3.1 Conformal Transformation

Consider the problem of computing the flow around a two-dimensional airfoil. A simple way to generate a body-fitted grid for the airfoil is to use the method of conformal mapping.

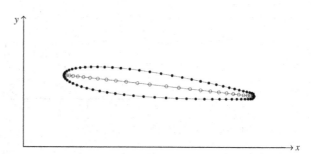

Figure 15.15. Mapping of an airfoil into a slit by conformal mapping. Open circles are sources points on the camber line. Black circles are enforcement points.

The method consists of two essential steps as schematically illustrated in Figures 15.15, 15.16, and 15.17. The first step is a mapping of the airfoil in the physical $x - y$ plane into a slit in the w or the $\xi - \eta$ plane. This is accomplished by conformal transformation using a multitude of point sources or doublets distributed along the camber line of the airfoil as shown in Figure 15.15. If point sources are used, the appropriate form of the conformal transformation is

$$w = z + \sum_{j=1}^{N} Q_j \log\left(\frac{z - z_j}{z - z_0}\right) + i\Omega, \qquad (15.17)$$

where $w = \xi + i\eta$ is the complex coordinates of a point in the mapped plane, $z = x + iy$ is a point in the physical plane, Q_j, z_j $(j = 1, 2, \ldots, N)$ are the source strengths and locations, and Ω is just a constant. The inclusion of Ω in the mapping function is important. Its inclusion makes it possible to map the airfoil onto a slit on the real axis of the w plane (see Figure 15.16). z_0 is the location of a reference source. A good location to place z_0 is in the middle of the camber line, such that all the lines joining z_j to z_0 lie within the airfoil. In this way, the branch cuts associated with the log functions of Eq. (15.17) would not present a problem in the computation.

In implementing the mapping, the coordinates z_j $(j = 1, 2, 3, \ldots, N)$ of the source points are first assigned. A useful rule in assigning the source locations is that the distance $|z_j - z_{j-1}|$ should be in some way proportional to the local thickness of the airfoil. That is to say, the sources should be closer together at where the airfoil is thin. Now the source strengths Q_j $(j = 1, 2, \ldots, N)$ are to be chosen to accomplish the mapping. For this purpose, a set of points, z_k $(k = 1, 2, \ldots, M)$, well distributed around the surface of the airfoil is chosen to enforce the mapping (see Figure 15.15). z_k are the enforcement points, and M should be much larger than N. On imposing the conditions that the surface of the airfoil be mapped onto a slit on the real axis of

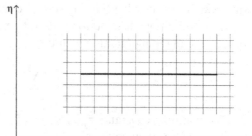

Figure 15.16. Complex w plane. The slit is the airfoil.

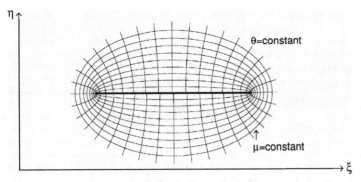

Figure 15.17. Elliptic coordinates $\mu - \theta$ is the body-fitted coordinates of the airfoil.

the w plane, Eq. (15.17) leads to the following system of linear equations for Q_j and Ω:

$$y_k + \sum_{j=1}^{N} \text{Im}\left[Q_j \log\left(\frac{z_k - z_j}{z_k - z_0}\right)\right] + \Omega = 0, \qquad k = 1, 2, \ldots, M. \quad (15.18)$$

This system of equations may be written in a matrix form (real variable) as follows:

$$\mathbf{Ax = b}, \qquad (15.19)$$

where x is a column vector of length $(2N + 1)$. The elements of x are

$$x_{2j-1} = \text{Re}(Q_j), \quad x_{2j} = \text{Im}(Q_j), \quad j = 1, 2, \ldots, N$$
$$x_{2N+1} = \Omega.$$

$\mathbf{A}$ is a $M \times (2N + 1)$ matrix. The elements of $\mathbf{A}$ are

$$a_{i,2j} + ia_{i,2j-1} = \log\left[\frac{z_i - z_j}{z_i - z_0}\right]$$
$$a_{i,2N+1} = 1, i = 1, 2, \ldots, M; \quad j = 1, 2, 3, \ldots, N.$$

$\mathbf{b}$ is a column vector of length M. The elements of $\mathbf{b}$ are

$$b_k = -y_k, \qquad k = 1, 2, 3, \ldots, M.$$

With $M \gg 2N + 1$ Eq. (15.19) is an overdetermined system. A least squares solution may be obtained by first premultiplying Eq. (15.19) by the transpose of $\mathbf{A}$. This leads to the normal equation,

$$\mathbf{A^T A x = A^T b}, \qquad (15.20)$$

which can be easily solved by any standard matrix solver.

If a point dipole distribution is used to map an airfoil into a slit instead of point sources, the mapping function would be

$$w = z + \sum_{j=1}^{N} \frac{A_j}{z - z_j} + iC. \qquad (15.21)$$

The mapping constants $z_j, A_j (j = 1, 2, \ldots)$ and C are determined in the same way as those for point sources.

The second step in developing a body-fitted grid to the airfoil is to take the midpoint of the slit in the $\xi - \eta$ plane as the center of an elliptic coordinate system. The two end points of the slit are the singular points of the conformal transformation. They are given by

$$\frac{dw}{dz} = 0. \tag{15.22}$$

Let the width of the slit be a and $\bar{\xi}$ be the x coordinate measured from the midpoint of the slit. The elliptic coordinates (μ, θ) are related to $(\bar{\xi}, \eta)$ by

$$\bar{\xi} = \frac{a}{2} \cosh \mu \cos \theta, \qquad \eta = \frac{a}{2} \sinh \mu \sin \theta. \tag{15.23}$$

The elliptic coordinate system is shown in Figure 15.17. The elliptic coordinate system is orthogonal. The curve $\mu = 0$ is the surface of the airfoil in the physical plane. If the computation is for a flow past the airfoil, then the boundary layer lies in the region $\mu < \Delta$ in the $\mu - \theta$ plane. The wake is in the region $-\Delta_1 < \theta < \Delta_2$, where Δ, Δ_1, and Δ_2 are small quantities for a flow from left to right. In a simulation, these are the regions where very fine meshes are needed.

15.3.2 Inverse Mapping from the w Plane to the Physical Plane

Eq. (15.17) gives a direct mapping of the airfoil from the physical plane to the w plane. After the computation, it will be necessary to reverse the procedure to find the point z corresponding to a point w in the transformed plane. For the mapping function under consideration, it is easy to see that away from the airfoil $z \cong w - i\Omega$. Therefore, for these points, the inverse mapping may be found by simple Newton iteration.

Let $z_{n+1} = z_n + \Delta z_n$ be the point in the z plane corresponding to the point $w = w_0$ in the w plane after the $(n + 1)$th iteration. Eq. (15.17) gives

$$w_0 = z_n + \Delta z_n + \sum_{j=1}^{N} Q_j \ln \left(\frac{z_n + \Delta z_n - z_j}{z_n + \Delta z_n - z_0} \right) + i\Omega. \tag{15.24}$$

By expanding the terms on the right side of Eq. (15.24) for small Δz_n, and ignoring all higher-order terms, it is straightforward to find that

$$\Delta z_n = \frac{w_0 - z_n - \sum_{j=1}^{N} Q_j \ln \left(\frac{z_n - z_j}{z_n - z_0} \right) - i\Omega}{1 + \sum_{j=1}^{N} Q_j \left(\frac{1}{z_n - z_j} - \frac{1}{z_n - z_0} \right)}. \tag{15.25}$$

To start the iteration, the obvious choice is to take $z_1 = w_0 - i\Omega$.

For points closer to the airfoil, it is advantageous to keep the second-order terms in expanding the right side of Eq. (15.24). Let

$$f(z) = \sum_{j=1}^{N} Q_j \ln \left(\frac{z - z_j}{z - z_0} \right). \tag{15.26}$$

By keeping the second-order terms in the expansion, it is easy to derive

$$\Delta z_n = \frac{-[1 + f'(z_n)] + [1 + 2f'(z_n) + (f'(z_n))^2 + 2f''(z_n)(w_0 - z_n - f(z_n) - i\Omega)]^{\frac{1}{2}}}{f''(z_n)},$$

(15.27)

where $' = d/dz$.

Points on the slit in the w plane are more difficult to invert. To invert these points, the following numerical method may be used. Let w_s be the point of which the inverse, say z_s, is to be found. Suppose the inverse of a point $\bar{w}$ denoted by $\bar{z}$ near w_s is already found. $\bar{w}$ may or may not be on the slit. Let w be a point on the line joining $\bar{w}$ to w_s so that

$$w - \bar{w} = re^{i\beta}.$$

(15.28)

Now, by Eq. (15.17) it is readily found that

$$\frac{dz}{dw} = \frac{1}{1 + \sum\limits_{j=1}^{N} Q_j \left[\frac{1}{z - z_j} - \frac{1}{z - z_0} \right]}.$$

(15.29)

On combining Eqs. (15.28) and (15.29), along the line joining $\bar{w}$ and w_s, the following differential equation is established:

$$\frac{dz}{dr} = \frac{e^{i\beta}}{1 + \sum\limits_{j=1}^{N} Q_j \left[\frac{1}{z - z_j} - \frac{1}{z - z_0} \right]}.$$

(15.30)

z_s may now be determined by integrating Eq. (15.30) from $r = 0$ to $r = |w_s - \bar{w}|$ with starting condition $r = 0$, $z = \bar{z}$.

15.3.3 NACA Four-Digit Airfoils

A very famous series of airfoil shapes is the NACA four-digit airfoils. These airfoils are completely defined by four parameters:

$\quad C$ the length of the chord line;

$\quad \varepsilon$ the maximum camber ratio;

$\quad p$ the chordwise position of the maximum camber ratio;

$\quad \tau$ the thickness ratio (maximum thickness/chord).

If x is the distance along the chord, the thickness distribution is

$$T(x) = 10\tau C \left[0.2969 \left(\frac{x}{C} \right)^{\frac{1}{2}} - 0.126 \frac{x}{C} - 0.3537 \left(\frac{x}{C} \right)^2 + 0.2843 \left(\frac{x}{C} \right)^3 - 0.1015 \left(\frac{x}{C} \right)^4 \right]$$

(15.31)

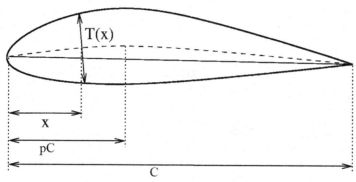

Figure 15.18 NACA four-digit airfoils.

and the camber line (see Figure 15.18) is given by two parabolas joined at the maximum camber point as follows:

$$\overline{Y}(x) = \begin{cases} \frac{\varepsilon x}{p^2}\left(2p - \frac{x}{C}\right), & \text{if } 0 < \frac{x}{C} < p; \\ \frac{\varepsilon(C-x)}{(1-p)^2}\left(1 + \frac{x}{C} - 2p\right), & \text{if } p < \frac{x}{C} < 1. \end{cases} \tag{15.32}$$

The first of the airfoil's four-digit designation is 100ε, the second digit is $10p$, and the last two are 100τ.

The upper part of the airfoil is given by

$$X_u(x) = x - \frac{T(x)\overline{Y}'(x)}{2[1 + (\overline{Y}'(x))^2]^{\frac{1}{2}}} \tag{15.33}$$

$$Y_u(x) = \overline{Y}(x) + \frac{T(x)}{2[1 + (\overline{Y}'(x))^2]^{\frac{1}{2}}} \tag{15.34}$$

and the lower part by

$$X_\ell(x) = x + \frac{T(x)\overline{Y}'(x)}{2[1 + (\overline{Y}'(x))^2]^{\frac{1}{2}}} \tag{15.35}$$

$$Y_\ell(x) = \overline{Y}(x) - \frac{T(x)}{2[1 + (\overline{Y}'(x))^2]^{\frac{1}{2}}} \tag{15.36}$$

where

$$\overline{Y}'(x) = \begin{cases} 2\frac{\varepsilon}{p^2}\left(p - \frac{x}{C}\right), & \text{if } 0 < \frac{x}{C} < p; \\ 2\frac{\varepsilon}{(1-p)^2}\left(p - \frac{x}{C}\right), & \text{if } p < \frac{x}{C} < 1. \end{cases} \tag{15.37}$$

Theoretically, one may have airfoils with sharp trailing edges. In reality, a sharp trailing edge is nearly impossible to make. Practical airfoils invariably have a blunt or slightly blunt trailing edge. A simple way to incorporate a blunt trailing edge to an airfoil is to replace the sharp tip by a semicircular arc as shown in Figure 15.19.

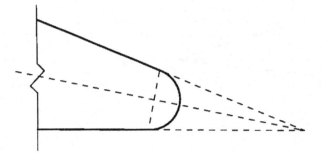

Figure 15.19. Blunt trailing edge of an airfoil.

Analytically, for $x > qc$ (q is the chordwise position of the circular arc), a blunt trailing edge airfoil has a thickness T and a camber line $\overline{Y}$ as follows:

$$T(x) = 2\left\{\frac{[T(qC)]^2}{4} - (x - qC)^2\left(1 + \left[\overline{Y}'(qC)\right]^2\right)\right\}^{\frac{1}{2}}. \tag{15.38}$$

$$\overline{Y}(x) = \overline{Y}'(qC)(x - qC) + \overline{Y}(qC) \tag{15.39}$$

However, the airfoil is slightly shortened. The new chord length is given by

$$C^* = qC + \frac{T(qC)}{2[1 + (\overline{Y}'(qC))^2]^{\frac{1}{2}}}. \tag{15.40}$$

15.4 Example I: Direct Numerical Simulation of the Generation of Airfoil Tones at Moderate Reynolds Number

It is known experimentally that an airfoil immersed in a uniform stream at moderate Reynolds number would emit tones. This phenomenon has been studied by a number of investigators. Surprisingly, there are significant differences in their experimental results. Furthermore, there is no clear consensus on the tone generation mechanism. There are two objectives in including this problem in this chapter. The first objective is to demonstrate how to develop a CAA code. The second objective is to illustrate how direct numerical simulation can assist in resolving or clarifying experimental disagreement. In addition, numerical simulation provides a full set of space-time data (not always available from an experiment), which may be used to investigate the many details of the phenomenon so as to shed light on the underlying tone generation mechanism. It is said that numerical simulation is used most often for prediction. However, the most useful contribution of numerical simulation is to provide a better understanding of the physics and mechanisms of a phenomenon under consideration.

Paterson et al. (1972) were the first to report the observation of tone genera-tion associated with uniform flow past a NACA0012 airfoil at moderate Reynolds number. Figure 15.20 represents the principal result of their work. In this figure, the frequencies of the tones detected are plotted as a function of flow velocity. The ladder structure of the data was quickly recognized (Tam, 1974) as an indication that the observed tones were related to a feedback loop. A feedback loop must satisfy

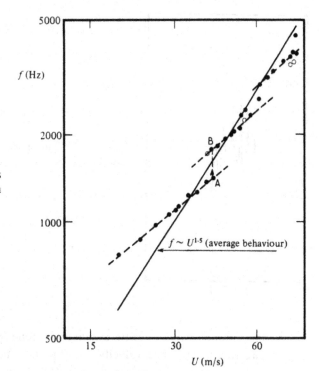

Figure 15.20. Tone frequency versus flow velocity measured by Paterson et al.

certain integral wave number requirements. The ladder structure is a consequence of a change in the quantization number of the feedback loop. At certain flow velocities, two tones were measured simultaneously. As velocity increases, the dominant tone frequency jumps to the next ladder step as indicated by A to B in Figure 15.20. This is usually referred to as staging. Further velocity increase causes the subdominant tone to vanish. By carefully analyzing their data, Paterson et al. found that the dominant tone followed a $U^{1.5}$ power law (see Figure 15.20) where U is the free stream velocity. Based on their extensive measurements, they proposed the following formula for the dominant frequency, f:

$$f = \frac{0.011 U^{1.5}}{(\nu C)^{\frac{1}{2}}}, \tag{15.41}$$

where C is the chord width and ν is the kinematic viscosity. Eq. (15.41) is referred to as the Paterson formula.

Arbey and Bataille (1983) repeated the airfoil experiment in an open wind tunnel as Paterson et al. did. In all their experiments, the airfoil was at a zero degree angle of attack. Just as Paterson et al. they observed tones. In fact, there is a multitude of them. Figure 15.21 shows a typical spectrum of their measurements. The tones are separated by about 110 Hz. Arbey and Bataille discovered that the frequency of the dominant tone (f_s in the center of the spectrum of Figure 15.21) was in good agreement with the Paterson formula (see Figure 15.22). However, there are distinct differences between the two experiments. First of all, Arbey and Bataille measured many tones, but Paterson et al. had at most two tones simultaneously. Moreover, the frequencies of the two tones in the Paterson et al. experiment are separated by a

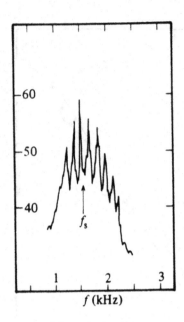

Figure 15.21. The spectrum of airfoil tones measured by Arbey and Bataille.

magnitude of about five times that of the tones of Arbey and Bataille. When plotted in a frequency versus velocity diagram, the tones of Arbey and Bataille also form a ladder structure suggesting again that the tones are related to a feedback loop.

Unlike Paterson et al. and Arbey and Bataille, Nash, Lowson, and McAlpine (1999) performed their airfoil experiment inside a wind tunnel. They quickly realized that the cross-duct modes and the Parker modes of the wind tunnel could interfere with the airfoil tone phenomenon. To eliminate this possibility, they lined the wind

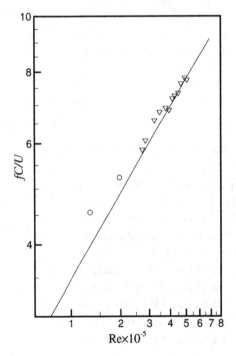

Figure 15.22. Strouhal number of dominant airfoil tone versus Reynolds number measured by Arbey and Bataille. Straight line is the Paterson formula.

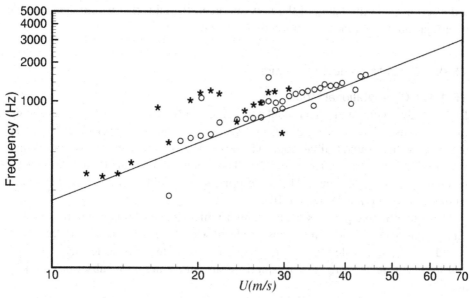

Figure 15.23. Airfoil tones measured by Nash et al. Straight line is the Paterson formula.

tunnel walls with sound-absorbing materials. Under such an experimental condition, they found that, for each free stream velocity, there was a single tone and no ladder structure. They suggested that the ladder structure data of Paterson et al. and Arbey and Bataille were influenced by their experimental facilities. The feedback and ladder structure of the data were not characteristics of the airfoil tone phenomenon. The feedback loop apparently was locked on something in the experimental facility outside the open wind tunnel. At a low angle of attack, the tone frequency-flow velocity relation measured by Nash et al. is also in good agreement with the Paterson formula (see Figure 15.23). At a high angle of attack, because of the blockage of the wind tunnel by the model, the measured relationship deviates from the Paterson formula, as would be expected.

It should be clear at this time that there is little agreement among the most prominent airfoil experiments reported in the literature. One major issue is whether the basic tone generation mechanism is related to a feedback loop, which would generate multiple tones or, as in the observations of Nash et al., feedback is spurious and there should only be a single tone. Even among investigators who observed multiple feedback tones, the details of the observed phenomenon are quite different. Therefore, it is fair to say that, up to the present time, no two experiments have the same results. The only agreement of all the major experiments is that the Paterson formula is a good prediction formula for the dominant tone frequency. This is true over a wide range of Reynolds numbers.

One way to resolve the disagreement is to perform highly accurate numerical simulations of the phenomenon. Numerical simulation has a number of advantages. First of all, when properly carried out, it can avoid facility-related feedback. Another advantage is that there is no background wind tunnel noise inherent in an experiment. It is no exaggeration to say that numerical simulation is, perhaps, the simplest way to reproduce the tones of a truly isolated airfoil. Finally, numerical simulation provides a full set of space-time data of the flow and acoustic fields that would facilitate the investigation of the tone generation mechanisms.

15.4.1 Direct Numerical Simulation

15.4.1.1 Computational Model
In this study, a computational model consisting of a NACA0012 airfoil at zero angle of attack is used. The chord width, C, is 0.1 m, but the sharp trailing edge is truncated and fitted with a round trailing edge. The airfoil is placed in a uniform stream with a velocity U ranging from 30 m/s to 75 m/s. The kinematic viscosity of air, v at 15°C, is taken to be 0.145×10^{-4} m^2/s. The corresponding Reynolds number, based on chord width, varies from 2×10^5 to 5×10^5.

The governing equations are the dimensionless Navier-Stokes equations in two dimensions. Here the length scale is C (chord width), velocity scale is a_∞ (ambient sound speed), time scale is C/a_∞, density scale is ρ_∞ (ambient gas density), pressure and stresses scale is $\rho_\infty a_\infty^2$. These equations (in Cartesian tensor subscript notation) are

$$\frac{\partial \rho}{\partial t} + \rho \frac{\partial u_j}{\partial x_j} + u_j \frac{\partial \rho}{\partial x_j} = 0 \tag{15.42}$$

$$\frac{\partial u_i}{\partial t} + u_j \frac{\partial u_i}{\partial x_j} = -\frac{1}{\rho} \frac{\partial p}{\partial x_i} + \frac{1}{\rho} \frac{\partial \tau_{ij}}{\partial x_j} \tag{15.43}$$

$$\frac{\partial p}{\partial t} + u_j \frac{\partial p}{\partial x_j} + \gamma p \frac{\partial u_j}{\partial x_j} = 0, \tag{15.44}$$

$$\text{where } \tau_{ij} = \frac{M}{R_e} \left(\frac{\partial u_i}{\partial x_j} + \frac{\partial u_j}{\partial x_i} \right), \qquad R_e = \frac{CU}{v}, M = \frac{U}{a_\infty}. \tag{15.45}$$

U is the free stream velocity, M is the Mach number, and R_e is the Reynolds number. Both viscous dissipation and heat conduction are neglected in the energy equation. Away from the airfoil, viscous effects are unimportant. In this part of the computational domain, viscous terms are dropped, that is, the Euler equations replace the Navier-Stokes equations as the governing equations.

15.4.1.2 Grid Design and Computational Domain
A well-designed grid is crucial to an accurate numerical simulation. For flow past an airfoil, accurate resolution of the boundary layer and wake flow is very important. To be able to do so, the use of a body-fitted grid is most advantageous.

The problem at hand is a multiscales problem. The smallest length scale of the problem that needs to be resolved is in the viscous boundary layer and in the wake. In this work, a multisize mesh is used in the time marching computation. Figure 15.24 shows the overall computational domain. It extends to a distance of 10 chordlengths in front of the leading edge of the airfoil and 30 chords downstream. The lateral dimension is 20 chords. The reason for choosing such a large computational domain is to provide sufficient distance for the wake flow and vortices to develop fully and to decay sufficiently before exiting the outflow boundary. A body-fitted elliptic coordinate system (μ, θ) and mesh, as discussed in the Section 15.3, is used around the

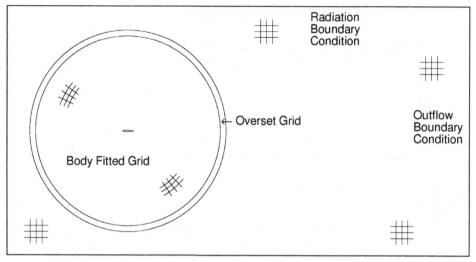

Figure 15.24. Computational domain in physical space for uniform flow past an airfoil.

airfoil. This region extends to a radius of 8 chords (for large μ, the elliptic coordinate system evolves into a polar coordinate system). Outside this nearly circular region, a Cartesian grid is used. This is illustrated in Figure 15.24. An overset grid region is created to encircle the body-fitted mesh. It is used to transfer computational information between the Cartesian grid and the body-fitted grid. The overset grid method of Chapter 13 is implemented here.

The boundary layer flow adjacent to the airfoil and the wake flow are believed to be responsible for the generation of airfoil tones. It is, therefore, important to provide sufficient spatial resolution in computing the flow in these thin shear layers. For this reason, in the grid design, the smallest size meshes are assigned to these regions. The grid layout in the $\mu - \theta$ plane (the computational domain for the elliptic coordinate region) is as shown in Figure 15.25. In this figure, the range of θ is from $-\pi$ to π with periodic boundary conditions imposed on the left and right boundaries. The airfoil surface is on the line $\mu = 0$. This computational domain is divided into seventeen subdomains, symmetric about the line $\theta = 0$. The subdomains are labeled by a capital letter "A," "B," "C," etc. In the physical plane, these subdomains are shown in Figures 15.26 and 15.27. Subdomains "B" and "C" contain the boundary layer. Subdomains "A" and "D" contain the near wake. Subdomains "F," "H," and "M" contain the intermediate wake. The far wake is in subdomain "N." Subdomain "N" extends to the edge of the Cartesian grid region. Table 15.1 provides the mesh size assigned to each of the computational subdomains with $\Delta\theta_0 = 0.00218$ and $\Delta\mu_0 = 0.0015$. The mesh size of the Cartesian grid is uniform with $\Delta x = \Delta y = 0.053$.

15.4.1.3 Computational Algorithm and Boundary Conditions
For time marching computation in the elliptic coordinates region of the computational domain, Eqs. (15.42) to (15.45) are first rewritten in the $\mu - \theta$ coordinates as in Section 5.6. The derivatives are then discretized according to the 7-point stencil DRP scheme. At the mesh-size change boundaries, the multi-size-mesh multi-time-step

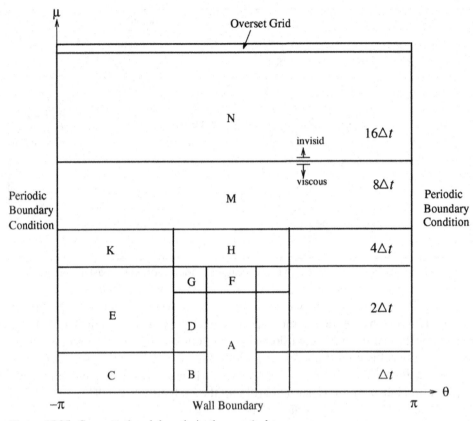

Figure 15.25. Computational domain in the $\mu - \theta$ plane.

DRP scheme of Chapter 12 is implemented. This scheme allows a smooth connection and transfer of the computed results using two different size meshes on the two sides of the boundary. It also allows a change in the size of the time step (see Figure 15.25). This makes it possible to maintain numerical stability without being forced to use very small time steps in subdomains where the mesh size is large. This results in a very efficient computation without loss of accuracy as noted in Chapter 12.

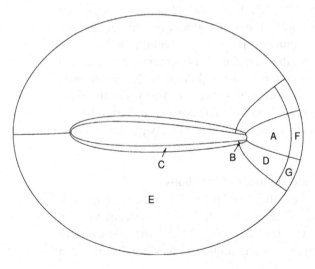

Figure 15.26. Various subdomains in the physical space.

Table 15.1. *Mesh size in different subdomains*

Subdomain	$\Delta\theta$	$\Delta\mu$
A	$\Delta\theta_0$	$\Delta\mu_0$
B, D, F, G	$2\,\Delta\theta_0$	$\Delta\mu_0$
C	$4\,\Delta\theta_0$	$\Delta\mu_0$
E, K	$4\,\Delta\theta_0$	$2\,\Delta\mu_0$
H, M, N	$2\,\Delta\theta_0$	$2\,\Delta\mu_0$

On the left, the top and bottom boundaries of the computation domain shown in Figure 15.24, the radiation boundary conditions of Section 6.5 are imposed. On the right boundary, outflow boundary conditions of Section 6.5 are enforced. On the airfoil surface, the no-slip boundary conditions are used. This, however, will create an overdetermined system of equations if all the discretized governing equations are required to be satisfied at the boundary points. To avoid this problem, the two momentum equations of the Navier-Stokes equations are not enforced at the wall boundary points. The continuity and energy equations are, nevertheless, computed to provide values of density, ρ, and pressure, p, at the airfoil surface. This way of enforcing the no-slip boundary conditions was discussed at the end of Section 6.7.

15.4.1.4 Distribution of Artificial Selective Damping

The addition of artificial selective damping is absolutely necessary when a central difference time marching scheme is used for computation. A central difference scheme has no intrinsic damping. It has to rely on artificial damping to eliminate spurious waves and to maintain numerical stability. To eliminate spurious waves requires background damping over the entire computational domain. The idea is to damp out spurious waves as they propagate across the computational domain. The

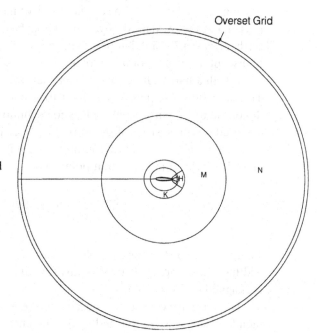

Figure 15.27. Subdomains H, K, M, and N in physical space.

damping terms to be added to the governing equations in the computation plane has the following form:

$$\frac{\partial u}{\partial t} + \cdots = -\frac{1}{R_{\Delta\mu}} \sum_{j=-3}^{3} d_j \left(u_{\ell+j,m} + u_{\ell,m+j} \right),$$ (15.46)

where t is nondimensionalized by $\Delta\mu/a_\infty$. $R_{\Delta\mu}$ is the mesh Reynolds number as discussed in Chapter 7.

To maintain numerical stability, a very different strategy is required. Spurious waves are often generated and amplified at boundaries and surfaces of discontinuities. They include solid surfaces, external boundaries of a computational domain, and mesh-size-change interfaces. To suppress the generation and amplification of spurious waves, especially grid-to-grid oscillations, extra damping is imposed along narrow strips of the computational domain bordering the boundaries and surfaces of discontinuities.

Background damping is usually imposed through the use of the damping curve with a half-width $\sigma = 0.2\pi$ (see Section 7.2). This damping curve targets the high wave numbers of the spectrum and introduces minimal damping to the low wave numbers. The amount of damping imposed is specified by the value assigned to the inverse mesh Reynolds number $R_\Delta^{-1} = (a_0 \Delta / \nu_a)^{-1}$. A recommended value to use is $R_\Delta^{-1} \cong 0.05$. The artificial selective damping stencil is designed to automatically adjust to the mesh size used, so there is no need to make provision to adjust the value of R_Δ^{-1}, because of the different mesh size used in different subdomains.

Near a surface of discontinuity such as a wall, a typical stencil arrangement is to impose extra damping as shown in Figure 7.7. Since all damping stencils are symmetric, the first mesh points to apply artificial selective damping are those on the first row immediately away from the surface of discontinuity. Here, a 3-point damping stencil is used. For the next row of mesh points, a 5-point damping stencil is used. For the third mesh row from the surface of discontinuity and for mesh points further away, a 7-point damping stencil is used. In the direction parallel to the surface of discontinuity, a 7-point damping stencil is used whenever possible. The damping curve with a half-width equal to 0.3π (see Section 7.2) would be most appropriate. To make sure damping is confined to a narrow strip adjacent to the surface of discontinuity, the inverse mesh Reynolds number is taken to be a Gaussian function centered on the surface of discontinuity. Specifically, if d is the distance of a mesh point from the surface of discontinuity, then it is recommended to take the inverse mesh Reynolds number at that point to be

$$\frac{1}{R_\Delta} = A e^{-(\ln 2)(d/\kappa\Delta)^2},$$ (15.47)

where for the airfoil tone problem, A is taken to be 0.1 and κ, which controls the width of the damping strip, is usually set equal to 3 or 4. For the problem at hand, κ is assigned the value of 4.

Near a sharp corner of a surface of discontinuity, additional damping is recommended. This is accomplished by adding an extra artificial selective damping term

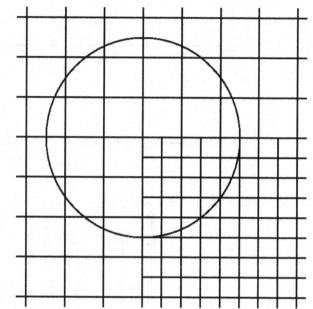

Figure 15.28. Extra damping at the corner of two subdomains with different-size meshes.

to the equations of motion. Let the coordinates of the corner point be (μ_0, θ_0), the inverse mesh Reynolds number of the extra damping term is taken to be

$$\frac{1}{R_\Delta} = Be^{-(\ln 2)\left[(\mu-\mu_0)^2+(\theta-\theta_0)^2\right]/(\kappa\Delta)^2}, \qquad (15.48)$$

where $\Delta^2 = (\Delta\mu^2 + \Delta\theta^2)$. The extra damping imposed at a corner point of a mesh size discontinuity is illustrated in Figure 15.28.

For the airfoil problem, the distribution of extra damping is shown in Figure 15.29. According to this figure, extra damping is added along the surface of the airfoil and along mesh-size-change boundaries. Additional damping is used at corner points. At the overset grid region, extra damping is applied to the mesh points just inside the $\mu = \mu_{max}$ boundary, where μ_{max} is the largest value of μ in the computational domain. No extra damping is imposed outside in the Cartesian grid domain.

15.4.1.5 Selection of Time Step

The time step used in the computation is selected to meet the stability and accuracy requirements of the subdomain with the smallest size mesh. In the computation domain, the governing partial differential equations have variable coefficients. It is impractical to perform numerical stability analysis for equations with variable coefficients. Here, the frozen coefficient approximation is invoked. Under this approximation, the coefficients are treated as constants. To further simplify the analysis, the viscous terms that are diffusive in nature will be neglected. In the $\mu - \theta$ plane, the linearized governing equations are Eq. (5.61), which may be written in the following form:

$$\frac{\partial \mathbf{u}}{\partial t} + \mathbf{A}\frac{\partial \mathbf{u}}{\partial \mu} + \mathbf{B}\frac{\partial \mathbf{u}}{\partial \theta} = 0, \qquad (15.49)$$

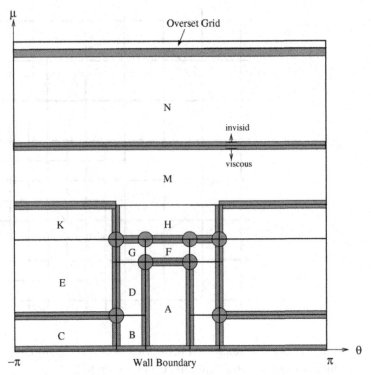

Figure 15.29. Distribution of extra artificial selective damping (shown shaded).

where

$$
\mathbf{u} = \begin{bmatrix} \rho \\ u \\ v \\ p \end{bmatrix}, \quad
\mathbf{A} = \begin{bmatrix}
\bar{u}\mu_x + \bar{v}\mu_y & \bar{\rho}\mu_x & \bar{\rho}\mu_y & 0 \\
0 & \bar{u}\mu_x + \bar{v}\mu_y & 0 & \mu_x/\bar{\rho} \\
0 & 0 & \bar{u}\mu_x + \bar{v}\mu_y & \mu_y/\bar{\rho} \\
0 & \gamma\bar{p}\mu_x & \gamma\bar{p}\mu_y & \bar{u}\mu_x + \bar{v}\mu_y
\end{bmatrix}
$$

$$
\mathbf{B} = \begin{bmatrix}
\bar{u}\theta_x + \bar{v}\theta_y & \bar{\rho}\theta_x & \bar{\rho}\theta_y & 0 \\
0 & \bar{u}\theta_x + \bar{v}\theta_y & 0 & \theta_x/\bar{\rho} \\
0 & 0 & \bar{u}\theta_x + \bar{v}\theta_y & \theta_y/\bar{\rho} \\
0 & \gamma\bar{p}\theta_x & \gamma\bar{p}\theta_y & \bar{u}\theta_x + \bar{v}\theta_y
\end{bmatrix},
$$

where subscripts x and y indicate partial derivatives. $\bar{u}, \bar{v}, \bar{\rho}$, and $\bar{p}$ are local values. Coefficient matrices $\mathbf{A}$ and $\mathbf{B}$ are to be regarded as constants in the analysis.

Numerical stability analysis follows the procedure of Section 5.3. For this purpose, the Fourier-Laplace transform is applied to Eq. (15.49) locally. The Fourier-Laplace transform in the $\mu - \theta$ plane is defined by

$$
\tilde{u}(\alpha, \beta, \omega) = \frac{1}{(2\pi)^3} \int \int \int_{-\infty}^{\infty} u(\mu, \theta, t)\, e^{-i(\alpha\mu + \beta\theta - \omega t)}\, d\mu\, d\theta\, dt.
$$

The Fourier-Laplace transform of Eq. (15.49) is

$$
\mathbf{K}\tilde{\mathbf{u}} = \tilde{\mathbf{G}}, \tag{15.50}
$$

where

$$
\mathbf{K} = \begin{bmatrix}
\omega - (\hat{\alpha}\bar{u} + \hat{\beta}\bar{v}) & -\bar{\rho}\hat{\alpha} & -\bar{\rho}\hat{\beta} & 0 \\
0 & \omega - (\hat{\alpha}\bar{u} + \hat{\beta}\bar{v}) & 0 & -\hat{\alpha}/\bar{\rho} \\
0 & 0 & \omega - (\hat{\alpha}\bar{u} + \hat{\beta}\bar{v}) & -\hat{\beta}/\bar{\rho} \\
0 & -\gamma\bar{p}\hat{\alpha} & -\gamma\bar{p}\hat{\beta} & \omega - (\hat{\alpha}\bar{u} + \hat{\beta}\bar{v})
\end{bmatrix},
$$

where $\hat{\alpha} = \alpha\mu_x + \beta\mu_y$, $\hat{\beta} = \alpha\theta_x + \beta\theta_y$. The eigenvalues of matrix $\mathbf{K}$ are as follows:

$$
\lambda_1 = \lambda_2 = [\omega - (\hat{\alpha}u + \hat{\beta}v)]
$$

$$
\lambda_{3,4} = \omega - (\hat{\alpha}\bar{u} + \hat{\beta}\bar{v}) \pm (\gamma\bar{p}/\bar{\rho})^{1/2}(\hat{\alpha}^2 + \hat{\beta}^2)^{1/2}.
$$

The dispersion relation that imposes the most stringent stability requirement is that of the acoustic wave $\lambda_4 = 0$. On proceeding as in Section 5.3, it is straightforward to derive the following maximum time step formula should the 7-point stencil DRP scheme be used for the computation.

$$
\Delta t_{max} = \frac{0.4}{\frac{|\bar{u}\mu_x + \bar{v}\mu_y|}{\Delta\mu} + \frac{|\bar{u}\theta_x + \bar{v}\theta_y|}{\Delta\theta} + \left(\frac{\gamma\bar{p}}{\bar{\rho}}\right)^{1/2}\left[\left(\frac{\mu_x}{\Delta\mu} + \frac{\theta_x}{\Delta\theta}\right)^2 + \left(\frac{\mu_y}{\Delta\mu} + \frac{\theta_y}{\Delta\theta}\right)^2\right]^{1/2}}. \tag{15.51}
$$

15.4.1.6 Starting Conditions

In the course of performing the airfoil tone simulations, it has been found that the following simple starting conditions have been very successful:

$$
t = 0, \quad u = U_{incoming}, \quad v = 0, \quad \rho = 1, \quad p = \frac{1}{\gamma}.
$$

The uniform starting condition is applied throughout the entire computational domain. In order to comply with the no-slip boundary conditions on the airfoil surface, the boundary conditions on the airfoil surface, $\mu = 0$, is taken to be

$$
\begin{bmatrix} u \\ v \end{bmatrix} = \begin{cases} \begin{bmatrix} U \\ 0 \end{bmatrix}\left[\frac{3(\tau-t)^2}{\tau^2} - \frac{2(\tau-t)^3}{\tau^3}\right], & t \leq \tau; \\ \begin{bmatrix} 0 \\ 0 \end{bmatrix}, & t > \tau. \end{cases}
$$

Here, τ is recommended to be five or six times of the tone period. The no-slip boundary conditions are attained at $t = \tau$.

15.4.1.7 Grid Refinement

To ensure a numerical simulation has adequate resolution, ideally, grid refinement should be performed. Essentially, one would like to show that the computed results are grid size independent. However, for large-scale simulation this is not always feasible. Often, this is because of the complexity of the computer code. Another reason is that some simulations may require exceedingly long computer run time. This invariably discourages the performance of grid refinement. For the airfoil tone simulations, it is a medium-size simulation so that grid refinements in relation to the numerical results discussed below have been carried out. In the grid refinement exercise, all the mesh sizes in the elliptical region of the computational domain are reduced by half. Two test cases are chosen with Reynolds numbers equal to 2×10^5

Table 15.2. *Airfoil dimensions and trailing edge thicknesses*

Airfoil	% truncation	Width of chord in cm	Trailing edge thickness in cm
NACA0012	0	10	0
#1	0.5	9.956	0.014
#2	2.0	9.82	0.055
#3	10.0	9.1	0.248

and 4×10^5. For the low Reynolds number case, the tone frequency is found to remain essentially the same. For the higher Reynolds number case, the tone frequency reduces by 2 percent with the finer mesh. On assuming that the difference is proportional to the Reynolds number of a simulation, it is estimated that the numerical results should be accurate to about 3 percent for the highest Reynolds number simulations.

15.4.2 Numerical Results

A computer code based on the methodologies described above has been developed. The code is then used to investigate airfoil tones of three NACA0012 airfoils with truncated trailing edges as proposed in Section 15.3.3. The dimensions and trailing edge thicknesses of the airfoils are given in Table 15.2. The trailing edge profiles of these airfoils are shown in Figure 15.30.

In this study, the original airfoil chordwidth is 0.1 m. The airfoil is placed in a uniform stream with a velocity U. The Reynolds number range of the present study is from 2×10^5 to 5×10^5. For a flat plate with the same length, the displacement thickness at the trailing edge at Reynolds number of 2×10^5 is $\delta^* = 1.72(Cv/U)^{\frac{1}{2}} = 3.85 \times 10^{-4}$ m. Thus, R_{δ^*} is equal to 770. For this Reynolds number, the wake flow is laminar in the absence of high-level ambient turbulence. In the present simulations,

(a)

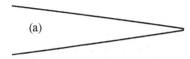

(b)

Figure 15.30. Trailing edge profiles of the three NACA0012 airfoils under study. (a) 0.5 percent truncation, (b) 2 percent truncation, (c) 10 percent truncation.

(c)

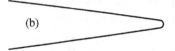

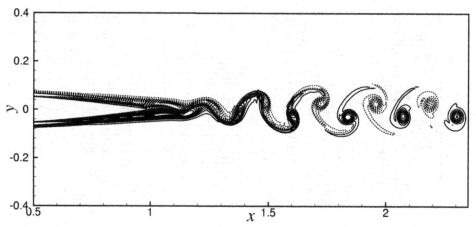

Figure 15.31. Vorticity field in the wake of Airfoil #2 at Re $= 2 \times 10^5$ showing the rolling up of the wake into discrete vortices.

this remains true even at the highest Reynolds number of the study. However, it may not be true for the three-dimensional simulation. In this study, the angle of attack is set at zero degree. This is to keep the flow and, hence, the tone generation as simple as possible. At a high angle of attack, separation bubble and open separation often occur on the suction side of the airfoil. Under such conditions, other tone generation mechanisms may become possible, but this is beyond the scope of this study.

The governing equations are the dimensionless Navier-Stokes equations in two dimensions and the energy equation. Here, the length scale is C (chord width), velocity scale is a_∞ (ambient sound speed), time scale is C/a_∞, density scale is ρ_∞ (ambient gas density), pressure and stress scales are $\rho_\infty a_\infty^2$. These equations are as follows:

$$\frac{\partial \rho}{\partial t} + \rho \frac{\partial u_j}{\partial x_j} + u_j \frac{\partial \rho}{\partial x_j} = 0 \tag{15.52}$$

$$\frac{\partial u_i}{\partial t} + u_j \frac{\partial u_i}{\partial x_j} = -\frac{1}{\rho} \frac{\partial p}{\partial x_i} + \frac{1}{\rho} \frac{\partial \tau_{ij}}{\partial x_j} \tag{15.53}$$

$$\frac{\partial p}{\partial t} + u_j \frac{\partial p}{\partial x_j} + \gamma p \frac{\partial u_j}{\partial x_j} = 0, \tag{15.54}$$

$$\text{where } \tau_{ij} = \frac{M}{\text{Re}} \left(\frac{\partial u_i}{\partial x_j} + \frac{\partial u_j}{\partial x_i} \right); \quad \text{Re} = \frac{CU}{\upsilon}; \quad M = \frac{U}{a_\infty}. \tag{15.55}$$

U is the free stream velocity, M is the Mach number, Re is the Reynolds number based on chord width. Both viscous dissipation and heat conduction are neglected in the energy equation as the Mach number is small.

15.4.2.1 Flow Field

At zero degree angle of attack, the computed boundary layer flow on the airfoil remains attached almost to the trailing edge. The boundary layer flow is steady while the flow in the wake is oscillatory. Figure 15.31 shows a plot of the computed instantaneous vorticity field around Airfoil #2 at a Reynolds number of 2×10^5.

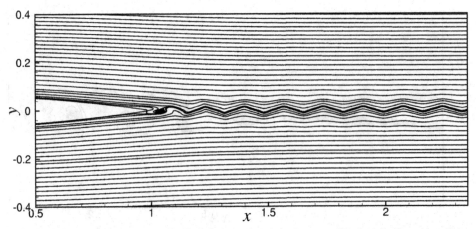

Figure 15.32. Instantaneous streamline pattern showing strong oscillations in the wake of the airfoil. Airfoil #2, Re $= 4 \times 10^5$.

The leading edge of the airfoil is placed at $x = 0$. This is typical of all the cases investigated. The flow leaves the trailing edge of the airfoil smoothly. There is no vortex shedding at the trailing edge, regardless of previous speculation (Paterson et al., 1973). The oscillations in the wake intensify in the downstream direction until it rolls up to form discrete vortices further downstream.

Figure 15.32 shows an instantaneous streamline pattern of the wake of Airfoil #2 at a Reynolds number of 4×10^5. The wake oscillation is prominent in this figure. The wavelengths of the oscillations appear to be fairly regular. This suggests that there is flow instability in the wake.

The mean velocity deficit profile in the wake resembles a Gaussian function. The half-width of the wake, b, near the airfoil trailing edge undergoes large variation in the flow direction. Figure 15.33 shows the variation of the half-width as a function of downstream distance for Airfoil #2 at a Reynolds number of 2×10^5. The width of the wake begins to shrink right downstream of the airfoil trailing edge until a

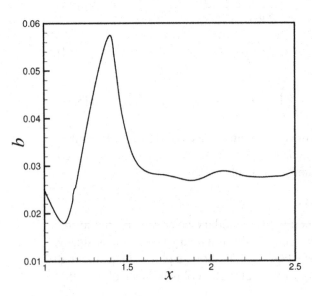

Figure 15.33. Spatial distribution of the half-width of the wake for Airfoil #2 at Re $= 2 \times 10^5$.

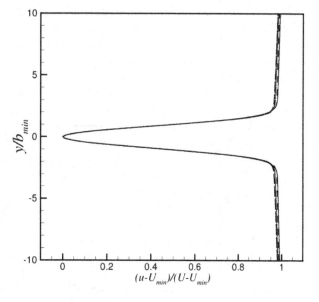

Figure 15.34. Similarity of velocity deficit profile at $b = b_{min}$ in the wake of Airfoil #2. Reynolds number varies from 2×10^5 to 5×10^5 at 0.5×10^5 interval.

minimum is reached at $b = b_{min}$. It grows rapidly further downstream, reaching a maximum before decreasing and attaining a nearly asymptotic value.

The velocity deficit profiles of the wake exhibits self-similarity. Figure 15.34 shows a collapse of the velocity profiles at minimum half-width over a large range of Reynolds numbers. In this figure, the variable $(u - U_{min})/(U - U_{min})$ is plotted against y/b_{min}, where U_{min} is the minimum velocity of the mean flow profile and b is the half-width. In Figure 15.34, data from Reynolds number of 2×10^5 to 5×10^5 at 0.5×10^5 intervals are included. There is a nearly perfect match of the set of seven profiles.

The similarity property of the wake velocity profiles suggests the possibility of the existence of a simple correlation of the minimum half-width of the wake with flow Reynolds number (Re). Figure 15.35 shows a plot of $(b_{min}/C)\mathrm{Re}^{1/2}$ against Re (C is the chord width) for Airfoil #2. It is seen that the measured values of this quantity from the numerical simulations are nearly independent of Reynolds number over the entire range of Reynolds number of the present study. Thus, the minimum half-width of the wake, b_{min}, of Airfoil #2 has the following form of Reynolds number dependence:

$$\frac{b_{min}}{C} \mathrm{Re}^{\frac{1}{2}} \approx 8.2. \qquad \text{(Airfoil #2).} \qquad (15.56)$$

A similar analysis has been performed for Airfoil #1. Figure 15.36 shows that the quantity $(b_{min}/C)\mathrm{Re}^{1/2}$ again is approximately a constant. In this case, the value b_{min} and Reynolds number are related by,

$$\frac{b_{min}}{C} \mathrm{Re}^{\frac{1}{2}} \approx 7.45. \qquad \text{(Airfoil #1).} \qquad (15.57)$$

15.4.2.2 Acoustics

In all the simulations that have been performed for Airfoils #1, #2, and #3 (see Table 15.2). only a single tone is detected. Figure 15.37 shows a typical airfoil trailing

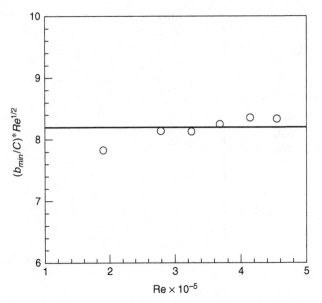

Figure 15.35. Relationship between b_{min} of the wake and Reynolds number (Airfoil #2).

edge noise spectrum. This is the noise spectrum of Airfoil #1 at a Reynolds number of 4×10^5. The measurement point is at (1.08, 0.5). Since there is only one single tone per computer simulation, it makes the finding the same as Nash et al. (1999).

Figure 15.38 shows the computed tone frequencies versus flow velocities for Airfoil #1 according to the results of our numerical simulations. This airfoil has a 0.5 percent truncation. It is nearly a sharp trailing edge airfoil. As shown in this figure, all the data points lie practically on a straight line parallel to the Paterson formula (the full line). The difference between the simulation results and the Paterson formula

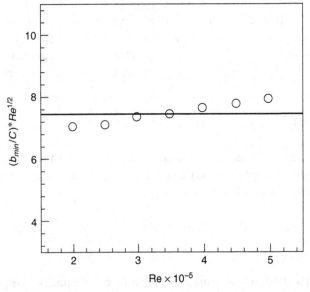

Figure 15.36. Relationship between b_{min} of the wake and Reynolds number (Airfoil #1).

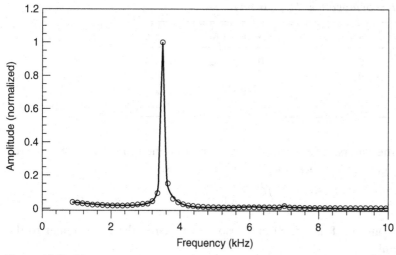

Figure 15.37. The measured noise spectrum for Airfoil #1 at Re $= 4 \times 10^5$.

is very small. On the basis of what is shown in Figure 15.38, it is believed that the simulations, for all intents and purposes, reproduce the empirical Paterson formula. The good agreement with the Paterson formula not only is a proof of the validity of the simulations, but also is an assurance that numerical simulations do contain the essential physics of airfoil tone generation.

Figure 15.39 shows the variations of the Strouhal numbers of the computed tones with Reynolds number for the three airfoils under consideration. The data for each airfoil lies approximately on a straight line. The straight lines are parallel to each

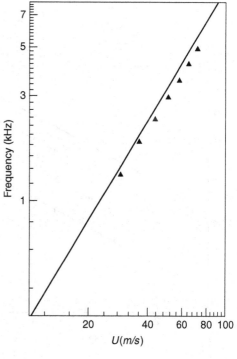

Figure 15.38. Measured tone frequencies at different flow velocities for Airfoil #1. Straight line is the Paterson formula.

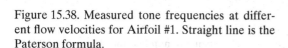

Table 15.3. *Values of proportionality constant* K

Airfoil	% truncation	Value of K
NACA0012	0	0.011 (Paterson formula)
#1	0.5	0.00973
#2	2.0	0.00835
#3	10.0	0.00717

other and to the Paterson formula (shown as the full line in Figure 15.39). All the four lines in Figure 15.39 fit a single formula as follows:

$$\frac{fC}{U} = K\mathrm{Re}^{\frac{1}{2}}, \tag{15.58}$$

where K is a constant. Eq. (15.58) may also be written in the form similar to the Paterson formula, i.e.,

$$f = K\frac{U^{\frac{3}{2}}}{(C\upsilon)^{\frac{1}{2}}}. \tag{15.59}$$

The values of constant K for each airfoil have been determined by best fit to the simulation data. They are given in Table 15.3.

Based on the computed results shown in Figure 15.39, the effect of trailing edge thickness at a given Reynolds number is to lower the tone frequency, that is, the thicker the blunt trailing edge the lower is the tone frequency. This dependence turns out to be similar to that of vortex shedding tones, such as that emitted by a long circular cylinder in a uniform flow. For the long cylinder vortex shedding problem, the tone Strouhal number is approximately equal to 0.2, i.e.,

$$\frac{fD}{U} \approx 0.2 \quad (D \text{ is the diameter}). \tag{15.60}$$

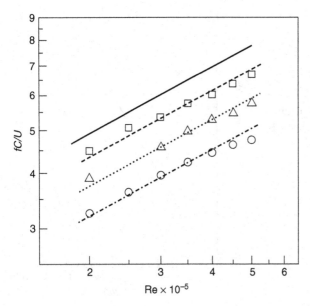

Figure 15.39. Dependence of tone Strouhal number on Reynolds number for Airfoils #1, #2, and #3. – – – –, Airfoil #1, · · · · · · Airfoil #2; — · —, Airfoil #3; ——, Paterson formula.

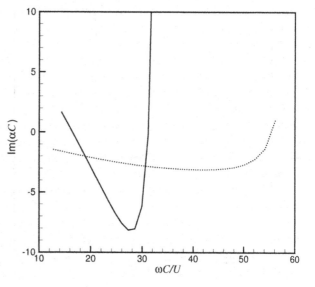

Figure 15.40. Comparison between the spatial growth rates of Kelvin-Helmholtz instability waves at the minimum half-width point in the wake (———) and at a point in the fully developed wake (· · · · · ·).

Therefore, the thicker the blunt object or trailing edge, the lower the tone frequency is. However, note that the physics and tone generation mechanisms are quite different.

15.4.3 Energy Source of Airfoil Tones

A number of investigators, e.g., Tam (1974), Fink (1975), Arbey and Bataille (1983), Lowson et al. (1994), Nash et al. (1999), Desquesnes et al, (2007), Sandberg et al. (2007), Chong and Joseph (2009), and Kingan and Pearse (2009), have suggested that airfoil tones are driven by the instabilities of the boundary layer. Various computations of Tollmien-Schlichting-type instabilities have been performed. The objective is to show that the frequency of the airfoil tone is the same as that of the most amplified instability wave. An examination of the past computation reveals that the spatial growth rate of the Tollmien-Schlichting wave is very small. Only in the separated flow region is there a substantial growth rate. However, all the instability calculations were carried out using the locally parallel flow approximation. Unlike attached boundary layers, the thickness of a separated boundary layer increases rapidly. This makes the appropriateness of the parallel flow approximation questionable. Thus, one should not be too ready to accept the results of past instability calculations without an examination of the approximation adopted in the computation.

In the present study, the NACA0012 airfoil is set at zero degree angle of attack on purpose. At a zero degree angle of attack, there is practically no flow separation on the airfoil until almost the trailing edge. Thus, the spatial growth rate of the Tollmien-Schlichting wave is much smaller than that of the Kelvin-Helmholtz instability in the wake of the airfoil. It is known that the growth rate of Kelvin-Helmholtz instability is inversely proportional to the half-width of the shear layer. In other words, a thin wake is more unstable than a thicker wake. That this is true is shown in Figure 15.40. This figure shows the computed spatial growth rate (antisymmetric mode) over the distance of a chord for Airfoil #1 at different instability wave frequencies. The computations are done using the nonparallel flow instability theory of Saric and

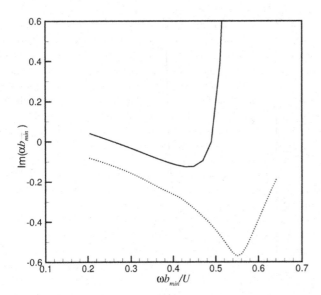

Figure 15.41. Comparison between the spatial growth rates of Kelvin-Helmholtz instability waves at the minimum half-width point using locally parallel ($\cdots\cdots$) and nonparallel (——) instability theories. Airfoil #2, Re $= 3 \times 10^5$.

Nayfeh (1975). The flow in the calculation has a Reynolds number of 3×10^5. The solid curve is for the minimum half-width point of the wake at $x/C = 1.088$ measured from the leading edge of the airfoil. The dotted curve is for a point in the developed wake region ($x/C = 1.556$). The maximum growth rate at the thinnest point of the wake is many times larger than that in the developed wake.

By performing instability computation, it is easy to establish that the maximum growth rate occurs in the wake at the location where the wake half-width is minimum. It is believed that because the growth rate is very large, the instability initiated at this region is ultimately responsible for the generation of airfoil tones. The necessary computed instability results to support this hypothesis will now be provided. Because of the rapid growth of the thickness of the wake at the minimum half-width location, Tam and Ju (2012) recognized that the nonparallel flow instability wave theory of Saric and Nayfeh (1975) should be used to perform all instability computations. Their method is a perturbation method. The zeroth order is the locally parallel flow solution. A first-order correction, to account for the spreading of the mean flow, is then computed. The combined growth rate is taken as the nonparallel spatial growth rate. Figure 15.41 shows the difference between parallel and nonparallel spatial growth rates at the minimum wake half-width point for Airfoil #2 at a Reynolds number of 3×10^5. It is straightforward to see from this figure that the nonparallel flow growth rate is smaller. Furthermore, the frequencies of the most amplified wave computed using parallel and nonparallel flow theories are quite different.

Figure 15.42 shows the angular frequency of the Kelvin-Helmholtz instability wave with the highest amplification rate at the narrowest point of the wake of Airfoil #2 as a function of Reynolds number. When nondimensionalized by minimum half-width $b_{\min}$ and free stream velocity U, the quantity ($\omega\, b_{\min}/U$) at the maximum growth rate is nearly independent of the Reynolds number. A good fit to the data is as follows"

$$\left(\frac{\omega b_{\min}}{U}\right)_{\text{maximum growth}} \approx 4.3. \qquad \text{(Airfoil \#2).} \qquad (15.61)$$

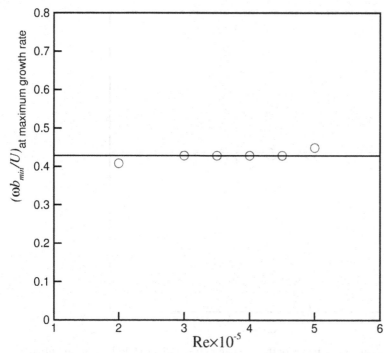

Figure 15.42. The angular frequency of the most amplified Kelvin-Helmholtz instability wave at the minimum half-width point of the wake as a function of Reynolds number. Airfoil #2.

Now, for this airfoil, b_{min} and Reynolds number are related by Eq. (15.56). By eliminating b_{min} from Eq. (15.56) and Eq. (15.61), it is found that

$$f = \frac{0.00835U^{1.5}}{(Cv)^{\frac{1}{2}}}. \tag{15.62}$$

Eq. (15.62) is exactly the same as Eq. (15.59) for Airfoil #2. The K value obtained directly from numerical simulations, displayed in Table 15.3, is in agreement with that from hydrodynamic instability consideration.

A similar nonparallel instability analysis for Airfoil #1 has also been carried out. Figure 15.43 shows the angular frequencies corresponding to that at maximum growth rate nondimensionalized by b_{min} and U as a function of the Reynolds number. An approximate relationship similar to Eq. (15.61) is

$$\frac{\omega b_{min}}{U} \approx 4.25. \qquad \text{(Airfoil \#1).} \tag{15.63}$$

Thus, by eliminating b_{min} from Eqs. (15.57) and (15.63), a tone frequency formula based on wake instability consideration for this airfoil is

$$f = \frac{0.00910U^{1.5}}{(Cv)^{\frac{1}{2}}}. \tag{15.64}$$

The proportionality constant in Eq. (15.64) differs slightly from that found from direct numerical simulation as given in Table 5.3. Figure 15.44 shows a direct comparison between the tone frequencies measured directly from numerical simulations

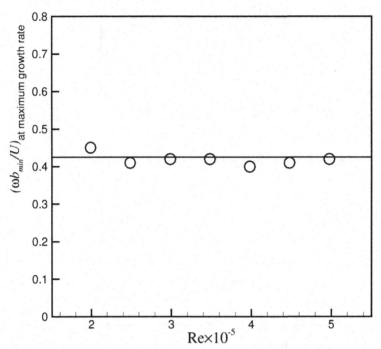

Figure 15.43. The angular frequency of the most amplified Kelvin-Helmholtz instability wave at the minimum half-width point of the wake as a function of Reynolds number. Airfoil #1.

(triangles) and the most unstable frequencies in the wake at minimum half-width (straight line). The agreement is quite good. It is, therefore, believed that the good agreements for Airfoils #1 and #2 lend support to the validity of the proposition that the energy source of airfoil tones at zero degree angle of attack is near wake flow instability.

15.4.4 Sound Generation Processes

In Section 15.4.3, it was demonstrated that the energy source responsible for the generation of airfoil tones is near-wake instability. These are antisymmetric instabilities. However, the flow is at a low subsonic Mach number. It is known that low subsonic flow instabilities are, by themselves, not strong or efficient noise radiators. In this section, the results of a study of the space-time data of the numerical simulations are reported. The objective of the study is to identify the tone generation processes. To facilitate this effort, the fluctuating pressure field p', defined by,

$$p' = p - \overline{p}, \tag{15.65}$$

where $\overline{p}$ is the time-averaged pressure is first computed. Positive p' indicates compression. Negative p' corresponds to rarefactions of the acoustic disturbances. Of special interest is the contour $p' = 0$. Following Tam and Ju (2012) this contour will be used as an indicator of the acoustic wave front. The propagation of this contour from the airfoil wake to the far field is tracked. The creation and spreading of the wave front provides the clues needed for identifying the tone generation processes.

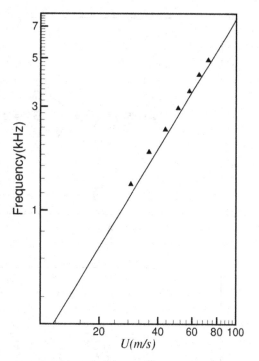

Figure 15.44. Comparison between DNS tone frequencies (▲) and most amplified instability wave frequencies (straight line). Airfoil #1.

A study of the space-time data reveals that the tone generation processes are rather complex. There is a dominant mechanism involving the interaction between the near-wake oscillations driven by wake instability and the airfoil trailing edge. However, there are also secondary mechanisms arising from flow and vortex adjustments in the near wake. Details of these tone generation processes can be understood by following the evolution of the wave front contour $p' = 0$ in time. Figure 15.45 illustrates the dominant tone generation process in space and time. Airfoil #2 (2 percent truncation) at $Re = 4 \times 10^5$ is used in this simulation. Figure 15.45a may be regarded as the beginning of a tone generation cycle. The trailing edge of the airfoil is shown on the left center of the figure. The flow is from left to right. Thus, the wake, defined by two $p' = 0$ contours, lies to the right of the airfoil. The near wake is highly unstable. The instability causes the wake to oscillate as can easily be seen in this figure. The wake is flanked by two rows of vortices in a staggered pattern. The spinning motion of the vortices creates a low-pressure region inside the vortices. Hence, all the vortices are located in the $p' < 0$ regions. These regions are labeled by a $(-)$ symbol. Regions with $p' > 0$ are labeled by a $(+)$ symbol. Shown in Figure 15.45a are wave front contours of $p' = 0$. The near-wake instability is antisymmetric with respect to the x-axis. This leads to an antisymmetric pressure field as shown in Figures 15.45a–15.45h. The pressure field shown in Figure 15.45a corresponds to the beginning of the downward motion of the wake in the region just downstream of the airfoil trailing edge. Since the airfoil is stationary, the downward movement of the near wake leads to the creation of a high-pressure region on the top side of the airfoil trailing edge and a corresponding low-pressure region on the mirror image bottom side. This is shown as two small circular regions right at the trailing edge of the airfoil in Figure 15.45b. A small arrow is inserted there to indicate the direction

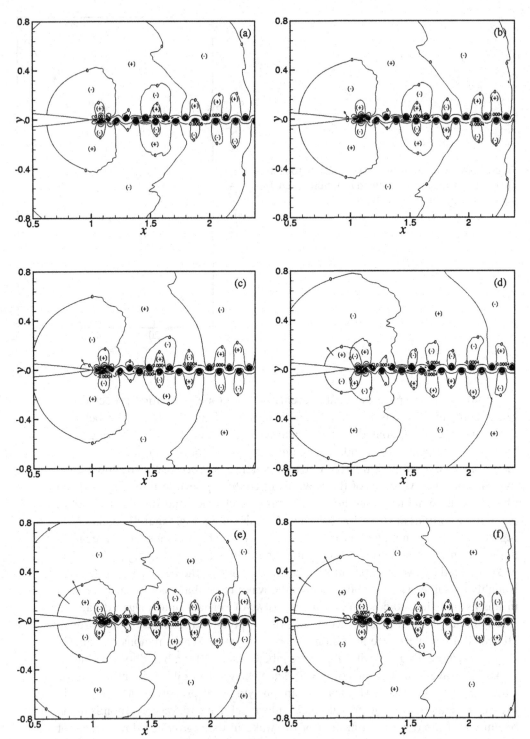

Figure 15.45A. Instantaneous positions of contour $p' = 0$ in the proximity of the near wake for Airfoil #2 at Re $= 4 \times 10^5$. (a) $t = 36.367$T, (b) $t = 36.402$T, (c) $t = 36.526$T, (d) $t = 36.69$T, (e) $t = 36.789$T, (f) $t = 36.974$T; T = period.

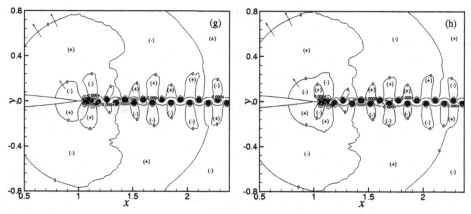

Figure 15.45B. Instantaneous positions of contour $p' = 0$ in the proximity of the near wake for Airfoil #2 at Re= 4×10^5. (g) $t = 37.183T$, (h) $t = 37.209T$; T = period.

of motion of the wave front. The high- and low-pressure regions grow rapidly in time as shown in Figures 15.45c–15.45e. The time difference between Figure 15.45a and Figure 15.45e is nearly half a cycle. Hence, Figure 15.45e to Figure 15.45h repeats a similar sequence of wave front motion, but with positive and negative fluctuating pressure fields interchanged. Each wave front contour in each half-cycle shows its creation at the trailing edge of the airfoil, its expansion in space and time, and its propagation to the far field in all directions.

The dominant tone generation process is observed in all the simulations that have been carried out. In some cases, other less dominant tone generation processes have been observed. These secondary processes may reinforce the primary process or may be important only in certain directions of radiation. By and large, the secondary mechanisms are related to the adjustment of the flow from boundary layer type to free shear wake flow and also the alignment of the wake vortices from their creation to that of a vortex street.

Figure 15.46 illustrates a secondary tone generation process. Figure 15.46a may be taken as the distribution of the wave front contours at the beginning of a cycle for Airfoil #2 at Reynolds number = 2×10^5. At the instant corresponding to Figure 15.46a, the near wake begins to move upward. This creates a high-pressure region on the bottom side of the airfoil trailing edge and, at the same time, a low-pressure region on the mirror image location on the top side. This is shown in Figure 15.46b. Figure 15.46c shows the expansion of the high (bottom)- and low (top)-pressure regions near the trailing edge. At the same time, this figure also shows the expansion of the next two loops; negative-pressure regions on the top half of the figure and positive-pressure regions in the lower half. The three negative-pressure regions that have expanded are labeled as 1, 2, and 3 in this figure. The locations of loops 2 and 3 coincide with the region where the wake oscillations are strongest with largest lateral displacements (see Figure 15.31). However, the wake has not rolled up to form vortices yet. Thus, this secondary tone generation process does not involve the formation of discrete vortices. Figures 15.46d–15.46g show the merging of the loops and the propagation of the wave front to the far field. Therefore, for this case, the adjustment of the near-wake flow assists and contributes to the generation of the observed airfoil tone.

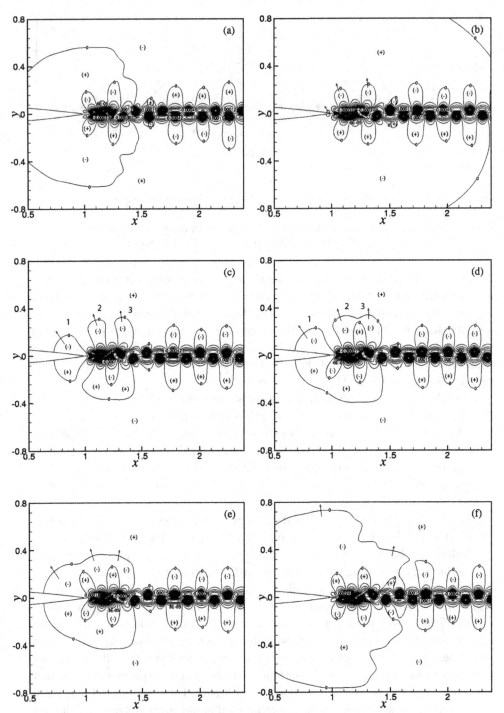

Figure 15.46A. Motion of wave fronts for Airfoil #2 at 2×10^5 Reynolds number illustrating a secondary tone generation process. (a) $t = 23.80T$, (b) $t = 24.02T$, (c) $t = 24.18T$, (d) $t = 24.201T$, (e) $t = 24.218T$, (f) $t = 24.35T$; T = period.

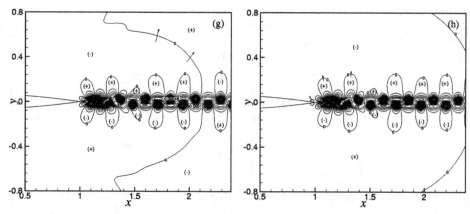

Figure 15.46B. Motion of wave fronts for Airfoil #2 at 2×10^5 Reynolds number illustrating a secondary tone generation process. (g) $t = 24.424T$, (h) $t = 24.53T$; T = period.

For airfoils with a sharp trailing edge, especially when operating at the high end of the moderate Reynolds number range, the wake rolls up into discrete vortices very close to the trailing edge (see Figure 15.47). In these cases, the formation of discrete vortices does contribute to the generation of airfoil tones. This secondary tone generation process is illustrated in the motion of the wave fronts in Figure 15.48. Figure 15.48a may be considered as the beginning of a tone generation cycle. To observe the secondary tone generation process, attention is directed to the shaded area of this figure. Figure 15.48b (at a later time) shows the expansion of the shaded region. In comparison with Figure 15.47, it is seen that the location of the expanded shaded region coincides with the beginning of the wake rolling up into isolated vortices. It is believed, therefore, that the formation of the vortices is what drives the expansion of the shaded region. Figures 15.48c–15.48h track the continued spreading of the sound pulse after it is generated. It appears that the dominant tone generation mechanism due to the interaction of near-wake instability and the airfoil trailing edge is responsible for the radiating sound to the upstream and sideline directions.

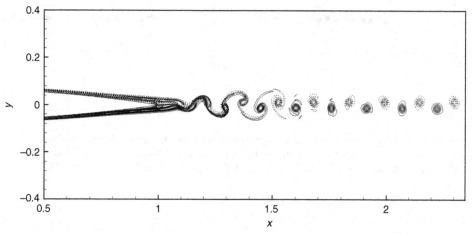

Figure 15.47. Vorticity contours showing the rolling up of the wake into discrete vortices near the sharp trailing edge of an airfoil. Airfoil #1 at $Re = 4 \times 10^5$.

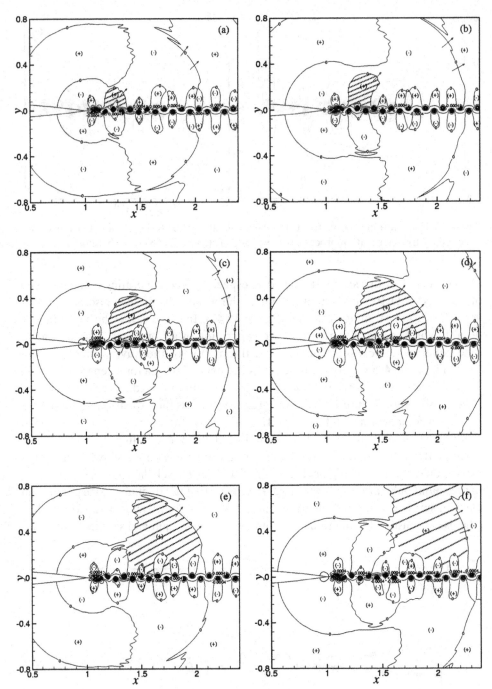

Figure 15.48A. A secondary tone generation mechanism due to the formation of discrete vortices in the wake illustrated by the motion of wave front contours. Airfoil #1 at Re = 4 × 10^5. (a) $t = 50.264T$, (b) $t = 50.41T$, (c) $t = 50.51T$, (d) $t = 50.643T$, (e) $t = 50.795T$, (f) $t = 51.002T$.

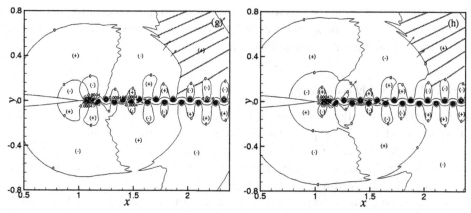

Figure 15.48B. A secondary tone generation mechanism due to the formation of discrete vortices in the wake illustrated by the motion of wave front contours. Airfoil #1 at Re $= 4 \times 10^5$. (g) $t = 51.188T$, (h) $t = 51.254T$.

The secondary tone generation mechanism due to the formation of discrete vortices is responsible for radiating sound to the downstream direction. The two mechanisms are perfectly coordinated and synchronized.

15.5 Computation of Turbulent Flows

In aeroacoustics, turbulence is a principal source of broadband noise. Therefore, research and development of turbulence modeling and turbulence simulation are an integral part of CAA.

Because of the availability of more and more powerful and faster and faster computers, turbulence, nowadays, becomes a favorite activity of large-scale computation. It is known that direct numerical simulation (DNS) of high Reynolds number turbulent flows requires an exceedingly large number of mesh points and long CPU time. On account of such requirements, DNS is presently not considered feasible for solving practical CAA problems. Recently, attention has turned to LES. However, owing to the three-dimensional nature of turbulence, realistically, LES can be carried out only in relatively small computational domains. Simple estimates of mesh and computer requirements would convince even the most ardent proponents of LES that it would be sometime in the future, when much more powerful and faster computers become available, before LES would become a design tool in CAA.

The appeal of DNS and LES is that they can, in principle, compute the entire or a large part of the turbulence spectrum. But, for noise prediction, it is highly plausible that it is not necessary to know everything about turbulence or the entire turbulence spectrum. How much does one really need to know about turbulence before one can calculate turbulence noise is an open question. It appears that if one's primary concern is on the dominant part of the noise spectrum, it is very likely that only the resolution of the most energetic part of the turbulence spectrum is necessary.

Calculating the mean flow velocity profile and other mean quantities of turbulent flows is, in general, important for engineering applications. For mean flow calculation, a practical way is to use the RANS equations with a two-equation turbulence model (e.g., the $k - \varepsilon$ or the $k - \omega$ model) or, if desired, a more advanced model.

The question pertinent to CAA is whether models of a similar level of sophistication could be developed for noise calculation. There is no clear answer to this question. It is a matter of intense research at this time.

15.5.1 Large Eddy Simulation

It was mentioned at the beginning of this chapter that, at present, there are two approaches to LES. One approach uses a subgrid scale model. The idea is to cut off and not to compute the very high wave number part of the turbulence spectrum. To account for the effect of the cutoff spectrum on the larger-scale turbulent motion, a model of the stress-strain rate relation is used. It turns out that a subgrid scale model based on a stress-strain rate relation does not simulate the effect very well. Such a model effectively provides only damping terms to the fluid motion, especially the high wave number components. This prevents the accumulation of small-scale turbulence energy due to the cascade process (energy continues to cascade to the shorter and shorter scales or higher and higher wave number components). The alternative approach is to use numerical damping to perform the same task. A natural way, in the later case, is to use artificial selective damping. For this purpose, the $\sigma = 0.3\pi$ damping curve of Section 7.2 may be used. There is no sharp cutoff for this damping curve. A reasonable choice is to take the cutoff wave number to be approximately at $\alpha \Delta x = 1.6$.

Instead of using artificial selective damping, one may use filtering. Filtering is a way to eliminate high wave number components with minimal effect on the long waves. A one-dimensional numerical filter may be constructed as follows.

Let f_ℓ be the value of a dynamical variable at the ℓ th mesh point. Consider a 7-point stencil filter. A filter should be symmetric to avoid any chance of amplifying the filtered value. Suppose $\hat{f}_\ell$ is the filtered value. The filtered value is related to the original values on the computational stencil by the following relation (assuming a linear filter is used).

$$\hat{f}_\ell = A(f_{\ell+3} + f_{\ell-3}) + B(f_{\ell+2} + f_{\ell-2}) + C(f_{\ell+1} + f_{\ell-1}) + Df_\ell. \quad (15.66)$$

The Fourier transform of Eq. (15.66) is

$$\hat{\tilde{f}} = [2A\cos(3\alpha\Delta x) + 2B\cos(2\alpha\Delta x) + 2C\cos(\alpha\Delta x) + D]\tilde{f} \equiv F(\alpha\Delta x)\tilde{f}. \quad (15.67)$$

On inverting the Fourier transform, the filtered function becomes

$$\hat{f}(x) = \int_{-\infty}^{+\infty} F(\alpha\Delta x)\tilde{f}e^{i\alpha x}d\alpha. \quad (15.68)$$

It is clear that the function $F(\alpha\Delta x)$ in Eq. (15.68) regulates the band of wave numbers that contributes to the filtered value of the original function. For this reason, $F(\alpha\Delta x)$ is referred to as the filter function.

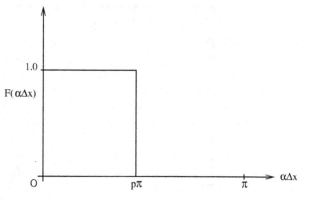

Figure 15.49. An ideal filter with a cutoff at $\alpha \Delta x = p\pi$.

Ideally, if all the wave numbers larger than $\alpha \Delta x \geq p\pi$ associated with a turbulence field are to be discarded or ignored, then the desirable filter function would be

$$F(\alpha \Delta x) = \begin{cases} 1 & \alpha \Delta x \leq p\pi \\ 0 & \alpha \Delta x > p\pi \end{cases}. \tag{15.69}$$

This ideal filter function is shown in Figure 15.49. When a finite size stencil is used, it is not possible to construct an ideal filter function. Thus, the objective is to make the filter as close to the ideal filter as possible.

For a 7-point stencil filter function, there are four free parameters. These parameters will now be determined by requiring the following conditions be satisfied.

1. $F(0) = 1.0$
2. $\left. \dfrac{\partial^2 F}{\partial (\alpha \Delta x)^2} \right|_{\alpha \Delta x = 0} = 0$
3. $\left. \dfrac{\partial^4 F}{\partial (\alpha \Delta x)^4} \right|_{\alpha \Delta x = 0} = 0$
4. $F(\pi) = 0$

It is easy to show that the values of the coefficients that satisfy these constraints are as follows:

$$A = \tfrac{1}{64}, \quad B = \tfrac{3}{32}, \quad C = \tfrac{15}{64}, \quad D = \tfrac{11}{16}. \tag{15.70}$$

Thus the filtering formula is

$$\hat{f}_\ell = \tfrac{1}{64}(f_{\ell+3} + f_{\ell-3}) - \tfrac{3}{32}(f_{\ell+2} + f_{\ell-2}) + \tfrac{15}{16}(f_{\ell+1} + f_{\ell-1}) + \tfrac{11}{16} f_\ell. \tag{15.71}$$

The filter function $F(\alpha \Delta x)$ is shown as the $n = 1$ curve in Figure 15.50. This figure does not look like the ideal filter function. To make the filter function resembling more of the ideal filter, one may perform repeated filtering. It is straightforward to show if the filtering process is performed n number of times, the filtered function becomes

$$f(x) = \int_{-\infty}^{\infty} F^n(\alpha \Delta x) \tilde{f}(\alpha) e^{i\alpha x} \, d\alpha. \tag{15.72}$$

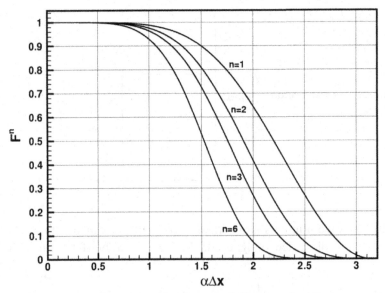

Figure 15.50. Profile of several 7-point stencil filter functions in wave number space.

That is, repeated filtering amounts to using a high power of the original filter function. Figure 15.50 shows a plot of $F^2(\alpha\Delta x)$, $F^3(\alpha\Delta x)$, $F^6(\alpha\Delta x)$, i.e., $n = 2, 3, 6$. It is clear that the filter function to a high power has a better resemblance to the ideal filter. However, it is not advisable to perform many filtering operations in one time step, because it will degrade the solution in the long wave range. If $n = 3$ filter is used, the approximate cutoff wave number, taken to be $F^3(\alpha_{\text{cutoff}}\Delta x) = 0.5$, is given by $\alpha_{\text{cutoff}}\Delta x \approx 1.8$.

15.5.2 RANS Computation: the $k - \varepsilon$ Model

For engineering applications, sometimes it is sufficient to compute the mean velocity profile of a turbulent flow. For this purpose, it has been recognized that the effect of turbulence on the mean flow may be adequately accounted for by including a turbulence-induced stress field. This is the Boussinesq approach. In this approach, the stress field is modeled by relating it to the strain rate field of the mean flow similar to that of laminar flow. This turbulence modeling approach is simple and relatively easy to implement on a computer. For this reason, it has become a wildly popular tool for solving practical problems for which only the mean flow is required.

Turbulence modeling is by now a well-established subject. An in-depth discussion of turbulence modeling is beyond the scope of this book. The objective here is to provide an introductory presentation of the subject. There are two most popular two-equation turbulence models. They are the $k - \varepsilon$ model and the $k - \omega$ model (Wilcox, 1998). A one-equation model by Spalart and Allmaras (1994) is also often used. It is now known that the $k - \varepsilon$ model works well for free shear turbulent flows, while the $k - \omega$ model works well for turbulent fluid layers adjacent to a solid surface. Motivated by this observation, Mentor (1994a, 1994b, 1997) has developed combined models, which is a clever way of blending the two models together for computing boundary layer type flows. In this book, only the $k - \varepsilon$ model is discussed and only to a limited extent.

The $k - \varepsilon$ turbulence model has been widely used in association with the RANS equations for turbulent mean flow calculations. However, it has been recognized that the applicability of the original $k - \varepsilon$ model is quite limited. This is because the model contains only a bare minimum of turbulence physics. Also, it is because the unknown constants of the original model were calibrated primarily by using low Mach number boundary layer and two-dimensional mixing layer flow data [see Hanjalic and Launder (1972), Launder and Spalding (1974), Launder and Reece (1975) and Hanjalic and Launder (1976)].

The useful range of the $k - \varepsilon$ model has since been extended. The extensions were carried out in two ways. First, a number of correction terms, intended to incorporate additional turbulence physics in the model, were proposed. Notable model corrections are the Pope (1978) correction developed for use in three-dimensional jets, the Sarkar and Lakshmanan (1991) correction developed for use when the convective Mach number is not too small, and the Tam and Ganesan (2003) correction developed for nonuniform high-temperature flows. Second, for application to a specific class of turbulent flows, the empirical constants of the original model were recalibrated using a large set of more appropriate data. The motivation for recalibration is the recognition that these constants are not really universal. The model would have a much better chance to be successful if it were applied to a restricted class of flows with similar turbulent mixing characteristics. For each class of flows, a new but more suitable set of constants is used. For instance, for calculating jet and free shear layer mean flows, Thies and Tam (1996) recalibrated the unknown model constants by using a large set of jet flow data covering a wide range of Mach numbers. Their computed jet mean flow velocity profiles for ambient temperature jets were found to be in excellent agreement with experimental measurements. More recently, applications of the recalibrated model to jets in simulated forward flight, coaxial jets, and jets with inverted velocity profile (see Tam, Pastouchenko, and Auriault (2001)) have been carried out with equal success.

The RANS equations including the $k - \varepsilon$ model in dimensionless form may be written as follows. Here, for jet flows, dimensionless variables with D, u_j, ρ_j, T_j (nozzle exit diameter, jet velocity, density, and temperature) as the length, velocity, density, and temperature scales will be used. Time, pressure, and the turbulence quantities k and ε will be nondimensionalized by D/u_j, $\rho_j u_j^2$, u_j^2, and u_j^3/D, respectively. Turbulent stresses τ_{ij} and eddy viscosity υ_T will be nondimensionalized by u_j^2 and u_j/D. In Cartesian tensor notation, the Favre-averaged equations of motion including the $k - \varepsilon$ model, as well as the Pope, Sarkar, and Tam and Ganesan correction terms (in dimensionless form) are as follows:

Continuity

$$\frac{\partial \rho}{\partial t} + \frac{\partial \rho u_\ell}{\partial x_\ell} = 0. \tag{15.73}$$

Momentum

$$\rho \left[\frac{\partial u_i}{\partial t} + v_\ell \frac{\partial u_i}{\partial x_\ell} \right] = -\frac{\partial p}{\partial x_i} - \frac{\partial (\rho \tau_{i\ell})}{\partial x_\ell} \tag{15.74}$$

Thermal energy

$$\rho \left[\frac{\partial T}{\partial t} + u_\ell \frac{\partial T}{\partial x_\ell} \right] = -\gamma(\gamma - 1) M_j^2 p \frac{\partial u_\ell}{\partial x_\ell} + \gamma(\gamma - 1) M_j^2 \rho \varepsilon$$

$$+ \frac{\gamma}{\mathrm{Pr}} \frac{\partial}{\partial x_\ell} \left(\rho v_T \frac{\partial T}{\partial x_\ell} \right). \tag{15.75}$$

The k and ε equations

$$\rho \left[\frac{\partial k}{\partial t} + u_\ell \frac{\partial k}{\partial x_\ell} \right] = -\rho \tau_{i\ell} \frac{\partial u_i}{\partial x_\ell} - \rho \varepsilon + \frac{1}{\sigma_k} \frac{\partial}{\partial x_\ell} \left(\rho v_T \frac{\partial k}{\partial x_\ell} \right) \tag{15.76}$$

$$\rho \left[\frac{\partial \varepsilon_s}{\partial t} + u_\ell \frac{\partial \varepsilon_s}{\partial x_\ell} \right] = -C_{\varepsilon 1} \frac{\varepsilon_s}{k} \rho \tau_{i\ell}^s \frac{\partial u_i}{\partial x_\ell} - (C_{\varepsilon 2} - C_{\varepsilon 3} \chi) \rho \frac{\varepsilon_s^2}{k} + \frac{1}{\sigma_\varepsilon} \frac{\partial}{\partial x_\ell} \left(\rho v_t^s \frac{\partial \varepsilon_s}{\partial x_\ell} \right). \tag{15.77}$$

Equation of state

$$p = \frac{1}{\gamma M_j^2} \rho T.$$

Auxiliary relations

$$v_T = v_t + v_\rho, \quad v_T^s = v_t^s + v_\rho$$

$$v_t = C_\mu \frac{k^2}{\varepsilon}, \quad v_t^s = C_\mu \frac{k^2}{\varepsilon_s}, \quad v_\rho = \begin{cases} C_\rho \frac{k^{7/2}}{\varepsilon^2} \frac{1}{\rho} \frac{|\nabla \rho \nabla V|}{|\nabla V|}, & \text{if } (\nabla \rho \cdot \nabla V) \text{ is negative} \\ 0, & \text{otherwise} \end{cases}$$

where V is the flow velocity

$$\varepsilon = \varepsilon_s \left(1 + \alpha_1 M_t^2 \right), \quad M_t^2 = \frac{2k}{T} M_j^2$$

$$\tau_{i\ell} = \frac{2}{3} k \delta_{i\ell} - v_T \left(\frac{\partial u_i}{\partial x_\ell} + \frac{\partial u_\ell}{\partial x_i} - \frac{2}{3} \frac{\partial u_k}{\partial x_k} \delta_{i\ell} \right)$$

$$\tau_{i\ell}^s = \frac{2}{3} k \delta_{i\ell} - v_T^s \left(\frac{\partial u_i}{\partial x_\ell} + \frac{\partial u_\ell}{\partial x_i} - \frac{2}{3} \frac{\partial u_k}{\partial x_k} \delta_{i\ell} \right)$$

$$\chi = \omega_{ik} \omega_{k\ell} \delta_{i\ell}, \quad \omega_{i\ell} = \frac{1}{2} \frac{k}{\varepsilon_s} \left(\frac{\partial u_i}{\partial x_\ell} - \frac{\partial u_\ell}{\partial x_i} \right),$$

where γ is the ratio of specific heats, Pr is the Prandtl number, and M_j is the jet Mach number. There are nine empirical constants in the preceding system. The values recommended by Thies and Tam (1996) and Tam and Ganesan (2004) for jet and similar type of flows are as follows:

$$C_\mu = 0.0874, \quad C_{\varepsilon 1} = 1.40, \quad C_{\varepsilon 2} = 2.02, \quad C_{\varepsilon 3} = 0.822, \quad C_\rho = 0.035,$$

$$\gamma \sigma_T = \mathrm{Pr} = 0.422, \quad \sigma_k = 0.324, \quad \sigma_\varepsilon = 0.377, \quad \alpha_1 = 0.518$$

The RANS equations and the $k - \varepsilon$ model were originally developed with the intention for computing the mean velocity profiles of turbulent flows. The question arises as to whether they can be used for computing unsteady flow and noise. For unsteady flow application, the set of equations is, generally, referred to as Unsteady Reynolds Averaged Navier-Stokes (URANS). Clearly, there are limitations to such

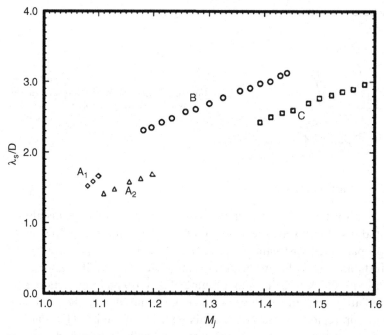

Figure 15.51. The different modes of jet screech tones. Data from Ponton and Seiner (1992).

usage. However, the system of URANS equations is relatively simple to use. It requires computational resources that are not too demanding. At present, there is no consensus as to whether it is appropriate to use such a formulation for noise prediction. Evidently, for certain types of problems, its use can be justified, whereas for other types it is not. One must decide based largely on the physics of the problem.

15.6 Example II: Numerical Simulation of Axisymmetric Jet Screech

Experimentally, it is found that an imperfectly expanded supersonic jet invariably emits strong tones called screech tones. Jet screech is a fairly complex phenomenon involving several modes of oscillations. When a convergent nozzle is used, it is observed experimentally that the screech modes are axisymmetric at low supersonic Mach numbers. There are two axisymmetric modes. They are usually designated as the A_1 and A_2 modes (see Figure 15.51). At Mach number 1.3 or higher, the jet screech switches to flapping or helical modes. They are designated as the B and C modes. In Figure 15.51 λ_s is the wavelength of the screech tone. Mode switching or staging is quite abrupt. The staging Mach number is found to be very sensitive to ambient experimental environment and also sensitive to upstream conditions of the jet flow. Largely because of this sensitivity, it is known that the staging Mach numbers differ slightly from experiment to experiment. Even in the same facility, they tend to differ somewhat when the experiment is repeated at a later time.

It is known, since the early work of Powell (1953), that screech tones are generated by a feedback loop. Recent works show that the feedback loop is driven by the instability waves of the jet flow. In the plume of an imperfectly expanded jet is a quasiperiodic shock cell structure. Figure 15.52 shows schematically the feedback loop. Near the nozzle lip where the jet mixing layer is thin and most receptive to

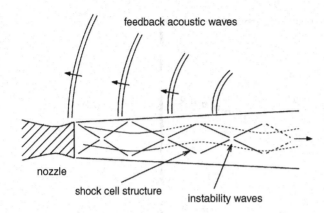

Figure 15.52. Schematic diagram of the feedback loop of jet screech.

external excitation, acoustic disturbances impinging on this area excite the instability waves. The excited instability waves, extracting energy from the mean flow, grow rapidly as they propagate downstream. After propagating a distance of four to five shock cells, the instability wave, having acquired a large enough amplitude, interacts with the quasiperiodic shock cells in the jet plume. The unsteady interaction generates acoustic radiation, part of which propagates upstream outside the jet. Upon reaching the nozzle lip region, they excite the mixing layer of the jet. This leads to the generation of new instability waves. In this way, the feedback loop is closed.

15.6.1 Computation of Axisymmetric Jet Screech

For the purpose of illustrating the essential steps and considerations needed to simulate jet screech computationally, only axisymmetric jet screech is considered (see Shen and Tam, 1998). This is to keep the discussion to a reasonable length. For axisymmetric jet screech associated with jets issued from a convergent nozzle, the jet Mach number is restricted to the range of 1.0 to 1.25. Numerical simulation of jet noise generation is not a straightforward undertaking. Tam (1995) had discussed some of the major computational difficulties anticipated in such an effort. First of all, the problem is characterized by very disparate length scales. For instance, the acoustic wavelength of the screech tone is more than 20 times larger than the initial thickness of the jet mixing layer that supports the instability waves. Furthermore, there is also a large disparity between the magnitude of the fluid particle velocity of the radiated sound and the velocity of the jet flow. Typically, they are five to six orders of magnitude different. To be able to compute accurately the instability waves and the radiated sound, a highly accurate CAA algorithm with shock-capturing capability as well as a set of high-quality numerical boundary conditions are required.

15.6.1.1 Computational Model

For an accurate simulation of jet screech generation, it is essential that the feedback loop be modeled and computed correctly. The important elements that form the feedback loop are the shock cell structure, the large-scale instability wave, and the feedback acoustic waves. In the mixing layer of the jet, turbulence is responsible for its spreading. The spreading rate of the jet affects the spatial growth and decay of the instability wave. Thus, turbulence in the jet plays an indirect but still important role in the feedback loop. The length scales of the instability wave, the shock cells,

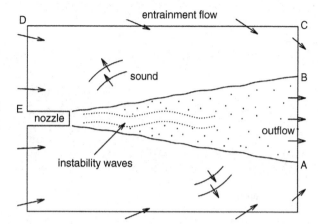

Figure 15.53. A sketch of the physical domain to be simulated.

as well as the feedback acoustic waves are much longer than that of the fine-scale turbulence in the mixing layer of the jet. Because of this disparity in length scales, no attempt is made here to resolve the fine-scale turbulence computationally. Based on these considerations, an unsteady RANS model is regarded as adequate. To provide the necessary jet spreading induced by turbulence, the $k - \varepsilon$ turbulence model is adopted. Here, the $k - \varepsilon$ model simulates the effect of the fine-scale turbulence on the jet mean flow. In the computation described below, the modified $k - \varepsilon$ model of Thies and Tam (1996), optimized for jet flows, is used.

Figure 15.53 shows the physical domain to be simulated. The following scales are used in the computations; length scale $= D$ (nozzle exit diameter), velocity scale $= a_\infty$ (ambient sound speed), time scale $= D/a_\infty$, density scale $= \rho_\infty$ (ambient gas density), pressure scale $\rho_\infty a_\infty^2$, temperature scale $= T_\infty$ (ambient gas temperature); scales for k, ε, and υ_t are a_∞^2, a^3/D, and $a_\infty D$, respectively. The dimensionless governing equations in Cartesian tensor notation in conservation form are as follows:

$$\frac{\partial \rho}{\partial t} + \frac{\partial (\rho u_j)}{\partial x_j} = 0 \tag{15.78}$$

$$\frac{\partial \rho u_i}{\partial t} + \frac{\partial}{\partial x_j}(\rho u_i u_j) = -\frac{\partial p}{\partial x_i} - \frac{\partial}{\partial x_j}(\rho \tau_{ij}) \tag{15.79}$$

$$\frac{\partial \rho E}{\partial t} + \frac{\partial}{\partial x_j}(\rho E u_j) = -\frac{\partial}{\partial x_j}(\rho u_j) - \frac{\partial}{\partial x_j}(\rho u_i \tau_{ij})$$
$$+ \frac{1}{P_r(\gamma - 1)} \frac{\partial}{\partial x_j}\left(\rho \upsilon_t \frac{\partial T}{\partial x_j}\right) + \frac{1}{\sigma_k} \frac{\partial}{\partial x_j}\left(\rho \upsilon_t \frac{\partial k}{\partial x_j}\right) \tag{15.80}$$

$$\frac{\partial \rho k}{\partial t} + \frac{\partial}{\partial x_j}(\rho k u_j) = -\rho \tau_{ij} \frac{\partial u_i}{\partial x_j} - \rho \varepsilon + \frac{1}{\sigma_k} \frac{\partial}{\partial x_j}\left(\rho \upsilon_t \frac{\partial k}{\partial x_j}\right) \tag{15.81}$$

$$\frac{\partial \rho \varepsilon}{\partial t} + \frac{\partial}{\partial x_j}(\rho \varepsilon u_j) = -C_{\varepsilon 1} \frac{\varepsilon}{(k + k_0)} \rho \tau_{ij} \frac{\partial u_j}{\partial x_j} - C_{\varepsilon 2} \frac{\rho \varepsilon^2}{(k + k_0)}$$
$$+ \frac{1}{\sigma_k} \frac{\partial}{\partial x_j}\left(\rho \upsilon_t \frac{\partial \varepsilon}{\partial x_j}\right) \tag{15.82}$$

$$\gamma p = \rho T, \tag{15.83}$$

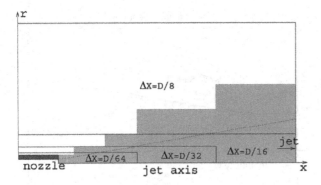

Figure 15.54. Computational domain in the $r - x$ plane showing the subdomains and mesh sizes. The RANS equations are solved in the shaded region. The Euler equations are solved in the unshaded region.

where,

$$E = \frac{T}{\gamma(\gamma - 1)} + \frac{1}{2}u_i^2 + k$$

$$\tau_{ij} = \frac{2}{3}k\delta_{ij} - \upsilon_t\left(\frac{\partial u_i}{\partial x_j} + \frac{\partial u_j}{\partial x_i} - \frac{2}{3}\frac{\partial u_k}{\partial x_k}\delta_{ij}\right)$$

$$\upsilon_t = C_\mu\frac{k^2}{(\varepsilon + \varepsilon_0)} + \frac{\upsilon}{a_\infty D}.$$

In these equations, γ is the ratio of specific heats, υ is the molecular kinematic viscosity. $k_0 = 10^{-6}$ and $\varepsilon_0 = 10^{-4}$ are small positive numbers to prevent division by zero. The model constants are given in Section 15.5.2. The inverse molecular Reynolds number $\upsilon/(a_\infty D)$ is assigned a value of 1.7×10^{-6} in the computation. Note that, for the range of Mach numbers and jet temperatures considered, the Pope and Sarkar corrections often added to the $k - \varepsilon$ model are not necessary and are omitted. Here, only cold jets are considered. For this reason, the Tam and Ganesan hot-jet correction is not needed. Outside the jet flow both k and ε are zero. On neglecting the molecular viscosity terms, the governing equations become the Euler equations.

15.6.1.2 Grid Design and Computational Scheme

All the computations are carried out in the $x - r$ plane. The size of the computational domain is $35D \times 17D$. By extending the computational domain to $35D$ downstream, all the screech tone noise sources are effectively included in the computation. Also, the amplitude of the excited instability waves would have decayed significantly at the outflow boundary, thus lessening the likelihood of reflection of unsteady disturbances back into the computational domain. By extending the computational domain to $17D$ in the radial direction, it is believed that the outer boundary is approximately in the far field. By measuring the radiation angle from the location in the jet where tone emission is the strongest, accurate screech tone directivity can be determined.

The computational domain is divided into four subdomains each having a different mesh size as shown in Figure 15.54. The subdomain immediately downstream of the nozzle exit, where the jet mixing layer is thin and the jet plume contains a shock cell structure, has the finest mesh : $\Delta x = \Delta r = D/64$. The mesh size of the next subdomain increases by a factor of 2. This continues on so that, outside the jet where the mesh is the coarsest, has a mesh size of $D/8$. The governing equations in cylindrical coordinates are discretized by the multi-size-mesh multi-time-step method (see Chapter 12). The use of multiple time steps is crucial to the computational effort. It greatly reduces the run time of the computer code.

15.6.1.3 Numerical Boundary Conditions
Numerical boundary conditions play a crucial role in the simulation of the jet screech phenomenon. An important requirement of this problem is that the pressure at the nozzle exit and that of the ambient condition are different and must be maintained. It is this pressure difference that leads to the formation of shock cells in the jet plume. Thus, both the outflow and radiation boundary condition must have the capability of imposing a desired ambient pressure to the mean flow solution. For the present problem, several additional types of numerical boundary conditions are required. In Figure 15.53, outflow boundary conditions (see Section 9.3) are imposed along boundary AB. Along boundary $BCDE$, radiation condition with entrainment flow is implemented (see Section 9.2). On the nozzle wall, the solid wall boundary condition is imposed. The jet flow is supersonic, so the inflow boundary condition can be prescribed at the nozzle exit plane. Finally, the equations of motion in cylindrical coordinates centered on the x-axis have an apparent singularity at the jet axis ($r \to 0$). A special treatment, discussed in Section 9.4, is used to avoid the singularity computationally.

15.6.1.3 Artificial Selective Damping
The DRP scheme is a central difference scheme and, therefore, has no intrinsic dissipation. For the purpose of eliminating spurious short waves and to improve numerical stability, artificial selective damping terms are added to the discretized finite difference equations. For the present problem, artificial selective damping is also critically needed for shock-capturing purposes.

In the interior region, the 7-point damping stencil with a half-width of $\sigma = 0.2\pi$ is used. An inverse mesh Reynolds number ($R_\Delta^{-1} = \upsilon_a/(a_\infty \Delta x)$), where υ_a is the artificial kinematic viscosity, of 0.05 is prescribed over the entire computation domain. This is to provide general background damping to eliminate possible propagating spurious waves. Near the boundaries of the computational domain where a 7-point stencil does not fit, the 5- and 3-point damping stencils given in Section 7.2 are used.

Spurious numerical waves are usually generated at the boundaries of a computational domain. The boundaries are also favorite sites for the occurrence of numerical instability. This is true for both exterior and interior boundaries such as the nozzle walls and buffer regions where there is a change of mesh size. To suppress both the generation of spurious numerical waves and numerical instability, additional artificial selective damping is imposed along these boundaries. Along the inflow, radiation, and outflow boundaries, a distribution of inverse mesh Reynolds number in the form of a Gaussian function with a half-width of 4 mesh points (normal to the boundary) and a maximum value of 0.1 right at the outmost mesh points is incorporated into the time marching scheme. Adjacent to the jet axis, a similar addition of artificial selective damping is implemented with a maximum value of the inverse mesh Reynolds number at the jet axis set equal to 0.35. On the nozzle wall, the use of a maximum value of additional inverse mesh Reynolds number of 0.2 has been found to be very satisfactory.

The two sharp corners of the nozzle lip and the transition point between the use of the outflow and the radiation boundary condition on the right side of the computational domain are locations requiring stronger numerical damping. This is carried out by adding a Gaussian distribution of damping around these special points.

As shown in Figure 15.54, the four computational subdomains are separated by buffer regions. Here, additional artificial selective damping is added to the finite difference scheme. In the supersonic region downstream of the nozzle exit, a shock cell structure develops in the jet flow. In order to provide the DRP scheme with shock-capturing capability, the variable stencil Reynolds number method discussed in Section 8.3 is used. The jet mixing layer in this region has a very large velocity gradient in the radial direction. Because of this, the U_{stencil} of the variable stencil Reynolds number method is determined by searching over the 7-point stencil in the axial direction. In the radial direction only the 2 mesh points immediately adjacent to the computational point are included in the search. An inverse stencil mesh Reynolds number distribution of the following form:

$$R_{\text{stencil}}^{-1} = 4.5 F(x) G(r),$$

where

$$F(x) = \begin{cases} 1, & 0 \le x \le 9 \\ \exp\left[-\frac{\ln 2}{(8\Delta x)^2}(x-9)^2\right], & 9 < x \end{cases}$$

$$G(r) = \begin{cases} 1, & 0 \le r \le 0.8 \\ \exp\left[-\frac{\ln 2}{(4\Delta x)^2}(r-0.8)^2\right], & 0.8 < r \end{cases}$$

is used. It is possible to show, based on the damping curve ($\sigma = 0.3\pi$), that the variable damping has minimal effect on the instability wave of the feedback loop. Also, extensive numerical experimentations indicate that the method used can, indeed, capture the shocks in the jet plume and that the time-averaged shock cell structure compares favorably with experimental measurements.

15.6.2 Numerical Results and Comparisons with Experiments

It has been found that numerical algorithm described above is capable of reproducing the jet screech phenomenon computationally. The nozzle configuration in the experiment of Ponton and Seiner (1992) is used in the simulation. Figure 15.55 shows the computed density field in the x–r plane at one instance after the initial transient disturbances have propagated out of the computational domain. The screech feedback loop locks itself into a periodic cycle without external interference. As can be seen, sound waves of the screech tones are radiated out in a region around the fourth and fifth shock cell downstream of the nozzle exit. Most of the prominent features of the numerically simulated jet screech phenomenon are in good agreement with physical experiment.

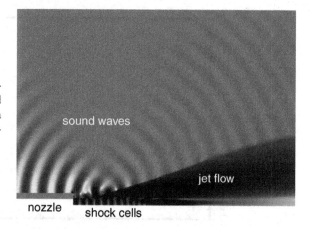

Figure 15.55. Density field from numerical simulation showing the generation and radiation of screech tone associated with a Mach 1.13 cold supersonic jet from a convergent nozzle.

15.6.2.1 Mean Velocity Profiles and Shock Cell Structure

To demonstrate that numerical simulation can, indeed, reproduce the physical experiment, the accuracy of the computed mean flow velocity profiles of the simulated jet is tested by comparison with experimental measurements. For this purpose, the time-averaged velocity profiles of the axial velocity of a Mach 1.2 cold jet from one jet diameter downstream of the nozzle exit to seven diameters downstream at one diameter intervals are measured from the numerical simulations. They are shown as a function of $\eta^* = (r - r_{0.5})/x$ in Figure 15.56, where $r_{0.5}$ is the radial distance from the jet axis to the location where the axial velocity is equal to one half of the fully expanded jet velocity. Numerous experiments have shown that the mean velocity profile when plotted as a function of η^* would nearly collapse into a single curve. The single curve is well represented by an error function in the following form:

$$\frac{U}{U_j} = 0.5[1 - erf(\sigma \eta^*)], \tag{15.85}$$

Figure 15.56. Comparison between computed mean velocity profiles at $M_j = 1.2$ and experiment (Lau, 1981). Simulation results: ———, $x/D = 1.0$; – – – –, $x/D = 2.0$; — · —, $x/D = 3.0$; — – – —, $x/D = 4.0$; — · · —, $x/D = 5.0$; - - - -, $x/D = 6.0$; · · · · · · ·, $x/D = 7.0$. Experiment $U/U_j = 0.5[1 - erf(\sigma \eta^*)]$; ○, $\sigma = 17.0$.

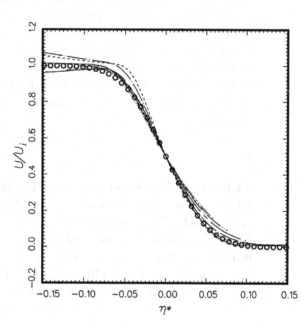

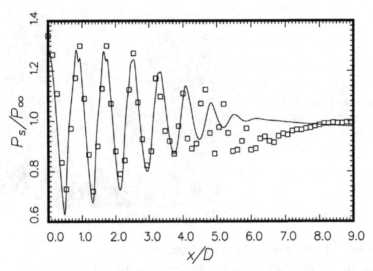

Figure 15.57. Comparison between calculated time-averaged pressure distribution along the centerline of a Mach 1.2 cold jet and the measurement of Norum and Brown (1993). ———, simulation; □, experiment.

where σ is the spreading parameter and U_j is the fully expanded jet exit velocity. Extensive jet mean velocity data had been measured by Lau (1981). By interpolating the data of Lau to Mach 1.2, it is found that experimentally σ is nearly equal to 17.0. The empirical formula with $\sigma = 17.0$ is also plotted in Figure 15.56 (the circles). It is clear that there is good agreement between measured mean velocity profile and those of the numerical simulation.

One important component of the screech feedback loop is the shock cell structure inside the jet plume. To ensure that the simulated shock cells are the same as those in an actual experiment, the time-averaged pressure distribution along the centerline of the simulated jet at Mach 1.2 is compared with the experimental measurements of Norum and Brown (1993). Figure 15.57 is a plot of the simulated and measured pressure distribution as a function of downstream distance. It is clear from this figure that the first five shocks of the simulation are in good agreement with experimental measurements both in terms of shock cell spacing and amplitude. Beyond the fifth shock cell, the agreement is less good. However, it is known that screech tones are generated around the fourth shock cell. Therefore, any minor discrepancies downstream of the fifth shock cell would not invalidate the screech tone simulation.

15.6.2.2 Screech Tone Frequency and Intensity

It is well known that at low supersonic jet Mach numbers, there are two axisymmetric screech modes. They are referred to as the A_1 and A_2 modes. Earlier, Norum (1983) had compared the frequencies of the A_1 and A_2 modes measured by a number of investigators. His comparison indicated that the screech frequencies and the Mach number at which transition from one mode to the other takes place (staging) could vary slightly from experiment to experiment. Experimentalists generally agree that

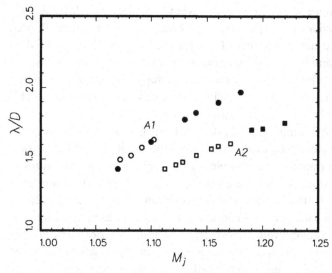

Figure 15.58. Comparison between acoustic wavelengths of simulated screech phenomenon and the experimental measurements of Ponton and Seiner (1992). ∘, □, Measurements; •, ■, simulations.

the screech phenomenon is extremely sensitive to minor details of the experimental facility and jet operation conditions.

By using this numerical method, it is found that numerical simulation does reproduce both the A_1 and the A_2 axisymmetric screech modes. Figure 15.58 shows the variation of the computed λ/D, where λ is the acoustic wavelength of the tone,

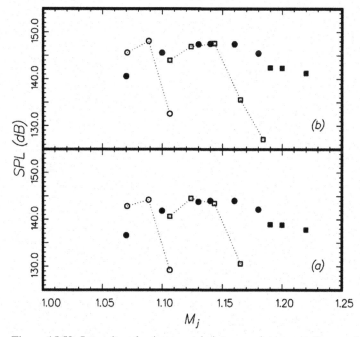

Figure 15.59. Intensity of axisymmetric jet screech tones at the nozzle exit plane. (a) $r/D = 0.889$, (b) $r/D = 0.642$. Experiment (Ponton and Seiner, 1992): ∘, A_1 mode; □, A_2 mode. Numerical simulation: •, A_1 mode; ■, A_2 mode.

with jet Mach number. Since $\lambda/D = a_\infty/(fD)$, where f is the screech frequency, this figure essentially provides the tone frequency Mach number relation. Plotted on this figure also are the measurements of Ponton and Seiner (1992). The data from both the numerical simulation and experiment fall on the same two curves, one for the A_1 mode and the other for the A_2 mode. This suggests that the computed screech frequencies are in good agreement with experimental measurements. This is so, although the staging Mach number is not the same.

Ponton and Seiner (1992) mounted two pressure transducers at a radial distance of $0.642D$ and $0.889D$, respectively, on the surface of the nozzle lip in their experiment. By means of these transducers, they were able to measure the intensities of screech tones. Their measured values are plotted in Figure 15.59. The transducer of Figure 15.59b is closer to the jet axis and, hence, shows a higher decibel level. Plotted on these figures also are the corresponding tone intensities measured in the numerical simulation. The peak levels of both physical and numerical experiments are nearly equal. Thus, except for the difference in the staging Mach number, the numerical simulation is, indeed, capable of providing accurate screech tone intensity prediction as well.

APPENDIX A

Fourier and Laplace Transforms

A.1 Fourier Transform

Let $f(x)$ be an absolutely integratable function; i.e., $\int_{-\infty}^{\infty} |F(x)|dx < \infty$, then $F(x)$ and its Fourier transform $\tilde{F}(\alpha)$ are related by

$$\tilde{F}(\alpha) = \frac{1}{2\pi} \int_{-\infty}^{\infty} F(x)e^{-i\alpha x}dx, \quad F(x) = \int_{-\infty}^{\infty} \tilde{F}(\alpha)e^{i\alpha x}d\alpha.$$

Derivative Theorem

By means of integration by parts, it is easy to find

$$\frac{\partial \tilde{F}}{\partial x} = \frac{1}{2\pi} \int_{-\infty}^{\infty} \frac{\partial F}{\partial x}e^{-i\alpha x}dx = \frac{i\alpha}{2\pi} \int_{-\infty}^{\infty} F(x)e^{-i\alpha x}dx = i\alpha\tilde{F}(\alpha).$$

Shifting Theorem

$$F(\tilde{x}+\lambda) = \frac{1}{2\pi} \int_{-\infty}^{\infty} F(x+\lambda)e^{-i\alpha x}dx = \frac{1}{2\pi} \int_{-\infty}^{\infty} F(\eta)e^{-i\alpha\eta+i\alpha\lambda}d\eta = e^{i\alpha\lambda}\tilde{F}(\alpha).$$

A.2 Laplace Transform

Let $f(t)$ satisfy the boundedness condition $\int_{-\infty}^{\infty} e^{-ct}|f(t)|dt < \infty$ for some $c > 0$, then $f(t)$ and its Laplace transform $\tilde{f}(\omega)$ are related by

$$\tilde{f}(\omega) = \frac{1}{2\pi} \int_{0}^{\infty} f(t)e^{i\omega t}dt, \quad f(t) = \frac{1}{2\pi} \int_{\Gamma} \tilde{f}(\omega)e^{-i\omega t}d\omega.$$

The inverse contour Γ is to be placed on the upper-half ω-plane above all poles and singularities of the integrand (see Figure A1). This condition is necessary in order to satisfy causality.

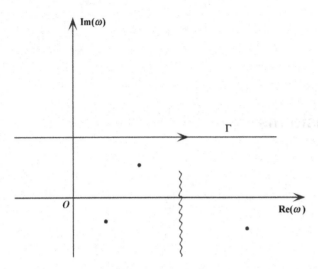

Figure A1. Complex ω-plane showing the inverse contour Γ, • poles, ~~~ branch cut.

Derivative Theorem

By applying integration by parts to the integral below, it is straightforward to show

$$\frac{\partial \tilde{f}}{\partial t} = \frac{1}{2\pi} \int_0^\infty \frac{\partial f}{\partial t} e^{i\omega t} dt = \frac{1}{2\pi} f(t) e^{i\omega t} \Big|_0^\infty - \frac{i\omega}{2\pi} \int_0^\infty f(t) e^{i\omega t} dt = -\frac{f(0)}{2\pi} - i\omega \tilde{f}(\omega).$$

Shifting Theorem

Let the initial conditions be

$$f(t) = \begin{cases} \Phi, & 0 \le t < \lambda \\ 0, & t < 0 \end{cases}$$

then,

$$\tilde{f}(t+\lambda) = \frac{1}{2\pi} \int_0^\infty f(t+\lambda) e^{i\omega t} dt = \frac{1}{2\pi} \int_\lambda^\infty f(\eta) e^{i\omega \eta - i\omega\lambda} d\eta$$

$$= e^{-i\omega\lambda} \left[\tilde{f}(\omega) - \frac{1}{2\pi} \int_0^\lambda f(\eta) e^{i\omega\eta} d\eta \right] = e^{-i\omega\lambda} \left[\tilde{f}(\omega) + \frac{\Phi}{2\pi i\omega} (1 - e^{i\omega\lambda}) \right].$$

Similarly,

$$\tilde{f}(t-\lambda = e^{i\omega\lambda} \tilde{f}(\omega).$$

The Method of Stationary Phase

Consider evaluating the following integral in the limit $\lambda \to \infty$.

$$I(\lambda) = \int_a^b H(x)e^{i\lambda\Phi(x)}dx. \tag{B1}$$

It was pointed out by Lord Kelvin (1887) that for large λ the function $e^{i\lambda\Phi(x)}$ will oscillate wildly with almost complete cancellation. Significant contribution to the integral comes from small intervals of x where $\Phi(x)$ varies slowly. This happens when $\Phi'(x) = 0$ or at the stationary points of $\Phi(x)$.

Suppose $\Phi(x)$ has only one stationary point in (a, b), say at $x = c$. That is

$$\Phi(x) = \Phi(c) + \frac{\Phi''(c)}{2}(x - c^2) + \dots. \tag{B2}$$

Since the major contribution to (B1) comes around $x = c$, asymptotically the integral is given by

$$I(\lambda) \sim \int_b^a H(c)e^{i\lambda\left[\Phi(c) + \frac{\Phi''(c)}{2}(x-c)^2 + \dots\right]}dx$$

$$\sim H(c)e^{i\lambda\Phi(c)} \int_{c-\delta}^{c+\delta} e^{i\frac{\lambda}{2}\Phi''(c)(x-c)^2}dx. \tag{B3}$$

It is possible to change the limits of integration to $-\infty$ to ∞ as the error of extending δ to ∞ is asymptotically small (nearly complete cancellation). On accounting for that $\Phi''(c)$ may be positive or negative, the integral of (B3) may be evaluated to give

$$\lim_{\lambda \to \infty} \int_a^b H(x)e^{i\lambda\Phi(x)}dx \sim \left(\frac{2\pi}{\lambda|\Phi''(c)|}\right)^{\frac{1}{2}} H(c)e^{i\left[\lambda\Phi(c)+\mathrm{sgn}(\Phi''(c))\frac{\pi}{4}\right]}, \tag{B4}$$

where sgn( ) is the sign of ( ).

APPENDIX C

The Method of Characteristics

The one-dimensional continuity, momentum, and energy equations of a compressible, inviscid fluids are

$$\frac{\partial \rho}{\partial t} + u\frac{\partial \rho}{\partial x} + \rho\frac{\partial u}{\partial x} = 0 \tag{C1}$$

$$\frac{\partial u}{\partial t} + u\frac{\partial u}{\partial x} + \frac{1}{\rho}\frac{\partial p}{\partial x} = 0 \tag{C2}$$

$$\frac{\partial p}{\partial t} + u\frac{\partial p}{\partial x} + \gamma p\frac{\partial u}{\partial x} = 0. \tag{C3}$$

This system of three equations supports three sets of characteristics. These characteristics may be found by combining the three equations in a special way. For this purpose, multiply (C1) by λ and (C2) by β and add to (C3), where λ and β are functions of the dependent variables ρ, u, and p.

$$\lambda\left[\frac{\partial \rho}{\partial t} + u\frac{\partial \rho}{\partial x}\right] + \beta\left[\frac{\partial u}{\partial t} + \left(u + \frac{\lambda\rho}{\beta} + \frac{\gamma p}{\beta}\right)\frac{\partial u}{\partial x}\right] + \frac{\partial p}{\partial t} + \left(u + \frac{\beta}{\rho}\right)\frac{\partial p}{\partial x} = 0. \tag{C4}$$

Now λ and β may be chosen such that all the derivatives of the dependent variables are in the form of a common convective derivative as follows:

$$\frac{\partial}{\partial t} + V\frac{\partial}{\partial x}.$$

There are three possible choices. They are

$$1. \quad \beta = 0, \quad \lambda = -\frac{\gamma p}{\rho}.$$

In this case, (C4) becomes

$$\frac{\partial p}{\partial t} + u\frac{\partial p}{\partial x} - \frac{\gamma p}{\rho}\left(\frac{\partial \rho}{\partial t} + u\frac{\partial \rho}{\partial x}\right) = 0. \tag{C5}$$

Now, in the x–t plane, along the curve

$$\frac{dx}{dt} = u, \tag{C6}$$

which will be referred to as the P-characteristic, (C5) may be rewritten as

$$\frac{dp}{dt} - \frac{\gamma p}{\rho}\frac{d\rho}{dt} = 0, \tag{C7}$$

where

$$\frac{d}{dt} = \frac{\partial}{\partial t} + \frac{dx}{dt}\frac{\partial}{\partial x} = \frac{\partial}{\partial t} + u\frac{\partial}{\partial x}$$

along the P-characteristic. (C7) may be expressed in a differential form as follows:

$$dp - \frac{\gamma p}{\rho}d\rho = 0. \tag{C8}$$

This may be integrated to yield

$$d\left(\frac{p}{\rho^\gamma}\right) = 0 \quad \text{or} \quad \frac{p}{\rho^\gamma} = \text{constant along a } P\text{-characteristic.} \tag{C9}$$

In other words, the flow is isentropic following the motion of a fluid element.

$$2. \quad \lambda = 0, \quad \beta = \rho\left(\frac{\gamma p}{\rho}\right)^{1/2} = \rho a$$

where a is the speed of sound. For this choice of λ and β, (C4) becomes

$$\frac{\partial p}{\partial t} + (u + a)\frac{\partial p}{\partial x} + \rho a\left[\frac{\partial u}{\partial t} + (u + a)\frac{\partial u}{\partial x}\right] = 0. \tag{C10}$$

Thus, along the C^+ characteristic,

$$\frac{dx}{dt} = u + a. \tag{C11}$$

(C10) reduces to

$$\frac{dp}{dt} + \rho a\frac{du}{dt} = 0 \quad \text{or} \quad dp + \rho a\, du = 0. \tag{C12}$$

$$3. \quad \lambda = 0, \quad \beta = -\rho\left(\frac{\gamma p}{\rho}\right)^{\frac{1}{2}} = -\rho a.$$

In this case, (C4) becomes

$$\frac{\partial p}{\partial t} + (u - a)\frac{\partial p}{\partial x} - \rho a\left[\frac{\partial u}{\partial t} + (u - a)\frac{\partial u}{\partial x}\right] = 0. \tag{C13}$$

Therefore, along the C^- characteristic,

$$\frac{dx}{dt} = u - a. \tag{C14}$$

(C13) reduces to

$$\frac{dp}{dt} - \rho a\frac{du}{dt} = 0 \quad \text{or} \quad dp - \rho a\, du = 0. \tag{C15}$$

Under certain special conditions one of the differential relations (C12) or (C15) may be integrated in a closed form. In this case, there are two integrals ((C9) is an integral) to the original system of equations (C1) to (C3). By means of these two integrals, the governing equations may be reduced to a single first-order partial differential equation.

APPENDIX D

Diffusion Equation

The dispersion-relation-preserving (DRP) scheme is well suited to compute numerical solutions of diffusion equations. To illustrate this, consider computing the solution of the initial value problem governed by the heat equation as follows:

$$\frac{\partial u}{\partial t} = \kappa \left(\frac{\partial^2 u}{\partial x^2} + \frac{\partial^2 u}{\partial y^2} \right) \tag{D1}$$

$$\text{at } t = 0, \quad u = f(x, y). \tag{D2}$$

Before implementing the DRP scheme, it is advantageous to cast (D1) into a system of first-order equations. This may be done by introducing two auxiliary variables defined by

$$v = \frac{\partial u}{\partial x}, \tag{D3}$$

$$w = \frac{\partial u}{\partial y}. \tag{D4}$$

(D1) becomes

$$\frac{\partial u}{\partial t} = \kappa \left(\frac{\partial v}{\partial x} + \frac{\partial w}{\partial y^2} \right). \tag{D5}$$

Now, the system of equations (D3) to (D5) is equivalent to (D1).

Upon applying the DRP scheme to (D3) to (D5), the finite difference system is

$$v_{\ell,m}^{(n)} = \frac{1}{\Delta x} \sum_{j=-3}^{3} a_j u_{\ell+j,m}^{(n)} \tag{D6}$$

$$w_{\ell,m}^{(n)} = \frac{1}{\Delta y} \sum_{j=-3}^{3} a_j u_{\ell,m+j}^{(n)} \tag{D7}$$

$$E_{\ell,m}^{(n)} = \kappa \sum_{j=-3}^{3} a_j \left(\frac{1}{\Delta x} v_{\ell+j,m}^{(n)} + \frac{1}{\Delta y} w_{\ell,m+j}^{(n)} \right) \tag{D8}$$

$$u_{\ell,m}^{(n+1)} = u_{\ell,m}^{(n)} + \Delta t \sum_{j=0}^{3} b_j E_{\ell,m}^{(n-j)}. \tag{D9}$$

The time step Δt that can be used in (D9) is constrained by stability considera-tion. To determine the largest time step without triggering numerical instability, the first step is to find the dispersion relation of the system of finite difference equations (D6) to (D9). This is done by generalizing the discretized system to a set of finite difference equations with continuous variables and then take their Fourier-Laplace transforms. This leads to (a $\tilde{\Phi}$ will be used to denote the Fourier-Laplace transform of Φ) the following:

$$\tilde{v} = i\bar{\alpha}\tilde{u} \tag{D10}$$

$$\tilde{w} = i\bar{\beta}\tilde{u} \tag{D11}$$

$$\tilde{E} = \kappa(i\bar{\alpha}\tilde{v} + i\bar{\beta}\tilde{w}) \tag{D12}$$

$$-i\bar{\omega}\tilde{u} - \frac{\tilde{f}(\alpha, \beta)}{2\pi} = \tilde{E}, \tag{D13}$$

where $\tilde{f}(\alpha, \beta)$ is the Fourier transform of the initial condition $f(x, y)$. α and β are the Fourier transform variables in the x and y directions, and ω is the Laplace transform variable. The solution of (D10) to (D13) is

$$\tilde{u} = \frac{i}{2\pi} \frac{\tilde{f}(\alpha, \beta)}{\bar{\omega} + i\kappa(\bar{\alpha}^2 + \bar{\beta}^2)}. \tag{D14}$$

The dispersion relation is given by setting the denominator of (D14) to zero, i.e.,

$$\bar{\omega} = -i\kappa(\bar{\alpha}^2 + \bar{\beta}^2). \tag{D15}$$

Note that $\bar{\omega}$ is purely imaginary for real α and β. According to the stability dia-gram of the DRP scheme (Figure 3.2) the stability limit along the negative imaginary axis of the $\bar{\omega}\Delta t$ plane is $|\bar{\omega}\Delta t| < 0.29$. Thus, by dispersion relation (D15), in order to maintain numerical stability, it is required that

$$\kappa(\bar{\alpha}^2 + \bar{\beta}^2)\Delta t < 0.29. \tag{D16}$$

(D16) may be rewritten as

$$\frac{\kappa\Delta t}{\Delta x^2}\left[(\bar{\alpha}\Delta x)^2 + (\bar{\beta}\Delta y)^2\left(\frac{\Delta x}{\Delta y}\right)^2\right] < 0.29. \tag{D17}$$

For the 7-point stencil DRP scheme, $\bar{\alpha}\Delta x$ and $\bar{\beta}\Delta y$ are less than 1.75. Thus, by replacing $\bar{\alpha}\Delta x$ and $\bar{\beta}\Delta y$ by 1.75 in (D17), the largest permissible time step, Δt_{Max} , is given by

$$\Delta t_{Max} = \frac{0.29(\Delta x)^2}{1.75^2\kappa\left[1 + \left(\frac{\Delta x}{\Delta y}\right)^2\right]}.$$

It is a fact that diffusion current always flows outward from the source. Math-ematically speaking, this means that the solution at a point, away from the source depends only on the solution at points closer to the source. Therefore, at the bound-ary of a finite computation domain, the appropriate boundary treatment is to use backward difference stencils in the boundary region. There is no need to impose special boundary conditions.

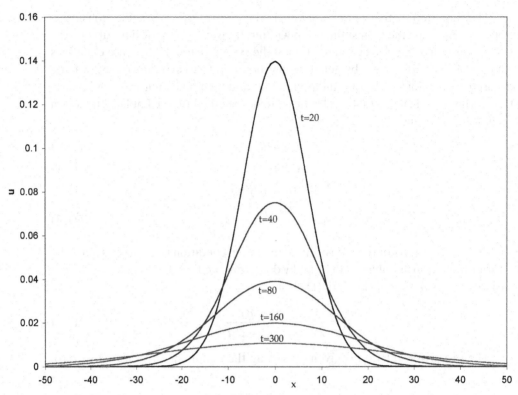

Figure D1. Diffusion of initial Gaussian pulse ($y = 0$).

As a simple numerical example, consider the case of an initial Gaussian pulse, i.e.,

$$t = 0, \quad u(x, y) = e^{-(\ln 2)\left(\frac{x^2 + y^2}{b^2}\right)}, \tag{D18}$$

where b is the half-width of the pulse. The exact solution is

$$u(x, y, t) = \frac{1}{1 + 4(\ln 2)\dfrac{\kappa t}{b^2}} e^{-\left(\frac{x^2 + y^2}{\left[\frac{b^2}{(\ln 2)} + 4\kappa t\right]}\right)}. \tag{D19}$$

Figure D1 shows the result of a computation using the 7-point stencil DRP scheme for the case $b = 3$. The length scale is $\Delta x = \Delta y$. The time scale is $\frac{(\Delta x)^2}{\kappa}$. A 100×100 mesh was used in the computation. The numerical solutions at $t = 20$, 40, 80, 160, and 300 are displayed in the figure. The exact solution is shown by a dotted line. However, it cannot be easily seen because it differs from the numerical solution by less than the thickness of the lines of the figure. The numerical solution is extremely accurate.

APPENDIX E

Accelerated Convergence to Steady State

In many aeroacoustics problems, it is sometimes advantageous to compute the mean flow first and then to turn the noise sources on and compute the sound field using the same computer code. This becomes a necessity if the sound intensity is very low; smaller than the error of the numerical mean flow solution (in comparison with the exact solution). One drawback of this approach is that it could take a long time to converge to a steady state when using an almost nondissipative computation scheme such as the dispersion-relation-preserving (DRP) scheme. On the other hand, using a more dissipative scheme might damp out the sound. Because of the nearly nondissipative nature of the scheme, the only way by which the residual associated with the initial condition can be reduced to a very low level is to have it expelled slowly through the boundaries of the computational domain. This sometimes could take a very long computation time. A method to achieve accelerated convergence to steady state without compromising accuracy would, therefore, be very useful.

To illustrate a way to attain accelerated convergence to steady state, consider a two-part problem of computing the mean flow created by a localized sink/source and then the sound field created by tiny oscillations of the sink/source strength. The problem is as shown in Figure E1. In two dimensions, the governing equations are as follows:

$$\frac{\partial \rho}{\partial t} + \rho \left(\frac{\partial u}{\partial x} + \frac{\partial v}{\partial y} \right) + u \frac{\partial \rho}{\partial x} + v \frac{\partial \rho}{\partial y} = S \tag{E1}$$

$$\frac{\partial u}{\partial t} + u \frac{\partial u}{\partial x} + v \frac{\partial u}{\partial y} + \frac{1}{\rho} \frac{\partial p}{\partial x} = -\frac{u}{\rho} S$$

$$\frac{\partial v}{\partial t} + u \frac{\partial v}{\partial x} + v \frac{\partial v}{\partial y} + \frac{1}{\rho} \frac{\partial p}{\partial y} = -\frac{v}{\rho} S \tag{E3}$$

$$\frac{\partial p}{\partial t} + u \frac{\partial p}{\partial x} + v \frac{\partial p}{\partial y} + \gamma p \left[\frac{\partial u}{\partial x} + \frac{\partial v}{\partial y} \right] = \frac{\gamma p}{\rho} S, \tag{E4}$$

where S is the sink/source distribution. Let the time-independent part of the source be

$$S(x, y) = \begin{cases} -Q, & (x - x_0)^2 + (y - y_0)^2 \le r_0^2 \\ -Q e^{-\sigma \left\{ \left[(x-x_0)^2 + (y-y_0)^2 \right]^{\frac{1}{2}} - r_0 \right\}^2}, & \text{otherwise} \end{cases} \tag{E5}$$

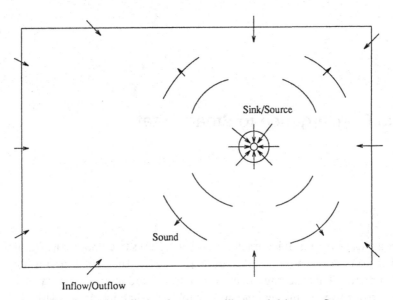

Figure E1. Acoustic radiation from an oscillatory sink/source flow.

In Eq. (E5) (x_0, y_0) are the coordinates of the center of the sink/source. Nondimensional variables with a_0 (sound speed) as the velocity scale, Δx as the length scale, $\Delta x / a_0$ as the time scale, ρ_0 as the density scale, and $\rho_0 a_0^2$ as the pressure scale are used in all computations.

In this illustration, the computational domain of Figure E1 is discretized by using a square mesh with $\Delta x = \Delta y$. A 120×100 computational grid centered at the origin of coordinates is used. The following values are assigned to the source parameters:

$$\sigma = \frac{\ln 2}{36}, \quad r_0 = 12, \quad Q = 0.0456, \quad x_0 = 20, \quad y_0 = 0.$$

To start the mean flow computation, a zero initial condition is used. The solution is time marched to a steady state by means of the 7-point stencil DRP scheme. Artificial selective damping with a mesh Reynolds number $(R_{\Delta x} = \frac{a_0 \Delta x}{\nu_a})$ equal to 10 is added to the discretized equations. With such a small amount of damping, in a trial computation, the residual is able to settle down over time to a magnitude of the order of 10^{-13} (the residual is effectively the value of the time derivative or its finite difference approximation). This is just the machine truncation accuracy (single precision). This very low background noise level is, however, a necessary condition for performing the subsequent computation of the very low intensity sound (of the order of 10^{-10}) produced by extremely tiny oscillations of the sink/source strength.

Through numerical experimentation, the following technique referred to as the "canceling the residuals" has been found to be very effective in promoting accelerated convergence. Since the difference between the numerical and the exact solution is of the order of a percent, further time marching calculation is a waste of effort in terms of improving the accuracy of the numerical solution, once the residuals reach a level of order 10^{-5}. With this in mind, it is possible to add terms that are exactly equal in magnitude, but opposite in sign to the residuals to the right side of

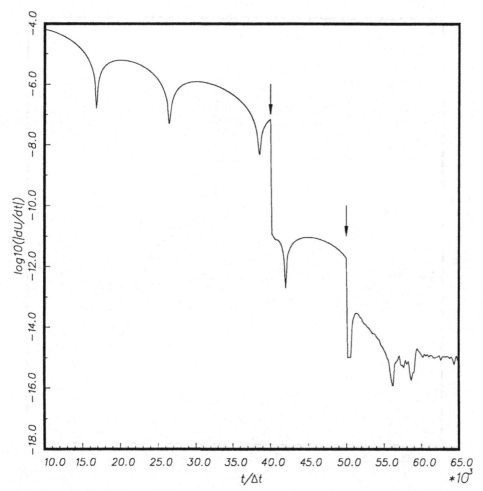

Figure E2. Time history of the residual of the u-velocity component at the grid point $(-20, -25)$. Arrows indicate the application of the canceling-the-residuals technique for accelerated convergence.

each of the governing equation. These are terms of order 10^{-5} or less and would, therefore, not materially affect the accuracy of the steady-state numerical solution. The added terms may be regarded as minute artificial distributed sources of fluid, momentum, or heat. The consequence of adding these minute source terms is to render the residuals instantaneously to zero. Of course, for a multilevel marching scheme, the residuals of the scheme are greatly reduced in this way, but they would not be exactly equal to zero at the next time level. Numerical experiments indicate that when this canceling-the-residuals procedure is performed, the overall residual of the computation drops almost instantaneously by several orders of magnitude. This method can be applied repeatedly until machine accuracy is achieved.

Figure E2 shows the time history of the residual, $\log_{10}|\frac{du}{dt}|$, at the point $(-20,-25)$. The arrows in this figure indicate when the canceling-the-residuals technique is applied. It is clear that there is a dramatic decrease (of 3 to 4 orders of magnitude) in the residual immediately after each application of the method.

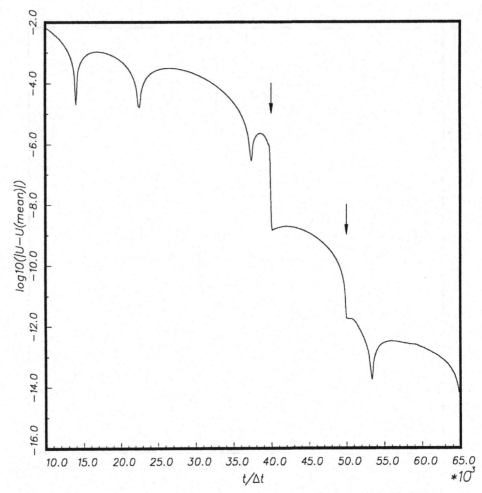

Figure E3. Convergence history of the *u*-velocity component at the point (+20, −25). Arrows indicate the application of the canceling-the-residuals technique for accelerated convergence.

Figure E3 shows the corresponding convergence history of the *u*-velocity component at the mesh point (+20, −25). The computed solution rapidly approaches the eventual mean value of the velocity component right after each application of the accelerated convergence technique. It is worthwhile to mention that the method works in all the cases that have been tested. The total computer run time needed to attain a steady-state numerical solution is drastically reduced in each instance.

APPENDIX F

Generation of Broadband Sound Waves with a Prescribed Spectrum by an Energy-Conserving Discretization Method

In this appendix, a method to generate broadband acoustic waves with a prescribed spectrum is considered. Suppose the sound waves to be generated in the time domain is to have a spectrum $S(\omega)$ as shown in Figure F1. For plane waves propagating in the x direction, the random-pressure field may be written as an inverse Fourier transform in the following form:

$$p(x, t) = \int_{-\infty}^{\infty} A(\omega) \cos\left[\omega\left(\frac{x}{a_0} - t\right) + \chi(\omega)\right] d\omega, \qquad (F1)$$

where $\chi(\omega)$ is a random-phase function, a_0 is the speed of sound, and $A(\omega)$ is the amplitude function.

For numerical computation, the spectrum $S(\omega)$ has to be discretized. A simple way is to divide the spectrum into narrow bands with bandwidth $\Delta\omega_j$ for the jth band. The bandwidths need not be constant. Let the center frequency of the jth band be ω_j. To avoid harmonic interaction, it will be required that no center frequency is the harmonic of another, i.e., $\omega_j \neq K\omega_k$, $j \neq k$, $K = $ integer.

A spectrum is a distribution of acoustic energy with respect to frequency. The energy in the jth band is $S(\omega_j)\Delta\omega_j$ as $\Delta\omega_j \to 0$. Now, for small $\Delta\omega_j$, a reasonable approximation is to assume that the energy in the jth band is concentrated in the center frequency ω_j. That is, the jth band is approximated by a single wave in the following form:

$$p_j = A_j \cos\left[\omega_j\left(\frac{x}{a_0} - t\right) + \chi_j\right].$$

A more formal way of stating the approximation is to replace the amplitude function $A(\omega)$ of Eq. (F1) by a row of delta functions, i.e.,

$$A(\omega) = \sum_{j=-\infty}^{\infty} A_j \delta(\omega - \omega_j) \qquad (F2)$$

The energy of a discrete frequency wave is equal to $\frac{1}{2}A_j^2$. Thus, to conserve energy, it is necessary to require

$$\tfrac{1}{2}A_j^2 = S(\omega)\Delta\omega_j.$$

$S(\omega)$

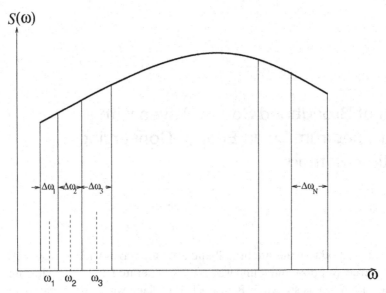

Figure F1. A prescribed energy spectrum of the broadband sound waves to be generated.

Thus,

$$A_j = [2S(\omega_j)\Delta\omega_j]^{1/2}. \tag{F3}$$

The energy-conserving discretization form of Eq. (F1) may now be written as

$$p(x,t) = \sum_{j=-N}^{N} [2S(\omega_j)\Delta\omega_j]^{1/2} \cos\left[\omega_j\left(\frac{x}{a_0} - t\right) + \chi_j\right], \tag{F4}$$

where χ_j is a random number, specifically $\chi_{-j} \neq \chi_j$.

Now, the two-point space-time correlation function of the sound wave field given by Eq. (F4) is

$$\overline{p(x',t')p(x'',t'+\tau)}$$

$$= \sum_{j=-N}^{N}\sum_{k=-N}^{N} 2[S(\omega_j)S(\omega_k)\Delta\omega_j\Delta\omega_k]^{1/2}$$

$$\overline{\times \cos\left[\omega_j\left(\frac{x'}{a_0} - t'\right) + \chi_j\right]\cos\left[\omega_k\left(\frac{x''}{a_0} - t' - \tau\right) + \chi_k\right]} \tag{F5}$$

where the overbar denotes a time average. Consider, for the moment, the time-averaged terms on the right side of Eq. (F5).

$$\overline{\cos\left[\omega_j\left(\frac{x'}{a_0} - t'\right) + \chi_j\right]\cos\left[\omega_k\left(\frac{x''}{a_0} - t' - \tau\right) + \chi_k\right]}$$

$$= \lim_{T\to\infty}\frac{1}{2T}\int_{-T}^{T} \cos\left[\omega_j\left(\frac{x'}{a_0} - t'\right) + \chi_j\right]\cos\left[\omega_k\left(\frac{x''}{a_0} - t' - \tau\right) + \chi_k\right] dt'$$

$$= \lim_{T\to\infty}\frac{1}{2T}\int_{-T}^{T} \cos\left[\omega_j\left(\frac{x'}{a_0} - t'\right) + \chi_j\right]\cos\left[\omega_k\left(\frac{x'}{a_0} - t'\right) + \chi_k + \omega_k\left(\frac{x'' - x'}{a_0} - \tau\right)\right] dt'$$

This time average is equal to zero except when $k = j$. In this special case, by expanding the second cosine term on the right side by compound angle formula, it is readily found that

$$\overline{\cos\left[\omega_j\left(\frac{x'}{a_0} - t'\right) + \chi_j\right]\cos\left[\omega_k\left(\frac{x''}{a_0} - t' - \tau\right) + \chi_k\right]} = \frac{1}{2}\cos\left[\omega_k\left(\frac{x'' - x'}{a_0} - \tau\right)\right]\delta_{jk},$$

$$(F6)$$

where δ_{jk} is the Kronecker delta. Therefore, Eq. (F5) reduces to

$$\overline{p\left(x', t'\right)p\left(x'', t' + \tau\right)} = \sum_{k=-N}^{N} S\left(\omega_k\right)\Delta\omega_k \cos\left[\omega_k\left(\frac{x'' - x'}{a_0} - \tau\right)\right].\qquad(F7)$$

The autocorrelation function of the wave field may be found by setting $x'' = x'$ in Eq. (F7). This yields

$$\overline{p(x', t')p(x', t' + \tau)} = \sum_{k=-N}^{N} S(\omega_k)\Delta\omega_k \cos(\omega_k\tau).\qquad(F8)$$

The power of the acoustic field is given by setting $\tau = 0$. Thus,

$$\overline{p^2(x', t')} = \sum_{k=-N}^{N} S(\omega_k)\Delta\omega_k \rightarrow \int_{-\infty}^{\infty} S(\omega)d\omega; \text{ in the limit } \Delta\omega_k \rightarrow 0.\qquad(F9)$$

Eq. (F9) indicates that $S(\omega)$ is the spectrum of the wave field.

Another way to find the spectrum is to make use of the fact that the spectrum is the Fourier transform of the autocorrelation function. That is, if $\overline{S}(\omega)$ is the spectrum of the wave field given by Eq. (F4), then it is equal to the Fourier transform of the right side of Eq. (F8) as follows:

$$\overline{S}(\omega) = \frac{1}{2\pi}\int_{-\infty}^{\infty} \sum_{k=-N}^{N} S(\omega_k)\Delta\omega_k \cos(\omega_k\tau)e^{i\omega\tau}\,d\tau.$$

In the limit $\Delta\omega_k \rightarrow 0$,

$$\overline{S}(\omega) = \frac{1}{2\pi}\int_{-\infty}^{\infty}\int_{-\infty}^{\infty} S(\Omega)\cos(\Omega\tau)e^{i\omega\tau}\,d\tau\,d\Omega$$

$$= \frac{1}{2}[S(\omega) + S(-\omega)] = S(\omega).\qquad(F10)$$

The usual symmetry assumption, i.e., $S(-\omega) = S(\omega)$ is assumed here.

A special case of interest is the sound field associated with a broadband monopole source. Let (Y, θ, ϕ) be the spherical coordinates system centered at the monopole. Y is the radial coordinate (see Figure 14.12). The sound field according to the energy conserving discretization procedure is

$$p(Y, t) = \sum_{j=-N}^{N}\left[2S(\omega_j)\Delta\omega_j\right]^{\frac{1}{2}}\frac{1}{Y}\cos\left[\omega_j\left(\frac{Y}{a_0} - t\right) + \chi_j\right].\qquad(F11)$$

APPENDIX G

Sample Computer Programs

Three sample programs are provided in this appendix. The first set of programs is for solving the one-dimensional convective wave equation. There are two programs. The first program uses the DRP scheme. The second program uses the standard 2nd, 4th and 6th order central difference scheme for spatial discretization and the Runge-Kutta scheme for time marching. The second set of programs is for solving the linearized two-dimensional Euler equations. The DRP scheme is used. The third set of programs is for solving the non-linear three-dimensional Euler equations. The DRP scheme is used.

G.1 Sample Program for Solving the One-Dimensional Convective Wave Equation

The numerical solution to the following problem in an infinite domain is considered. The convective equation in dimensionless form is

$$\frac{\partial u}{\partial t} + \frac{\partial u}{\partial x} = 0$$

Two types of initial conditions are used in this example

1. A Gaussian function.
2. A box car function i.e.,

$$u(x, 0) = H(x + L) - H(x - L)$$

where $H(x)$ is the unit step function.

The Fortran 90 program below computes the solution using the DRP scheme.

```
program DRP_1D
!
!-----------------------------------------------------------------
!
! This code solves the 1-D convective wave equation on the real line
!    using the DRP scheme
!   -equation is non-dimensionalized by mesh spacing and wave speed
```

```
! -equation is discretized with 7-point spatial stencils and
!  4 level multi-step time advancement
! -the solution is advanced from the initial conditions to the
!  desired time and then output to file
!
! Variables:
!       id    -number of mesh points to the left and right of origin
!       nmax    -maximum number of time steps
!       t     -time
!       delt    -time step
!       u     -solution values
!       k     -time derivatives, dudt, at different time levels
!       k1    -converts time level to index in k array
!       a     -spatial stencil coefficients
!       b     -multi-step advancement coefficients
!
!----------------------------------------------------------------
!
implicit none
integer, parameter::rkind=8,id=800
integer::n,l,nmax,bh
integer, dimension(-3:0)::k1
real(kind=rkind)::delt,h,t
real(kind=rkind), dimension(-id-3:id+3)::u
real(kind=rkind), dimension(-3:0,-id:id)::k
real(kind=rkind), dimension(-3:3)::a
real(kind=rkind), dimension(-3:0)::b
integer::outdata=20
!
!    Open output data file 'out.dat'
!
open(unit=outdata, file="outDRP.dat", status="replace")
!
!    Optimized 7-point spatial stencil
!
a(-3)=-0.02003774846106832_rkind
a(-2)=0.163484327177606692_rkind
a(-1)=-0.766855408972008323_rkind
a(0)=0.0_rkind
a(1)=0.766855408972008323_rkind
a(2)=-0.163484327177606692_rkind
a(3)=0.02003774846106832_rkind
!
!    Optimized multi-step
!
b(0)=2.3025580888_rkind
```

```
b(-1)=-2.4910075998_rkind
b(-2)=1.5743409332_rkind
b(-3)=-0.3858914222_rkind
!
!    Set simulation parameters
!
delt=0.1_rkind ! time step size (stability/accuracy limit is 0.2111)
nmax=3000 ! maximum number of time steps
!
!    Set 'boundary' conditions (assume u=0 far from origin)
!
u(-id-3)=0.0_rkind
u(-id-2)=0.0_rkind
u(-id-1)=0.0_rkind
u(id+1)=0.0_rkind
u(id+1)=0.0_rkind
u(id+3)=0.0_rkind
!
!    Set initial conditions
!
t=0.0_rkind
!
!    Gaussian function with half-width 'bh' and height 'h'
!
!bh=3
!h=0.5_rkind
!do l=-id,id
! u(l)=h*exp(-log(2.0_rkind)*(real(l,rkind)/real(bh,rkind))**2)
!end do
!
!    Boxcar
!
u=0.0_rkind
u(-50:50)=1.0_rkind
!
!    Set solution to zero for t<0
!
k=0.0_rkind
k1(-3)=-3
k1(-2)=-2
k1(-1)=-1
k1(0)=0
!
!----------------------------------------------------------------
!
!    Main loop - march nmax steps with multi-level DRP
```

```
!
do n=1,nmax
!
t=t+delt
write (*,*)"n: ",n," t: ",real(t,4)
!
!     Update k
!
do l=-id,id
 k(kl(0),l)=a(-3)*u(l-3)+a(-2)*u(l-2)+a(-1)*u(l-1)+a(1)*u(l+1)+a(2)*u(l+2)+a(3)*u(l+3)
end do
!
!     Update u to time level n
!
do l=-id,id
 u(l)=u(l)-delt*(b(0)*k(kl(0),l)+b(-1)*k(kl(-1),l)+b(-2)*k(kl(-2),l)+b(-3)*k(kl(-3),l))
end do
!
!     Shift k array back
!
kl=cshift(kl,1)
!
end do
!
!     Output u results
!
50 format(i8,e16.7e3)
do l=-id,id
 write (outdata,50)l,u(l)
end do
!
stop
end program DRP_1D
```

The Fortran 90 program below computes the solution to the convective wave equation by means of the standard 2nd, 4th and 6th order central difference scheme for spatial discretization and the Runge-Kutta scheme for time marching.

```
program RK4_1D
!
!-----------------------------------------------------------------
!
!     This code solves the 1-D convective wave equation on the real line
!        using the standard fourth order Runge-Kutta scheme
!         -equation is non-dimensionalized by mesh spacing and wave speed
!         -equation can be discretized with either 6th,4th,2nd order spatial stencils
```

```
!    -the solution is advanced from the initial conditions to the
!     desired time and then output to file
!
!    Variables:
!         id   -number of mesh points to the left and right of origin
!         nmax   -maximum number of time steps
!         t    -time
!         delt -time step
!         u,ut  -solution values
!         k,k1,k2 -time derivatives, dudt
!         a2    -2nd order spatial stencil coefficients
!         a4    -4th order spatial stencil coefficients
!         a6    -6th order spatial stencil coefficients
!
!-------------------------------------------------------------
!
implicit none
integer, parameter::rkind=8,id=800
integer::n,l,nmax,bh,iorder
real(kind=rkind)::delt,dth,dt6,h,t
real(kind=rkind), dimension(-id-3:id+3)::u,ut
real(kind=rkind), dimension(-id:id)::k,k1,k2
real(kind=rkind), dimension(-1:1)::a2
real(kind=rkind), dimension(-2:2)::a4
real(kind=rkind), dimension(-3:3)::a6
integer::outdata=20
!
!    Open output data file 'out.dat'
!
open(unit=outdata, file="outRK4.dat", status="replace")
!
!    2nd order spatial stencil
!
a2(-1)=-0.5_rkind
a2(0)=0.0_rkind
a2(1)=0.5_rkind
!
!    4th order spatial stencil
!
a4(-2)=1.0_rkind/12.0_rkind
a4(-1)=-2.0_rkind/3.0_rkind
a4(0)=0.0_rkind
a4(1)=2.0_rkind/3.0_rkind
a4(2)=-1.0_rkind/12.0_rkind
!
!    6th order stencil
```

```
!
a6(-3)=-1.0_rkind/60.0_rkind
a6(-2)=3.0_rkind/20.0_rkind
a6(-1)=-3.0_rkind/4.0_rkind
a6(0)=0.0_rkind
a6(1)=3.0_rkind/4.0_rkind
a6(2)=-3.0_rkind/20.0_rkind
a6(3)=1.0_rkind/60.0_rkind
!
!    Set simulation parameters
!
delt=0.4_rkind ! time step size
dth=delt*0.5_rkind
dt6=delt/6.0_rkind
nmax=750 ! maximum number of time steps
iorder=6 ! order of the spatial stencil (2,4,6)
!
!    Set 'boundary' conditions (assume u=0 far from origin)
!
u(-id-3)=0.0_rkind
u(-id-2)=0.0_rkind
u(-id-1)=0.0_rkind
u(id+1)=0.0_rkind
u(id+1)=0.0_rkind
u(id+3)=0.0_rkind
!
ut=0.0_rkind
!
!    Set initial conditions
!
t=0.0_rkind
!
!    Gaussian function with half-width 'bh' and height 'h'
!
!bh=3
!h=0.5_rkind
!do l=-id,id
! u(l)=h*exp(-log(2.0_rkind)*(real(l,rkind)/real(bh,rkind))**2)
!end do
!
!    Boxcar
!
u=0.0_rkind
u(-50:50)=1.0_rkind
!
!    Initial du/dt
```

```
!
call dudt(u,k,iorder)
!
!----------------------------------------------------------------
!
!      Main loop - march nmax steps in time using RK4 integration
!
do n=1,nmax
!
t=t+delt
write (*,*)"n: ",n," t: ",real(t,4)
!
!      Second step
!
do l=-id,id
 ut(l)=u(l)+dth*k(l)
end do
call dudt(ut,k1,iorder)
!
!      Third step
!
do l=-id,id
 ut(l)=u(l)+dth*k1(l)
end do
call dudt(ut,k2,iorder)
!
!      Fourth step
!
do l=-id,id
 ut(l)=u(l)+delt*k2(l)
 k2(l)=k2(l)+k1(l)
end do
call dudt(ut,k1,iorder)
!
!      Update u
!
do l=-id,id
 u(l)=u(l)+dt6*(k(l)+k1(l)+2.0_rkind*k2(l))
end do
!
!      Update dudt
!
call dudt(u,k,iorder)
!
end do
!
```

```fortran
!     Output u results
!
50 format(i8,e16.7e3)
do l=-id,id
 write (outdata,50)l,u(l)
end do
!
stop
!
contains
!
subroutine dudt(f,kf,iflag)
!
!!!!!!!!!!!!!!!!!!!!!!!!!!!!!!!!!!!!!!!!!!!!!!!!!!!!!!!!!!!!!!!!!!!!!!!
!
!     Calculate the derivative du/dt=-du/dx of the convectice wave eqn
!     -iflag is a flag for the spatial stencil (2nd,4th,6th order)
!
!!!!!!!!!!!!!!!!!!!!!!!!!!!!!!!!!!!!!!!!!!!!!!!!!!!!!!!!!!!!!!!!!!!!!!!!
!
implicit none
integer, intent(in)::iflag
real(kind=rkind), intent(in), dimension(-id-3:id+3)::f
real(kind=rkind), intent(out), dimension(-id:id)::kf
integer::l
!
select case (iflag)
  case (2)
  do l=-id,id
    kf(l)=-(a2(-1)*f(l-1)+a2(1)*f(l+1))
  end do
  case (4)
  do l=-id,id
    kf(l)=-(a4(-2)*f(l-2)+a4(-1)*f(l-1)+a4(1)*f(l+1)+a4(2)*f(l+2))
  end do
  case (6)
  do l=-id,id
    kf(l)=-(a6(-3)*f(l-3)+a6(-2)*f(l-2)+a6(-1)*f(l-1)+a6(1)*f(l+1)+a6(2)*f(l+2)+a6(3)*f(l+3))
  end do
end select
!
return
end subroutine dudt
!
end program RK4_1D
```

Samples of computed results and the exact solutions are shown below.

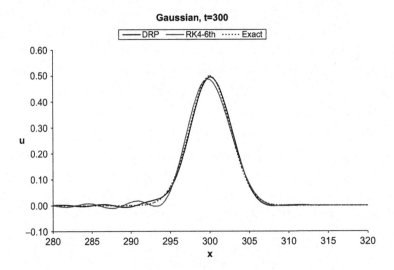

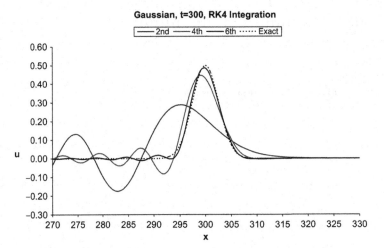

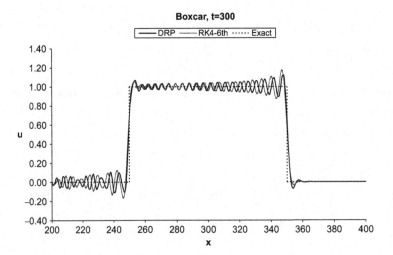

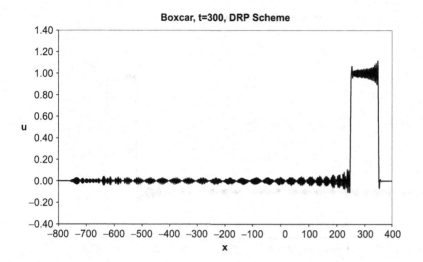

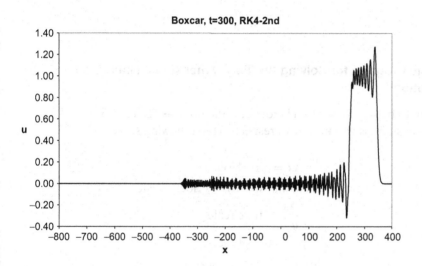

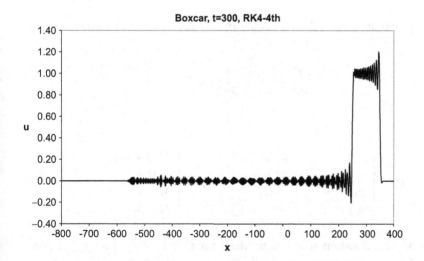

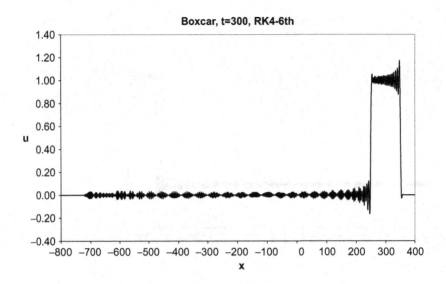

G.2 Sample Program for Solving the Two-Dimensional Linearized Euler Equations

The problem to be solved is stated below. It is the same as Exercise 6.3b.

Use dimensionless variables with respect to the following scales

$$\Delta x = \text{length scale}$$
$$a_\infty \, (\text{ambient sound speed}) = \text{velocity scale}$$
$$\frac{\Delta x}{a_\infty} = \text{time scale}$$
$$\rho_\infty = \text{density scale}$$
$$\rho_\infty a_\infty^2 = \text{pressure scale}$$

The linearized two-dimensional Euler equations on a uniform mean flow are to be solved

$$\frac{\partial \mathbf{U}}{\partial t} + \frac{\partial \mathbf{E}}{\partial x} + \frac{\partial \mathbf{F}}{\partial y} = 0$$

where

$$\mathbf{U} = \begin{bmatrix} \rho \\ u \\ v \\ p \end{bmatrix}, \quad \mathbf{E} = \begin{bmatrix} M_x \rho + u \\ M_x u + p \\ M_x v \\ M_x p + u \end{bmatrix}, \quad \mathbf{F} = \begin{bmatrix} M_y \rho + v \\ M_y u \\ M_y v + p \\ M_y p + v \end{bmatrix},$$

M_x and M_y are constant mean flow Mach number in the x and y direction, respectively. Use a computation domain $-100 \le x \le 100$, $-100 \le y \le 100$ embedded in free space.

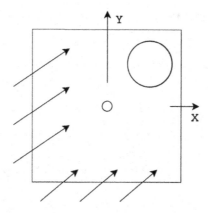

Let $M_x = M_y = 0.5\cos(\pi/4)$. Solve the initial value problem, $t = 0$.

$$p = \exp\left[-(\ln 2)\left(\frac{x^2 + y^2}{9}\right)\right]$$

$$\rho = \exp\left[-(\ln 2)\left(\frac{x^2 + y^2}{9}\right)\right] + 0.1\exp\left[-(\ln 2)\left(\frac{(x - 67)^2 + (y - 67)^2}{25}\right)\right]$$

$$u = 0.04\,(y - 67)\exp\left[-(\ln 2)\left(\frac{(x - 67)^2 + (y - 67)^2}{25}\right)\right]$$

$$v = -0.04\,(x - 67)\exp\left[-(\ln 2)\left(\frac{(x - 67)^2 + (y - 67)^2}{25}\right)\right]$$

Note: The mean flow is in the direction of the diagonal of the computational domain.
Give the distributions of p, ρ and u at $t = 60, 80, 100, 200$ and 600.

The Fortran 77 program below computes the solution using the DRP scheme.
Note: All Fortran 77 statements should start on the 7th column except comments
and the continuation symbol. The continuation symbol should be in the 6th column.

```
c   This code is designed to solve Problem 6.3(b)


    Program Test_radiation_boundary_conditions


c   Solves an initial value problem governed by the linearized two
c   dimensional Euler equationi on a uniform mean flow with Mach number
c   (Mx,My)


c   A computational domain -M < x < M, -N < y < N embedded in
c   free space with dx = dy is used.


c   Radiation boundary conditions are imposed at the bottom
c   and the left boundaries of the computational domain.
c   Outflow boundary conditions are imposed on the top and right
c   side of the computational domain.


c   To compute the solution, the DRP scheme is used in the interior
c   of the computational domain. On the boundaries optimized seven
```

```
c   ponts backward differences are used.


c   Variables computed:


c   ro(-M:M,-N:N) - density
c   p(-M:M,-N:N) - pressure
c   u(-M:M,-N:N) - u component of velocity
c   v(-M:M,-N:N) - v component of velocity


c   Input parameters: Time - time at which solution is to be computed
c   e.g. Time = 80.d0


c   Output is written to files ro.diag.dat (unit 22) - The spatial
c   distribution of density along the diagonal line x = y
c   ro.dat (unit 23) - density as a function of coordinates (x,y)




c-----------------------------------------------------------------
c    Main program
c-----------------------------------------------------------------


     Implicit none
     Double precision Time, dx, dy, dt, Mx, My
     Integer N, M, N_Time, i, j, k

c    Set the size of the domain

     Parameter( N = 100, M = 100)

     Double precision ro(-M:M,-N:N), p(-M:M,-N:N)
     Double precision u(-M:M,-N:N), v(-M:M,-N:N)
     Double precision drodt(-M:M,-N:N,0:3), dpdt(-M:M,-N:N,0:3)
     Double precision dudt(-M:M,-N:N,0:3), dvdt(-M:M,-N:N,0:3)
     Double precision pin, vin, uin, roin
     External pin, vin, uin, roin
     common Mx, My, dx, dy, dt

c    Mach number

     Mx = 0.35355339D+00
     My = 0.35355339D+00


c    Delta x and delta y

     dx = 1.0000000000D+00
     dy = 1.0000000000D+00


c    Time step
```

```fortran
      dt = (0.190D+00*dx)/(1.750D+00*(Mx + My*dx/dy +
     &  sqrt(1.00D+00 + (dx/dy)**2.00D+00)))


      open( unit = 22, file = 'ro.diag.dat')
      open( unit = 23, file = 'ro.dat')


      write(*,*) 'Time = '
      read(*,*) Time
      N_Time = int(Time/dt)


      write(*,*) N_Time
      write(*,*) dt

c    Initial conditions

      call init(ro,u,v,p,drodt,dudt,dvdt,dpdt)

c    Compute Solution

      Do k=1,N_Time
       call tderiv(ro,u,v,p,drodt,dudt,dvdt,dpdt)
      ENDDO

c    Write output to the file

      Do i=-M,M
       write(22,*) ro(i,i)
      ENDDO
      Do j=-N,N
       write(23,*) ( ro(i,j),i=-M,M )
      ENDDO


      END
c=====================================================
c  End of the main program
c=====================================================


c    Function specifying initial density

      Double precision function roin(x,y)
      Implicit none
      Double precision x, y
      roin = dexp(-dlog(2.00D+00)*((x**2.0D+00+y**2.0D+00)/9.0D+00))
     & + 0.100D+00*dexp(-dlog(2.00D+00)*
     & ((x - 67.0D+00)**2.00D+00+(y - 67.0D+00)**2.00D+00)/25.00D+00)
      RETURN
      END
```

```
c  Function specifying initial u component of velocty

   Double precision function uin(x,y)
   Implicit none
   Double precision x, y
   uin = 0.040D+00*(y - 67.00D+00)*
  & dexp(-dlog(2.00D+00)*
  & (((x - 67.0D+00)**2.00D+00+(y - 67.0D+00)**2.00D+00)/25.00D+00))
   RETURN
   END


c  Function specifying initial v component of velocity

   Double precision function vin(x,y)
   Implicit none
   Double precision x, y
   vin = -0.040D+00*(x - 67.00D+00)*
  & dexp(-dlog(2.00D+00)*
  & ((x - 67.0D+00)**2.00D+00+(y - 67.0D+00)**2.00D+00)/25.00D+00)
   RETURN
   END


c  Function specifying initial pressure

   Double precision function pin(x,y)
   Implicit none
   Double precision x, y
   pin = dexp(-dlog(2.00D+00)*((x**2.0D+00+y**2.0D+00)/9.0D+00))
   RETURN
   END


c  Set initial conditions

   subroutine init(ro,u,v,p,drodt,dudt,dvdt,dpdt)
   Implicit none
   common Mx, My, dx, dy, dt
   Integer N, M, i, j, k
   Parameter( N = 100, M = 100)
   Double precision ro(-M:M,-N:N), p(-M:M,-N:N)
   Double precision u(-M:M,-N:N), v(-M:M,-N:N)
   Double precision drodt(-M:M,-N:N,0:3), dpdt(-M:M,-N:N,0:3)
   Double precision dudt(-M:M,-N:N,0:3), dvdt(-M:M,-N:N,0:3)
   Double precision dx, dy, dt,x,y, roin, uin, vin, pin, Mx, My

   Do i = -M,M
    Do j = -N,N
```

```
    x = float(i)*dx
    y = float(j)*dy
    Do k=0,3
      drodt(i,j,k) = 0.00D+00
      dudt(i,j,k) = 0.00D+00
      dvdt(i,j,k) = 0.00D+00
      dpdt(i,j,k) = 0.00D+00
    ENDDO
    ro(i,j) = roin(x,y)
    u(i,j) = uin(x,y)
    v(i,j) = vin(x,y)
    p(i,j) = pin(x,y)
   ENDDO
  ENDDO
  RETURN
  END


c  Compute x derivative at the boundaries

   subroutine bnd_xderiv(p,i,j,a,i1,i2,dpdx)
   Implicit none
   common Mx, My, dx, dy, dt
   Integer N, M, i, j, i1, i2, k
   Parameter( N = 100, M = 100)
   Double precision p(-M:M,-N:N), a(-6:6), dx, dpdx
   Double precision Mx, My, dy, dt
   dpdx = 0.00D+00
   Do k=i1,i2
    dpdx = dpdx + a(k)*p(i+k,j)/dx
   ENDDO
   RETURN
   END


c  Compute y derivative at the boundaries

   subroutine bnd_yderiv(p,i,j,b,j1,j2,dpdy)
   Implicit none
   common Mx, My,dx, dy, dt
   Integer N, M, i, j, j1, j2, k
   Parameter( N = 100, M = 100)
   Double precision p(-M:M,-N:N),b(-6:6),dy,dpdy,dx,dt,Mx,My
   dpdy = 0.00D+00
   Do k=j1,j2
    dpdy = dpdy + b(k)*p(i,j+k)/dy
   ENDDO
   RETURN
   END
```

```
c   Compute central x derivative

      subroutine centr_xderiv(p,i,j,dpdx)
      Implicit none
      common Mx, My, dx, dy, dt
      Integer N, M, i, j, k
      Parameter( N = 100, M = 100)
      Double precision p(-M:M,-N:N),a(3),dx,dpdx,dy,Mx,My,dt
      a(1) = 0.770882380518D+00
      a(2) = -0.166705904415D+00
      a(3) = 0.0208431427703D+00
      dpdx = 0.00D+00
      Do k=1,3
       dpdx = dpdx + (a(k)*(p(i+k,j) - p(i-k,j)))/dx
      ENDDO
      RETURN
      END


c   Compute central y derivative

      subroutine centr_yderiv(p,i,j,dpdy)
      Implicit none
      common Mx, My, dx, dy, dt
      Integer N, M, i, j, k
      Parameter( N = 100, M = 100)
      Double precision p(-M:M,-N:N),a(3),dy,dpdy,dx,dt,Mx,My
      a(1) = 0.770882380518D+00
      a(2) = -0.166705904415D+00
      a(3) = 0.0208431427703D+00
      dpdy = 0.00D+00
      Do k=1,3
       dpdy = dpdy + (a(k)*(p(i,j+k) - p(i,j-k)))/dy
      ENDDO
      RETURN
      END


c   Compute radiation boundary conditions

      subroutine radiation_bc(p,i,j,dpdx,dpdy,dpdt)
      Implicit none
      common Mx, My, dx, dy, dt
      Integer N, M, i, j
      Parameter( N = 100, M = 100)
      Double precision p(-M:M,-N:N), dpdt(-M:M,-N:N,0:3)
      Double precision Vth, r, cs, sn, dpdx, dpdy
      Double precision Mx, My, dx, dy, dt, x, y
      x = float(i)*dx
      y = float(j)*dy
```

```fortran
      r = dsqrt(x**2.000D+00 + y**2.000D+00)
      cs = x/r
      sn = y/r
      Vth = Mx*cs + My*sn + dsqrt(1.00D+00 -
     &     (My*cs - Mx*sn)**2.00D+00)
      dpdt(i,j,0) = -Vth*(cs*dpdx + sn*dpdy + p(i,j)/(2.00D+00*r))
      RETURN
      END


c   Compute outflow boundary conditions for velocity

      subroutine vel_outbc(u,v,i,j,dudx,dudy,dvdx,dvdy,
     &               dpdx,dpdy,dudt,dvdt)
      Implicit none
      common Mx, My, dx, dy, dt
      Integer N, M, i, j
      Parameter( N = 100, M = 100)
      Double precision u(-M:M,-N:N), v(-M:M,-N:N)
      Double precision dudt(-M:M,-N:N,0:3), dvdt(-M:M,-N:N,0:3)
      Double precision dt,dudx,dudy,dvdx,dvdy,dpdx,dpdy
      Double precision Mx, My, dx, dy

      dudt(i,j,0) = -(Mx*dudx + My*dudy + dpdx)
      dvdt(i,j,0) = -(Mx*dvdx + My*dvdy + dpdy)
      RETURN
      END


c   Compute outflow boundary conditions for density

      subroutine ro_outbc(ro,i,j,dpdx,dpdy,dpdt,
     &               drodx,drody,drodt)
      Implicit none
      common Mx, My, dx, dy, dt
      Integer N, M, i, j
      Parameter( N = 100, M = 100)
      Double precision ro(-M:M,-N:N),Mx,My,dx,dy,
     &drodt(-M:M,-N:N,0:3), dpdt(-M:M,-N:N,0:3)
      Double precision dt,dpdx,dpdy,drodx,drody

      drodt(i,j,0) = -Mx*drodx - My*drody + dpdt(i,j,0) +
     &          Mx*dpdx + My*dpdy
      RETURN
      END


c   Set appropriate coefficients for derivatives

      subroutine change_ab(a,b,i,j)
      Implicit none
```

```
      Integer i, j, k
      Double precision a(-6:6), b(i:j)
      Do k=i,j
       a(k) = b(k)
      ENDDO
      RETURN
      END


C=========================================================
c  Advance to the next time level
C=========================================================

      subroutine tderiv(ro,u,v,p,drodt,dudt,dvdt,dpdt)
      implicit none
      common Mx, My, dx, dy, dt
      Integer N, M, i1, i2, j1, j2, i, j, k
      Parameter( N = 100, M = 100)
      Double precision ro(-M:M,-N:N), p(-M:M,-N:N), b(0:3), a(3)
      Double precision u(-M:M,-N:N), v(-M:M,-N:N)
      Double precision drodt(-M:M,-N:N,0:3), dpdt(-M:M,-N:N,0:3)
      Double precision dudt(-M:M,-N:N,0:3), dvdt(-M:M,-N:N,0:3)
      Double precision dx, dy, dt, ac(-6:6), bc(-6:6),
     & x,y,cs,sn,Vth,r, Mx, My, sum1, sum2, sum3, sum4
      Double precision drody, drodx, dudx, dpdy, dpdx,
     & dvdy, dvdx, dudy
      Double precision a06(0:6), a15(-1:5), a24(-2:4),
     &          a60(-6:0), a51(-5:1), a42(-4:2)

      b(0) = 2.30255809D+00
      b(1) = -2.49100760333333238D+00
      b(2) = 1.57434093666666630D+00
      b(3) = -0.385891423333333039D+00


      a(1) = 0.770882380518D+00
      a(2) = -0.166705904415D+00
      a(3) = 0.0208431427703D+00


C-------------------------------------------------------------

      a06(0) = -2.19228033900D+00
      a06(1) = 4.74861140100D+00
      a06(2) = -5.10885191500D+00
      a06(3) = 4.46156710400D+00
      a06(4) = -2.83349874100D+00
      a06(5) = 1.12832886100D+00
      a06(6) = -0.20387637100D+00
```

```
    a15(-1) = -0.20933762200D+00
    a15(0) = -1.08487567600D+00
    a15(1) = 2.14777605000D+00
    a15(2) = -1.38892832200D+00
    a15(3) = 0.76894976600D+00
    a15(4) = -0.28181465000D+00
    a15(5) = 0.48230454000D-01


    a24(-2) = 0.49041958000D-01
    a24(-1) = -0.46884035700D+00
    a24(0) = -0.47476091400D+00
    a24(1) = 1.27327473700D+00
    a24(2) = -0.51848452600D+00
    a24(3) = 0.16613853300D+00
    a24(4) = -0.26369431000D-01


c---------------------------------------------------------------

    Do k=-6,6
     ac(k) = 0.00D+00
     bc(k) = 0.00D+00
    ENDDO
    Do k=-6,0
     a60(k) = -a06(-k)
    ENDDO
    Do k=-5,1
     a51(k) = -a15(-k)
    ENDDO
    Do k=-4,2
     a42(k) = -a24(-k)
    ENDDO


c---------------------------------------------------------------
c  Lower left hand corner
c---------------------------------------------------------------

    Do j=0,2
     IF(j.EQ.0) THEN
      j1 = 0
      j2 = 6
      call change_ab(bc,a06,0,6)
     ELSE IF(j.EQ.1) THEN
      j1 = -1
      j2 = 5
      call change_ab(bc,a15,-1,5)
     ELSE
      j1 = -2
```

```
       j2 = 4
       call change_ab(bc,a24,-2,4)
     ENDIF
    Do i=0,2
     IF(i.EQ.0) THEN
      i1 = 0
      i2 = 6
      call change_ab(ac,a06,0,6)
     ELSE IF(i.EQ.1) THEN
      i1 = -1
      i2 = 5
      call change_ab(ac,a15,-1,5)
     ELSE
      i1 = -2
      i2 = 4
      call change_ab(ac,a24,-2,4)
     ENDIF
     call bnd_xderiv(p,-M+i,-N+j,ac,i1,i2,dpdx)
     call bnd_yderiv(p,-M+i,-N+j,bc,j1,j2,dpdy)
     call radiation_bc(p,-M+i,-N+j,dpdx,dpdy,dpdt)
     call bnd_xderiv(u,-M+i,-N+j,ac,i1,i2,dudx)
     call bnd_yderiv(u,-M+i,-N+j,bc,j1,j2,dudy)
     call radiation_bc(u,-M+i,-N+j,dudx,dudy,dudt)
     call bnd_xderiv(v,-M+i,-N+j,ac,i1,i2,dvdx)
     call bnd_yderiv(v,-M+i,-N+j,bc,j1,j2,dvdy)
     call radiation_bc(v,-M+i,-N+j,dvdx,dvdy,dvdt)
     call bnd_xderiv(ro,-M+i,-N+j,ac,i1,i2,drodx)
     call bnd_yderiv(ro,-M+i,-N+j,bc,j1,j2,drody)
     call radiation_bc(ro,-M+i,-N+j,drodx,drody,drodt)
    ENDDO
    ENDDO


c---------------------------------------------------------------
c
c  Upper left hand corner
c---------------------------------------------------------------
    Do j=0,2
     IF(j.EQ.0) THEN
      j1 = -6
      j2 = 0
      call change_ab(bc,a60,-6,0)
     ELSE IF(j.EQ.1) THEN
      j1 = -5
      j2 = 1
      call change_ab(bc,a51,-5,1)
     ELSE
      j1 = -4
```

```
      j2 = 2
      call change_ab(bc,a42,-4,2)
     ENDIF
    Do i=0,2
     IF(i.EQ.0) THEN
      i1 = 0
      i2 = 6
      call change_ab(ac,a06,0,6)
     ELSE IF(i.EQ.1) THEN
      i1 = -1
      i2 = 5
      call change_ab(ac,a15,-1,5)
     ELSE
      i1 = -2
      i2 = 4
      call change_ab(ac,a24,-2,4)
     ENDIF
    call bnd_xderiv(p,-M+i,N-j,ac,i1,i2,dpdx)
    call bnd_yderiv(p,-M+i,N-j,bc,j1,j2,dpdy)
    call radiation_bc(p,-M+i,N-j,dpdx,dpdy,dpdt)
    call bnd_xderiv(u,-M+i,N-j,ac,i1,i2,dudx)
    call bnd_yderiv(u,-M+i,N-j,bc,j1,j2,dudy)
    call radiation_bc(u,-M+i,N-j,dudx,dudy,dudt)
    call bnd_xderiv(v,-M+i,N-j,ac,i1,i2,dvdx)
    call bnd_yderiv(v,-M+i,N-j,bc,j1,j2,dvdy)
    call radiation_bc(v,-M+i,N-j,dvdx,dvdy,dvdt)
    call bnd_xderiv(ro,-M+i,N-j,ac,i1,i2,drodx)
    call bnd_yderiv(ro,-M+i,N-j,bc,j1,j2,drody)
    call radiation_bc(ro,-M+i,N-j,drodx,drody,drodt)
    ENDDO
    ENDDO


c----------------------------------------------------------------
c  Lower right hand corner
c----------------------------------------------------------------

    Do j=0,2
     IF(j.EQ.0) THEN
      j1 = 0
      j2 = 6
      call change_ab(bc,a06,0,6)
     ELSE IF(j.EQ.1) THEN
      j1 = -1
      j2 = 5
      call change_ab(bc,a15,-1,5)
     ELSE
      j1 = -2
```

```
          j2 = 4
          call change_ab(bc,a24,-2,4)
        ENDIF
       Do i=0,2
        IF(i.EQ.0) THEN
         i1 = -6
         i2 = 0
         call change_ab(ac,a60,-6,0)
        ELSE IF(i.EQ.1) THEN
         i1 = -5
         i2 = 1
         call change_ab(ac,a51,-5,1)
        ELSE
         i1 = -4
         i2 = 2
         call change_ab(ac,a42,-4,2)
        ENDIF
        call bnd_xderiv(p,M-i,-N+j,ac,i1,i2,dpdx)
        call bnd_yderiv(p,M-i,-N+j,bc,j1,j2,dpdy)
        call radiation_bc(p,M-i,-N+j,dpdx,dpdy,dpdt)
        call bnd_xderiv(u,M-i,-N+j,ac,i1,i2,dudx)
        call bnd_yderiv(u,M-i,-N+j,bc,j1,j2,dudy)
        call radiation_bc(u,M-i,-N+j,dudx,dudy,dudt)
        call bnd_xderiv(v,M-i,-N+j,ac,i1,i2,dvdx)
        call bnd_yderiv(v,M-i,-N+j,bc,j1,j2,dvdy)
        call radiation_bc(v,M-i,-N+j,dvdx,dvdy,dvdt)
        call bnd_xderiv(ro,M-i,-N+j,ac,i1,i2,drodx)
        call bnd_yderiv(ro,M-i,-N+j,bc,j1,j2,drody)
        call radiation_bc(ro,M-i,-N+j,drodx,drody,drodt)
       ENDDO
      ENDDO


c----------------------------------------------------------------
c  Upper right hand corner
c----------------------------------------------------------------

      Do j=0,2
        IF(j.EQ.0) THEN
         j1 = -6
         j2 = 0
         call change_ab(bc,a60,-6,0)
        ELSE IF(j.EQ.1) THEN
         j1 = -5
         j2 = 1
         call change_ab(bc,a51,-5,1)
        ELSE
         j1 = -4
         j2 = 2
```

```
      call change_ab(bc,a42,-4,2)
     ENDIF
     Do i=0,2
     IF(i.EQ.0) THEN
      i1 = -6
      i2 = 0
      call change_ab(ac,a60,-6,0)
     ELSE IF(i.EQ.1) THEN
       i1 = -5
       i2 = 1
       call change_ab(ac,a51,-5,1)
     ELSE
       i1 = -4
       i2 = 2
       call change_ab(ac,a42,-4,2)
     ENDIF
     call bnd_xderiv(p,M-i,N-j,ac,i1,i2,dpdx)
     call bnd_yderiv(p,M-i,N-j,bc,j1,j2,dpdy)
     call radiation_bc(p,M-i,N-j,dpdx,dpdy,dpdt)
     call bnd_xderiv(u,M-i,N-j,ac,i1,i2,dudx)
     call bnd_yderiv(u,M-i,N-j,bc,j1,j2,dudy)
     call bnd_xderiv(v,M-i,N-j,ac,i1,i2,dvdx)
     call bnd_yderiv(v,M-i,N-j,bc,j1,j2,dvdy)
     call bnd_xderiv(ro,M-i,N-j,ac,i1,i2,drodx)
     call bnd_yderiv(ro,M-i,N-j,bc,j1,j2,drody)
     call vel_outbc(u,v,M-i,N-j,dudx,dudy,dvdx,dvdy,
    & dpdx,dpdy,dudt,dvdt)
     call ro_outbc(ro,M-i,N-j,dpdx,dpdy,dpdt,
    & drodx,drody,drodt)
     ENDDO
     ENDDO


c----------------------------------------------------------------
c  Lower boundary
c----------------------------------------------------------------

     Do j=0,2
     IF(j.EQ.0) THEN
      j1 = 0
      j2 = 6
      call change_ab(bc,a06,0,6)
     ELSE IF(j.EQ.1) THEN
      j1 = -1
      j2 = 5
      call change_ab(bc,a15,-1,5)
     ELSE
      j1 = -2
      j2 = 4
```

```
        call change_ab(bc,a24,-2,4)
      ENDIF
    Do i=-M+3,M-3
      call bnd_yderiv(p,i,-N+j,bc,j1,j2,dpdy)
      call centr_xderiv(p,i,-N+j,dpdx)
      call radiation_bc(p,i,-N+j,dpdx,dpdy,dpdt)
      call bnd_yderiv(u,i,-N+j,bc,j1,j2,dudy)
      call centr_xderiv(u,i,-N+j,dudx)
      call radiation_bc(u,i,-N+j,dudx,dudy,dudt)
      call bnd_yderiv(v,i,-N+j,bc,j1,j2,dvdy)
      call centr_xderiv(v,i,-N+j,dvdx)
      call radiation_bc(v,i,-N+j,dvdx,dvdy,dvdt)
      call bnd_yderiv(ro,i,-N+j,bc,j1,j2,drody)
      call centr_xderiv(ro,i,-N+j,drodx)
      call radiation_bc(ro,i,-N+j,drodx,drody,drodt)
    ENDDO
    ENDDO


c-----------------------------------------------------------------
c  Left boundary
c-----------------------------------------------------------------

    Do i=0,2
      IF(i.EQ.0) THEN
       i1 = 0
       i2 = 6
       call change_ab(ac,a06,0,6)
      ELSE IF(i.EQ.1) THEN
       i1 = -1
       i2 = 5
       call change_ab(ac,a15,-1,5)
      ELSE
       i1 = -2
       i2 = 4
       call change_ab(ac,a24,-2,4)
      ENDIF
      Do j=-N+3,N-3
       call bnd_xderiv(p,-M+i,j,ac,i1,i2,dpdx)
       call centr_yderiv(p,-M+i,j,dpdy)
       call radiation_bc(p,-M+i,j,dpdx,dpdy,dpdt)
       call bnd_xderiv(u,-M+i,j,ac,i1,i2,dudx)
       call centr_yderiv(u,-M+i,j,dudy)
       call radiation_bc(u,-M+i,j,dudx,dudy,dudt)
       call bnd_xderiv(v,-M+i,j,ac,i1,i2,dvdx)
       call centr_yderiv(v,-M+i,j,dvdy)
       call radiation_bc(v,-M+i,j,dvdx,dvdy,dvdt)
       call bnd_xderiv(ro,-M+i,j,ac,i1,i2,drodx)
```

```
        call centr_yderiv(ro,-M+i,j,drody)
        call radiation_bc(ro,-M+i,j,drodx,drody,drodt)
      ENDDO
     ENDDO


c----------------------------------------------------------------
c  Right boundary
c----------------------------------------------------------------

     Do i=0,2
      IF(i.EQ.0) THEN
       i1 = -6
       i2 = 0
       call change_ab(ac,a60,-6,0)
      ELSE IF(i.EQ.1) THEN
       i1 = -5
       i2 = 1
       call change_ab(ac,a51,-5,1)
      ELSE
       i1 = -4
       i2 = 2
       call change_ab(ac,a42,-4,2)
      ENDIF
      Do j=-N+3,N-3
       call bnd_xderiv(p,M-i,j,ac,i1,i2,dpdx)
       call centr_yderiv(p,M-i,j,dpdy)
       call radiation_bc(p,M-i,j,dpdx,dpdy,dpdt)
       call bnd_xderiv(u,M-i,j,ac,i1,i2,dudx)
       call centr_yderiv(u,M-i,j,dudy)
       call bnd_xderiv(v,M-i,j,ac,i1,i2,dvdx)
       call centr_yderiv(v,M-i,j,dvdy)
       call bnd_xderiv(ro,M-i,j,ac,i1,i2,drodx)
       call centr_yderiv(ro,M-i,j,drody)
       call vel_outbc(u,v,M-i,j,dudx,dudy,dvdx,dvdy,
     &    dpdx,dpdy,dudt,dvdt)
       call ro_outbc(ro,M-i,j,dpdx,dpdy,dpdt,
     &    drodx,drody,drodt)
      ENDDO
     ENDDO


c----------------------------------------------------------------
c  Upper boundary
c----------------------------------------------------------------

     Do j=0,2
      IF(j.EQ.0) THEN
       j1 = -6
```

```
        j2 = 0
        call change_ab(bc,a60,-6,0)
      ELSE IF(j.EQ.1) THEN
        j1 = -5
        j2 = 1
        call change_ab(bc,a51,-5,1)
      ELSE
        j1 = -4
        j2 = 2
        call change_ab(bc,a42,-4,2)
      ENDIF
      Do i=-M+3,M-3
        call bnd_yderiv(p,i,N-j,bc,j1,j2,dpdy)
        call centr_xderiv(p,i,N-j,dpdx)
        call radiation_bc(p,i,N-j,dpdx,dpdy,dpdt)
        call bnd_yderiv(u,i,N-j,bc,j1,j2,dudy)
        call centr_xderiv(u,i,N-j,dudx)
        call bnd_yderiv(v,i,N-j,bc,j1,j2,dvdy)
        call centr_xderiv(v,i,N-j,dvdx)
        call bnd_yderiv(ro,i,N-j,bc,j1,j2,drody)
        call centr_xderiv(ro,i,N-j,drodx)
        call vel_outbc(u,v,i,N-j,dudx,dudy,dvdx,dvdy,
     &  dpdx,dpdy,dudt,dvdt)
        call ro_outbc(ro,i,N-j,dpdx,dpdy,dpdt,
     &  drodx,drody,drodt)
       ENDDO
      ENDDO


c----------------------------------------------------------------
c  Interior region
c----------------------------------------------------------------

      Do i=-M+3,M-3
       Do j=-N+3,N-3

        dpdx = 0.00D+00
        dpdy = 0.00D+00
        dudx = 0.00D+00
        dudy = 0.00D+00
        dvdx = 0.00D+00
        dvdy = 0.00D+00
        drodx = 0.00D+00
        drody = 0.00D+00

        Do k=1,3
          dpdx = dpdx + (a(k)*(p(i+k,j) - p(i-k,j)))/dx
          dpdy = dpdy + (a(k)*(p(i,j+k) - p(i,j-k)))/dy
          dudx = dudx + (a(k)*(u(i+k,j) - u(i-k,j)))/dx
```

```
       dudy = dudy + (a(k)*(u(i,j+k) - u(i,j-k)))/dy
       dvdx = dvdx + (a(k)*(v(i+k,j) - v(i-k,j)))/dx
       dvdy = dvdy + (a(k)*(v(i,j+k) - v(i,j-k)))/dy
      drodx = drodx + (a(k)*(ro(i+k,j) - ro(i-k,j)))/dx
      drody = drody + (a(k)*(ro(i,j+k) - ro(i,j-k)))/dy
      ENDDO
      dpdt(i,j,0) = -(Mx*dpdx + My*dpdy + dudx + dvdy)
      dvdt(i,j,0) = -(Mx*dvdx + My*dvdy + dpdy)
      dudt(i,j,0) = -(Mx*dudx + My*dudy + dpdx)
      drodt(i,j,0) = -(Mx*drodx + My*drody + dudx + dvdy)
      ENDDO
      ENDDO


      Do i=-M,M
      Do j=-N,N
      sum1 = 0.00D+00
      sum2 = 0.00D+00
      sum3 = 0.00D+00
      sum4 = 0.00D+00


      Do k=0,3
       sum1 = sum1 + b(k)*dpdt(i,j,k)
       sum2 = sum2 + b(k)*dudt(i,j,k)
       sum3 = sum3 + b(k)*dvdt(i,j,k)
       sum4 = sum4 + b(k)*drodt(i,j,k)
      ENDDO
      p(i,j) = p(i,j) + dt*sum1
      u(i,j) = u(i,j) + dt*sum2
      v(i,j) = v(i,j) + dt*sum3
      ro(i,j) = ro(i,j) + dt*sum4
      Do k=0,2
       dpdt(i,j,3-k) = dpdt(i,j,2-k)
       dudt(i,j,3-k) = dudt(i,j,2-k)
       dvdt(i,j,3-k) = dvdt(i,j,2-k)
       drodt(i,j,3-k) = drodt(i,j,2-k)
      ENDDO
      ENDDO
      ENDDO


      RETURN
      END
```

G.3 Sample Program for Solving the Three-Dimensional Euler Equations

The problem to be solved is as stated below. It is the same as Exercise 6.7.

Consider the time history of an acoustic pulse initiated at time $t = 0$ by a Gaussian pressure distribution which is spherically symmetric with respect to the origin of the coordinate system. To find the exact linearized solution, let R be the radial distance of the spherical coordinate system. The radial momentum and energy equations are:

$$\frac{\partial v}{\partial t} = -\frac{1}{\rho_0}\frac{\partial p}{\partial R} \tag{1}$$

$$\frac{\partial p}{\partial t} + \gamma p_0 \frac{1}{R^2}\frac{\partial}{\partial R}(R^2 v) = 0 \tag{2}$$

The initial conditions at $t = 0$ are:

$$p = \varepsilon \exp\left[-(\ln(2))\left\{\frac{R}{b}\right\}^2\right] \tag{3}$$

$$v = 0 \tag{4}$$

By recasting the problem in p alone, the problem becomes,

$$\frac{\partial^2 p}{\partial t^2} - a_0^2 \frac{1}{R}\frac{\partial^2}{\partial R^2}(Rp) = 0 \tag{5}$$

At $t = 0$,

$$p = \varepsilon \exp\left[-(\ln(2))\left\{\frac{R}{b}\right\}^2\right] \tag{6}$$

$$\frac{\partial p}{\partial t} = 0 \tag{7}$$

Also, it is necessary to require that p is finite at $R \to 0$ \hfill (8)

The exact solution is,

$$p = \varepsilon \left\{\frac{R - a_0 t}{2R}e^{-(\ln 2)\left(\frac{R-a_0 t}{b}\right)^2} + \frac{R + a_0 t}{2R}e^{-(\ln 2)\left(\frac{R+a_0 t}{b}\right)^2}\right\} \tag{9}$$

At $R \to 0$,

$$p(0, t) = \varepsilon e^{-(\ln 2)\left(\frac{a_0 t}{b}\right)^2}\left[1 - 2\frac{(\ln 2)}{b^2}(a_0 t)^2\right] \tag{10}$$

The solution is to be computed in three-dimensional Cartesian coordinates.

The Fortran 90 program below computes the three-dimensional nonlinear solution of the Euler equations using the DRP scheme.

```
program DRP_3D
!-----------------------------------------------------------------
!
!   This code solves the 3D non-linear Euler's equations using the DRP scheme
!   -the domain is rectangular with constant mesh spacing dx=dy=dz in
!    the three coordinate directions
!     -equations are non-dimensionalized by the mesh spacing, speed of sound,
!    and freestream density
!   -radiation boundary conditions are used on all boundary faces
!   -equations are discretized with 7-point spatial stencils and
!   4 level multi-step time advancement
```

```
!   -artificial selective damping stencils are added to the discretized equations
!   -the solution is advanced from the initial conditions to the
!    desired time and then output to file
!
!   Main variables and arrays:
!   ld,md,kd    -1/2 the number of mesh points in the x,y,z directions
!   nmax        -maximum number of time steps
!   t           -time
!   dt          -time step
!   gamma       -ratio of specific heats
!   Xmean       -mean field values
!   R_inv       -inverse of artificial mesh Reynolds number
!   u(:,:,:) -velocity in x-dir
!   v(:,:,:) -velocity in y-dir
!   w(:,:,:) -velocity in z-dir
!   p(:,:,:) -pressure
!   rho(:,:,:) -density
!   ku(:,:,:,-3:0)  -time derivatives, dudt, at different time levels
!   kv(:,:,:,-3:0)  -time derivatives, dvdt, at different time levels
!   kw(:,:,:,-3:0)   -time derivatives, dwdt, at different time levels
!   kp(:,:,:,-3:0)  -time derivatives, dpdt, at different time levels
!   krho(:,:,:,-3:0)-time derivatives, drhodt, at different time levels
!   kl(-3:0)        -converts time level to index in k array
!   aXX()                   -spatial stencil coefficients
!   b            -multi-step advancement coefficients
!   dXX()            -artificial selective damping stencil coefficients
!
!-----------------------------------------------------------------
!
implicit none
integer, parameter::rkind=8,ld=30,md=30,kd=30
integer::n,l,m,k,nmax
integer, dimension(-3:0)::kl
real(kind=rkind)::dt,h,bh,t,gamma,umean,vmean,wmean,pmean,rhomean,R_inv,pexact,r
real(kind=rkind), dimension(-ld-3:ld+3,-md-3:md+3,-kd-3:kd+3)::u,v,w,p,rho
real(kind=rkind), dimension(-3:0,-ld-3:ld+3,-md-3:md+3,-kd-3:kd+3)::ku,kv,kw,kp,krho
real(kind=rkind), dimension(-3:3)::a33,d33
real(kind=rkind), dimension(-2:4)::a24
real(kind=rkind), dimension(-1:5)::a15
real(kind=rkind), dimension(0:6)::a06
real(kind=rkind), dimension(-3:0)::b
real(kind=rkind), dimension(-2:2)::d22
real(kind=rkind), dimension(-1:1)::d11
integer::outdata=20,tecplot1=22,tecplot2=24,tecplot3=26,trace1=28
!
!   Open output data files
!
```

```
open(unit=outdata, file="outDRP3D.dat", status="replace")
open(unit=tecplot1, file="tecplot1.dat", status="replace")
open(unit=tecplot2, file="tecplot2.dat", status="replace")
open(unit=tecplot3, file="tecplot3.dat", status="replace")
open(unit=trace1, file="trace1.dat", status="replace")
!
!-------Initialization----------------------------------------
!
!   Optimized 7-point spatial stencil
!
a33(-3)=-0.02084314277031176_rkind
a33(-2)=0.166705904414580469_rkind
a33(-1)=-0.77088238051822552_rkind
a33(0)=0.0_rkind
a33(1)=0.77088238051822552_rkind
a33(2)=-0.166705904414580469_rkind
a33(3)=0.02084314277031176_rkind
!
a24(-2)=0.049041958_rkind
a24(-1)=-0.468840357_rkind
a24(0)=-0.474760914_rkind
a24(1)=1.273274737_rkind
a24(2)=-0.518484526_rkind
a24(3)=0.166138533_rkind
a24(4)=-0.026369431_rkind
!
a15(-1)=-0.209337622_rkind
a15(0)=-1.084875676_rkind
a15(1)=2.147776050_rkind
a15(2)=-1.388928322_rkind
a15(3)=0.768949766_rkind
a15(4)=-0.281814650_rkind
a15(5)=0.048230454_rkind
!
a06(0)=-2.192280339_rkind
a06(1)=4.748611401_rkind
a06(2)=-5.108851915_rkind
a06(3)=4.461567104_rkind
a06(4)=-2.833498741_rkind
a06(5)=1.128328861_rkind
a06(6)=-0.203876371_rkind
!
!   Optimized multi-step
!
b(0)=2.3025580888_rkind
b(-1)=-2.4910075998_rkind
b(-2)=1.5743409332_rkind
```

```
b(-3)=-0.3858914222_rkind
!
!    Damping Stencils
!
d33(-3)=-0.023853048191278_rkind
d33(-2)=0.10630357876989_rkind
d33(-1)=-0.22614695180872_rkind
d33(0)=0.28739284246022_rkind
d33(1)=-0.22614695180872_rkind
d33(2)=0.10630357876989_rkind
d33(3)=-0.023853048191278_rkind
!
d22(-2)=0.0625_rkind
d22(-1)=-0.25_rkind
d22(0)=0.375_rkind
d22(1)=-0.25_rkind
d22(2)=0.0625_rkind
!
d11(-1)=-0.25_rkind
d11(0)=0.5_rkind
d11(1)=-0.25_rkind
!
!    Set simulation parameters
!
dt=0.1_rkind ! time step size
nmax=350 ! maximum number of time steps
gamma=1.4_rkind ! ratio of specific heats
!
!    Mean field values
!
umean=0.0_rkind
vmean=0.0_rkind
wmean=0.0_rkind
pmean=1/gamma
rhomean=1.0_rkind
!
!    Inverse of the artificial mesh Reynolds number
!
R_inv=0.05_rkind
!
!    Set initial conditions
!
t=0.0_rkind
!
u=umean
v=vmean
w=wmean
```

```fortran
rho=rhomean
!
!    Initial pressure pulse is a Gaussian function with half-width 'bh' and height 'h'
!
bh=3.0_rkind ! half width is three mesh spacings
h=0.001_rkind ! wave amplitude
do k=-kd-3,kd+3
 do m=-md-3,md+3
  do l=-ld-3,ld+3
   p(l,m,k)=pmean+h*exp(-log(2.0_rkind)*(sqrt(real(l**2,rkind)+real(m**2,rkind)+ &
       real(k**2,rkind))/bh)**2)
  end do
 end do
end do
!
!    Set solution to zero for t<0
!
ku=0.0_rkind
kv=0.0_rkind
kw=0.0_rkind
kp=0.0_rkind
krho=0.0_rkind
kl(-3)=-3
kl(-2)=-2
kl(-1)=-1
kl(0)=0
!
!-------Execution--------------------------------------------------
!
!    Main loop - march 'nmax' steps in time with the DRP scheme
!
do n=1,nmax
!
t=t+dt
write (*,*)"n: ",n," t: ",real(t,4)
!
!    Compute ku,kv,kw,kp,krho at interior points (Euler's equations)
!
call euler(u,v,w,p,rho,ku,kv,kw,kp,krho)
!
!    Compute ku,kv,kw,kp,krho at boundary points (Radiation and Outflow equations)
!
call bcond(u,v,w,p,rho,ku,kv,kw,kp,krho)
!
!    Update fields to time level n+1
!
do k=-kd-3,kd+3
```

```fortran
  do m=-md-3,md+3
    do l=-ld-3,ld+3
      u(l,m,k)=mladvance(l,m,k,u,ku)
      v(l,m,k)=mladvance(l,m,k,v,kv)
      w(l,m,k)=mladvance(l,m,k,w,kw)
      p(l,m,k)=mladvance(l,m,k,p,kp)
     rho(l,m,k)=mladvance(l,m,k,rho,krho)
    end do
  end do
end do
!
!   Circle shift k array
!
kl=cshift(kl,1)
!
!   Output trace
!
pexact=h*exp(-log(2.0_rkind)*(t/bh)**2)*(1.0_rkind- &
2.0_rkind*log(2.0_rkind)*(t/bh)**2)
if (abs(pexact) .lt. 1.0E-8) then
 pexact=0.0_rkind
end if
write (trace1,52)t,p(0,0,0)-pmean,pexact
!
end do
!
!-------Output--------------------------------------------------
!
50 format(2i5,2e16.7e2)
51 format(i5,2e16.7e2)
52 format(3e16.7e2)
!
!   Line plots
!
write (outdata,*)"x line at m=0, k=0"
do l=-ld-3,ld+3
 r=abs(real(l,rkind))
 if (r .gt. 10.0_rkind*tiny(r)) then
  pexact=h*(((r-t)/(2.0_rkind*r))*exp(-log(2.0_rkind)*((r-t)/bh)**2)+ &
      ((r+t)/(2.0_rkind*r))*exp(-log(2.0_rkind)*((r+t)/bh)**2))
 else
  pexact=h*exp(-log(2.0_rkind)*(t/bh)**2)*(1.0_rkind- &
2.0_rkind*log(2.0_rkind)*(t/bh)**2)
 end if
 if (abs(pexact) .lt. 1.0E-8) then
  pexact=0.0_rkind
 end if
```

```
 write (outdata,51)l,p(l,0,0)-pmean,pexact
end do
write (outdata,*)"y line at l=0, k=0"
do m=-md-3,md+3
 r=abs(real(m,rkind))
 if (r .gt. 10.0_rkind*tiny(r)) then
  pexact=h*(((r-t)/(2.0_rkind*r))*exp(-log(2.0_rkind)*((r-t)/bh)**2)+ &
      ((r+t)/(2.0_rkind*r))*exp(-log(2.0_rkind)*((r+t)/bh)**2))
 else
  pexact=h*exp(-log(2.0_rkind)*(t/bh)**2)*(1.0_rkind- &
2.0_rkind*log(2.0_rkind)*(t/bh)**2)
 end if
 if (abs(pexact) .lt. 1.0E-8) then
  pexact=0.0_rkind
 end if
 write (outdata,51)m,p(0,m,0)-pmean,pexact
end do
write (outdata,*)"z line at l=0, m=0"
do k=-kd-3,kd+3
 r=abs(real(k,rkind))
 if (r .gt. 10.0_rkind*tiny(r)) then
  pexact=h*(((r-t)/(2.0_rkind*r))*exp(-log(2.0_rkind)*((r-t)/bh)**2)+ &
      ((r+t)/(2.0_rkind*r))*exp(-log(2.0_rkind)*((r+t)/bh)**2))
 else
  pexact=h*exp(-log(2.0_rkind)*(t/bh)**2)*(1.0_rkind- &
2.0_rkind*log(2.0_rkind)*(t/bh)**2)
 end if
 if (abs(pexact) .lt. 1.0E-8) then
  pexact=0.0_rkind
 end if
 write (outdata,51)k,p(0,0,k)-pmean,pexact
end do
!
!   Tecplot files for contour plots
!
!     x-y plane at z=0
!
write (tecplot1,*)"VARIABLES= x y p pe"
write (tecplot1,*)"ZONE I=",2*ld+7," J=",2*md+7
do m=-md-3,md+3
 do l=-ld-3,ld+3
  r=sqrt(real(l**2,rkind)+real(m**2,rkind))
  if (r .gt. 10.0_rkind*tiny(r)) then
   pexact=h*(((r-t)/(2.0_rkind*r))*exp(-log(2.0_rkind)*((r-t)/bh)**2)+&
       ((r+t)/(2.0_rkind*r))*exp(-log(2.0_rkind)*((r+t)/bh)**2))
  else
```

```fortran
    pexact=h*exp(-log(2.0_rkind)*(t/bh)**2)*(1.0_rkind- &
2.0_rkind*log(2.0_rkind)*(t/bh)**2)
  end if
  if (abs(pexact) .lt. 1.0E-8) then
   pexact=0.0_rkind
  end if
  write (tecplot1,50)l,m,p(l,m,0)-pmean,pexact
 end do
end do
!
!    x-y plane at z=30
!
write (tecplot2,*)"VARIABLES= x y p pe"
write (tecplot2,*)"ZONE I=",2*ld+7," J=",2*md+7
do m=-md-3,md+3
 do l=-ld-3,ld+3
  r=sqrt(real(l**2,rkind)+real(m**2,rkind)+real(30**2,rkind))
  if (r .gt. 10.0_rkind*tiny(r)) then
   pexact=h*(((r-t)/(2.0_rkind*r))*exp(-log(2.0_rkind)*((r-t)/bh)**2)+ &
        ((r+t)/(2.0_rkind*r))*exp(-log(2.0_rkind)*((r+t)/bh)**2))
  else
    pexact=h*exp(-log(2.0_rkind)*(t/bh)**2)*(1.0_rkind- &
2.0_rkind*log(2.0_rkind)*(t/bh)**2)
  end if
  if (abs(pexact) .lt. 1.0E-8) then
   pexact=0.0_rkind
  end if
  write (tecplot2,50)l,m,p(l,m,30)-pmean,pexact
 end do
end do
!
!    y-z plane at x=0
!
write (tecplot1,*)"VARIABLES= y z p pe"
write (tecplot1,*)"ZONE I=",2*md+7," J=",2*kd+7
do k=-kd-3,kd+3
 do m=-md-3,md+3
   r=sqrt(real(m**2,rkind)+real(k**2,rkind))
   if (r .gt. 10.0_rkind*tiny(r)) then
    pexact=h*(((r-t)/(2.0_rkind*r))*exp(-log(2.0_rkind)*((r-t)/bh)**2)+ &
         ((r+t)/(2.0_rkind*r))*exp(-log(2.0_rkind)*((r+t)/bh)**2))
   else
    pexact=h*exp(-log(2.0_rkind)*(t/bh)**2)*(1.0_rkind- &
2.0_rkind*log(2.0_rkind)*(t/bh)**2)
   end if
   if (abs(pexact) .lt. 1.0E-8) then
    pexact=0.0_rkind
```

```
     end if
   write (tecplot1,50)m,k,p(0,m,k)-pmean,pexact
  end do
 end do
!
stop
!
contains
!
subroutine euler(u,v,w,p,rho,ku,kv,kw,kp,krho)
!
!!!!!!!!!!!!!!!!!!!!!!!!!!!!!!!!!!!!!!!!!!!!!!!!!!!!!!!!!!!!!!!!!!!!!!!!!
!
!   Compute the time derivatives ku,kv,kw,kp,krho at every interior
!     point using the non-linear Euler's equations
!
!!!!!!!!!!!!!!!!!!!!!!!!!!!!!!!!!!!!!!!!!!!!!!!!!!!!!!!!!!!!!!!!!!!!!!!!!
!
implicit none
real(kind=rkind), intent(in), dimension(-ld-3:ld+3,-md-3:md+3,-kd-3:kd+3)::u,v,w,p,rho
real(kind=rkind), intent(in out), dimension(-3:0,-ld-3:ld+3,-md-3:md+3,-kd- &
3:kd+3)::ku,kv,kw,kp,krho
integer::l,m,k
real(kind=rkind)::dpdx,dpdy,dpdz,xpdmp,ypdmp,zpdmp,drhodx,drhody,drhodz, &
          xrhodmp,yrhodmp,zrhodmp,dudx,dudy,dudz,xudmp,yudmp,zudmp, &
dvdx,dvdy,dvdz,xvdmp,yvdmp,zvdmp,dwdx,dwdy,dwdz,xwdmp,ywdmp,zwdmp
!
do k=-kd,kd
 do m=-md,md
  do l=-ld,ld
   !
   !  dudt
   !
   dudx=xdiff(l,m,k,3,u)
   dudy=ydiff(l,m,k,3,u)
   dudz=zdiff(l,m,k,3,u)
   dpdx=xdiff(l,m,k,3,p)
   xudmp=xdamp(l,m,k,3,u)
   yudmp=ydamp(l,m,k,3,u)
   zudmp=zdamp(l,m,k,3,u)
   call dudt(l,m,k,dudx,dudy,dudz,dpdx,xudmp,yudmp,zudmp,rho,u,v,w,ku)
   !
   !  dvdt
   !
   dvdx=xdiff(l,m,k,3,v)
   dvdy=ydiff(l,m,k,3,v)
```

```fortran
    dvdz=zdiff(l,m,k,3,v)
    dpdy=ydiff(l,m,k,3,p)
    xvdmp=xdamp(l,m,k,3,v)
    yvdmp=ydamp(l,m,k,3,v)
    zvdmp=zdamp(l,m,k,3,v)
    call dvdt(l,m,k,dvdx,dvdy,dvdz,dpdy,xvdmp,yvdmp,zvdmp,rho,u,v,w,kv)
    !
    !  dwdt
    !
    dwdx=xdiff(l,m,k,3,w)
    dwdy=ydiff(l,m,k,3,w)
    dwdz=zdiff(l,m,k,3,w)
    dpdz=zdiff(l,m,k,3,p)
    xwdmp=xdamp(l,m,k,3,w)
    ywdmp=ydamp(l,m,k,3,w)
    zwdmp=zdamp(l,m,k,3,w)
    call dwdt(l,m,k,dwdx,dwdy,dwdz,dpdz,xwdmp,ywdmp,zwdmp,rho,u,v,w,kw)
    !
    !  dpdt
    !
    xpdmp=xdamp(l,m,k,3,p)
    ypdmp=ydamp(l,m,k,3,p)
    zpdmp=zdamp(l,m,k,3,p)
    call dpdt(l,m,k,dpdx,dpdy,dpdz,dudx,dvdy,dwdz,xpdmp,ypdmp,zpdmp,u,v,w,p,kp)
    !
    !  drhodt
    !
    drhodx=xdiff(l,m,k,3,rho)
    drhody=ydiff(l,m,k,3,rho)
    drhodz=zdiff(l,m,k,3,rho)
    xrhodmp=xdamp(l,m,k,3,rho)
    yrhodmp=ydamp(l,m,k,3,rho)
    zrhodmp=zdamp(l,m,k,3,rho)
    call drhodt(l,m,k,drhodx,drhody,drhodz,dudx,dvdy,dwdz,xrhodmp,yrhodmp,zrhodmp, &
            u,v,w,rho,krho)
    !
    end do
  end do
end do
!
return
end subroutine euler
!
subroutine bcond(u,v,w,p,rho,ku,kv,kw,kp,krho)
!
```

```
!!!!!!!!!!!!!!!!!!!!!!!!!!!!!!!!!!!!!!!!!!!!!!!!!!!!!!!!!!!!!!!!!!!!!!!!!!
!
!   The domain is a 3D rectangular box. The mean flow speed is zero.
!    Radiation equations are used on all faces. The boundary
!    region is 3 mesh lines thick.
!
!!!!!!!!!!!!!!!!!!!!!!!!!!!!!!!!!!!!!!!!!!!!!!!!!!!!!!!!!!!!!!!!!!!!!!!!!!!!!!
!
implicit none
integer::l,m,k,fptsx,fptsy,fptsz
real(kind=rkind), intent(in), dimension(-ld-3:ld+3,-md-3:md+3,-kd-3:kd+3)::u,v,w,p,rho
real(kind=rkind), intent(in out), dimension(-3:0,-ld-3:ld+3,-md-3:md+3,-kd- &
3:kd+3)::ku,kv,kw,kp,krho
real(kind=rkind)::r,sinphi,cosphi,sintheta,costheta,dpdx,dpdy,dpdz,xpdmp,ypdmp,zpdmp, &
drhodx,drhody,drhodz,xrhodmp,yrhodmp,zrhodmp,dudx,dudy,dudz,xudmp,yudmp,zudmp,dvdx, &
dvdy,dvdz,xvdmp,yvdmp,zvdmp,dwdx,dwdy,dwdz,xwdmp,ywdmp,zwdmp
!
!   'kd' faces
!
do l=-ld-3,ld+3
 do m=-md-3,md+3
  !
  ! Bottom
  !
  do k=-kd-3,-kd-1
  !
  if (l .lt. -ld) then
  fptsx=3-ld-l
  else if (l .gt. ld) then
   fptsx=3+ld-l
  else
   fptsx=3
  end if
  if (m .lt. -md) then
   fptsy=3-md-m
  else if (m .gt. md) then
   fptsy=3+md-m
  else
   fptsy=3
  end if
   fptsz=3-kd-k
  !
  call polar(l,m,k,r,sinphi,cosphi,sintheta,costheta)
  !
  dpdx=xdiff(l,m,k,fptsx,p)
  dpdy=ydiff(l,m,k,fptsy,p)
  dpdz=zdiff(l,m,k,fptsz,p)
```

```
xpdmp=xdamp(l,m,k,fptsx,p)
ypdmp=ydamp(l,m,k,fptsy,p)
zpdmp=zdamp(l,m,k,fptsz,p)
call radiation_bc(l,m,k,r,sinphi,cosphi,sintheta,costheta,pmean,dpdx,dpdy,dpdz, &
          xpdmp,ypdmp,zpdmp,p,kp)
!
drhodx=xdiff(l,m,k,fptsx,rho)
drhody=ydiff(l,m,k,fptsy,rho)
drhodz=zdiff(l,m,k,fptsz,rho)
xrhodmp=xdamp(l,m,k,fptsx,rho)
yrhodmp=ydamp(l,m,k,fptsy,rho)
zrhodmp=zdamp(l,m,k,fptsz,rho)
call radiation_bc(l,m,k,r,sinphi,cosphi,sintheta,costheta,rhomean,drhodx,drhody, &
          drhodz,xrhodmp,yrhodmp,zrhodmp,rho,krho)
!
dudx=xdiff(l,m,k,fptsx,u)
dudy=ydiff(l,m,k,fptsy,u)
dudz=zdiff(l,m,k,fptsz,u)
xudmp=xdamp(l,m,k,fptsx,u)
yudmp=ydamp(l,m,k,fptsy,u)
zudmp=zdamp(l,m,k,fptsz,u)
call radiation_bc(l,m,k,r,sinphi,cosphi,sintheta,costheta,umean,dudx,dudy,dudz, &
          xudmp,yudmp,zudmp,u,ku)
!
dvdx=xdiff(l,m,k,fptsx,v)
dvdy=ydiff(l,m,k,fptsy,v)
dvdz=zdiff(l,m,k,fptsz,v)
xvdmp=xdamp(l,m,k,fptsx,v)
yvdmp=ydamp(l,m,k,fptsy,v)
zvdmp=zdamp(l,m,k,fptsz,v)
call radiation_bc(l,m,k,r,sinphi,cosphi,sintheta,costheta,vmean,dvdx,dvdy,dvdz, &
          xvdmp,yvdmp,zvdmp,v,kv)
!
dwdx=xdiff(l,m,k,fptsx,w)
dwdy=ydiff(l,m,k,fptsy,w)
dwdz=zdiff(l,m,k,fptsz,w)
xwdmp=xdamp(l,m,k,fptsx,w)
ywdmp=ydamp(l,m,k,fptsy,w)
zwdmp=zdamp(l,m,k,fptsz,w)
call radiation_bc(l,m,k,r,sinphi,cosphi,sintheta,costheta,wmean,dwdx,dwdy,dwdz, &
          xwdmp,ywdmp,zwdmp,w,kw)
!
end do
!
! Top
!
do k=kd+1,kd+3
```

```
!
if (l .lt. -ld) then
 fptsx=3-ld-l
else if (l .gt. ld) then
 fptsx=3+ld-l
else
 fptsx=3
end if
if (m .lt. -md) then
 fptsy=3-md-m
else if (m .gt. md) then
 fptsy=3+md-m
else
 fptsy=3
end if
fptsz=3+kd-k
!
call polar(l,m,k,r,sinphi,cosphi,sintheta,costheta)
!
dpdx=xdiff(l,m,k,fptsx,p)
dpdy=ydiff(l,m,k,fptsy,p)
dpdz=zdiff(l,m,k,fptsz,p)
xpdmp=xdamp(l,m,k,fptsx,p)
ypdmp=ydamp(l,m,k,fptsy,p)
zpdmp=zdamp(l,m,k,fptsz,p)
call radiation_bc(l,m,k,r,sinphi,cosphi,sintheta,costheta,pmean,dpdx,dpdy,dpdz, &
          xpdmp,ypdmp,zpdmp,p,kp)
!
drhodx=xdiff(l,m,k,fptsx,rho)
drhody=ydiff(l,m,k,fptsy,rho)
drhodz=zdiff(l,m,k,fptsz,rho)
xrhodmp=xdamp(l,m,k,fptsx,rho)
yrhodmp=ydamp(l,m,k,fptsy,rho)
zrhodmp=zdamp(l,m,k,fptsz,rho)
call radiation_bc(l,m,k,r,sinphi,cosphi,sintheta,costheta,rhomean,drhodx,drhody, &
          drhodz,xrhodmp,yrhodmp,zrhodmp,rho,krho)
!
dudx=xdiff(l,m,k,fptsx,u)
dudy=ydiff(l,m,k,fptsy,u)
dudz=zdiff(l,m,k,fptsz,u)
xudmp=xdamp(l,m,k,fptsx,u)
yudmp=ydamp(l,m,k,fptsy,u)
zudmp=zdamp(l,m,k,fptsz,u)
call radiation_bc(l,m,k,r,sinphi,cosphi,sintheta,costheta,umean,dudx,dudy,dudz, &
          xudmp,yudmp,zudmp,u,ku)
!
dvdx=xdiff(l,m,k,fptsx,v)
```

```
    dvdy=ydiff(l,m,k,fptsy,v)
    dvdz=zdiff(l,m,k,fptsz,v)
    xvdmp=xdamp(l,m,k,fptsx,v)
    yvdmp=ydamp(l,m,k,fptsy,v)
    zvdmp=zdamp(l,m,k,fptsz,v)
    call radiation_bc(l,m,k,r,sinphi,cosphi,sintheta,costheta,vmean,dvdx,dvdy,dvdz, &
            xvdmp,yvdmp,zvdmp,v,kv)
    !
    dwdx=xdiff(l,m,k,fptsx,w)
    dwdy=ydiff(l,m,k,fptsy,w)
    dwdz=zdiff(l,m,k,fptsz,w)
    xwdmp=xdamp(l,m,k,fptsx,w)
    ywdmp=ydamp(l,m,k,fptsy,w)
    zwdmp=zdamp(l,m,k,fptsz,w)
    call radiation_bc(l,m,k,r,sinphi,cosphi,sintheta,costheta,wmean,dwdx,dwdy,dwdz, &
            xwdmp,ywdmp,zwdmp,w,kw)
    !
   end do
   !
  end do
 end do
 !
 !   'md' faces
 !
 do l=-ld-3,ld+3
  do k=-kd,kd
   !
   !  Bottom
   !
   do m=-md-3,-md-1
    !
    if (l .lt. -ld) then
     fptsx=3-ld-l
    else if (l .gt. ld) then
     fptsx=3+ld-l
    else
     fptsx=3
    end if
    fptsy=3-md-m
    fptsz=3
    !
    call polar(l,m,k,r,sinphi,cosphi,sintheta,costheta)
    !
    dpdx=xdiff(l,m,k,fptsx,p)
    dpdy=ydiff(l,m,k,fptsy,p)
    dpdz=zdiff(l,m,k,fptsz,p)
```

```
xpdmp=xdamp(1,m,k,fptsx,p)
ypdmp=ydamp(1,m,k,fptsy,p)
zpdmp=zdamp(1,m,k,fptsz,p)
call radiation_bc(1,m,k,r,sinphi,cosphi,sintheta,costheta,pmean,dpdx,dpdy,dpdz, &
         xpdmp,ypdmp,zpdmp,p,kp)

!
drhodx=xdiff(1,m,k,fptsx,rho)
drhody=ydiff(1,m,k,fptsy,rho)
drhodz=zdiff(1,m,k,fptsz,rho)
xrhodmp=xdamp(1,m,k,fptsx,rho)
yrhodmp=ydamp(1,m,k,fptsy,rho)
zrhodmp=zdamp(1,m,k,fptsz,rho)
call radiation_bc(1,m,k,r,sinphi,cosphi,sintheta,costheta,rhomean,drhodx,drhody, &
         drhodz,xrhodmp,yrhodmp,zrhodmp,rho,krho)
!
dudx=xdiff(1,m,k,fptsx,u)
dudy=ydiff(1,m,k,fptsy,u)
dudz=zdiff(1,m,k,fptsz,u)
xudmp=xdamp(1,m,k,fptsx,u)
yudmp=ydamp(1,m,k,fptsy,u)
zudmp=zdamp(1,m,k,fptsz,u)
call radiation_bc(1,m,k,r,sinphi,cosphi,sintheta,costheta,umean,dudx,dudy,dudz, &
         xudmp,yudmp,zudmp,u,ku)
!
dvdx=xdiff(1,m,k,fptsx,v)
dvdy=ydiff(1,m,k,fptsy,v)
dvdz=zdiff(1,m,k,fptsz,v)
xvdmp=xdamp(1,m,k,fptsx,v)
yvdmp=ydamp(1,m,k,fptsy,v)
zvdmp=zdamp(1,m,k,fptsz,v)
call radiation_bc(1,m,k,r,sinphi,cosphi,sintheta,costheta,vmean,dvdx,dvdy,dvdz, &
         xvdmp,yvdmp,zvdmp,v,kv)
!
dwdx=xdiff(1,m,k,fptsx,w)
dwdy=ydiff(1,m,k,fptsy,w)
dwdz=zdiff(1,m,k,fptsz,w)
xwdmp=xdamp(1,m,k,fptsx,w)
ywdmp=ydamp(1,m,k,fptsy,w)
zwdmp=zdamp(1,m,k,fptsz,w)
call radiation_bc(1,m,k,r,sinphi,cosphi,sintheta,costheta,wmean,dwdx,dwdy,dwdz, &
         xwdmp,ywdmp,zwdmp,w,kw)
!
end do
!
! Top
!
```

```
do m=md+1,md+3
!
if (1 .lt. -1d) then
 fptsx=3-1d-1
else if (1 .gt. 1d) then
 fptsx=3+1d-1
else
 fptsx=3
end if
fptsy=3+md-m
fptsz=3
!
call polar(1,m,k,r,sinphi,cosphi,sintheta,costheta)
!
dpdx=xdiff(1,m,k,fptsx,p)
dpdy=ydiff(1,m,k,fptsy,p)
dpdz=zdiff(1,m,k,fptsz,p)
xpdmp=xdamp(1,m,k,fptsx,p)
ypdmp=ydamp(1,m,k,fptsy,p)
zpdmp=zdamp(1,m,k,fptsz,p)
call radiation_bc(1,m,k,r,sinphi,cosphi,sintheta,costheta,pmean,dpdx,dpdy,dpdz, &
          xpdmp,ypdmp,zpdmp,p,kp)
!
drhodx=xdiff(1,m,k,fptsx,rho)
drhody=ydiff(1,m,k,fptsy,rho)
drhodz=zdiff(1,m,k,fptsz,rho)
xrhodmp=xdamp(1,m,k,fptsx,rho)
yrhodmp=ydamp(1,m,k,fptsy,rho)
zrhodmp=zdamp(1,m,k,fptsz,rho)
call radiation_bc(1,m,k,r,sinphi,cosphi,sintheta,costheta,rhomean,drhodx,drhody, &
          drhodz,xrhodmp,yrhodmp,zrhodmp,rho,krho)
!
dudx=xdiff(1,m,k,fptsx,u)
dudy=ydiff(1,m,k,fptsy,u)
dudz=zdiff(1,m,k,fptsz,u)
xudmp=xdamp(1,m,k,fptsx,u)
yudmp=ydamp(1,m,k,fptsy,u)
zudmp=zdamp(1,m,k,fptsz,u)
call radiation_bc(1,m,k,r,sinphi,cosphi,sintheta,costheta,umean,dudx,dudy,dudz, &
          xudmp,yudmp,zudmp,u,ku)
!
dvdx=xdiff(1,m,k,fptsx,v)
dvdy=ydiff(1,m,k,fptsy,v)
dvdz=zdiff(1,m,k,fptsz,v)
xvdmp=xdamp(1,m,k,fptsx,v)
yvdmp=ydamp(1,m,k,fptsy,v)
zvdmp=zdamp(1,m,k,fptsz,v)
```

```fortran
      call radiation_bc(l,m,k,r,sinphi,cosphi,sintheta,costheta,vmean,dvdx,dvdy,dvdz, &
               xvdmp,yvdmp,zvdmp,v,kv)
      !
      dwdx=xdiff(l,m,k,fptsx,w)
      dwdy=ydiff(l,m,k,fptsy,w)
      dwdz=zdiff(l,m,k,fptsz,w)
      xwdmp=xdamp(l,m,k,fptsx,w)
      ywdmp=ydamp(l,m,k,fptsy,w)
      zwdmp=zdamp(l,m,k,fptsz,w)
      call radiation_bc(l,m,k,r,sinphi,cosphi,sintheta,costheta,wmean,dwdx,dwdy,dwdz, &
               xwdmp,ywdmp,zwdmp,w,kw)
      !
      end do
      !
     end do
    end do
    !
    !   'ld' faces
    !
    do m=-md,md
     do k=-kd,kd
      !
      ! Bottom
      !
      do l=-ld-3,-ld-1
      !
      fptsx=3-ld-1
      fptsy=3
      fptsz=3
      !
      call polar(l,m,k,r,sinphi,cosphi,sintheta,costheta)
      !
      dpdx=xdiff(l,m,k,fptsx,p)
      dpdy=ydiff(l,m,k,fptsy,p)
      dpdz=zdiff(l,m,k,fptsz,p)
      xpdmp=xdamp(l,m,k,fptsx,p)
      ypdmp=ydamp(l,m,k,fptsy,p)
      zpdmp=zdamp(l,m,k,fptsz,p)
      call radiation_bc(l,m,k,r,sinphi,cosphi,sintheta,costheta,pmean,dpdx,dpdy,dpdz, &
               xpdmp,ypdmp,zpdmp,p,kp)
      !
      drhodx=xdiff(l,m,k,fptsx,rho)
      drhody=ydiff(l,m,k,fptsy,rho)
      drhodz=zdiff(l,m,k,fptsz,rho)
      xrhodmp=xdamp(l,m,k,fptsx,rho)
      yrhodmp=ydamp(l,m,k,fptsy,rho)
      zrhodmp=zdamp(l,m,k,fptsz,rho)
```

```fortran
call radiation_bc(l,m,k,r,sinphi,cosphi,sintheta,costheta,rhomean,drhodx,drhody, &
         drhodz,xrhodmp,yrhodmp,zrhodmp,rho,krho)
!
dudx=xdiff(l,m,k,fptsx,u)
dudy=ydiff(l,m,k,fptsy,u)
dudz=zdiff(l,m,k,fptsz,u)
xudmp=xdamp(l,m,k,fptsx,u)
yudmp=ydamp(l,m,k,fptsy,u)
zudmp=zdamp(l,m,k,fptsz,u)
call radiation_bc(l,m,k,r,sinphi,cosphi,sintheta,costheta,umean,dudx,dudy,dudz, &
         xudmp,yudmp,zudmp,u,ku)
!
dvdx=xdiff(l,m,k,fptsx,v)
dvdy=ydiff(l,m,k,fptsy,v)
dvdz=zdiff(l,m,k,fptsz,v)
xvdmp=xdamp(l,m,k,fptsx,v)
yvdmp=ydamp(l,m,k,fptsy,v)
zvdmp=zdamp(l,m,k,fptsz,v)
call radiation_bc(l,m,k,r,sinphi,cosphi,sintheta,costheta,vmean,dvdx,dvdy,dvdz, &
         xvdmp,yvdmp,zvdmp,v,kv)
!
dwdx=xdiff(l,m,k,fptsx,w)
dwdy=ydiff(l,m,k,fptsy,w)
dwdz=zdiff(l,m,k,fptsz,w)
xwdmp=xdamp(l,m,k,fptsx,w)
ywdmp=ydamp(l,m,k,fptsy,w)
zwdmp=zdamp(l,m,k,fptsz,w)
call radiation_bc(l,m,k,r,sinphi,cosphi,sintheta,costheta,wmean,dwdx,dwdy,dwdz, &
         xwdmp,ywdmp,zwdmp,w,kw)
!
end do
!
!  Top
!
do l=ld+1,ld+3
!
fptsx=3+ld-1
fptsy=3
fptsz=3
!
call polar(l,m,k,r,sinphi,cosphi,sintheta,costheta)
!
dpdx=xdiff(l,m,k,fptsx,p)
dpdy=ydiff(l,m,k,fptsy,p)
dpdz=zdiff(l,m,k,fptsz,p)
xpdmp=xdamp(l,m,k,fptsx,p)
ypdmp=ydamp(l,m,k,fptsy,p)
```

```fortran
zpdmp=zdamp(l,m,k,fptsz,p)
call radiation_bc(l,m,k,r,sinphi,cosphi,sintheta,costheta,pmean,dpdx,dpdy,dpdz, &
          xpdmp,ypdmp,zpdmp,p,kp)
!
drhodx=xdiff(l,m,k,fptsx,rho)
drhody=ydiff(l,m,k,fptsy,rho)
drhodz=zdiff(l,m,k,fptsz,rho)
xrhodmp=xdamp(l,m,k,fptsx,rho)
yrhodmp=ydamp(l,m,k,fptsy,rho)
zrhodmp=zdamp(l,m,k,fptsz,rho)
call radiation_bc(l,m,k,r,sinphi,cosphi,sintheta,costheta,rhomean,drhodx,drhody, &
          drhodz,xrhodmp,yrhodmp,zrhodmp,rho,krho)
!
dudx=xdiff(l,m,k,fptsx,u)
dudy=ydiff(l,m,k,fptsy,u)
dudz=zdiff(l,m,k,fptsz,u)
xudmp=xdamp(l,m,k,fptsx,u)
yudmp=ydamp(l,m,k,fptsy,u)
zudmp=zdamp(l,m,k,fptsz,u)
call radiation_bc(l,m,k,r,sinphi,cosphi,sintheta,costheta,umean,dudx,dudy,dudz, &
          xudmp,yudmp,zudmp,u,ku)
!
dvdx=xdiff(l,m,k,fptsx,v)
dvdy=ydiff(l,m,k,fptsy,v)
dvdz=zdiff(l,m,k,fptsz,v)
xvdmp=xdamp(l,m,k,fptsx,v)
yvdmp=ydamp(l,m,k,fptsy,v)
zvdmp=zdamp(l,m,k,fptsz,v)
call radiation_bc(l,m,k,r,sinphi,cosphi,sintheta,costheta,vmean,dvdx,dvdy,dvdz, &
          xvdmp,yvdmp,zvdmp,v,kv)
!
dwdx=xdiff(l,m,k,fptsx,w)
dwdy=ydiff(l,m,k,fptsy,w)
dwdz=zdiff(l,m,k,fptsz,w)
xwdmp=xdamp(l,m,k,fptsx,w)
ywdmp=ydamp(l,m,k,fptsy,w)
zwdmp=zdamp(l,m,k,fptsz,w)
call radiation_bc(l,m,k,r,sinphi,cosphi,sintheta,costheta,wmean,dwdx,dwdy,dwdz, &
          xwdmp,ywdmp,zwdmp,w,kw)
!
end do
!
end do
end do
!
return
end subroutine bcond
```

```fortran
!
function xdiff(lc,mc,kc,fpts,f)
!
!!!!!!!!!!!!!!!!!!!!!!!!!!!!!!!!!!!!!!!!!!!!!!!!!!!!!!!!!!!!!!!!!!!!!!!!!!!!
!
!   Calculate a difference stencil in the x-direction
!   -stencil is calculated at the point (lc,mc,kc)
!   -fpts is a key that gives the number of forward stencil points
!     (3 for central difference)
!
!!!!!!!!!!!!!!!!!!!!!!!!!!!!!!!!!!!!!!!!!!!!!!!!!!!!!!!!!!!!!!!!!!!!!!!!!!!!!
!
implicit none
real(kind=rkind)::xdiff
integer, intent(in)::lc,mc,kc,fpts
real(kind=rkind), intent(in), dimension(-ld-3:ld+3,-md-3:md+3,-kd-3:kd+3)::f
!
select case (fpts)
  case (3) ! central difference
    xdiff=(a33(3)*(f(lc+3,mc,kc)-f(lc-3,mc,kc))+a33(2)*(f(lc+2,mc,kc)-f(lc-2,mc,kc))+ &
        a33(1)*(f(lc+1,mc,kc)-f(lc-1,mc,kc)))
  case (4) ! '24' stencil
    xdiff=(a24(-2)*f(lc-2,mc,kc)+a24(-1)*f(lc-1,mc,kc)+a24(0)*f(lc,mc,kc)+ &
        a24(1)*f(lc+1,mc,kc)+a24(2)*f(lc+2,mc,kc)+a24(3)*f(lc+3,mc,kc)+ &
        a24(4)*f(lc+4,mc,kc))
  case (2) ! '42' stencil
    xdiff=-(a24(-2)*f(lc+2,mc,kc)+a24(-1)*f(lc+1,mc,kc)+a24(0)*f(lc,mc,kc)+ &
        a24(1)*f(lc-1,mc,kc)+a24(2)*f(lc-2,mc,kc)+a24(3)*f(lc-3,mc,kc)+ &
        a24(4)*f(lc-4,mc,kc))
  case (5) ! '15' stencil
    xdiff=(a15(-1)*f(lc-1,mc,kc)+a15(0)*f(lc,mc,kc)+a15(1)*f(lc+1,mc,kc)+ &
        a15(2)*f(lc+2,mc,kc)+a15(3)*f(lc+3,mc,kc)+a15(4)*f(lc+4,mc,kc)+ &
        a15(5)*f(lc+5,mc,kc))
  case (1) ! '51' stencil
    xdiff=-(a15(-1)*f(lc+1,mc,kc)+a15(0)*f(lc,mc,kc)+a15(1)*f(lc-1,mc,kc)+ &
        a15(2)*f(lc-2,mc,kc)+a15(3)*f(lc-3,mc,kc)+a15(4)*f(lc-4,mc,kc)+ &
        a15(5)*f(lc-5,mc,kc))
  case (6) ! '06' stencil
    xdiff=(a06(0)*f(lc,mc,kc)+a06(1)*f(lc+1,mc,kc)+a06(2)*f(lc+2,mc,kc)+ &
        a06(3)*f(lc+3,mc,kc)+a06(4)*f(lc+4,mc,kc)+a06(5)*f(lc+5,mc,kc)+ &
        a06(6)*f(lc+6,mc,kc))
  case (0) ! '60' stencil
    xdiff=-(a06(0)*f(lc,mc,kc)+a06(1)*f(lc-1,mc,kc)+a06(2)*f(lc-2,mc,kc)+ &
        a06(3)*f(lc-3,mc,kc)+a06(4)*f(lc-4,mc,kc)+a06(5)*f(lc-5,mc,kc)+ &
        a06(6)*f(lc-6,mc,kc))
end select
!
```

```fortran
      return
      end function xdiff
!
      function ydiff(lc,mc,kc,fpts,f)
!
      !!!!!!!!!!!!!!!!!!!!!!!!!!!!!!!!!!!!!!!!!!!!!!!!!!!!!!!!!!!!!!!!!!!!!!!!
!
!    Calculate a difference stencil in the y-direction
!    -stencil is calculated at the point (lc,mc,kc)
!    -fpts is a key that gives the number of forward stencil points
!        (3 for central difference)
!
      !!!!!!!!!!!!!!!!!!!!!!!!!!!!!!!!!!!!!!!!!!!!!!!!!!!!!!!!!!!!!!!!!!!!!!!!
!
      implicit none
      real(kind=rkind)::ydiff
      integer, intent(in)::lc,mc,kc,fpts
      real(kind=rkind), intent(in), dimension(-ld-3:ld+3,-md-3:md+3,-kd-3:kd+3)::f
!
      select case (fpts)
        case (3) ! central difference
          ydiff=(a33(3)*(f(lc,mc+3,kc)-f(lc,mc-3,kc))+a33(2)*(f(lc,mc+2,kc)-f(lc,mc-2,kc))+ &
              a33(1)*(f(lc,mc+1,kc)-f(lc,mc-1,kc)))
        case (4) ! '24' stencil
          ydiff=(a24(-2)*f(lc,mc-2,kc)+a24(-1)*f(lc,mc-1,kc)+a24(0)*f(lc,mc,kc)+ &
              a24(1)*f(lc,mc+1,kc)+a24(2)*f(lc,mc+2,kc)+a24(3)*f(lc,mc+3,kc)+ &
              a24(4)*f(lc,mc+4,kc))
        case (2) ! '42' stencil
          ydiff=-(a24(-2)*f(lc,mc+2,kc)+a24(-1)*f(lc,mc+1,kc)+a24(0)*f(lc,mc,kc)+ &
              a24(1)*f(lc,mc-1,kc)+a24(2)*f(lc,mc-2,kc)+a24(3)*f(lc,mc-3,kc)+ &
              a24(4)*f(lc,mc-4,kc))
        case (5) ! '15' stencil
          ydiff=(a15(-1)*f(lc,mc-1,kc)+a15(0)*f(lc,mc,kc)+a15(1)*f(lc,mc+1,kc)+ &
              a15(2)*f(lc,mc+2,kc)+a15(3)*f(lc,mc+3,kc)+a15(4)*f(lc,mc+4,kc)+ &
              a15(5)*f(lc,mc+5,kc))
        case (1) ! '51' stencil
          ydiff=-(a15(-1)*f(lc,mc+1,kc)+a15(0)*f(lc,mc,kc)+a15(1)*f(lc,mc-1,kc)+ &
              a15(2)*f(lc,mc-2,kc)+a15(3)*f(lc,mc-3,kc)+a15(4)*f(lc,mc-4,kc)+ &
              a15(5)*f(lc,mc-5,kc))
        case (6) ! '06' stencil
          ydiff=(a06(0)*f(lc,mc,kc)+a06(1)*f(lc,mc+1,kc)+a06(2)*f(lc,mc+2,kc)+ &
              a06(3)*f(lc,mc+3,kc)+a06(4)*f(lc,mc+4,kc)+a06(5)*f(lc,mc+5,kc)+ &
              a06(6)*f(lc,mc+6,kc))
        case (0) ! '60' stencil
          ydiff=-(a06(0)*f(lc,mc,kc)+a06(1)*f(lc,mc-1,kc)+a06(2)*f(lc,mc-2,kc)+ &
              a06(3)*f(lc,mc-3,kc)+a06(4)*f(lc,mc-4,kc)+a06(5)*f(lc,mc-5,kc)+ &
              a06(6)*f(lc,mc-6,kc))
```

```fortran
end select
!
return
end function ydiff
!
function zdiff(lc,mc,kc,fpts,f)
!
!!!!!!!!!!!!!!!!!!!!!!!!!!!!!!!!!!!!!!!!!!!!!!!!!!!!!!!!!!!!!!!!!!!!!!!!!!!!!
!
!  Calculate a difference stencil in the z-direction
!   -stencil is calculated at the point (lc,mc,kc)
!   -fpts is a key that gives the number of forward stencil points
!      (3 for central difference)
!
!!!!!!!!!!!!!!!!!!!!!!!!!!!!!!!!!!!!!!!!!!!!!!!!!!!!!!!!!!!!!!!!!!!!!!!!!!!!!!!!
!
implicit none
real(kind=rkind)::zdiff
integer, intent(in)::lc,mc,kc,fpts
real(kind=rkind), intent(in), dimension(-ld-3:ld+3,-md-3:md+3,-kd-3:kd+3)::f
!
select case (fpts)
  case (3) ! central difference
  zdiff=(a33(3)*(f(lc,mc,kc+3)-f(lc,mc,kc-3))+a33(2)*(f(lc,mc,kc+2)-f(lc,mc,kc-2))+ &
     a33(1)*(f(lc,mc,kc+1)-f(lc,mc,kc-1)))
  case (4) ! '24' stencil
  zdiff=(a24(-2)*f(lc,mc,kc-2)+a24(-1)*f(lc,mc,kc-1)+a24(0)*f(lc,mc,kc)+ &
     a24(1)*f(lc,mc,kc+1)+a24(2)*f(lc,mc,kc+2)+a24(3)*f(lc,mc,kc+3)+ &
     a24(4)*f(lc,mc,kc+4))
  case (2) ! '42' stencil
  zdiff=-(a24(-2)*f(lc,mc,kc+2)+a24(-1)*f(lc,mc,kc+1)+a24(0)*f(lc,mc,kc)+ &
     a24(1)*f(lc,mc,kc-1)+a24(2)*f(lc,mc,kc-2)+a24(3)*f(lc,mc,kc-3)+  &
     a24(4)*f(lc,mc,kc-4))
  case (5) ! '15' stencil
  zdiff=(a15(-1)*f(lc,mc,kc-1)+a15(0)*f(lc,mc,kc)+a15(1)*f(lc,mc,kc+1)+ &
     a15(2)*f(lc,mc,kc+2)+a15(3)*f(lc,mc,kc+3)+a15(4)*f(lc,mc,kc+4)+ &
     a15(5)*f(lc,mc,kc+5))
  case (1) ! '51' stencil
  zdiff=-(a15(-1)*f(lc,mc,kc+1)+a15(0)*f(lc,mc,kc)+a15(1)*f(lc,mc,kc-1)+ &
     a15(2)*f(lc,mc,kc-2)+a15(3)*f(lc,mc,kc-3)+a15(4)*f(lc,mc,kc-4)+ &
     a15(5)*f(lc,mc,kc-5))
  case (6) ! '06' stencil
  zdiff=(a06(0)*f(lc,mc,kc)+a06(1)*f(lc,mc,kc+1)+a06(2)*f(lc,mc,kc+2)+ &
     a06(3)*f(lc,mc,kc+3)+a06(4)*f(lc,mc,kc+4)+a06(5)*f(lc,mc,kc+5)+ &
     a06(6)*f(lc,mc,kc+6))
  case (0) ! '60' stencil
```

```fortran
        zdiff=-(a06(0)*f(lc,mc,kc)+a06(1)*f(lc,mc,kc-1)+a06(2)*f(lc,mc,kc-2)+ &
            a06(3)*f(lc,mc,kc-3)+a06(4)*f(lc,mc,kc-4)+a06(5)*f(lc,mc,kc-5)+ &
            a06(6)*f(lc,mc,kc-6))
  end select
!
  return
  end function zdiff
!
  function xdamp(lc,mc,kc,fpts,f)
!
!!!!!!!!!!!!!!!!!!!!!!!!!!!!!!!!!!!!!!!!!!!!!!!!!!!!!!!!!!!!!!!!!!!!!!!!!!!!!!
!
!   Calculate an artificial selective damping stencil in the x-direction
!     -stencil is calculated at the point (lc,mc,kc)
!     -fpts is a key that gives the number of forward stencil points
!
!!!!!!!!!!!!!!!!!!!!!!!!!!!!!!!!!!!!!!!!!!!!!!!!!!!!!!!!!!!!!!!!!!!!!!!!!!!!!!
!
  implicit none
  real(kind=rkind)::xdamp
  integer, intent(in)::lc,mc,kc,fpts
  real(kind=rkind), intent(in), dimension(-ld-3:ld+3,-md-3:md+3,-kd-3:kd+3)::f
!
  select case (fpts)
    case (3) ! 7 point stencil
      xdamp=R_inv*(d33(3)*(f(lc+3,mc,kc)+f(lc-3,mc,kc))+d33(2)*(f(lc+2,mc,kc)+ &
          f(lc-2,mc,kc))+d33(1)*(f(lc+1,mc,kc)+f(lc-1,mc,kc))+d33(0)*f(lc,mc,kc))
    case (2,4) ! 5 point stencil
      xdamp=R_inv*(d22(2)*(f(lc+2,mc,kc)+f(lc-2,mc,kc))+d22(1)*(f(lc+1,mc,kc)+ &
          f(lc-1,mc,kc))+d22(0)*f(lc,mc,kc))
    case (1,5) ! 3 point stencil
      xdamp=R_inv*(d11(1)*(f(lc+1,mc,kc)+f(lc-1,mc,kc))+d11(0)*f(lc,mc,kc))
    case (0,6)
      xdamp=0.0_rkind
  end select
!
  return
  end function xdamp
!
  function ydamp(lc,mc,kc,fpts,f)
!
!!!!!!!!!!!!!!!!!!!!!!!!!!!!!!!!!!!!!!!!!!!!!!!!!!!!!!!!!!!!!!!!!!!!!!!!!!!!!!
!
!   Calculate an artificial selective damping stencil in the y-direction
!     -stencil is calculated at the point (lc,mc,kc)
!     -fpts is a key that gives the number of forward stencil points
!
!!!!!!!!!!!!!!!!!!!!!!!!!!!!!!!!!!!!!!!!!!!!!!!!!!!!!!!!!!!!!!!!!!!!!!!!!!!!!!
```

```
!
implicit none
real(kind=rkind)::ydamp
integer, intent(in)::lc,mc,kc,fpts
real(kind=rkind), intent(in), dimension(-ld-3:ld+3,-md-3:md+3,-kd-3:kd+3)::f
!
select case (fpts)
  case (3) ! 7 point stencil
    ydamp=R_inv*(d33(3)*(f(lc,mc+3,kc)+f(lc,mc-3,kc))+d33(2)*(f(lc,mc+2,kc)+ &
      f(lc,mc-2,kc))+d33(1)*(f(lc,mc+1,kc)+f(lc,mc-1,kc))+d33(0)*f(lc,mc,kc))
  case (2,4) ! 5 point stencil
    ydamp=R_inv*(d22(2)*(f(lc,mc+2,kc)+f(lc,mc-2,kc))+d22(1)*(f(lc,mc+1,kc)+ &
      f(lc,mc-1,kc))+d22(0)*f(lc,mc,kc))
  case (1,5) ! 3 point stencil
    ydamp=R_inv*(d11(1)*(f(lc,mc+1,kc)+f(lc,mc-1,kc))+d11(0)*f(lc,mc,kc))
  case (0,6)
    ydamp=0.0_rkind
end select
!
return
end function ydamp
!
function zdamp(lc,mc,kc,fpts,f)
!
!!!!!!!!!!!!!!!!!!!!!!!!!!!!!!!!!!!!!!!!!!!!!!!!!!!!!!!!!!!!!!!!!!!!!!!!!!!
!
!   Calculate an artificial selective damping stencil in the z-direction
!    -stencil is calculated at the point (lc,mc,kc)
!    -fpts is a key that gives the number of forward stencil points
!
!!!!!!!!!!!!!!!!!!!!!!!!!!!!!!!!!!!!!!!!!!!!!!!!!!!!!!!!!!!!!!!!!!!!!!!!!!!
!
implicit none
real(kind=rkind)::zdamp
integer, intent(in)::lc,mc,kc,fpts
real(kind=rkind), intent(in), dimension(-ld-3:ld+3,-md-3:md+3,-kd-3:kd+3)::f
!
select case (fpts)
  case (3) ! 7 point stencil
    zdamp=R_inv*(d33(3)*(f(lc,mc,kc+3)+f(lc,mc,kc-3))+d33(2)*(f(lc,mc,kc+2)+ &
      f(lc,mc,kc-2))+d33(1)*(f(lc,mc,kc+1)+f(lc,mc,kc-1))+d33(0)*f(lc,mc,kc))
  case (2,4) ! 5 point stencil
    zdamp=R_inv*(d22(2)*(f(lc,mc,kc+2)+f(lc,mc,kc-2))+d22(1)*(f(lc,mc,kc+1)+ &
      f(lc,mc,kc-1))+d22(0)*f(lc,mc,kc))
  case (1,5) ! 3 point stencil
    zdamp=R_inv*(d11(1)*(f(lc,mc,kc+1)+f(lc,mc,kc-1))+d11(0)*f(lc,mc,kc))
```

```fortran
   case (0,6)
    zdamp=0.0_rkind
end select
!
return
end function zdamp
!
function mladvance(lc,mc,kc,f,k)
!
!!!!!!!!!!!!!!!!!!!!!!!!!!!!!!!!!!!!!!!!!!!!!!!!!!!!!!!!!!!!!!!!!!!!!!!!!!!!!
!
!   Optimized multi-level time advancement to new time level
!
!!!!!!!!!!!!!!!!!!!!!!!!!!!!!!!!!!!!!!!!!!!!!!!!!!!!!!!!!!!!!!!!!!!!!!!!!!!!!
!
implicit none
integer, intent(in)::lc,mc,kc
real(kind=rkind), intent(in), dimension(-ld-3:ld+3,-md-3:md+3,-kd-3:kd+3)::f
real(kind=rkind), intent(in), dimension(-3:0,-ld-3:ld+3,-md-3:md+3,-kd-3:kd+3)::k
real(kind=rkind)::mladvance
!
mladvance=f(lc,mc,kc)+dt*(b(0)*k(kl(0),lc,mc,kc)+b(-1)*k(kl(-1),lc,mc,kc)+ &
    b(-2)*k(kl(-2),lc,mc,kc)+b(-3)*k(kl(-3),lc,mc,kc))
!
return
end function mladvance
!
subroutine polar(lc,mc,kc,r,sinphi,cosphi,sintheta,costheta)
!
!!!!!!!!!!!!!!!!!!!!!!!!!!!!!!!!!!!!!!!!!!!!!!!!!!!!!!!!!!!!!!!!!!!!!!!!!!!!!
!
!   Calculates r,sinphi,cosphi,sintheta,costheta at given boundary node
!
!!!!!!!!!!!!!!!!!!!!!!!!!!!!!!!!!!!!!!!!!!!!!!!!!!!!!!!!!!!!!!!!!!!!!!!!!!!!!
!
implicit none
integer, intent(in)::lc,mc,kc
real(kind=rkind), intent(out)::r,sinphi,cosphi,sintheta,costheta
real(kind=rkind)::theta,d
!
!   Radius from center of domain
!
r=sqrt(real(lc**2,rkind)+real(mc**2,rkind)+real(kc**2,rkind))
!
!   Angles in spherical coordinates (phi is azimuthal, theta is polar)
!
theta=acos(real(kc,rkind)/r)
```

```
costheta=cos(theta)
sintheta=sin(theta)
d=r*abs(sintheta)
if (d .gt. 10.0_rkind*tiny(d)) then
 sinphi=real(mc,rkind)/d
 cosphi=real(lc,rkind)/d
else
 sinphi=0.0_rkind
 cosphi=0.0_rkind
end if
!
return
end subroutine polar
!
subroutine
radiation_bc(lc,mc,kc,r,sinphi,cosphi,sintheta,costheta,fmean,dfdx,dfdy,dfdz, &
     xdmp,ydmp,zdmp,f,k)
!
!!!!!!!!!!!!!!!!!!!!!!!!!!!!!!!!!!!!!!!!!!!!!!!!!!!!!!!!!!!!!!!!!!!!!!!!
!
!   Calculates time derivative 'K' for radiation boundary node
!   -This radiation equation passes a spherically symmetric outgoing
!    wave in a fluid with zero mean velocity
!
!!!!!!!!!!!!!!!!!!!!!!!!!!!!!!!!!!!!!!!!!!!!!!!!!!!!!!!!!!!!!!!!!!!!!!!!
!
implicit none
integer, intent(in)::lc,mc,kc
real(kind=rkind), intent(in)::r,sinphi,cosphi,sintheta,costheta,dfdx,dfdy,dfdz, &
    fmean,xdmp,ydmp,zdmp
real(kind=rkind), intent(in), dimension(-ld-3:ld+3,-md-3:md+3,-kd-3:kd+3)::f
real(kind=rkind), intent(in out), dimension(-3:0,-ld-3:ld+3,-md-3:md+3,-kd-3:kd+3)::k
!
k(kl(0),lc,mc,kc)=-(cosphi*sintheta*dfdx+sinphi*sintheta*dfdy+costheta*dfdz+xdmp+ydmp+zdmp)
!
return
end subroutine radiation_bc
!
subroutine dudt(lc,mc,kc,dudx,dudy,dudz,dpdx,xudmp,yudmp,zudmp,rho,u,v,w,ku)
!
!!!!!!!!!!!!!!!!!!!!!!!!!!!!!!!!!!!!!!!!!!!!!!!!!!!!!!!!!!!!!!!!!!!!!!!!
!
!   Calculate dudt at given node point (Non-linear Eulers equations)
!
!!!!!!!!!!!!!!!!!!!!!!!!!!!!!!!!!!!!!!!!!!!!!!!!!!!!!!!!!!!!!!!!!!!!!!!!
!
implicit none
integer, intent(in)::lc,mc,kc
```

```fortran
real(kind=rkind), intent(in)::dudx,dudy,dudz,dpdx,xudmp,yudmp,zudmp
real(kind=rkind), intent(in), dimension(-ld-3:ld+3,-md-3:md+3,-kd-3:kd+3)::rho,u,v,w
real(kind=rkind), intent(out), dimension(-3:0,-ld-3:ld+3,-md-3:md+3,-kd-3:kd+3)::ku
!
ku(kl(0),lc,mc,kc)=-u(lc,mc,kc)*dudx-v(lc,mc,kc)*dudy-w(lc,mc,kc)*dudz &
    -dpdx/rho(lc,mc,kc)-xudmp-yudmp-zudmp
!
return
end subroutine dudt
!
subroutine dvdt(lc,mc,kc,dvdx,dvdy,dvdz,dpdy,xvdmp,yvdmp,zvdmp,rho,u,v,w,kv)
!
!!!!!!!!!!!!!!!!!!!!!!!!!!!!!!!!!!!!!!!!!!!!!!!!!!!!!!!!!!!!!!!!!!!!!!!!!
!
! Calculate dvdt at given node point (Non-linear Eulers equations)
!
!!!!!!!!!!!!!!!!!!!!!!!!!!!!!!!!!!!!!!!!!!!!!!!!!!!!!!!!!!!!!!!!!!!!!!!!!
!
implicit none
integer, intent(in)::lc,mc,kc
real(kind=rkind), intent(in)::dvdx,dvdy,dvdz,dpdy,xvdmp,yvdmp,zvdmp
real(kind=rkind), intent(in), dimension(-ld-3:ld+3,-md-3:md+3,-kd-3:kd+3)::rho,u,v,w
real(kind=rkind), intent(out), dimension(-3:0,-ld-3:ld+3,-md-3:md+3,-kd-3:kd+3)::kv
!
kv(kl(0),lc,mc,kc)=-u(lc,mc,kc)*dvdx-v(lc,mc,kc)*dvdy-w(lc,mc,kc)*dvdz &
    -dpdy/rho(lc,mc,kc)-xvdmp-yvdmp-zvdmp
!
return
end subroutine dvdt
!
subroutine dwdt(lc,mc,kc,dwdx,dwdy,dwdz,dpdz,xwdmp,ywdmp,zwdmp,rho,u,v,w,kw)
!
!!!!!!!!!!!!!!!!!!!!!!!!!!!!!!!!!!!!!!!!!!!!!!!!!!!!!!!!!!!!!!!!!!!!!!!!!
!
! Calculate dwdt at given node point (Non-linear Eulers equations)
!
!!!!!!!!!!!!!!!!!!!!!!!!!!!!!!!!!!!!!!!!!!!!!!!!!!!!!!!!!!!!!!!!!!!!!!!!!
!
implicit none
integer, intent(in)::lc,mc,kc
real(kind=rkind), intent(in)::dwdx,dwdy,dwdz,dpdz,xwdmp,ywdmp,zwdmp
real(kind=rkind), intent(in), dimension(-ld-3:ld+3,-md-3:md+3,-kd-3:kd+3)::rho,u,v,w
real(kind=rkind), intent(out), dimension(-3:0,-ld-3:ld+3,-md-3:md+3,-kd-3:kd+3)::kw
!
kw(kl(0),lc,mc,kc)=-u(lc,mc,kc)*dwdx-v(lc,mc,kc)*dwdy-w(lc,mc,kc)*dwdz &
    -dpdz/rho(lc,mc,kc)-xwdmp-ywdmp-zwdmp
!
```

```fortran
        return
      end subroutine dwdt
!
      subroutine drhodt(lc,mc,kc,drhodx,drhody,drhodz,dudx,dvdy,dwdz,xrhodmp,yrhodmp, &
          zrhodmp,u,v,w,rho,krho)
!
!!!!!!!!!!!!!!!!!!!!!!!!!!!!!!!!!!!!!!!!!!!!!!!!!!!!!!!!!!!!!!!!!!!!!!!!!!
!
!   Calculate drhodt at given node point (Non-linear Eulers equations)
!
!!!!!!!!!!!!!!!!!!!!!!!!!!!!!!!!!!!!!!!!!!!!!!!!!!!!!!!!!!!!!!!!!!!!!!!!!!
!
      implicit none
      integer, intent(in)::lc,mc,kc
      real(kind=rkind), intent(in)::drhodx,drhody,drhodz,dudx,dvdy,dwdz,xrhodmp, &
          yrhodmp,zrhodmp
      real(kind=rkind), intent(in), dimension(-ld-3:ld+3,-md-3:md+3,-kd-3:kd+3)::u,v,w,rho
      real(kind=rkind), intent(out), dimension(-3:0,-ld-3:ld+3,-md-3:md+3,-kd-3:kd+3)::krho
!
      krho(kl(0),lc,mc,kc)=-u(lc,mc,kc)*drhodx-v(lc,mc,kc)*drhody-w(lc,mc,kc)*drhodz &
          -rho(lc,mc,kc)*(dudx+dvdy+dwdz)-xrhodmp-yrhodmp-zrhodmp
!
        return
      end subroutine drhodt
!
      subroutine dpdt(lc,mc,kc,dpdx,dpdy,dpdz,dudx,dvdy,dwdz,xpdmp,ypdmp,zpdmp,u,v,w,p,kp)
!

!!!!!!!!!!!!!!!!!!!!!!!!!!!!!!!!!!!!!!!!!!!!!!!!!!!!!!!!!!!!!!!!!!!!!!!!!!
!
!   Calculate dpdt at given node point (Non-linear Eulers equations)
!
!!!!!!!!!!!!!!!!!!!!!!!!!!!!!!!!!!!!!!!!!!!!!!!!!!!!!!!!!!!!!!!!!!!!!!!!!!
!
      implicit none
      integer, intent(in)::lc,mc,kc
      real(kind=rkind), intent(in)::dpdx,dpdy,dpdz,dudx,dvdy,dwdz,xpdmp,ypdmp,zpdmp
      real(kind=rkind), intent(in), dimension(-ld-3:ld+3,-md-3:md+3,-kd-3:kd+3)::u,v,w,p
      real(kind=rkind), intent(out), dimension(-3:0,-ld-3:ld+3,-md-3:md+3,-kd-3:kd+3)::kp
!
      kp(kl(0),lc,mc,kc)=-u(lc,mc,kc)*dpdx-v(lc,mc,kc)*dpdy-w(lc,mc,kc)*dpdz &
          -gamma*p(lc,mc,kc)*(dudx+dvdy+dwdz)-xpdmp-ypdmp-zpdmp
!
        return
      end subroutine dpdt
!
      end program DRP_3D
```

Some Comments about the 3-D Code and Sample Computed Results

This code solves the full non-linear Eulers equations (non-dimensionalized) on a 3D Cartesian domain. Artificial selective damping terms are added to the equations. The boundary conditions are set to the spherical radiation equations on all faces (no mean flow in the domain). The data structure is a 3D array, and the number of mesh points is easily adjustable. The code contains functions and subroutines that can be used to construct more sophisticated 3D codes. The code is clean and very efficient, and is about 1200 lines in length.

A test case for the simulation code:

Currently the code uses $67 \times 67 \times 67 \sim 300,000$ nodes to compute a spherically symmetric pulse exiting the domain. The initial Gaussian pressure pulse has a half width of 3 mesh spacings and an amplitude of 0.001. The code advances the solution to times $t = 2, 4, 6, 8, 10, 15, 25, 35, 45$ and output the results. Attached are some line contour plots of the pressure on the $x - y$ plane ($z = 0$), profiles of the pressure along the x-axis, and a time trace of the pressure at the origin. In the contour plots and the time trace, the solid line is the computational solution. In the profiles along the x-axis, the open circles represent the computational solution at mesh points. The non-linear computational solution compares very well with the exact linear solution.

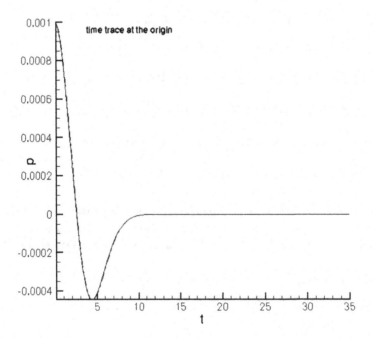

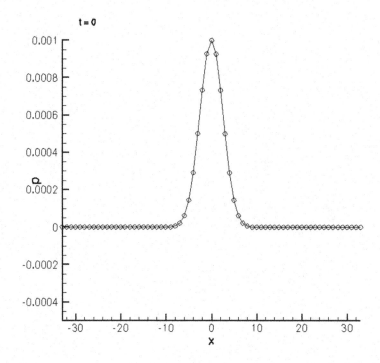

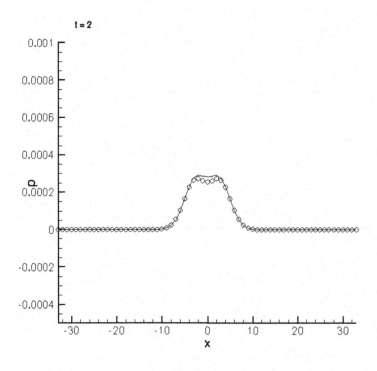

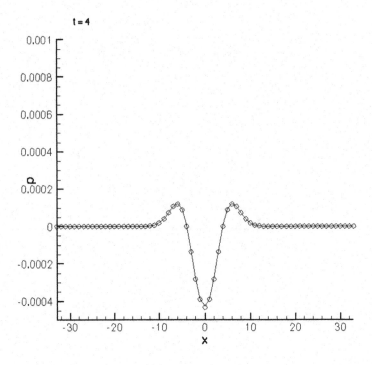

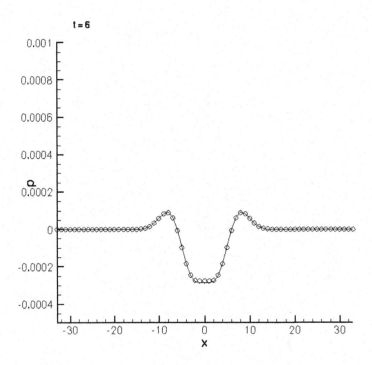

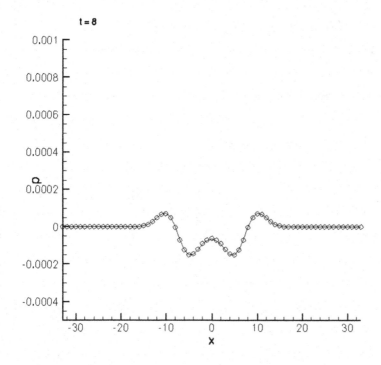

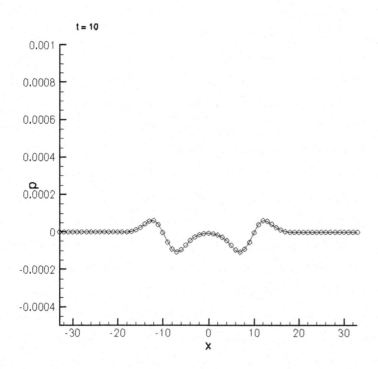

References

Abramowitz, M. and I.A. Stegun, I.A. (1964) Handbook of Mathematical Functions with Formulas, Graphs and Mathematical Tables, *National Bureau of Standards*, Dover Publications, Inc., New York.

Arbey, H. & Bataille, J. (1983) Noise generated by airfoil profiles placed in a uniform laminar flow. *Journal of Fluid Mechanics*, 134, 33–47.

Bayliss, A. and Turkel, E. (1980) Radiation Boundary Conditions for Wave-Like Equations. *Communications on Pure and Applied Mathematics*, 33, 707–725.

Bayliss, A. and Turkel, E. (1982) Far Field Boundary Conditions for Compressible Flows. *Journal of Computational Physics*, 48, 182–199.

Berenger. J.P. (1994) A perfectly matched layer for the absorption of electromagnetic waves, *Journal of Computational Physics*, 114, 185–200.

Berenger, J.P. (1996) Three dimensional perfectly matched layer for the absorption of electromagnetic waves, *Journal of Computational physics*, 127, 363–379.

Berger, M.J. and Colella, P. (1989) Local Adaptive Mesh Refinement for Shock Hydrodynamics. *Journal of Computational Physics*, 82, 64–84.

Bradury, L. J. S. (1965) The structure of a self-preserving turbulent plane jet. *Journal of Fluid Mechanics*, 23, 31–64.

Chandrasekhar, S. (1989) Adjoint differential systems in the theory of hydrodynamic stability. In *Selected Papers of S. Chandrasekhar*, Vol. 4, pp. 221–228. University of Chicago Press.

Cho, Y. C. (1980) Reciprocity principle in duct acoustics. *Journal of the Acoustical Society of America*, 67, 1421–1426.

Chong, T. P. & Joseph, P. (2009) An experimental study of tonal noise mechanism of laminar airfoils. *AIAA Paper* 2009–3345.

Dahl, Milo D. (2000) Proceedings of the Third Computational Aeroacoustics (CAA) Workshop on Benchmark Problems. NASA CP-2000-209790.

Desquesnes, G., Terracol, M. & Sagaut, P. (2007) Numerical investigation of the tone noise mechanism over laminar airfoils. *Jouranl of Fluid Mechanics*, 591, 155–182.

Dong, T.Z. (1997) On Boundary Conditions for Acoustic Computations in Non-uniform Mean Flows. *Journal of Computational Acoustics*, 5, 297–315.

Dowling, A. P. (1983) Flow-acoustic interaction near a flexible wall. *Journal of Fluid Mechanics*, 128, 181–198.

Eckhaus, W. (1965) *Studies in Nonlinear Stability Theory*. Springer.

Erdelyi, A., Magnus, W., Oberhettinger, F. and Tricomi, F. G. (1954) *Tables of Integral Transforms* Vol. 1. McGraw Hill.

Eversman, W. (1991) Theoretical models for duct acoustic propagation and radiation, in: H.H. Hubbard (Ed.), *Aeroacoustics of Flight Vehicles: Theory and Practice*, Vol. 2, NASA RP-1258, pp. 101–164.

Falle, S.A.E.G. and Giddings, J. (1993) Body Capturing using Adaptive Cartesian Grids. *Numerical Methods in Fluid Dynamics*, 4, 337–343.

Ffowcs Williams, J.E. and Hawkings, D.L. (1969) Sound Generation by Turbulence and Surface in Arbitrary Motion. *Philosophical Transactions of the Royal Society of London.* 264A, 321–342.

Giles, M.B. (1990) Nonreflecting Boundary Conditions for Euler Equation Calculations. *AIAA Journal*, 28, 2050–2058.

Givoli, D. (1991) Non-Reflecting Boundary Conditions. *Journal of Computational Physics*, 94, 1–29.

Hanjalic, K. & Launder, B.E. (1972) A Reynolds stress model of turbulence and its applications to thin shear flows. *Journal of Fluid Mechanics*, 52, 609–638.

Hanjalic, K. & Launder, B.E. (1976) Contribution towards a Reynolds stress closure for low Reynolds number turbulence. *Journal of Fluid Mechanics*, 74, 593–610.

Henderson, B.S. (2004) Problem 2: Sound Generation by Flow over a Cavity. *Fourth Computational Aeroacoustics (CAA) Workshop on Benchmark Problems.*NASA/CP-2004-212954. pp. 71–77.

Hill, D. C. (1995) Adjoint systems and their role in the receptivity problem for boundary layers. *Journal of Fluid Mechanics*, 292, 183–294.

Hixon, R., Shih, S.H. and Mankbadi, R.R. (1995) Evaluation of Boundary Conditions for Computational Aeroacoustics. *AIAA Journal*, 33, 2006–2012.

Howe, M. S. (1975) The generation of sound by aerodynamics sources in an inhomongeneous steady flow. *Journal of Fluid Mechanics*, 67, 597–610.

Howe, M. S. (1981) The displacement thickness theory of trailing edge noise. *Journal of Sound and Vibration*, 75, 239–250.

Hu, F.Q., Hussaini, M.Y. and Manthey, J.L. (1996) Low-Dissipation and Low-Dispersion Runge-Kutta Schemes for Computational Acoustics, *Journal of Computational Physics*, 124, 177–191.

Hu, F.Q. (1996) On Absorbing Boundary Conditions of Linearized Euler Equations by a Perfectly Matched Layer. *Journal of Computational Physics*, 129, 201–219.

Hu, F.Q. (2001) A Stable Perfectly Matched Layer for Linearized Euler Equations in Unsplit Physical Variables. *Journal of Computational Physics*, 173, 455–480.

Hu, F.Q. (2004) *Absorbing Boundary Conditions. International Journal of Computational Fluid Dynamics.* 18, 513–522.

Hu, F.Q. (2005) A Perfectly Matched Layer Absorbing Boundary Condition for Linearized Euler Equations with a Non-uniform Mean Flow. *Journal of Computational Physics*, 208, 469–492.

Kibens, V. (1968) The intermittent region of a turbulent boundary layer. Ph. D. thesis, The Johns Hopkins University.

Kingan, M. J. & Pearse, J. R. (2009) Laminar boundary layer instability noise produced by an aerofoil. *Journal of Sound and Vibration*, 322, 808–828.

Kurbatskii, K.A. and Tam, C.K.W. (1997) Cartesian Boundary Treatment of Curved Walls for High-Order Computational Aeroacoustics Schemes. *AIAA Journal*, 35, 133–140.

Lau, J.C. (1981) Effects of exit Mach number and temperature on mean flow and turbulence characteristics in round jets. *Journal of Fluid Mechanics*, 105, 193–218.

Launder, B.E. & Spalding, D.B. (1974) The numerical computation of turbulent flows. *Computer Methods in Applied Mechanics and Engineering.* 3, 269–289.

Launder, B.E., Reece, G.J. & Rodi, W. (1975) Progress in the development of a Reynolds stress turbulence closure. *Journal of Fluid Mechanics*, 68, 537–566.

Lele, S.K. (1992) Compact Finite Difference Schemes with Spectral-Like Resolution. *Journal of Computational Physics*, 103, 16–42.

Lele, S.K. (1997) Computational Aeroacoustics: A Review. *AIAA Paper 97-0018.*

Li, Y. (1997) Wave Number-extended High-order Upwind-biased Finite Difference Schemes for Convective Scalar Transport. *Journal of Computational Physics*, 133, 235–255.

Lockard, D.P., Brentner, K.S. and Atkins, H.L. (1995) High Accuracy Algorithms for Computational Aeroacoustics. *AIAA Journal*, 33, 246–251.

Lyrintzis, A.S. (2003) Integral acoustics methods: from the (CFD) near-field to the (acoustic) far-field. *International Journal of Aeroacoustics*, 2, 95–128.

Maestrello, L. (1973) On the relationship between acoustic energy density flux near the jet axis and far-field acoustic intensity. NASA TN D-7269.

Magnus, W. and Oberhettinger, F. (1949) *Formulas and Theorems for the Functions of Mathematical Physics*.Chelsea Publishing Company.

McAlpine, A., Wright, M.C.M., Batard, H. and Thezelais, S. (2003) Finite/boundary element assessment of a turbofan spliced intake liner at supersonic fan operating conditions. *AIAA Paper* 2003–3305.

Melling, T.H. (1993) The acoustic impedance of perforate at medium and high sound pressure levels. *Journal of Sound and Vibration*, 29, 1–65.

Menter, F.R. (1994a) Two-equation eddy-viscosity turbulence models for engineering applications. *AIAA Journal*, 32, 1598–1605.

Menter, F.R. (1994b) Eddy viscosity transport equations and their relation to the $k - \varepsilon$ model. NASA TM-108854.

Menter, F.R. (1997) Eddy viscosity transport equations and their relation to the $k - \varepsilon$ model. *Journal of Fluids Engineering*, 119, 876–884.

Mickens, R.E. (1990) Difference Equations: Theory and Applications. 2nd Edition, van Nostrand Reinhold, New York.

Morfey, C.L. (1971) Acoustic energy in non-uniform flows. *Journal of Sound and Vibration*, 14, 159–170.

Motsinger, R.E. and Kraft, R.E. (1991) Design and performance of duct acoustic treatment, in: H.H. Hubbard (Ed.), *Aeroacoustics of Flight Vehicles: Theory and Practice*, Vol. 2, NASA RP-1258, pp. 165–206.

Myers, M.K. (1980) On the Acoustic Boundary Condition in the Presence of Flow. *Journal of Sound and Vibration*, 71, 429–434.

Nash, E.C., Lowson, M.V. & McAlpine, A. (1999) Boundary-layer instability noise on airfoils. *Journal of Fluid Mechanics*, 382, 27–61.

Norum, T.D. (1983) Screech suppression in supersonic jets. *AIAA Journal*, 21, 235–240.

Norum, T.D. & Brown, M.C. (1993) Simulated high-speed flight effects of supersonic jet noise. *AIAA Paper 93-4388*.

Paterson, R. W., Vogt, P. G., Fink, M. R. & Munch, C. L. (1973) Vortex noise of isolated Airfoils. *Journal of Aircrafts*, 10, 296–302.

Phillips, O. M. (1955) The irrotational motion outside a free turbulent boundary. *Proceedings Cambridge Philosophical Society*, 51, 220–229.

Ponton, M.K. & Seiner, J.M. (1992) The effects of nozzle exit lip thickness on plume resonance. *Journal of Sound and Vibration*, 154, 531–549.

Pope, S.B. (1978) An explanation of the turbulent round-jet/plane-jet anomaly. *AIAA Journal*, 16, 279–281.

Powell, A. (1953) On the mechanism of choked jet noise. *Proceedings of the Physical Society, London*, 66, 1039–1056.

Regan, B. and Eaton, J. (1999) Modelling the influence of acoustic liner non-uniformities on duct modes. *Journal of Sound and Vibration* 219, 859–879.

Roberts, P. H. (1960) Characteristic value problems posed by differential equations arising in hydrodynamics and hydromagnetics. *Journal of Mathematical Analysis and Applications*, 1, 193–214.

Roe, P.L. (1992) Technical Prospectus for Computational Aeroacoustics. *AIAA Paper 92-02032*.

Sanberg, R. D., Sandham, N. D. & Joseph, P. F. (2007) Direct numerical simulations of trialing-edge noise generated by boundary-layer instabilities. *Journal of Sound and Vibration*, 304, 677–690.

Saric, W. S. & Nayfeh, A. H. (1975) Nonparallel stability of boundary layer flows. *Physics of Fluids*, 18, 945–950.

Sarkar, S. & Lakshmanan, B. (1991) Application of a Reynolds stress turbulence model to compressible shear layer. *AIAA Journal*, 29, 743–749.

Shen, H. and Tam, C.K.W. (1998) Numerical Simulation of the Generation of Axisymmetric Mode Jet Screech Tones. *AIAA Journal*, 36, 1801–1807.

Sherer, S. E. (2004) Acoustic Scattering from Multiple Circular Cylinders: Category 2, Problems 1 and 2, Analytical Solution. Fourth Computational Aeroacoustics (CAA) *Workshop on Benchmark Problems. NASA/CP-2004-212954.*

Spalart, P.R. & Allmaras, S.R. (1994) A one-equation turbulence model for aerodynamic flows. *La Recherche Aerospatiale,* n 1, 5–21.

Stewart, R. W. (1956) Irrotational motion associated with free turbulent flows. *Journal of Fluid Mechanics,* 51, 593–606.

Tam, C.K.W. 1974 Discrete tones of isolated airfoils. *Journal of the Acoustical Society of America,* 55, 1173–1177.

Tam, C.K.W. (1995) Computational Aeroacoustics: Issues and Methods. *AIAA Journal,* 33, 1788–1796.

Tam, C.K.W. (1998) Advances in Numerical Boundary Conditions for Computational Aeroacoustics. *Journal of Computational Acoustics,* 6, 377–402.

Tam, C.K.W. (2004) Computational Aeroacoustics: An Overview of Computational Challenges and Applications. *International Journal of Computational Fluid Dynamics,* 18, 547–567.

Tam, C.K.W. and Block, P.J.W. (1978) On the Tones and Pressure Oscillations Induced by Flows over Rectangular Cavities. *Journal of Fluid Mechanics,* 89, 373–399.

Tam, C.K.W. and Auriault, L. (1996) Time-Domain Impedance Boundary Conditions for Computational Aeroacoustics. *AIAA Journal,* 34, 917–923.

Tam, C.K.W. and Dong, Z. (1994) Wall Boundary Conditions for High-Order Finite Difference Schemes in Computational Aeroacoustics. *Theoretical and Computational Fluid Dynamics,* 6, 303–322.

Tam, C.K.W. and Dong, Z. (1996) Radiation and Outflow Boundary Conditions for Direct Computation of Acoustic and Flow Disturbances in Nonuniform Mean Flow. *Journal of Computational Acoustics,* 4, 175–201.

Tam, C. K. W., Golebiowski, M. and Seiner, J. M. (1996) Two components of turbulent mixing noise from supersonic jets. *AIAA Paper 96-1716.*

Tam, C.K.W. and Shen, H. (1993) Direct Computation of Nonlinear Acoustic Pulses using High-Order Finite Difference Schemes. *AIAA Paper 93-4325.*

Tam, C.K.W. and Webb, J.C. (1993) Dispersion-Relation-Preserving Finite difference Schemes for Computational Acoustics. *Journal of Computational Physics,* 107, 262–281.

Tam, C.K.W., Webb, J.C. and Dong, Z. (1993) A Study of the Short Wave Components in Computational Acoustics. *Journal Computational Acoustics,* 1, 1–30.

Tam, C. K. W. and Auriault, L. (1998) Mean flow refraction effects on sound radiated from localized sources in a jet. *Journal of Fluid Mechanics,* 370, 149–174.

Tam, C.K.W., Auriault, L and Cambuli, F. (1998) Perfectly Matched Layer as an Absorbing Boundary Condition for the Linearized Euler Equations in Open and Ducted Domains. *Journal Computational Physics,* 144, 213–234.

Tam, C.K.W., Fang, J. and Kurbatskii, K.A. (1998) Nonhomogeneous Radiation and Outflow Boundary Conditions Simulating Incoming Acoustic and Vorticity Waves for Exterior Computational Aeroacoustic Problems. *International Journal for Numerical Methods in Fluids,* 26, 1107–1123.

Tam, C.K.W. and Kurbatskii, K.A. (2000) A Wave number Based Extrapolation and Interpolation Method for Use in Conjunction with High-Order Finite Difference Schemes. *Journal of Computational Physics,* 157, 588–617.

Tam, C.K.W. and Kurbatskii, K.A. (2000) Microfluid dynamics and acoustics of resonant liners. *AIAA Journal,* 38, 1331–1339.

Tam, C.K.W., Kurbatskii, K.A., Ahuja, K.K. and Gaeta Jr., R.J. (2001) A numerical and experimental investigation of the dissipation mechanisms of resonant acoustic liners. *Journal of Sound and Vibration,* 245, 545–557.

Tam, C.K.W., Pastouchenko, N.N. and Auriault, L. (2001) Effects of forward flight on jet mixing noise from fine-scale turbulence. *AIAA Journal,* 39, 1261–1269.

Tam, C.K.W. and Ganesan, A. (2004) Modified $k - \varepsilon$ turbulence model for calculating hot jet mean flows and noise. *AIAA Journal*, 42, 26–34.

Tam, C.K.W. and Ju, H. (2009) Finite difference computation of acoustic scattering by small surface inhomogeneities and discontinuities. *Journal of Computational Physics*, 228, 5917–5932.

Tam, C.K.W., Pastouchenko, N.N. and Viswanathan, K. (2010) Continuation of the near acoustic field of a jet to the far field. Part I: theory. *AIAA Paper 2010-3728*.

Tam, C.K.W., Vishwanathan, K. and Pastouchenko, N.N. (2010) Continuation of the near acoustic field of a jet to the far field. Part II. Experimental validation and noise source characteristics. *AIAA Paper 2010-3729*.

Tam, C.K.W. and Ju, H. (2012) Airfoil tones at moderate Reynolds number. *Journal of Fluid Mechanics*, 690, 536–570.

Tam, C. K. W., Viswanathan, K., Ahuja, K. K. and Panda, J. (2008) The sources of jet noise: experimental evidence. *Journal of Fluid Mechanics*, 615, 253–292.

Tam, C.K.W., Ju, H. and Chien, E.W. (2008) Scattering Acoustic Duct Modes by Axial Liner Splices. *Journal of Sound and Vibration*, 310, 1014–1035.

Tester, B. J. (1973) The Propagation and Attenuation of Sound in Lined Ducts Containing Uniform or Plug Flow. *Journal of Sound and Vibration*, 28, 151–203.

Tester, B.J., Powles, C.J., Baker, N.J. and Kempton, A.J. (2006) Scattering of sound by liner splices: a Kirchhoff model with numerical verification. *AIAA Journal*, 44, 2009–2017.

Thies, A.T. and Tam, C.K.W. (1996) Computation of turbulent axisymmetric and nonaxisymmetric jet flows using the $k - \varepsilon$ model. *AIAA Journal*, 34, 309–316.

Thomas, J.P. and Roe, P.L. (1993) Development of Non-Dissipative Numerical Schemes for Computational Aeroacoustics. AIAA Paper 93–3382.

Thompson, K.W. (1989) Time Dependent Boundary Conditions for Hyperbolic Systems. *Journal of Computational Physics*, 68, 1–24.

Thompson, K.W. (1989) Time Dependent Boundary Conditions for Hyperbolic Systems, II. *Journal of Computational Physics*, 89, 439–461.

Trefethen, L.N. (1981) Group Velocity in Finite Difference Schemes. *Siam Review*, 24, 113–136.

Tyler, J. M. and Sofrin, T. G. (1962) Axial flow compressor noise studies. *SAE Transaction*, 70, 309–332.

Vichnevetsky, R. and Bowles, J.B. (1982) *Fourier Analysis of Numerical Approximations of Hyperbolic Equations*.SIAM, Philadelphia.

White, F.M. (1991) *Viscous Fluid Flow*. 2nd Edition, McGraw Hill.

Whitham, G.B. (1974) *Linear and Nonlinear Waves*. Wiley-Interscience.

Wilcox, D.C. (1998) *Turbulence Modeling for CFD*. 2nd Edition, DCW Industries Inc.

Zhuang, M. and Chen, R. (1998) Optimized Upwind Dispersion-Relation-Preserving Finite Difference Scheme for Computational Aeroacoustics. *AIAA Journal*, 36, 2146–2148.

Zhuang, M. and Chen, R. (2002) Applications of High-Order Optimized Upwind Schemes for Computational Aeroacoustics. *AIAA Journal*, 40, 443–449.

Zingg, D.W., Lomax, H. and Jurgens, H. (1993) An Optimized Finite Difference Scheme for Wave Propagation Problems. *AIAA Paper 93-0459*.

Zingg, D.W. (2000) Comparison of High-Accuracy Finite Difference Methods for Linear Wave Propagation. *SIAM Journal of Scientific Computing*, 22, 476–502.

Index

Accelerated convergence to steady state method, 403

Acoustic liner, 180
   splice, 192–193

Acoustic wave, 162, 164
   pulse, 287

Adjoint Green's function, 308
   adjoint equations and boundary conditions, 311
   computation of, 316
   conical surface, 313
   cylindrical surface, 312
   non-self-adjoint, 309–310
   reciprocity relation, 311
   self-adjoint, 308–309

Airfoil tone generation processes, 372–379
   primary, 373–375
   secondary, 377–379

Airfoil tones, 350–379
   airfoil wake at moderate Reynolds number, 363–365
   direct numerical simulation, 354
   energy source, 369–371
   feedback loop, 350
   multiple tones, 350–352
   Paterson formula for tone frequency, 351
   single tone, 352–353
   tone frequency dependence on Reynolds number, 367–368
   tone generation processes, 372–379

Airfoil wake instability, 369–372

Aliasing, 124

Analytic continuation, into the $r < 0$ region, 154

Artificial selective damping, 113
   at surfaces of discontinuity, 122
   basic concept, 114–115
   distribution of, 332
   excessive damping, 118
   stencils, 116, 118, 122, 126–127, 238, 239, 256–261
   variable, 137

Artificial viscous diffusion, 119

Asymptotic outflow boundary conditions, 150

Asymptotic radiation boundary condition, 249, 332

Asymptotic solutions, linearized Euler equations, 61

Autocorrelation function, 409

Axial liner splice, 193

Axis boundary treatment, cylindrical coordinates, 152

Axisymmetric jet screech, 385

Background damping, 245

Backward difference, 76
   stencils, 32

Basic elements of a CAA code, 329
   boundary conditions, 332
   computational algorithm, 331
   computational domain, 330
   computational grid, 331
   computational model, 329
   distribution of artificial selective damping, 332

Blasius solution, 249

Blunt trailing edge, of airfoil, 349–350

Body-fitted grid, 263, 344

Boundaries with discontinuities, 167

Boundary conditions
   asymptotic, 81–82, 150, 249, 332
   axis treatment, 152
   for curved surface, 102, 225
   for diffusion equation, 109, 401
   ghost point method, 89, 104–105
   in CAA code development, 332
   outflow, 81–82, 83, 88, 149–152, 249
   perfectly matched layer absorbing, 158, 159, 165
   radiation, 81, 83, 88, 106, 110, 147, 249, 332
   surface of discontinuity interface, 9–10
   three-parameter broadband impedance, 182
   time-domain impedance, 180
   wall treatment, 89, 102

Boundary data, overspecification, 301

Boundary instability mode, 213

Boundary modes, spurious, 15–17

Boxcar initial condition, 112
Broadband
  noise spectrum, energy-conserving
    discretization of, 322, 407–409
  random acoustic field, generation of, 321
  three-parameter impedance model, 182
  two-point space-time correlation function, 323,
    324, 326, 408
Burger's equation, 142

C_+ and C_- characteristic, 130
Canceling the residuals method, 404
Cartesian boundary treatment of curved wall
  surfaces, 102, 225
Characteristics
  C_+ and C_-, 130
  method of, 398
  P, 130
Circumferential liner splice, 193, 199
Coefficients of large optimized stencils, 34
Complex change of variables, 159
Computer code, 410–471
  sample 1D code, 410–417
  sample 2D code, 420–437
  sample 3D code, 438–467
Conformal transformation, 344
  elliptic coordinates, body-fitted grid for an
    airfoil, 346
  enforcement points, 345
  inverse mapping, 347–348
  source points, 345
Conservation form, 132
Constrained optimization, 208
Constraints (see Order constraints)
Continuation of near acoustic field to far field,
  298
  adjoint Green's function method, 308
  continuation problem, 299–301
  Ffowcs-Williams and Hawkings method, 300
  Kirchhoff method, 300
  surface Green's function method, 301
Contours of local interpolation error, 273–278
Convergence (see Accelerated convergence to
  steady state)
Curved walls
  Cartesian boundary treatment, 102, 225
  ghost point method, 104–105
  simplified enforcement method, 106

Damping (see Numerical damping)
Data transfer, between grids, 265
Diffusion equation, 400
Directional derivative, 155
Dispersion (see Numerical dispersion)
Dispersion relation, 46
  DRP scheme, 69
  linearized Euler equations, 61
  multidimensional, 59
Dispersion-relation-preserving (DRP) finite
  difference scheme, 50, 67
  accuracy consideration, 72

algorithmic analysis, 67–69
  curvilinear coordinates, 73
  dispersion relation, 69
  group velocity, 70–71
  numerical stability, 69
  starting condition, 68
Dispersive wave, 45, 48
  system, solution of, 46
Dissipation (see Numerical dissipation)

Energy cascade, 137
Energy flux (PWL), 195
Energy source of airfoil tones, 369
Energy spectrum, 48
Energy-conserving discretization of broadband
    noise spectrum, 322, 407–409
Entrainment flow, 146–149
Entropy wave, 162, 164
  pulse, 287
Equivalent noise source, 326
Error
  amplification mechanism, 210
  contours of local interpolation error, 273–278
  extrapolation, 206, 207
  impact of order constraints on, 274
  interpolation, 216, 268, 272–277
Euler equations (see Linearized Euler equations)
Extra damping, 245
Extrapolation, 203
  error in wave number space, 206
  formula, 206
  local error, 207
  numerical instability, 204–206
  optimized method, 206–213
  wave number analysis in one dimension, 206

Far field (see Continuation of near acoustic field
  to far field)
Ffowcs-Williams and Hawkings integral method,
  300
Filter function, 381
Finite difference equation, 1–4
  eigenvalue problem, 5–8
  exact solution, 2–4
  linear equation with constant coefficients, 2–4
  order of, 1
Forward difference, 76
Fourier transform, 395

Gaussian profile, 135
Ghost point method, 89, 249
  for time-domain impedance boundary
    condition, 188–189
Ghost values, 90
Gibbs phenomenon, 133
Global interpolation, 278
Grid search method for eigenvalues, 16
Grid-to-grid oscillations, 240, 332
Grids
  body-fitted, 263, 344
  data transfer between, 265

moving, 287
overset, 263, 286
refinement, 361
resolution requirement, 333
stationary, 287
Group velocity, 47, 48
  DRP scheme, 70–71
  of a finite difference scheme, 25

Half-period cosine function, 136
Helmholtz resonator, 180

Impedance, 180
Incoming disturbances, 144, 146
Initial condition, multilevel marching scheme, 49, 68
Instability
  airfoil wake, 369–372
  boundary mode, 213
  damping-induced, 121
  extrapolation, 204–206
  Kelvin-Helmholtz, 369
  nonparallel flow, 369–372
  Tollmien-Schlichting, 369
Integral representation, 301
Interpolation, 203
  coefficients, 268
  errors in wave number space, 272–277
  formula, 216
  global, 278
  irregular stencil, 267, 280
  local error, 216, 268, 273–278
  order constraints, 270
  regular stencil, 272–273, 280
  total error, 268
  two-dimensional, 267
Irregular interpolation stencil, 267, 280

Jet screech, 385–394
  axisymmetric mode, 385
  feedback loop, 385–386
  helical or flapping mode, 385
  shock cell structure, 391–392
  staging, 394
  tone frequency, 392–393
  tone intensity, 393–394

k–ε model, 382
  corrections to, 383–384
k–ω model, 382
Kelvin-Helmholtz instability, 369
Kirchhoff method, 300

Lagrange multiplier, 208
Lagrange polynomials method, 203, 206
Laplace transform, 395
Large stencils (*see* Stencils)
Large eddy simulation (LES), 330
Leading waves, 60
Linearized Euler equations, 61
  acoustic waves, 65–67

asymptotic solutions, 63–67
dispersion relation, 61–63
entropy wave, 63
numerical stability, 69
vorticity wave, 64
Local error
  extrapolation, 207
  interpolation, 216, 268
Long waves, finite difference scheme, 24
Lorentz line shape, 136
Low-dissipation low-dispersion Runge-Kutta (LDDRK) scheme, 39, 54

Maxwell equations, 75
Mentor's transport eddy viscosity model, 382
Mesh Reynolds number, 117
Mesh-size-change buffer region (*see* Stencils)
Mesh-size-change interface, 229
Method of characteristics, 398
Method of stationary phase, 397
Moving boundary problems, 293, 294
Moving grid, 287
Multi-size-mesh multi-time-step scheme, 230, 249
Multidimensional dispersion relation, 59
Multidimensional optimized interpolation method, 267, 291, 294
Multilevel time discretization method, 40
  spurious solutions, 41
  stability diagram, 41–43
  starting condition, 49, 68
Multiscales problems, 229
Multisize mesh, 229
Myers impedance boundary condition, 185

NACA four-digit airfoils, 348
  blunt trailing edge, 349–350
  camber ratio, 348–349
  chord length, 348
  thickness ratio, 348–349
Necessary condition for numerical stability, 58
Newton iteration method, 347
Nonlinear outflow boundary conditions, 151
Nonlinear simple wave equation, 132
  energy cascade, 137
  nonlinear steepening effects, 133, 137
  shocks, 132, 133
  spurious oscillations, 132, 137
Nonlinear simple waves, 130
Nonlinear steepening effects, 133
Nonparallel flow instability, 369–372
Normal acoustic modes, closed-end tube, 5–8
Normal incidence impedance tube, 189, 201
Numerical damping
  as cause of numerical instability, 121
  at surfaces of discontinuity, 122
  background, 245
  due to temporal discretization, 58
  excessive, 118
  extra, 245
  selective (*see* Artificial selective damping)

Numerical dispersion, 25
　due to spatial discretization, 50
　due to temporal discretization, 53, 55
　origin of, 48, 51
Numerical dissipation, origin of, 57
Numerical error amplification mechanism, 210
Numerical filtering method, 380–382
Numerical stability
　necessary condition for, 58
　of DRP scheme, 69–70
　of time-domain impedance boundary condition,
　　182
　stability diagram, multi time step method, 41–43

Oblique Cartesian coordinates, 316
Optimized methods
　extrapolation, 206–213
　finite difference, 22
　interpolation, 213, 286
　multidimensional interpolation, 266, 291
　time discretization, 40
Order constraints
　as interpolation requirement, 270
　impact on interpolation error, 274
Order of truncation, 271
Outflow boundary conditions, 81, 149, 249
　asymptotic solution based, 81–82, 88, 150
　implementation, 83
　nonlinear, 151
Outgoing waves, 146
Overlapping mesh, 263
Overset grids, 263, 286

P characteristic, 130
Perfectly matched layer (PML), 158
　derivation, 159
　in three dimensions, 165
Phase function, 47
Phase velocity, 47, 48
Point-sink/source approximation, 148
Pope's correction to the k–ε model, 383–384
Potential representation of velocity, 152
Pressure gradient transformation method,
　170–171

r derivative of a scalar variable near $r = 0$,
　156
r derivative of the velocity field near $r = 0$,
　156
Radiation boundary conditions
　asymptotic solution based, 81, 88
　entrainment flow, 147–149
　implementation, 83
　one-dimensional Euler equations, 110
　one-dimensional wave equation, 106
　spherical symmetric problem, 106
Random broadband acoustic field, generation of,
　321
Reactance, 180

Regular interpolation stencil, 272–273, 280
Resistance, 180
Reynolds Averaged Navier-Stokes (RANS)
　equations, 330
Runge-Kutta (RK) scheme, 38–40, 54

Sarkar and Lakshmanan correction to the k–ε
　model, 383–384
Scalar potential, 152
Shed vortices, 250
Shock tube problem, 143
Shock-capturing computation, 132–142
Short waves, 112
　finite difference scheme, 24
Single time step method, 38
Sliding interface, 285
Spalart Allmaras model, 382
Spatial discretization methods, 21
　optimized finite difference approximation,
　　22
　truncated Taylor series method, 21–22
Spatial resolution requirements, 333
　2D turbulent jet, 337–338
　2D turbulent mixing layer, 336–337
　laminar boundary layer, 338–340
　turbulent boundary layer, 340–344
　wake flow, 333–336
Split-variable method, 144
　for PML, 317
Spurious numerical scattering, 196
Spurious numerical solutions, 7
Stability (see Numerical stability)
Stationary grid, 287
Stationary point, 397
Steady state, accelerated convergence to,
　403
Stencil Reynolds number, 139–140
Stencils
　artificial selective damping, 116, 118, 122,
　　126–127, 238, 239, 256–261
　backward difference, 32
　finite difference, 24, 26, 30, 32, 34, 231, 233,
　　252–256
　irregular interpolation, 267, 280
　large, 30, 34, 118, 126–127, 252–261
　mesh-size-change buffer region, 231, 233, 235,
　　237, 238, 239, 252–261,
　regular interpolation, 272–273, 280
　time marching, 40, 235, 237
Stokes oscillatory boundary layer, 102
Subgrid scale model, 330
Surface Green's function, 301–308
　conical surface, 316
　cylindrical surface, 302, 304, 306, 307
　normal velocity as matching variable, 305
　pressure as matching variable, 301
Surface of discontinuity, finite difference model,
　11–15
　boundary modes, 15
　interface boundary conditions, 9–10

Tam and Ganesan correction to the k–ε model, 383–384
Three-parameter broadband impedance model, 182
Time-domain impedance boundary condition, 180
  equivalent Myers impedance boundary condition, 185
  stability, 182
  three-parameter broadband model, 182
Tollmien-Schlichting instability, 369
Tolmien-Schlichting waves, 247
Total internal refraction, 14
Total interpolation error, two-dimensional, 268
Trailing waves, 54, 60
Truncated Fourier series method, 198
Truncated unit step function, 194
Truncation, order of, 271
Two-point space-time correlation function, 323, 324, 326, 408

Unsteady Reynolds Averaged Navier-Stokes (URANS), 384
Upwind schemes, 58
Upwinding requirement, 58

Variable artificial selective damping, 137
Vector potential, 152
Very large eddy simulation (VLES), 330

Vortex shedding, at trailing edge of acoustic liner cavity, 250
Vorticity wave, 162, 164

Wall boundary conditions, 89
Wave energy, 48
Wave equation (see Nonlinear simple wave equation)
Wave number of finite difference scheme, 23
Wave number truncation method, 168, 194, 200, 201
Wave resolution, 333
Wave velocities, group and phase, 47, 48
Waves
  acoustic, 162, 164
  dispersive, 45, 48
  electromagnetic, 75
  entropy, 162, 164
  leading, 60
  long, 24
  nonlinear simple, 130
  outgoing, 146
  short, 24, 112
  Tollmien-Schlichting, 247
  trailing, 54, 60
  vorticity, 162, 164
Witham's equal area rule, 140, 143

Zero fluctuating pressure contour method, 372–379

Printed in the United States
By Bookmasters